Mechanics

Stefan Roth · Achim Stahl

Mechanics

Experimental Physics—Descriptively Explained

Stefan Roth
III. Physikalisches Institut B
RWTH Aachen University
Aachen, Germany

Achim Stahl
III. Physikalisches Institut B
RWTH Aachen University
Aachen, Germany

ISBN 978-3-662-68078-0   ISBN 978-3-662-68079-7 (eBook)
https://doi.org/10.1007/978-3-662-68079-7

This book is a translation of the original German edition "Mechanik und Wärmelehre" by Stefan Roth and Achim Stahl, published by Springer-Verlag GmbH, DE in 2016. The translation was done with the help of an artificial intelligence machine translation tool. A subsequent human revision was done primarily in terms of content, so that the book will read stylistically differently from a conventional translation. Springer Nature works continuously to further the development of tools for the production of books and on the related technologies to support the authors.

© The Editor(s) (if applicable) and The Author(s), under exclusive license to Springer-Verlag GmbH, DE, part of Springer Nature 2025

This work is subject to copyright. All rights are solely and exclusively licensed by the Publisher, whether the whole or part of the material is concerned, specifically the rights of translation, reprinting, reuse of illustrations, recitation, broadcasting, reproduction on microfilms or in any other physical way, and transmission or information storage and retrieval, electronic adaptation, computer software, or by similar or dissimilar methodology now known or hereafter developed.
The use of general descriptive names, registered names, trademarks, service marks, etc. in this publication does not imply, even in the absence of a specific statement, that such names are exempt from the relevant protective laws and regulations and therefore free for general use.
The publisher, the authors, and the editors are safe to assume that the advice and information in this book are believed to be true and accurate at the date of publication. Neither the publisher nor the authors or the editors give a warranty, expressed or implied, with respect to the material contained herein or for any errors or omissions that may have been made. The publisher remains neutral with regard to jurisdictional claims in published maps and institutional affiliations.

This Springer imprint is published by the registered company Springer-Verlag GmbH, DE, part of Springer Nature.
The registered company address is: Heidelberger Platz 3, 14197 Berlin, Germany

Paper in this product is recyclable

Preface

This book is aimed at all those who would like to learn the basics of experimental physics. It originated from the lecture "Experimental Physics 1", which we held for our freshmen in physics at RWTH Aachen. We have tried to capture and pass on the fun that this course with its many experiments gives us. We hope we have succeeded to some extent. You can do some of these experiments yourself. Give it a try! That's when it really becomes fun. But experiments alone are not enough. You have to deal with the model ideas and explanations of physics. This book also aims to support you in this.

This book is the first volume in a series on experimental physics. The topic of this volume is classical mechanics. What the term "classical" means in this context, you will see in ▶ Chap. 4. The book is divided into five parts. In the first edition, this volume also included thermodynamics. Although this may seem surprising, there are good arguments for counting thermodynamics as part of mechanics and treating them together. In the second edition, we have outsourced it to its own volume for space reasons. The volume on mechanics would otherwise have become too extensive. These five parts of the volume are

1. Introduction: Here we want to discuss some basic things about physics.
2. Mechanics of point masses, where we learn to know and explain movements, initially for the simple case of bodies that we approximate as points.
3. Mechanics of rigid bodies: Here we want to abandon the approximation of point masses and deal with extended bodies.
4. Elastic bodies, where we finally allow deformations of the formerly rigid bodies. In this part, we can also deal with liquids and gases.
5. Vibrations and waves: This is an important topic that is found in all areas of physics. Therefore, we have dedicated a separate part to it.

For understanding the book, special prior knowledge about physics is not required. Even school physics knowledge is largely repeated. However, we assume school knowledge in mathematics. If you have difficulties here, we offer you a little help in the appendix ("Mathematical Introduction").

Dealing with physics requires extensive mathematical knowledge. You must learn this thoroughly. If you do not possess it today, you must acquire it parallel to the study of experimental physics. We will use some of this knowledge in this book, possibly even before the thorough and systematic learning process is completed for you. This is unfortunately inevitable. Take it as motivation to engage even more intensively with mathematics!

This textbook presents you with the material of classical mechanics, but you have to work it out yourself. This is not a trick of the authors, no, it is in the nature of the matter. In the end, you want to be ready to research independently. But researching means entering new scientific territory, i.e., entering areas where there are no more books and no more people who could teach you what the new territory looks like. You will have to explore it yourself. To prepare for this, you have to learn to work out the new things yourself. It is still material that others already know, but in the end, it will be completely new.

Understand this textbook as an offer that should make it easier for you to work out the material. The actual work has to happen in your head. There you have to build your own building of physics with your ideas, explanations, and connections. This is a big change and often the biggest problem at the beginning.

The goal is to gain a deep understanding of physics. Knowledge of the facts is also important, but usually easier to achieve than a deep understanding. The question of *Why?* is of central importance. Why does a particular process happen this way and not otherwise? Why does this result occur and not another?

Here are some tips. Unfortunately, there are no magic formulas:
- Get used to constantly asking the *Why?* question when you read in this or other books, when you think about the experiments, when you talk to someone about physics, …
- Question the information that is offered to you. You have to expect that there are also errors in this book: typos, inaccurate or incorrect formulations, and perhaps also points that we have not yet fully understood. Every other textbook may contain errors and online articles usually contain even more. Think critically and recognize these errors!
- Physical statements are tied to conditions. Don't just learn the statements, but think about under what conditions they apply.
- Have you understood a particular fact? If so, you should be able to explain the fact. Try it! Imagine a friend asks you about it, and try (e.g., in your mind) to explain the fact in **your** own words.
- Revise the individual chapters. Make notes by writing down the essential content in your own words.
- Use different sources (e.g., several textbooks) on a topic, so that you get to know different approaches and explanations on the topic.
- Discuss physics. Address points that you do not yet fully understand, or discuss further questions and problems.

About the structure of the book: The five parts are divided into chapters. Each chapter contains a text with illustrations and equations that describe the essential material of the chapter. In addition, there are experiments and examples that supplement the text, as well as some remarks on the methodological approach in physics.

> **Experiment 0.1: Experiment**
>
> In this layout, the experiments are presented in the book. Some are cited in the text and are very important for understanding. Others serve more for illustration. You might be able to recreate some of them. Try it!

> **Example 0.1: Representation of an example**
>
> Furthermore, examples are integrated into the text that supplement the content. This text shows the formatting of an example. Examples are important. They show how you can apply the knowledge you have learned. Work through the examples. You can test your understanding with them.

> **Methodical Remark**
>
> In addition, texts are inserted at some points that point to the methodological approach in physics. They should make clear to you how physics works. You can recognize them by this formatting.

At the end of most chapters, there are also practice exercises. You can find a brief outline of the solutions in the appendix.

Have fun and success!

This book is not just our work. Behind the book are many helpers, to whom we would like to express our heartfelt thanks. Our thanks go to Beate Roth for proofreading, to our colleagues Prof. Lutz Feld, Dr. Katja Klein and Egon Schneevoigt, to whom we owe many of the great experiments (and the photos of them), further to Jennifer Merz, Franziska Scholz, Richard Brauer, Niklas Mohr, Jan Domenik Sammet, Rüdiger Jussen, Hendrik Jansen, Joschka Lingemann, Lukas Gromann, Julius Schniewind, Sarah Böhm and to all the students who have pointed out errors to us. Finally, we would like to thank Springer-Verlag for their excellent support.

Achim Stahl
Stefan Roth
Aachen
March 2025

Contents

I Introduction

1 **What is Physics?** 3

The Importance of Physics 5

2 **Physical Quantities** 9
2.1 Definition 10
2.2 The Basic Quantities 10
2.3 The Length 13
2.4 Time 16
2.5 The Mass 19
2.6 The Angle Measures 23
2.7 Notations 23

3 **Measurement Errors** 27
3.1 Measured Value 28
3.2 Statistical Interpretation 32
3.3 Error Propagation 33
3.4 Systematic Errors 35

4 **Methodology** 39
4.1 Physical Theories 40
4.2 Scientific Method 41
4.3 Scope of Classical Physics 43

II Mechanics of Point Masses

5 **Kinematics of the Point Mass** 47
5.1 The Point Mass 48
5.2 Reference Systems 49
5.3 Uniform Motion 56
5.4 Non-uniform Motion 59
5.5 Acceleration 61
5.6 Free fall 65
5.7 Projectile Motion 70
5.8 Circular Motion 75

6 **Dynamics of a Point Mass** 81
6.1 The Inertia Principle 82
6.2 The Basic Law of Mechanics 87
6.3 The Principle of Reaction 97

| 6.4 | Superposition of Forces | 101 |
| 6.5 | Measurement of Forces | 104 |

7 Work and Energy ... 107
7.1	Work and Power	108
7.2	Energy	116
7.3	Energy Conservation	119
7.4	Symmetries	131

8 Momentum ... 137
8.1	Conservation of Momentum	138
8.2	Center of Mass	140
8.3	Collision Processes	146
8.4	Systems with Variable Mass	155
8.5	The Impulse	158

9 Friction ... 163
9.1	Overview	164
9.2	Static Friction	166
9.3	Kinetic Friction	172
9.4	Rolling Friction	174

10 Fictitious Forces ... 177
10.1	Overview	178
10.2	Uniformly Accelerated Reference Systems	180
10.3	Centrifugal Force	183
10.4	Coriolis Force	189
10.5	Absolute Motion?	205

11 Celestial Mechanics ... 209
11.1	Kepler's First Law	210
11.2	Kepler's Second and Third Laws	214
11.3	Newton's Law of Universal Gravitation	222
11.4	Gravitational and Inertial Mass	239
11.5	Potential and Potential Energy	241

III The Rigid Body

12 Statics of Rigid Bodies ... 253
12.1	Definition	254
12.2	The Torque	256
12.3	The Center of Gravity of a Body	261
12.4	The Theorems of Statics	264
12.5	Statics of Rigid Bodies	268

13	**Rotations**	283
13.1	The Angular Momentum	284
13.2	Rotation Around a Fixed Axis	289
13.3	Conservation of Angular Momentum	300
13.4	Rolling	306
13.5	Gyroscopic Motion	313
13.6	The Inertia Tensor	327
13.7	Rotation around Free Axes	338
13.8	Comparison	340

IV Elastic Bodies

14	**Elastomechanics**	349
14.1	Strain	350
14.2	Bending	355
14.3	Compression	360
14.4	Shearing	362
14.5	Finite Element Method	364

15	**Hydro- and Aerostatics**	369
15.1	Pressure	370
15.2	Compressibility	377
15.3	Hadrostatic Pressure	381
15.4	Buoyancy	388
15.5	Interfaces	393

16	**Hydro- and Aerodynamics**	407
16.1	Description of Flows	408
16.2	The Continuity Equation	413
16.3	The Flow of Ideal Fluids	417
16.4	Internal Friction	432
16.5	Laminar Flows	437
16.6	Turbulent Flows	446
16.7	Drag	447
16.8	Dynamic Lift	455

V Oscillations and Waves

17	**Oscillations**	473
17.1	Harmonic Oscillations	474
17.2	Damped Oscillations	487
17.3	Forced Oscillations	495
17.4	Coupled Oscillations	507
17.5	Standing Waves	517

18	**Waves**	533
18.1	Harmonic Waves	534
18.2	Wave Equation	542
18.3	Wave Packets	544
18.4	Energy Density and Energy Transport	551
18.5	Reflection and Interference	553
19	**Acoustics**	559
19.1	Sound Waves	560
19.2	Perception of Sound	570
19.3	Moving Sound Sources	576
19.4	Musical Instruments	583

Supplementary Information

A1 List of Symbols	596
A2 Solutions to the Problems	599
A3 Mathematical Introduction	621

List of Experiments

Experiment 2.1: Scales. 15
Experiment 2.2: Clocks 18
Experiment 2.3: Scales. 21
Experiment 5.1: Projectile Speed. 57
Experiment 5.2: Stroboscopic Recording of a Falling Ball 60
Experiment 5.3: Free Fall in Vacuum 65
Experiment 5.4: Free Fall in the Slipstream 65
Experiment 5.5: Free Fall with the Drop Tower 68
Experiment 5.6: Superposition with the Train 70
Experiment 5.7: Projectile trajectory with a water jet. 72
Experiment 5.8: Superposition with the crossbow 72
Experiment 5.9: Superposition with the Ski Jump 73
Experiment 6.1: Inertia of a heavy ball. 86
Experiment 6.2: Inertia at the set table 86
Experiment 6.3: Inertia in Rotation 87
Experiment 6.4: Acceleration on the Air Track 94
Experiment 6.5: Reactive Forces on Skateboards. 98
Experiment 6.6: Recoil from the Medicine Ball 98
Experiment 6.7: Water Rocket. 99
Experiment 6.8: Addition of Forces 102
Experiment 7.1: Lifting with a Tackle. 108
Experiment 7.2: Energy Conversion with a Dynamo 120
Experiment 7.3: Conservation of Energy on the Pendulum. 121
Experiment 7.4: Drinking Duck 131
Experiment 8.1: Conservation of Momentum on
the Air Cushion Track 139
Experiment 8.2: Pendulum Cart 143
Experiment 8.3: Collisions on an Air Cushion Track. 150
Experiment 8.4: Newton's Cradle 153
Experiment 8.5: Bouncy Ball Pyramid 154
Experiment 8.6: Ballistic Pendulum 154
Experiment 8.7: Rocket Car 158
Experiment 8.8: Impulse with the Skateboard 160
Experiment 9.1: Static Friction on an Inclined Plane. 167
Experiment 9.2: Static Friction on the Inclined Plane 2. 169
Experiment 9.3: Static and Kinetic Friction on a Rod 172
Experiment 9.4: Measurement of the Coefficient of Kinetic Friction 173
Experiment 9.5: Sliding Rod 173
Experiment 9.6: Heat through Friction. 174
Experiment 10.1: Fictious Force in the Accelerated Reference System 182
Experiment 10.2: Measurement of the Centrifugal Force 184
Experiment 10.3: Centrifugal Force 186

Experiment 10.4: Earth Ellipsoid 186
Experiment 10.5: The Surface of Rotating Liquids 187
Experiment 10.6: Rotating Bucket 188
Experiment 10.7: Polar Bear Hunting with Difficulties 193
Experiment 10.8: Model Experiment on Foucault's Pendulum 199
Experiment 10.9: Foucault's Pendulum 200
Experiment 11.1: Gravitational Balance 229
Experiment 12.1: Torque Disk 259
Experiment 12.2: Torque on the Spool of Thread 260
Experiment 12.3: Determination of the Center of Gravity of Plates 261
Experiment 12.4: A Stair of Bricks 274
Experiment 13.1: Measurement of Moments of Inertia 295
Experiment 13.2: Parallel Axis Theorem 297
Experiment 13.3: Conservation of Angular Momentum with the Swivel Chair 302
Experiment 13.4: The Gimbal Gyroscope 304
Experiment 13.5: The Stubborn Suitcase 304
Experiment 13.6: Gyroscope 305
Experiment 13.7: Fire Tornado 305
Experiment 13.8: Cylinder rolls on an inclined plane 308
Experiment 13.9: Maxwell's Wheel 312
Experiment 13.10: Vertical Gyroscope—Part 1 314
Experiment 13.11: Humming Top 318
Experiment 13.12: Vertical Gyroscope—Part 2 321
Experiment 13.13: A Ball Gyroscope 324
Experiment 13.14: Levitron 324
Experiment 13.15: Magnus Gyrocompass 325
Experiment 13.16: Rotation around Free Axes 339
Experiment 13.17: Juggling with Cigar Boxes 339
Experiment 14.1: Straining a Copper Wire 350
Experiment 14.2: Lateral Strain of a Rubber Band 353
Experiment 14.3: Elongation beyond the Elastic Limit 353
Experiment 14.4: Visualization of Stresses 355
Experiment 15.1: Isotropy of Pressure 374
Experiment 15.2: Hydraulic Press 376
Experiment 15.3: Compressibility of Water 379
Experiment 15.4: Hydrostatic Pressure in Water 382
Experiment 15.5: Communicating Tubes 382
Experiment 15.6: Hydrostatic Paradox 383
Experiment 15.7: Density Balance 383
Experiment 15.8: Buoyancy 389
Experiment 15.9: Cartesian Diver 391
Experiment 15.10: Magdeburg Hemispheres 392
Experiment 15.11: Measurement of Surface Tension 395
Experiment 15.12: Floating Paperclip 396
Experiment 15.13: Minimal Surface through Surface Tension 396
Experiment 15.14: Pressure in a Soap Bubble 397

Experiment 15.15: Critical Angle in a Wedge Glass 400
Experiment 15.16: Capillarity 402
Experiment 16.1: Water Tunnel. 408
Experiment 16.2: Hydrodynamic Paradox 420
Experiment 16.3: A Bucket with a Hole 422
Experiment 16.4: Prandtl Tube 425
Experiment 16.5: Water Jet Pump. 426
Experiment 16.6: Floating Balls in an Air Stream 430
Experiment 16.7: Aerodynamic Paradox 431
Experiment 16.8: Hagen-Poiseuille 440
Experiment 16.9: Stokes's Law 444
Experiment 16.10: Drag 450
Experiment 16.11: Smoke Rings 450
Experiment 16.12: Dynamic Lift on a Wing. 456
Experiment 16.13: Magnus Effect 466
Experiment 17.1: Spring Pendulum 474
Experiment 17.2: Compound Pendulum. 478
Experiment 17.3: Mach's Pendulum Apparatus 478
Experiment 17.4: Torsion Pendulum. 479
Experiment 17.5: Simple Gravity Pendulum. 479
Experiment 17.6: Simple Gravity Pendulum 2 480
Experiment 17.7: Pohl's Wheel 488
Experiment 17.8: Forced Oscillations on the Pohl's Wheel 496
Experiment 17.9: Resonance with Spring Pendulum 503
Experiment 17.10: Tongue Frequency Meter 503
Experiment 17.11: Resonance Catastrophe with a Wine Glass 506
Experiment 17.12: Coupled Pendulums 507
Experiment 17.13: Coupled Metronomes 509
Experiment 17.14: Beating 515
Experiment 17.15: Eigenmodes. 518
Experiment 17.16: Pendulum Chain 521
Experiment 17.17: Wave Machine. 522
Experiment 17.18: Rope Waves in the Rubber Band 523
Experiment 17.19: Standing Waves with the Wave Machine 526
Experiment 17.20: Standing Waves on Rubber Rope 526
Experiment 17.21: Kundt's Tube. 527
Experiment 17.22: Chladni Patterns 528
Experiment 17.23: Vibrations of a Guitar String 529
Experiment 18.1: Water Waves 534
Experiment 18.2: Water Waves 2 537
Experiment 18.3: Fourier Analysis with Microphone 548
Experiment 18.4: Reflection of Water Waves 554
Experiment 18.5: Interference with Water Waves 557
Experiment 19.1: Flickering Candle Flame 563
Experiment 19.2: Rubens's Flame Tube 563
Experiment 19.3: Sound Propagation in Air. 564

Experiment 19.4: Sound in a Vacuum..................................... 565
Experiment 19.5: Sound Propagation in Wood 565
Experiment 19.6: Pin Hole Siren.. 566
Experiment 19.7: Interference with Sound 567
Experiment 19.8: Measurement of the Speed of Sound..................... 570
Experiment 19.9: Speed of Sound in Helium 571
Experiment 19.10: Voice Pitch in Helium................................ 571
Experiment 19.11: Speed of Sound in a Solid 571
Experiment 19.12: Doppler Effect with Whistle.......................... 581
Experiment 19.13: Monochord.. 588

Introduction

Contents

Chapter 1 What is Physics? – 3

Chapter 2 Physical Quantities – 9

Chapter 3 Measurement Errors – 27

Chapter 4 Methodology – 39

What is Physics?

© The Author(s), under exclusive license to Springer-Verlag GmbH, DE, part of Springer Nature 2025
S. Roth and A. Stahl, *Mechanics*,
https://doi.org/10.1007/978-3-662-68079-7_1

Fig. 1.1 Rudolf Mößbauer. © Nobel Foundation

Physics is an ancient science, whose origin lies in Greek philosophy. The name "Physics" also comes from Greek (φυσική θεωρα, physike theoria), which was coined by Aristotle and refers to the description and explanation of causes and relationships in nature.

Today, physics is considered one of the natural sciences. This classification has its reasons. Physics is a science of nature, meaning it deals with nature. This usually involves the inanimate nature, such as the laws of levers or polar lights. Physics is divided into different disciplines. This first volume is dedicated to mechanics, which can be described as the study of the motion of bodies.

Modern physics is an experimental science. Whether a particular model is right or wrong is determined by physicists using experiments. This was not always the case. In the Middle Ages, physicists studied ancient texts to find knowledge. But since Galileo Galilei (1564–1642), the experiment has been the basis of physics.

Physics is an exact science. Qualitative explanations of phenomena may be important for understanding physics, but they are not sufficient. Every physical model must be able to make quantitative predictions about processes in nature that can be verified by experiments. To do this, physics uses the language of mathematics. If you want to understand physics, you must learn this language.

Here is a more detailed article by Rudolf Mößbauer (Fig. 1.1), Nobel laureate in Physics 1961, on the importance of physics[1]:

---

1 © With kind permission C. Mößbauer.

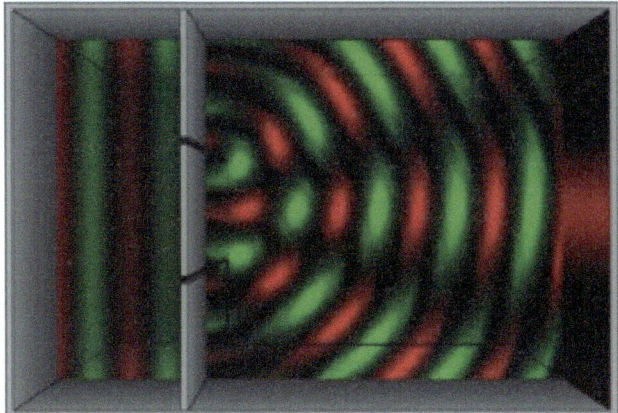

Fig. 1.2 Diffraction of light at the double slit

The Importance of Physics

Physics strives to trace the course of events back to universally valid laws - the laws of nature. These laws are of superior validity, nothing can escape them. While all material in this world is subject to constant change, the order of natural law is timeless. We can trust it fully, at any place and at any time, and build on it.

The roots of physics lie in antiquity. However, it was not until the seventeenth century that Johannes Kepler, Galileo Galilei and Isaac Newton established the methodology of modern physics by isolating individual processes from their context and quantitatively examining them with the help of experiments; that they finally formulated a basic physical law mathematically.

At the end of the nineteenth century, the electronic structure of matter began to be elucidated (Figs. 1.2, 1.3 and 1.4). Joseph Thomson discovered the electron, the first indivisible particle, in 1897. At the beginning of the twentieth century, Albert Einstein revolutionized our conception of space and time with his theory of relativity. He and Max Planck discovered the photon as the elementary particle of light. Werner Heisenberg, Erwin Schrödinger, Paul Dirac, and Wolfgang Pauli solved the

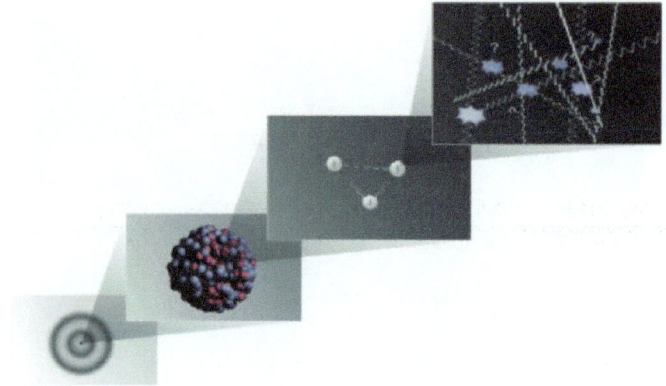

Fig. 1.3 Journey into the atomic nucleus

Fig. 1.4 From Earth to the cosmos

problem of wave-particle duality with the development of quantum theory. These discoveries mark the beginning of modern physics of the twentieth century. Since then, physicists have been discovering new, exciting phenomena and laws of nature.

Many discoveries of our time are significantly shaped by the use of physical methods. Examples include the decoding of deoxyribonucleic acid (DNA, ◘ Fig. 1.5), as well as the elucidation of sequences in the human genome. Physical research forms an indispensable element, especially of the natural and engineering sciences. It is the basis of modern technology. The semiconductor transistor, invented 50 years ago, has changed our entire electronic technology. Through it and other semiconductor components, whose miniaturization into highly integrated electronic circuits made it possible to develop modern computer and communication technologies.

Even at the beginning of the twenty-first century, physicists are researching fundamental problems of physics, chemistry, technology, and, increasingly, biology. The open field of exciting physical research in the future is enormous. Some examples: Physicists are now able to generate macroscopically coherent matter waves. These consist of a large number of atoms, but possess wave properties. The exploration of these coherent matter waves and their interaction with other matter and light promises to be very exciting. The exploration of smallest structures with diameters of a few nanometers (◘ Fig. 1.6), which consist of only a few atoms or molecules, will lead to numerous new technical applications in the so-called nanotechnology.

Methods of mathematical physics will remain indispensable for the analysis of complex dynamic processes and structure formation processes. Research continues in high-temperature superconductivity, among other things, with the aim of reducing losses in electrical power supply. Controlled nuclear fusion is one of the most ambitious future projects of physics (◘ Fig. 1.7).

The survey of the cosmos, e.g., with X-ray telescopes, will expand our knowledge about the origin, structure, and dynamics of the cosmos, as well as the analysis of invisible neutrino streams using elaborate underground detectors. Experiments and theoretical developments will lead to a new understanding of the elementary building blocks of the universe and a unified descrip-

DNA

◘ Fig. 1.5   DNA

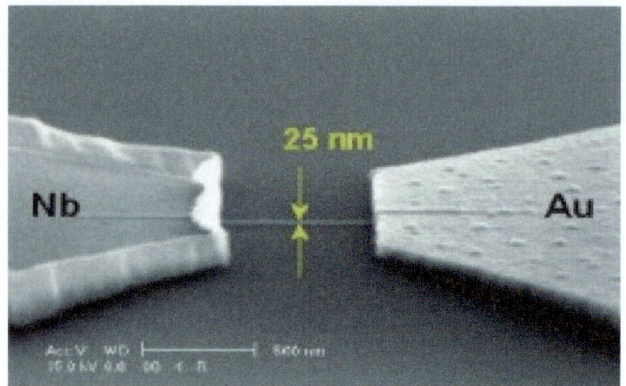

Fig. 1.6 Nanometer metal wire

Fig. 1.7 TOKAMAK experiment for the fusion of hydrogen

tion of the fundamental forces of nature. In the future, physicists will more intensively investigate questions of a physical nature in other natural sciences, such as biophysical elementary processes. Physics was, remains, and will continue to be a fundamental natural science, part of our culture, and the basis of our technology.

Munich 2002
Rudolf Mößbauer

Physical Quantities

Contents

2.1 Definition – 10

2.2 The Basic Quantities – 10

2.3 The Length – 13

2.4 Time – 16

2.5 The Mass – 19

2.6 The Angle Measures – 23

2.7 Notations – 23

© The Author(s), under exclusive license to Springer-Verlag GmbH, DE, part of Springer Nature 2025
S. Roth and A. Stahl, *Mechanics*,
https://doi.org/10.1007/978-3-662-68079-7_2

2.1 Definition

Physics describes objects and phenomena through certain properties, which are called "physical quantities". These physical quantities must be quantifiable. This can either be done through a measurement process, in which case we speak of a "measured quantity", or the quantity can be calculated from other measured quantities, in which case we speak of a "derived quantity". Physical laws convey the relationship between physical quantities.

To meaningfully state the value of a measured quantity, you must provide a numerical value, also called a magnitude, and a unit. So, if we describe a length as 1.5 kilometers, then 1.5 is the magnitude and "kilometer" is the unit. We use the unit "kilometer" as a scale and indicate with the magnitude 1.5 that the length is one and a half times as long as a kilometer.

2.2 The Basic Quantities

If scientific results are to be comparable worldwide, a uniform definition of physical quantities must be agreed upon. This uniform definition is based on a series of so-called basic or base quantities, from which the other physical quantities are derived. At the 10th General Conference on Weights and Measures (Conférence Générale des Poids et Mesures, CGPM) in Paris in 1954, seven base quantities were agreed upon, which, along with the associated base units, are recognized today in most countries. They are listed in ◘ Table 2.1.

◘ Table 2.1 Basic quantities according to the SI system

Basic quantity	Basic unit	Unit symbol
Length	Meter	m
Mass	Kilogram	kg
Time	Second	s
Electric current	Ampere	A
Temperature	Kelvin	K
Amount of substance	Mol	mol
Luminous intensity	Candela	cd

2.2 · The Basic Quantities

Since 1960, this agreement has been referred to as SI (SI = *Système International d'Unités*= International System of Units). The basic units of the SI are mandatory in the European Union and many other countries for official and business transactions and are also widely used in almost all areas. The imperial and US customary measurement systems go back to British standards. They are still in use, although today, the units are realted to the SI system.

In the SI system, the seven basic units are defined as follows:

Meter – The meter is the unit of length in the SI system. It has undergone a number of different definitions and is now indirectly traced back to the unit of time through the speed of light. A meter is the length of the path that light travels in vacuum in the 299,792,458th fraction of a second. The fraction was chosen so that this definition roughly corresponds to the previous ones.

Kilogram – The unit of mass in the SI system is the kilogram. Until 2019, it was defined by a standard mass body (prototype kilogram), the mass of which defined the kilogram. Today, it is derived from the Planck constant using a Kibble balance.

Second – The definition of the second has also gone through many versions. It is no longer derived from the astronomical definition of the day. Today, it refers to an oscillation process in cesium atoms of the nuclide ^{133}Cs, which is extremely stable. The second corresponds to 9,192,631,770 oscillation periods of the radiation emitted during the transition between the two hyperfine structure levels in the ground state of these atoms.

Ampere – The ampere is the basic electrical unit. It indicates the strength of the electric current. If a current flows through two parallel, straight and infinitely long conductors of negligibly small, circular cross-section, the conductors attract each other. For a long time, the ampere was defined by this force. If a current of 1 A flows through the conductors at a distance of 1 m in vacuum, a force of $2 \cdot 10^{-7}$ N per meter of conductor length results. Today, the ampere is traced back to the charge transported by moving electrons.

Kelvin – The Kelvin is used to measure temperatures. The definition is derived from the triple point of water. The Kelvin is the 273.16th part of the temperature of the triple point in relation to the absolute zero of temperature.

Today, the Kelvin is also traced back to a natural constant, namely the Boltzmann constant. We will discuss this in the volume on thermodynamics.

Mol – The unit Mol is used to measure the amount of substance in a system. It corresponds to the amount of substance composed of as many atoms, molecules, or ions as there are atoms in 12 g of the carbon nuclide ^{12}C. Like all other units today, this unit is also traced back to a natural constant. Here it is the Avogadro constant, which is set to the value $6.022\,140\,76 \cdot 10^{23}$. A mol is the amount of substance in a system that contains just as many atoms or molecules.

Candela – The unit Candela is finally used to indicate the intensity of a light source. It is based on a light source of monochromatic radiation of frequency $540 \cdot 10^{12}$ Hz. If the intensity of the radiation in a certain direction is exactly 1/683 W per steradian, this corresponds to a luminous intensity of one candela.

The 11th General Conference on Weights and Measures (CGPM) established two additional angular units in the SI system in 1960. They are listed in ◘ Table 2.2.

Radian – The radian indicates the angle in a plane. If you connect the legs of the angle with an arc, the radian is the ratio of the arc length to the length of the legs.

Steradian – The steradian is the corresponding solid angle. The definition is described in more detail in ▶ Sect. 2.6.

All other physical quantities and units can be traced back to the basic quantities and units of the SI system mentioned above. These additional quantities are traced back to basic quantities and supplementary quantities using definition equations. For example, the unit of force—the Newton—is not a basic unit. The force is traced back to the basic quantities of length, mass, and time through the Newton's fundamental law of mechanics.

◘ **Table 2.2** Supplementary SI Units

Quantity	Unit	Unit Symbol
Plane Angle	Radian	rad
Solid Angle	Steradian	sr

$$F = ma \tag{2.1}$$

From Eq. 2.1 it can be seen that for the Newton it must hold

$$1\,\text{N} = [F] = [m][a] = 1\,\text{kg}\frac{\text{m}}{\text{s}^2} \tag{2.2}$$

The square brackets around a quantity denote the unit of the enclosed quantity.

Please note that this selection of basic quantities, like the choice of standards, is a convention. For example, instead of time, one could have chosen speed as a basic quantity. Then time would appear as a derived quantity.

In fact, the SI system is not the only unit system still in use today. In the British Empire, the *imperial system* was officially in use for a long time and is still used on many occasions. It has many more basic quantities and units. In addition to length, basic quantities for area and volume are defined. This system is still officially used in the USA. It is almost the only country that has not joined the SI system. Another alternative unit system is the CGS system, which is sometimes used in science. The abbreviation stands for the three mechanical basic quantities in this system: "Centimeter-Gram-Second". It uses different standards in the field of mechanics than the SI system. The various extensions of the CGS system into electrodynamics even work without an electrical basic unit.

As we have seen, the number and selection of basic quantities is a convention. The SI system uses only seven basic quantities. But even these could be further reduced. For example, one could trace the amount of substance back to mass and therefore not treat it as a basic quantity, but as a derived quantity. Or one could trace the temperature back to thermal energy on m, kg, and s. In this book, we want to refer exclusively to the SI system as described above. The mechanical basic quantities are presented in more detail below.

2.3 The Length

The significance of length as a physical quantity should be intuitively clear. The unit of length is the meter. To realize a length measurement, one needs a length standard. In the simplest case, this is a rod whose length corresponds as closely as possible to one meter. To measure a length, one counts how often the length standard has to be applied until the length to be measured is reached. This gives the length of the distance in meters. However, for an accurate measurement, one also needs a way to divide the length standard into equal parts.

The meter was invented in Paris amidst the turmoil of the French Revolution to put an end to the chaos of the length measures of the time (Fig. 2.1). It was originally defined over the circumference of the Earth. The distance from the North Pole to the Equator, the so-called meridian quadrant, was assumed to be 10 million meters. In a six-year effort, the two astronomers Jean-Baptiste Delambre and Pierre Méchain measured the distance from Dunkirk in northern France to Barcelona using trigonometric methods. With the known curvature of the Earth, they were able to determine the length of the meter and represented it in the form of a scale, the so-called prototype meter. Based on this definition, several prototype meters were made from a platinum-iridium alloy. It was the standard in the SI system until 1960 (Fig. 2.2). However, even the elaborate metal alloy soon no longer met the requirements for stability. In addition, the comparison with the national standards of all the other countries was very laborious. They had to be regularly transported to the prototype meter in Paris and compared with it. Therefore, in 1960, it was decided to define the meter over the wavelength of a krypton laser. This can be reproduced in any laboratory, eliminating the laborious comparison with the prototype meter.

In 1983, it was finally agreed to completely abandon an independent definition of the meter. The meter has since been linked to the speed of light with the unit "sec-

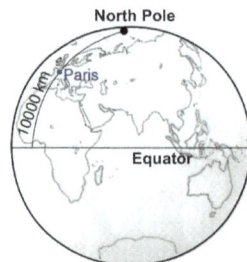

Fig. 2.1 Historical definition of the meter

Fig. 2.2 Two prototypes of the meter. At both ends there are three parallel markings. The middle ones indicate the beginning and end of the meter. © National Institute of Standards and Technology NIST

2.3 · The Length

ond". It is the distance that light travels in vacuum in the 299,792,458th part of a second. The meter is therefore no longer a base unit, but a unit derived from the second. We have mentioned it above for historical reasons nonetheless.

Experiment 2.1: Scales

The first experiment presents different measuring instruments for length measurement. The first three measuring instruments should be well known:

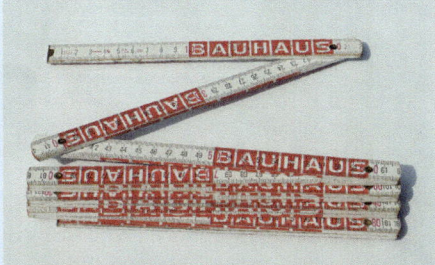

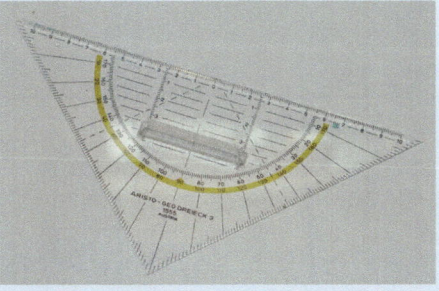

The following image shows a laser rangefinder measuring the width of a door way. It directly uses the definition of the meter based on the speed of light. The device emits a short laser light pulse, which is reflected at the end of the measuring distance. From the runtime Δt and the speed of light (in air) c the distance is calculated as $s = \frac{1}{2} c \Delta t$. Due to the very short times that are to be measured here (1 meter corresponds to about 3 ns), the accuracy is however limited. Even expensive devices only achieve an accuracy of typically 0.5 mm.

The last illustration shows a caliper, also called a slide gauge, for precise measurement of small distances. Equipped with a vernier scale or today more and more frequently with a digital display. It achieves an accuracy of up to 0.1 mm.

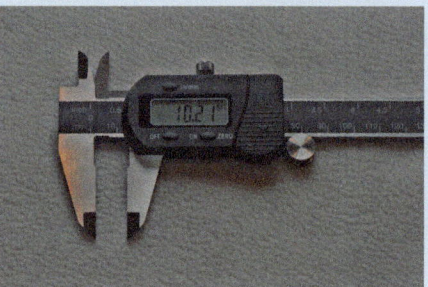

2.4 Time

Like length, time was long defined by astronomical processes, namely by the rotation of the Earth. A day is divided into 24 hours, an hour into 60 minutes, and a minute into 60 seconds. The second, the current base unit of time, was thus the $24 \times 60 \times 60 = 86{,}400$-th part of a day. The name "second" comes from Latin. It refers to the second division of the hour after the minute.

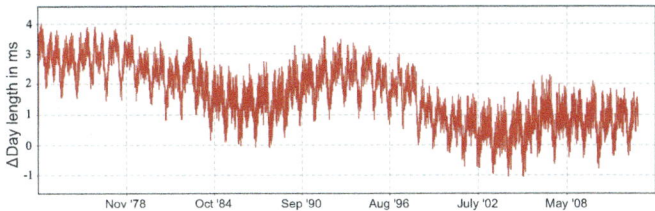

Fig. 2.3 Deviation of the actual day length from 86,400 s. © Federal Office for Cartography and Geodesy, International Earth Rotation and Reference Systems Service, Annual Report 2012

However, Earth's rotation and the length of a day vary from day to day. The deviations of the length of a day from 86,400 s are constantly recorded. They are shown in Fig. 2.3 for the years 1972 to 2013. One can see deviations of a few milliseconds. The fluctuations mainly arise because the Earth is not a completely rigid body. The Earth's core and mantle, but also the oceans and the atmosphere, move against each other and accelerate and decelerate each other in their rotations. This creates the fluctuations in the rotation of the Earth's crust, from which we measure the length of a day. The deviations accumulate over the days. As the cumulative deviation approaches a second, a leap second is inserted into the time calculation to keep the clocks synchronized with the course of the day.

While deviations of 3 ms in relation to the length of the day are of little significance for practical life, this accuracy is not sufficient for some scientific measurements. Therefore, as early as 1967, it was decided to derive the unit of time from atomic measurements. Today, it is defined by the period of a radiation transition in the ground state of ^{133}Cs atoms. If one compares the rate of several Cs atomic clocks with each other, one can assure that their fluctuations against each other are significantly lower than those of the Earth's rotation. The best Cs atomic clocks achieve frequency uncertainties of about 10^{-15}. This is better than 1 s in a million years. Further improvements and alternative clocks are being intensively researched.

> **Experiment 2.2: Clocks**
>
> Time is measured with clocks. Our photo shows a simple example. If very precise time pulses are needed in the laboratory, a good local clock can be synchronized to the radio signal of the National Metrology Institute or that of the GPS satellites. For the highest requirements, one must operate an atomic clock oneself.

The Bureau International des Poids et Mesures (BIPM) in Paris generates a universal time from the time cycle of several hundred atomic clocks of national authorities. It is called UTC for *Universal Time Coordinated*. From this, Central European Time (CET) is also derived (+1 hour). In Germany, the Physikalisch-Technische Bundesanstalt in Braunschweig is responsible for the official time. It disseminates the time cycle via radio, telephone, and the internet. So-called radio clocks synchronize themselves to this time signal.

> **Example 2.1: Fountain Atomic Clock**
>
> Cesium atoms are evaporated at room temperature forming a gas. In a cross of six opposing laser beams (optical molasses), about 10^6 cesium atoms can be collected in an atomic cloud and strongly cooled. The relative velocities between the atoms are reduced to values of a few centimeters per second. Now the atoms get a small "push" upwards. This also happens through momentum transfer from a laser. The atoms follow trajectories upwards until they, slowed down by gravity, fall back down. The movement resembles that of water molecules in a fountain, hence the name fountain clock. The trajectory reaches a height of about 1 m and lasts about 1 s.
>
> The atoms are in a uniform state before the flight. During the flight, they pass through a microwave resonator both when ascending and when falling back. It works like a Rabi apparatus. It stimulates transitions if the radiated frequency corresponds to the frequency of the transition in the atoms. When falling back, the transitions are detected below the resonator in the detection zone. With an electrical control loop, the microwave frequency is now adjusted to the value at which a maximum signal is generated. By definition, this is a fre-

quency of 9,192,631,770 Hz. One would now only have to count 9,192,631,770 oscillation periods and thus have determined a second. In practice, the second is obtained by frequency division from the quartz oscillator, from which the microwave signal for irradiating the atoms is generated.

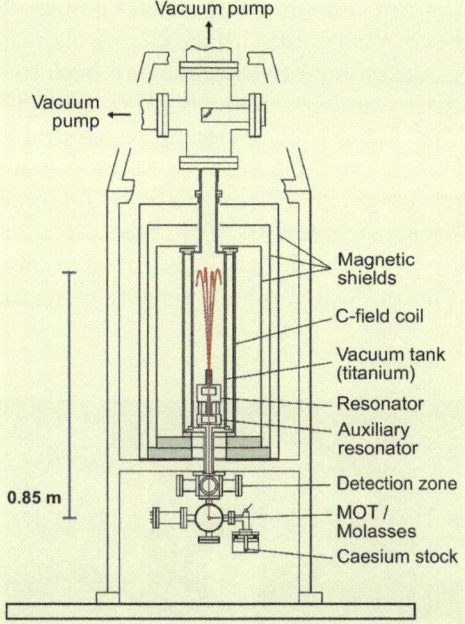

© Courtesy of the Physikalisch-Technischen Bundesanstalt

The photo shows the cesium fountains CSF1 and CSF2 of the Physikalisch-Technischen Bundesanstalt in Braunschweig.

© Courtesy of the Physikalisch-Technischen Bundesanstalt

2.5 The Mass

The kilogram was the last unit that was still defined by a material standard made by humans. It is the unit of mass. The International Bureau of Weights and Measures (BIPM) stores the prototype kilogram under three glass bells. It is shaped like a cylinder and is made of a particularly stable platinum-iridium alloy. ◘ Figure 2.4 shows a photo of it. The cylinder is 39 mm high and has a diameter of 39 mm as well. Its mass defined the kilogram. The national metrology institutes have their own standards, which they typically compare with working standards at the BIPM every ten years. The national standards deviate by up to one milligram from the original. The comparisons allowed the difference to be determined to within a few micrograms.

But even with the most careful handling, such a standard is not immune to minimal changes. Wear and chemical deposits on the surface change its mass. Therefore, ways were sought to trace the unit back to the measurement of natural constants. It is not enough to find a relation that defines the kilogram via natural constants, a measurement method must be developed that allows a standard weight to be realized with reasonable effort and sufficient accuracy. In 2018, measurement technology finally achieved the goal of a relative accuracy of 10^{-9}, so that on May 20, 2019, the day of metrology, the

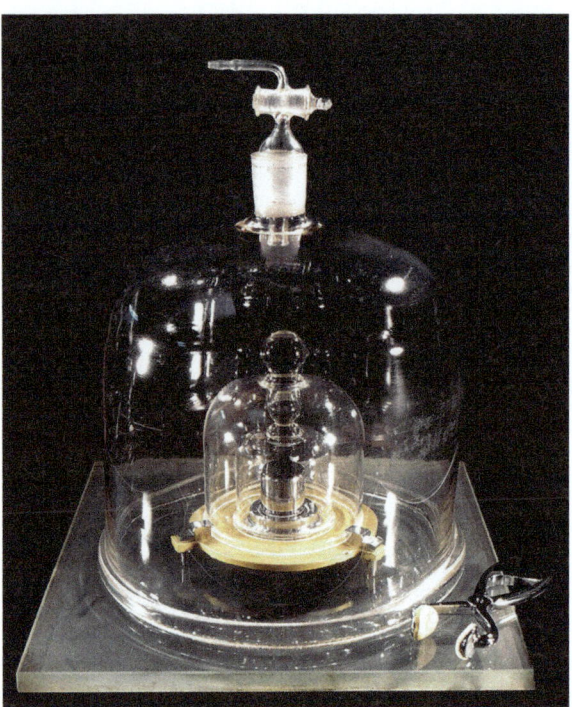

Fig. 2.4 The prototype kilogram of the International Bureau of Weights and Measures (BIPM) in Sèvres near Paris. © Courtesy of the International Bureau of Weights and Measures

2.5 · The Mass

new definition of the kilogram could come into force. The definition reads: "The kilogram, symbol kg, is the SI unit of mass. It is defined by taking the fixed numerical value of the Planck constant h to $6.62607015 \cdot 10^{-34}$ when expressed in the unit Js, which is equal to kg m/s², where the meter and the second are defined by c and Δv_{Cs}. Here, c denotes the speed of light and Δv_{Cs} refers to the frequency of the radiation transition described in ▶ Sect. 2.4 in ^{133}Cs atoms.

The definition does not specify a measurement method. Ultimately, any measurement method that achieves the required accuracy is acceptable. The BIPM has published a list of recognized measurement methods. The most well-known is the so-called Kibble balance, which compares the weight of a body with the magnetic force on a coil in a magnetic field. This force can be determined very precisely through quantum effects. In the volume on electricity and magnetism, we will discuss the Kibble balance, but the explanation of the quantum physical measurement will have to wait for later volumes of our series. In ◘ Fig. 2.5, the photo shows the Kibble balance of the National Institutes of Standards and Technology (NIST) in the USA. This measuring instrument is a beam balance, whose beam is designed in the form of a wheel. It can be seen in the photo at the top. Also, you can see on the right one of the arms of the balance, on which the weight to be determined is placed. The entire balance is located in a vacuum chamber to exclude the influence of buoyancy in air.

◘ **Fig. 2.5** Kibble balance for determining a weight via the Planck constant. © Jennifer Lauren Lee, for the U.S. National Institute of Standards and Technology (NIST)

> **Experiment 2.3: Scales**
>
> The mass of a body is determined with a scale. There are mechanical scales and electronic ones. Examples shown are a beam balance and a kitchen scale. In an electronic scale, the mass deforms a spring, a bending beam, or something similar due to its weight. The deformation is electronically detected and the mass is calculated from it.

© Wikimedia: Simon Speed

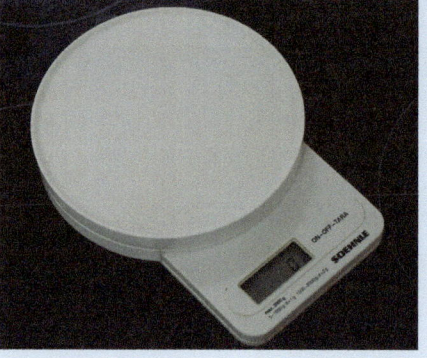

2.6 The Angle Measures

Angles are often measured in degrees (symbol °). A full circle corresponds to 360°. The degree is divided into arc minutes and seconds.
- 1° corresponds to 60′ (arc minutes)
- 1′ corresponds to 60″ (arc seconds)

In addition to the degree, the radian is also common as a unit. This is called the radian measure of the angle α. It can be traced back to length units (see ◻ Fig. 2.6), e.g. through the relation

$$\alpha = \frac{b}{r} \tag{2.3}$$

One can convert degrees and radians through the relation

$$2\pi \text{ rad} = 360 \text{ Grad} \tag{2.4}$$

The full circle has 2π rad.

The solid angle Ω is defined accordingly in three dimensions (see ◻ Fig. 2.7). The relation

$$\Omega = \frac{A}{r^2} \tag{2.5}$$

applies, where A is the area of the spherical section and r is the radius of the sphere.

The unit is called "steradian" (sr). The full sphere corresponds to a solid angle of 4π sr. The solid angle can also be determined from the opening angle of the cone in ◻ Fig. 2.7. It is

$$\Omega = 4\pi \sin^2 \frac{\alpha}{4} \tag{2.6}$$

where α is the full opening angle of the cone.

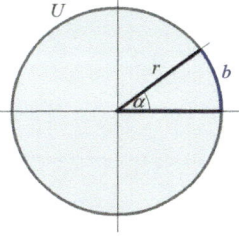

◻ **Fig. 2.6** Definition of the angle measure

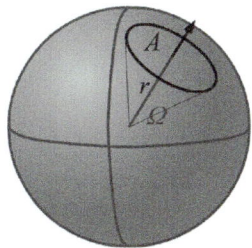

◻ **Fig. 2.7** Definition of the solid angle Ω

2.7 Notations

In science, both very large and very small numbers occur. In such cases, it can be advantageous to use exponential notation. In this notation, for example, the speed of light is

$$c = 2.99792458 \cdot 10^8 \frac{\text{m}}{\text{s}}. \tag{2.7}$$

For very large numbers, a small space or a comma is sometimes inserted after every third digit to separate the thousands. As an example, consider the average distance of the moon from the earth. It is 384,401.0 km.

The units are often provided with prefixes. The most important ones are given in ◘ Table 2.3.

? Problems

1. Which of the following physical quantities are scalars (a numerical value with a unit), and which are vectors (have a magnitude as well as a direction)?
   - Time
   - Momentum
   - Force
   - Temperature
   - Mass
   - Acceleration
   - Energy
   - Work
   - Speed

2. The Hagen-Poiseuille law describes the laminar flow of a fluid through a pipe:

$$I = \frac{\pi}{8\eta} \frac{r^4}{l} \Delta p$$

   Here, I is the volume throughput per unit of time, r is the radius and l is the length of the pipe and Δp is the pressure difference (unit N/m^2) between the ends of the pipe. Determine the physical unit of the material constant η, which is referred to as viscosity. Express this in terms of the base units of the SI system.

3. In the following equations, the distance s in given in meter (m), the time t in seconds (s) and the speed v in meters per second (m/s). Determine the physical unit of the constants c_i:

   a) $s = c_1 + c_2 t + \frac{1}{2} c_3 t^2$
   b) $v^2 - c_1^2 = 2c_2(s - c_3)^2$
   c) $t = \sqrt{\frac{2s}{c_1}}$
   d) $s = c_1 \cos(c_2 t) + \frac{c_3}{c_2} \sin(c_2 t)$

2.7 · Notations

Table 2.3 Metric prefixes of units

Prefix	Power	Abbreviation
Exa	10^8	E
Peta	10^{15}	P
Tera	10^{12}	T
Giga	10^9	G
Mega	10^6	M
Kilo	10^3	k
Hecto	10^2	h
Deca	10^1	da
	10^0	–
Deci	10^{-1}	d
Centi	10^{-2}	c
Milli	10^{-3}	m
Micro	10^{-6}	μ
Nano	10^{-9}	n
Pico	10^{-12}	p
Femto	10^{-15}	f
Atto	10^{-18}	a

4. Have you noticed that the height and diameter of the cylindrical standard of the kilogram coincide? This is no coincidence. Show that this results in the minimum surface area for a given volume.

Measurement Errors

Contents

3.1 Measured Value – 28

3.2 Statistical Interpretation – 32

3.3 Error Propagation – 33

3.4 Systematic Errors – 35

© The Author(s), under exclusive license to Springer-Verlag GmbH, DE, part of Springer Nature 2025
S. Roth and A. Stahl, *Mechanics*,
https://doi.org/10.1007/978-3-662-68079-7_3

3.1 Measured Value

"Measurement is an art" is a well-known saying about experimental physics. Even if this saying should not be taken literally, it indicates an important problem. Anyone who carries out a measurement may hope to obtain the true value of a quantity as a result, but unfortunately this is not the case. There are many reasons why the measured value will deviate from the true value: environmental influences disturb the measurement, the measuring devices only have limited accuracy, to name just two examples. The experimenter must live with the fact that the measurement result does not reflect the true value. The art lies in getting as close as possible to the true value.

We want to formalize this a bit: The deviation of a value obtained from measurements from the true value of the measured quantity is called measurement error Δx. The definition is

$$\Delta x = x_i - x_t \tag{3.1}$$

Here, x_i is the measured value and x_t is the true value. The experimenter's tasks include not only developing a measurement method that keeps the measurement error within acceptable limits, but also determining its magnitude. In many cases, this task is more difficult and time-consuming than the actual measurement. The definition from Eq. 3.1 is useless for determining the measurement error, as the true value is unknown. Therefore, one must try to estimate the measurement error in another way. How this is done in detail, you will learn in the laboratory courses. It would go beyond the scope of this book.

Example 3.1: Statistical Error

From the measurement of the oscillation period of a pendulum, one can determine the standard acceleration of gravity on Earth (see Experiment 17.5). Try yourself! Using a stopwatch, you determine the duration of the oscillations. To reduce the inaccuracy when starting and stopping the clock, you measure 20 oscillation periods. You deflect the pendulum and let it swing. Then you start the clock at maximum deflection (where the motion is slowest), count off 20 oscillation periods, and stop the clock again at maximum deflection. You carry

3.1 · Measured Value

out this measurement a total of 15 times. The table shows the measured values.

No.	Measured Value
1	43.4 s
2	41.0 s
3	42.8 s
4	40.6 s
5	41.2 s
6	41.8 s
7	41.8 s
8	45.6 s
9	39.0 s
10	44.8 s
11	41.4 s
12	41.8 s
13	41.2 s
14	37.4 s
15	41.2 s

From these measurements, you determine the average value $\overline{T}$ using the formula

$$\overline{T} = \frac{1}{N} \sum_{i=1}^{N} \frac{t_i}{20}$$

where t_i are the measurements from the table for 20 oscillation periods and N is the number of measurements (here 15). It results in $\overline{T} = 2.083$ s. However, the 15 measurements contain not only information about the average value, but the scattering of the measurements can also be used to determine the statistical error. Here, one must consider the deviation of the individual measurements from the average value. The so-called standard deviation of the individual measurement σ indicates how far these deviate from the average value. It is calculated according to

$$\sigma_{T_i} = \sqrt{\frac{1}{N-1} \sum_{i=1}^{N} \left(\frac{t_i}{20} - \overline{T}\right)^2}$$

With the values from the table, we obtain $\sigma = 0.010$ s. Now let's look at the first measurement from the table. It yields the measurement value $T_1 = (2.17 \pm 0.10)$ s, from the second $T_2 = (2.05 \pm 0.10)$ s etc. But you have already determined the average. This gives a more accurate result than a single measurement. Its statistical error can also be determined from the table. Now you need to determine the standard deviation of the average. This is done using the formula

$$\sigma_{\overline{T}} = \sqrt{\frac{1}{N-1}\frac{1}{N}\sum_{i=1}^{N}\left(\frac{t_i}{20} - \overline{T}\right)^2} = \frac{\sigma_{T_i}}{\sqrt{N}}$$

You will receive $\sigma_{\overline{T}} = 0.026$ s and thus as a result of the measurement

$$\overline{T} = (2.084 \pm 0.026)\text{ s}$$

The result of a measurement is given as

$$x = (A \pm \Delta A) \cdot \text{unit} \qquad (3.2)$$

with the measured value A and the (absolute) measurement error ΔA. The relative measurement error is $\Delta A / A$. Only the significant figures are given. It makes no sense to specify a measurement with an accuracy that significantly exceeds the measurement error. A sensible specification would be, for example,

$$L = (1.53 \pm 0.02)\text{ m} \quad \text{or} \quad L = (1.534 \pm 0.023)\text{ m} \qquad (3.3)$$

The measured value and measurement error are always given with the same number of decimal places.

> **Example 3.2: Planning a measurement: Measurement accuracy**
>
> Imagine the following situation: You want to buy paint at the hardware store to repaint your room. How many liters of paint do you need to get?
>
> First, you need to determine the wall area, but unfortunately, you don't have a measuring tape or a laser distance meter or anything similar. The room height is 2.5 m. You come up with the idea of measuring the width of the walls with your feet, but you wonder if this is accurate enough.

3.1 · Measured Value

Your measurement for the rectangular room gives a base area of 14.5 feet × 16.5 feet. You estimate that the length of your foot is between 20 cm and 30 cm and choose as a measure (25 ± 5) cm. This results in a wall area of $(2 \cdot 14.5 + 2 \cdot 16.5) \cdot 0.25\,\text{m} \cdot 2.5\,\text{m} = 38.75\,\text{m}^2$. You subtract $2\,\text{m}^2$ for the door and $1.5\,\text{m}^2$ for two windows, which results in $33.75\,\text{m}^2$. The paint has a yield of $10\,\text{m}^2/l$, so that a 10-liter bucket would be enough for two coats of paint. How large is the uncertainty of this estimate? You calculate again with 20 cm or 30 cm foot length and get between $26\,\text{m}^2$ and $41.5\,\text{m}^2$. In all cases, you come to the same conclusion: A 5-liter bucket is too little, a 10-liter bucket is large enough even in the worst case. You buy a 10-liter bucket.

Of course, it is possible to measure the walls of your room more accurately, but you would not benefit from it. The 10-liter bucket is the right amount in any case. A more precise measurement would have only caused more effort.

The example is intended to show how important it is to consider the accuracy of the measurement that is required before starting. This significantly determines the effort that must be made.

3.2 Statistical Interpretation

In the previous section, you learned that stating a measurement without error makes little sense: "The length of a distance was determined to be 1.53 m." What does this mean? Could it be that the true value of this distance is below 1.5 m? Or even below 1.4 m? Without stating an error, none of these questions can be answered.

Even with error indication, the answer to these questions is not obvious. Let's assume that the measurement resulted in $L = (1.53 \pm 0.02)$ m. Does this ensure that the true value is not below 1.5 m? Not at all! The error only indicates how far the measured value could be from the true value. However, it does not limit the deviation.

A correctly determined error is based on a statistical interpretation. It is imagined that a measurement is not only carried out once, but many times. It is important that the individual measurements are independent of each other, i.e., errors that lead to a deviation by a certain value in a certain direction in one measurement must not lead to the same (or correlated) deviation in other measurements. For example, when measuring length, one must imagine that a different ruler is used for each measurement, otherwise deviations in the actual length of the ruler would lead to the same error each time. The measurement results of these many measurements will scatter around a central value. The error is a measure of this scatter. The exact meaning depends on the probability distribution of the measured values.

In ◻ Fig. 3.1, such a probability distribution can be seen. It indicates the probability with which a certain measured value will occur with multiple repetitions of the measurement. As an example, a normal distribution (Gaussian curve) is chosen here.

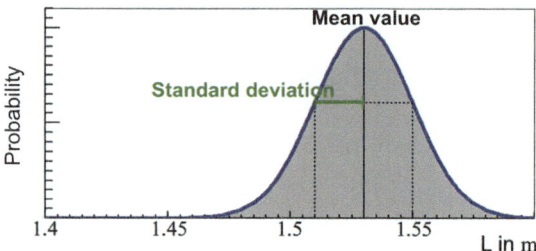

◻ **Fig. 3.1** Probability distribution of a simple length measurement with Gaussian error

$$\text{Prob}(x) = \frac{1}{\sigma\sqrt{2\pi}} e^{-\frac{1}{2}\left(\frac{x-\mu}{\sigma}\right)^2} \tag{3.4}$$

with the mean μ and the standard deviation σ. The mean μ, also called expectation value, is the most probable value. It will be given as the measured value. The standard deviation σ determines the width of the curve. It is given as the error, so Result $= \mu \pm \sigma$. If one carries out a single measurement, it can be determined from the curve how likely it is that the measured value will lie between $\mu - \sigma$ and $\mu + \sigma$. The result is 68.27%. The error specification $L = (1.53 \pm 0.02)$ m, implies that with a multiple repetition of this measurement, 68.27% of the measurement results will fall within 2 cm of the average value 1.53 m.

The number 68.27% is called the "confidence interval" of the measurement. This confidence interval is often used because it corresponds exactly to one standard deviation of the normal distribution. However, other confidence intervals can also be used. Strictly speaking, for a complete specification, the form of the probability distribution should also be specified in addition to the measured value and the error. But this is only rarely done due to the central limit theorem of statistics. It states that when many uncorrelated statistical scatterings are superimposed, a normal distribution results in the end, regardless of the form, which the superimposed probability distributions originally had. Since a measurement is usually a superposition of many statistical effects, the measurement results should follow a normal distribution. This is implicitly assumed when specifying a measurement as $L = (1.53 \pm 0.02)$ m.

3.3 Error Propagation

The previous section referred to the direct measurement of physical quantities. We had considered a length measurement as an example. Now we want to investigate the influence of measurement errors on the determination of derived quantities.

We consider again the rod of length $L = (1.53 \pm 0.02)$ m. Let's assume that the rod has a radius of $r = 2.5\,\text{cm}$. Then how large is the volume V? It applies

$$V = \pi r^2 L \tag{3.5}$$

The result is $3004.1\,\text{cm}^3$. How large is the error due to the inaccurate measurement of the length? It can be determined by using $L = (1.53 + 0.02)\,\text{m}$ and $L = (1.53 - 0.02)\,\text{m}$ as the length. The resulting volumes are $V_+ = 3043.4\,\text{cm}^3$ and from $V_- = 2964.9\,\text{cm}^3$ which leads to $V = (3004 \pm 39)\,\text{cm}^3$.

The error of the direct measurement of the length has an influence on the accuracy of the determination of the volume as a derived quantity. This is referred to as error propagation to the derived quantity. Above, the error on the volume was determined by inserting different values of the length, quasi by trial and error. Although this leads to the correct result, it is not exactly an elegant procedure. In error calculation, one learns that the error on a quantity x propagates to a derived quantity A as follows:

$$\sigma_A = \frac{\partial A(x)}{\partial x} \sigma_x \tag{3.6}$$

Here, σ_x the error on the quantity x and σ_A the propagated error on A. In between is the derivative of A with respect to x. Applied to our example, the result is

$$\sigma_V = \frac{\partial V(L)}{\partial L} \sigma_L = \pi r^2 \sigma_L = 19.6\,\text{cm}^2 \cdot 2\,\text{cm} = 39.2\,\text{cm}^3 \tag{3.7}$$

in accordance with the empirical determination by trying out above.

Finally, one can ask the question of how the error changes on the derived quantity when more than one error-prone quantity is involved in the determination. Then, the errors from the various input quantities must be individually propagated and added in quadrature to a total error. In the case of a quantity A, which depends on two input quantities x and y, this would be[1]

---

[1] Here, partial derivatives appear in the formula, e.g. $\frac{\partial A(x,y)}{\partial x}$. This notation means that the function A is differentiated with respect to the mentioned variable (here x) while all other variables are kept constant.

$$\sigma_A = \sqrt{\left(\frac{\partial A(x,y)}{\partial x}\right)^2 \sigma_x^2 + \left(\frac{\partial A(x,y)}{\partial y}\right)^2 \sigma_y^2} \quad (3.8)$$

or with any number of input quantities x_i with errors σ_i

$$\sigma_A = \sqrt{\sum_{i=1}^{n} \left(\frac{\partial A(x_1,\ldots,x_n)}{\partial x_i}\right)^2 \sigma_i^2} \quad (3.9)$$

Let's return to our example of determining the volume of the rod and assume that in addition to the length measurement, the measurement of the radius is also subject to error. Let $r = (2.5 \pm 0.1)$ cm. Then it is

$$\begin{aligned}\sigma_V &= \sqrt{\left(\frac{\partial V(L,r)}{\partial L}\right)^2 \sigma_L^2 + \left(\frac{\partial V(L,r)}{\partial r}\right)^2 \sigma_r^2} \\ &= \sqrt{(\pi r^2)^2 \sigma_L^2 + (2\pi rL)^2 \sigma_r^2} \\ &= \sqrt{(39.2 \text{ cm}^3)^2 + (240.3 \text{ cm}^3)^2} = 243.5 \text{ cm}^3\end{aligned} \quad (3.10)$$

The result is now $V = (3000 \pm 240)$ cm^3.

3.4 Systematic Errors

In the previous sections, you have learned about measurement errors that lead to statistical fluctuations of the measured value. In the example of measuring the length of a rod, the experimenter will apply the scale slightly differently in a repetition of the measurement, resulting in a different measured value. This would be an example of an effect that causes the measured value to scatter. The influence of this effect can be reduced by repeating the measurement several times and giving the average of all measurements instead of a single measurement. In the individual measurements, he will have applied the scale too far forward and too far back, which then averages out. Therefore, by repeating the measurement, this error can be reduced. Such measurement errors are therefore called statistical errors.

Not every measurement error is of this statistical nature. If the experimenter always uses the same scale to measure the length of the rod, an error occurs that leads to a systematic deviation of the measured values in a certain direction. This is called a systematic error. For example, if the scale is actually a little too long, the measured 1.52 m actually corresponds to a slightly greater length. The influence of this effect cannot be reduced by repeating the measurement, as the same error occurs anew with each measurement. Sometimes statistical and systematic errors are reported separately in the results. If, in the length measurement described above, an error of 1 cm were added due to the inaccuracy of the scale, one could write $L = (1.53 \pm 0.02_{stat} \pm 0.01_{sys})$ m.

The determination of the systematic error is often the greatest challenge for the experimenter and in many cases takes more time and energy than the determination of the measured value itself. In our example, the experimenter might be lucky and the accuracy of the scale was noted by the manufacturer on the scale. Then he will probably adopt this value. If not, he could determine the systematic error caused by the limited accuracy of the scale by obtaining several different scales. If he repeated the measurement 10 times with the first scale and stated the average value, he will now perform it 10 times with each of the new scales and compare the average values with each other. The scatter of the average values against each other gives him a measure of their accuracy. The difficulty with systematic errors is that they can be overlooked. Each measurement has different systematic errors. The experimenter must first recognize where systematic errors could be hidden before he can devise a method for their determination (or elimination).

❓ Problems

1. You buy a 125-g cup of yogurt. Your kitchen scale (accuracy 0.5 g) measures a weight of 137.6 g. After you have eaten the yogurt, you measure the rinsed cup. The scale shows 12.9 g. Is that okay?
2. Your wristwatch has a time stability of 10^{-5}. How often should you set the clock if it should not be more than half a minute wrong?
3. With a time-of-flight measurement of laser pulses, the distance from the Earth to a mirror left behind by the Apollo astronauts on the moon can be determined with a relative accuracy of $3 \cdot 10^{-10}$. What is the absolute accuracy of the measurement (distance Earth–Moon: 326,321 km)? With what accuracy was the time of flight measured?

3.4 · Systematic Errors

4. In Example 3.1, the gravitational acceleration g was to be determined from the measurement of the period T of the oscillation of a pendulum of length l, where the following applies:

$$g = \frac{4\pi^2 l}{T^2}$$

Show that the error propagation law gives:

$$\frac{\sigma_g}{g} = \sqrt{\left(\frac{\sigma_l}{l}\right)^2 + \left(2\frac{\sigma_T}{T}\right)^2}$$

A 1.1 m long pendulum was used, with the uncertainty of the pendulum length estimated at 2 mm. Determine the acceleration due to gravity g and its error, along with the measurement of the period $T = (2.095 \pm 0.022)\,\text{s}$. What is the dominant contribution to the error? Can any of the contributions be neglected?

Methodology

Contents

4.1 Physical Theories – 40

4.2 Scientific Method – 41

4.3 Scope of Classical Physics – 43

© The Author(s), under exclusive license to Springer-Verlag GmbH, DE, part of Springer Nature 2025
S. Roth and A. Stahl, *Mechanics*,
https://doi.org/10.1007/978-3-662-68079-7_4

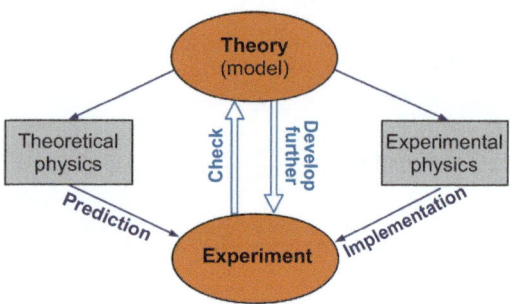

Fig. 4.1 Schematic representation of the interplay between theory and experiment

4.1 Physical Theories

The method of physics is an interplay between model building and experimental verification of the models. The models contain ideas about the nature of the universe within the validity range of the model. They are written in the language of mathematics. The models allow us to make predictions about the results of experiments. Likewise, everyday processes can be described and predicted with the models. The predictions are verified with experiments. The more frequently and precisely this is achieved, the more confidence physicists have in a model.

Verifying the models is not the only task of the experiments. It may be a dream to find a model one day that explains nature quantitatively on its own, but to this day all models contain a number of natural constants, whose values are not given by the models. They must be determined experimentally. Only when all natural constants occurring in a model have been measured can the model make quantitative predictions. The same applies to material constants, which appear in many models and specify the properties of certain materials[1].

Not every idea about physical processes can be accepted as a model. There are fixed criteria that a model must meet (Fig. 4.1). The most important are internal consistency and falsifiability. A model must be free

---

[1] For example, the melting temperature of iron is a material constant, while the gravitational constant is a natural constant. The melting temperature of iron is at least in principle calculable within the framework of the applicable model , the gravitational constant is not.

of internal contradictions, and it must be possible to derive predictions from it that can be validated experimentally. An idea that cannot be checked is not considered a model. Sometimes the term "model" is also used as "theory". Basically, both refer to the same thing. If the scope is rather specific and narrow, physicists usually use the term "model", while "theory" is reserved for cases with a large scope that have been extensively verified.

Physics can be divided into two disciplines: Theoretical physics studies theories and derives predictions for experiments. Experimental physics develops experiments and carries them out. The boundaries are fluid and it is important for a good physicist to master both at least to some extent.

4.2 Scientific Method

There is an important point that we need to examine more closely. It comes from epistemology (or philosophy of science), a field of philosophy. It deals with how we gain knowledge, among other things, in natural science.

Can one prove that a certain theory is true? Surprisingly, the answer is "no"[2]. Even if many experiments yield a consistent result when testing a theory, it does not necessarily mean that the theory must be true, because this is an inductive reasoning. Induction is the (logical) inference from individual cases to general statements. Such conclusions are only logically compelling under special boundary conditions (e.g., in the case of complete induction as a mathematical proof). An experiment always examines a particular individual case, while the theory makes statements about the general case. This is usually discussed in philosophy using a simple example of swans as the Hume-Kant-Popper's induction problem. The example is as follows: You observe (adult) swans and find that all observed swans have white plumage. Does this justify the statement (theory) "All swans are white"? The answer is "no", because observing a series of swans says nothing about the color of unobserved swans. The next one could be black (◻ Fig. 4.2). In the method of natural science, falsification replaces the proof of the validity of

---

[2] This approach to epistemology is called "Critical Rationalism", an important representative was Karl Popper.

◻ **Fig. 4.2** David Hume (1711–1776), Immanuel Kant (1724–1804), and Karl Popper (1902–1994) recognized that, although they had never observed a swan whose plumage was not white, one cannot conclude that all swans are white. The photo shows the counterproof, an Australian black swan. © With kind permission from Christian Träger

a theory (verification). The scientist does not try to prove a theory, but tries to refute it (falsify). If he still finds a match between his result and the theory, confidence in the theory increases, but a single experiment that demonstrates a contradiction is enough to refute the theory.

Unfortunately, this is not so clear in practice. There are almost always measurements for every theory that do not agree with it. Scientists discuss the reliability of the experiments or the theoretical predictions, speculate about solutions or misunderstood aspects of the theory. A classic example is Michelson's measurement of the speed of light in 1881 and 1887, which contradicted the then theory of light propagation. These measurements were only recognized as a falsification of the theory 30 years later, when Einstein designed the "correct" theory with the special theory of relativity in 1905. Such processes contain a lot of psychology[3].

Epistemology cannot be the subject of this book. However, we strongly recommend engaging with epistemology. It sharpens the thinking and helps to become a good physicist.

We would like to draw your attention to another aspect of the methodology of scientific thinking, which we believe is important for understanding physics, too. These thoughts go back to the critical philosophy of Imma-

---

[3] This was first outlined in the book by Thomas S. Kuhn: The Structure of Scientific Revolutions.

nuel Kant. He dealt with the concepts of "a priori" and "a posteriori". He refers to knowledge gained from experience, which in science is obtained through experiments and observations, as a posteriori. He refers to knowledge that precedes any experience as a priori. Such a priori are strictly universally valid and necessary for any science. They are prerequisites on which we build science. Here are some examples:

- Scientific thinking is based on a concept of space and time. Without such a concept, we could not practice physics.
- An integral part of this concept of space and time is the homogeneity of space. It refers to the assumption that all places in space are equal. Only in this way is it possible for scientific results to be verified by other scientists, as the test is likely to be conducted in a different place.
- The temporal constancy of natural laws is another prerequisite of physical science. Experiments must be repeatable and always lead to the same result, which implies that the underlying laws of nature do not change.
- We assume that the laws of nature are simple—we often speak of the elegance of a theory. If several, differently complex explanations compete with each other, we choose the simplest explanation, e.g., one that is based on a monocausal chain.

We believe it is important for you to be aware that there are such prerequisites for scientific thinking, which also mark the limits of scientific thinking.

4.3 Scope of Classical Physics

The subject of this volume is classical mechanics. It is part of classical physics. Like all physical theories, classical mechanics has a limited scope. It can be described by a triangle with three boundaries (◘ Fig. 4.3). At the boundaries of the triangle, classical mechanics loses its validity.

These are reached as soon as
1. Speeds approach the speed of light, i.e.

$$\frac{v}{c} \approx 1$$

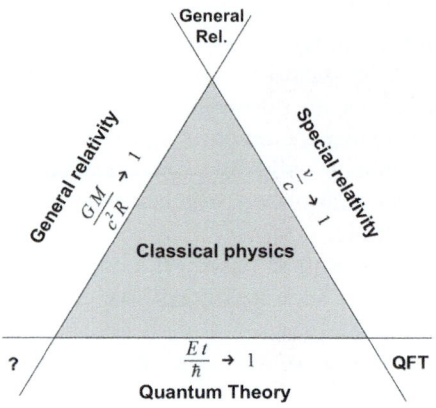

Fig. 4.3 Scope of classical physics

2. The influence of masses or energy densities causes a significant space-time curvature, i.e.

$$\frac{GM}{c^2 R} \approx 1$$

3. Effects are no longer large compared to Planck's quantum of action, i.e.

$$\frac{Et}{\hbar} \approx 1$$

Classical physics must then be replaced by a more general theory. These are respectively:
1. Special theory of relativity,
2. General theory of relativity,
3. Quantum theory.

If more than one of the above limit cases is reached, then one uses at

1. + 2. the General Theory of Relativity (includes the Special Theory of Relativity).
1. + 3. the Relativistic Quantum Field Theory (QFT).
2. + 3. the Quantum Theory of Gravity. However, this theory does not yet exist.

Mechanics of Point Masses

Contents

Chapter 5   Kinematics of the Point Mass – 47

Chapter 6   Dynamics of a Point Mass – 81

Chapter 7   Work and Energy – 107

Chapter 8   Momentum – 137

Chapter 9   Friction – 163

Chapter 10   Fictitious Forces – 177

Chapter 11   Celestial Mechanics – 209

Kinematics of the Point Mass

Contents

5.1 The Point Mass – 48

5.2 Reference Systems – 49

5.3 Uniform Motion – 56

5.4 Non-uniform Motion – 59

5.5 Acceleration – 61

5.6 Free fall – 65

5.7 Projectile Motion – 70

5.8 Circular Motion – 75

© The Author(s), under exclusive license to Springer-Verlag GmbH, DE, part of Springer Nature 2025
S. Roth and A. Stahl, *Mechanics*,
https://doi.org/10.1007/978-3-662-68079-7_5

5.1 The Point Mass

The point mass is a model representation of a body. It is primarily used to simplify the description of the body's motion. Properties of the body such as volume and shape are neglected. The body is considered as a mathematical point that has no extension, but a finite mass. The entire mass of the body is localized at this point. Usually, one would choose the center of gravity[1] of the body as this point.

Describing a body as a point mass is an approximation. In nature, there are no point masses. Every body has an extension. The approximation can only describe the behavior of the body incompletely. For example, a point mass has no aerodynamic drag, as it is proportional to the cross-sectional area of the body, which is zero for a point mass. Or consider the rotation of a body: It is meaningless to talk about the rotation of a point around itself or its orientation in space. One cannot assign a rotation to a point. Nevertheless, a point mass can be an excellent approximation. It does not necessarily depend on the actual size of the body. For example, one can determine the trajectory of ions through a mass spectrometer with point masses, just as one can calculate the motion of the Earth around the Sun assuming that the Earth is a point mass. It depends on whether the effects that are ignored are negligibly small. In the case of the Earth's orbital motion, for example, one ignores that the Sun attracts the sun-facing side of the Earth more strongly, as it is closer to the Sun than the opposite side. You have to decide in each individual case whether the point mass is an appropriate approximation or not.

This raises the question of what is an appropriate approximation. To answer this question, one must first realize that one can never solve a real problem exactly. Even with seemingly simple problems, such as the fall of a lead ball in a vacuum, one can discover influences that one has not taken into account in one's calculation. In the case of the lead ball, this would be, for example, the gravitational pull of the experimenter on the lead ball. The treatment of a real problem can only be done with finite accuracy. Anyone who approaches a problem

---

1   The exact definition of the center of gravity follows in ▶ Sect. 12.3.

should think in advance about the accuracy they want to achieve. All approximations, including that of the point mass, must be validated against this accuracy. If the effects neglected do not compromise the desired accuracy, the point mass is an appropriate approximation.

> **Example 5.1: Pendulum**
>
> In treating the pendulum's weight as a point mass, it is assumed that the forces act at the center of gravity of the weight and that a uniform pendulum length applies to all points of the weight. We want to illustrate this approximation with an estimate. We consider a spherical weight with a radius $R = 2$ cm and a mass $m = 200$ g, which hangs on a thread of length $l = 80$ cm. If it is deflected by 5° from the rest position, the restoring force is 0.17 N. If we let the weight go, it will swing and reach its maximum speed at the zero crossing. At this point, the air resistance is also maximum. As we will learn in ▶ Sect. 16.7, it is $4.8 \cdot 10^{-9}$ N. In treating the pendulum body as a point mass, we neglect this force compared to the restoring force.

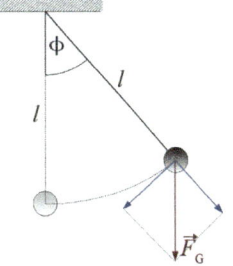

5.2 Reference Systems

To describe the motion of a body, we must choose a reference system. The reference system determines the location in relation to which the position of the body is specified, and it establishes the spatial directions in relation to which directions are given.

The motion of a body can appear very different from different reference systems (◘ Fig. 5.1). Imagine a train moving at a constant speed through a station. A passenger drops a wooden ball out of the window. From his frame of reference (coordinate origin anchored to the train, x axis upwards, y axis towards the platform, z axis in the direction of travel) the ball seems to fall straight down at first, until it is finally deflected backwards, i.e. in the negative z direction, by the airstream. From the frame of reference of an observer on the platform, a completely different picture emerges (coordinate origin stationary with the platform, x axis upwards, z axis in the direction of the train's movement, y axis perpendicular to both):

Fig. 5.1 View from a high-speed ICE train on a neighboring slower train. From the perspective of the ICE, the slower train seems to be moving backwards

The wooden ball initially flies at the same speed as the train in the positive z-direction. Its path points increasingly downwards due to the fall, and the speed in the z-direction decreases due to air drag.

Both descriptions of the trajectory are equivalent. One cannot say that one is better than the other or even that one is correct and the other is wrong. It is a question of practicality which reference system one chooses. Usually, we choose a reference system in which the description or calculation of the motion appears as simple as possible. Objective questions can be answered from any reference system. Imagine a basket was placed on the track bed under the train's window. Does the wooden ball hit the basket or not? This is an objective question. The same answer must be obtained in every reference system. The question of whether the ball shatters upon impact can also be answered objectively. Other questions may depend on the reference system, such as the question of what speed the ball has upon impact (in relation to the origin of coordinates).

To quantitatively describe movements, coordinate systems are used. The most common is the Cartesian or orthogonal coordinate system. Cylinder- or spherical coordinates are also frequently used. These three coordinate systems are shown in ◘ Fig. 5.2. In addition, there are many other coordinate systems, often adapted to specific problems.

5.2 · Reference Systems

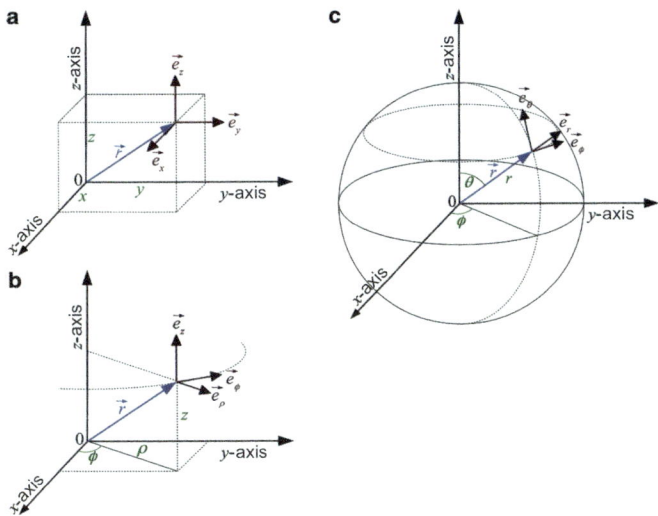

Fig. 5.2 Coordinate systems: Cartesian coordinates (**a**), Cylinder coordinates (**b**), Spherical coordinates (**c**)

A point in a Cartesian coordinate system is represented by the intersects with the x-, y- and z-axes. Expressed as a vector:

$$\vec{r} = (x, y, z) \tag{5.1}$$

In cylindrical coordinates, the position vector $\vec{r}$ is projected onto the x-y plane. The length of the projection is called ρ. The angle ϕ of the projection to the x axis and the height z of the point above the x-y plane are given. As a vector, we write

$$\vec{r} = (\rho, \phi, z) \tag{5.2}$$

In spherical coordinates, the length of the position vector $\vec{r}$, the angle ϕ, which we already know from cylindrical coordinates (azimuthal angle), and the angle θ of the position vector against the z-axis (polar angle) represent the location of the point:

$$\vec{r} = (r, \phi, \theta) \tag{5.3}$$

The axes of the coordinate systems are characterized by unit vectors along the axes. They are shown in red in **Fig. 5.2**. In these three examples, the coordinate axes are perpendicular to each other. They form a right-handed tripod. However, there are also coordinate systems where the axes are not perpendicular to each other. You can convert the coordinates of a point from one coordinate system to another. These so-called coordinate transformations can be derived from the figures. For the conversion between Cartesian coordinates and polar coordinates the result is:

$$\begin{aligned} x &= \rho \cos\phi & \rho &= \sqrt{x^2 + y^2} \\ y &= \rho \sin\phi & \phi &= \arctan\frac{y}{x} \\ z &= z & z &= z \end{aligned} \qquad (5.4)$$

In the calculation of the angle ϕ, the periodicity of the arc tangent function must be taken into account. For negative x, it may be necessary to add or subtract π. The conversion between Cartesian and spherical coordinates is obtained with the same remark about the angle ϕ as

$$\begin{aligned} x &= r\sin\theta\cos\phi & r &= \sqrt{x^2 + y^2 + z^2} \\ y &= r\sin\theta\sin\phi & \phi &= \arctan\frac{y}{x} \\ z &= r\cos\theta & \theta &= \arccos\frac{z}{\sqrt{x^2 + y^2 + z^2}} \end{aligned} \qquad (5.5)$$

With the example of a train passing through a station at the beginning of this section, we introduced two coordinate systems that are moving relative to each other, namely the coordinate system anchored to the platform and the coordinate system moving with the train (see ◐ Fig. 5.3). For simplicity, we chose Cartesian coordinates for both systems. How do we convert from one system to the other?

We denote the system on the platform with S and the corresponding coordinates with (x, y, z), the system that

◐ **Fig. 5.3** Moving coordinate systems on the platform and in the train

moves with the train is S' and the coordinates accordingly are (x', y', z'). Essential for the conversion between S and S' is the vector $\vec{r}(t)$, which connects the origin of coordinates of S with the origin of S'. With the movement of the train, this vector will change. Please note that we can specify it both in the coordinates of S and those of S'. For example, if the origin of S' at time t_0 is seen from the platform at location (x_0, y_0, z_0), then the origin of S from the perspective of the train at (x'_0, y'_0, z'_0), where $x'_0 = -x_0, y'_0 = -y_0$ and $z'_0 = -z_0$ apply, provided that the axes of the two coordinate systems are oriented the same way. A point $\vec{s}$, which in S has the coordinates (x_s, y_s, z_s), in S' the coordinates $(x_{s'}, y_{s'}, z_{s'})$, where for the x-coordinate $x_s = x_0 + x'_s$ must apply, as well as corresponding relations for the other two coordinates.

If the two coordinate systems S and S' move with constant velocity against each other, the vector $\vec{r}(t)$ can be expressed by a constant velocity $\vec{v} = (v_x, v_y, v_z)$ via $\vec{r}(t) = \vec{v}t$, where we have assumed that the coordinate systems coincide initially. Then it follows:

$$\begin{aligned} x_s &= v_x t + x'_s \\ y_s &= v_y t + y'_s \, . \\ z_s &= v_z t + z'_s \end{aligned} \qquad (5.6)$$

This is called the Galilean transformation between the coordinate systems. Often, the transformation is generalized to a group of transformations, also called the Galilean transformation, by adding the following transformations:

- The displacement of the coordinate origins against each other by a fixed vector $\vec{a} = (a_x, a_y, a_z)$.
- The offset between the clocks in the two coordinate systems by a value t_a, so that $t = t_a + t'$ applies.
- The rotation of the coordinate systems against each other, described by a rotation matrix M.

It generally applies:

$$\begin{pmatrix} x_s \\ y_s \\ z_s \end{pmatrix} = M \begin{pmatrix} v_x(t_a + t') + a_x + x_{s'} \\ v_y(t_a + t') + a_y + y_{s'} \\ v_z(t_a + t') + a_z + z_{s'} \end{pmatrix}. \qquad (5.7)$$

Example 5.2: Speed from the Perspective of a Car

An employee of a car insurance company approached us with the following situation:

A customer reported for the umpteenth time damage to his vehicle, an Audi A8, caused by rock fall. He was driving on the highway at a speed between 100 km/h and 120 km/h under a bridge when teenagers dropped stones onto the roadway. Three stones hit his car, two on the trunk lid, the third smashed the rear window. Is this believable?

Among the customer's information is the accumulation of hits in the rear area of the car. From the the perspective of the roadside, the stones fell straight down. We can determine their speed from the free fall. However, from the driver's perspective, the stones hit the vehicle at a shallow angle from above. Let's denote with $\vec{v}_v$ the vertical speed due to the fall and with $\vec{v}_h$ the speed of the vehicle, each from the perspective of the roadside, then the driver sees the stones approaching at a speed $\vec{v}' = \vec{v}_v - \vec{v}_h$. We have calculated the direction of this speed for $v_h = 100$ km/h and entered it as slanted lines in the illustration. Now let's assume that the teenagers dropped many stones, we can read from the picture the probability with which a stone hits the car in one of the areas A to F. The probability of hitting the rear window is only 2.2%! At slightly higher speeds, it becomes zero. The combined probability of two hits on the trunk lid and one hit on the rear window is less than 0.01%. We have sent the picture and the probabilities to the insurance company. We do not know how the insurance company and the customer came to an agreement.

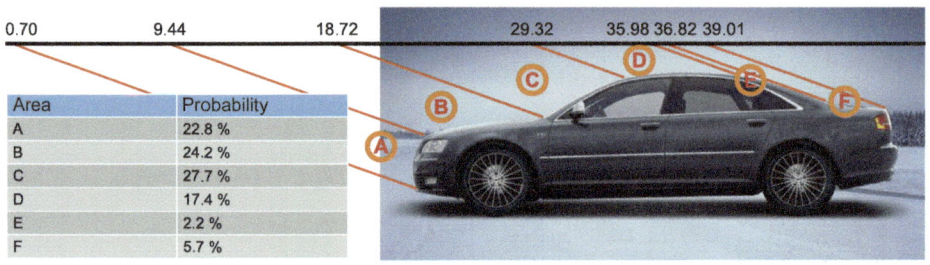

Area	Probability
A	22.8 %
B	24.2 %
C	27.7 %
D	17.4 %
E	2.2 %
F	5.7 %

5.2 · Reference Systems

Example 5.3: Moving Coordinate System

A person is riding on a roller coaster through a loop. She is holding a ball on a string in her hand. During the loop, the person observes how the pendulum moves out of its resting position. We use spherical coordinates with the origin at the suspension point of the string. This coordinate system moves along with the roller coaster. The z-axis points to the center of the loop, the x-axis in the direction of motion. The position of the ball in this system is given by $(r, \phi, \theta) = (l, 0, \theta(t))$, where l is the length of the string.

We do not want to calculate the motion, but only describe it. We choose a Cartesian coordinate system, whose origin is at the entry and exit point of the loop. The directions of the coordinate axes can be taken from the sketch. The path of the suspension point through the loop is given in the Cartesian coordinate system as follows:

$$\vec{s}_H(t) = (R\sin(2\pi t/T), Rt/10, R(1 - \cos(2\pi t/T))).$$

We obtain the coordinates of the ball in this system by adding (r, ϕ, θ) from the co-moving system to the coordinates of the holding point:

$$\vec{s}_K = \big(R\sin(2\pi t/T) + l\sin(\theta(t)), Rt/10,$$
$$R(1 - \cos(2\pi t/T)) + l\cos(\theta(t))\big).$$

This is the motion that a person standing next to the roller coaster observes. If we wanted to calculate the movements, we would have to determine the forces acting on the ball. This is, on the one hand, the weight, which always points downwards in the stationary system, and on the other hand, the centrifugal force, which always points away from the center of the loop in the moving system, i.e., in the moving system in the negative z direction. With the coordinate transformations, we can transform the forces into the respective other system. Give it a try!

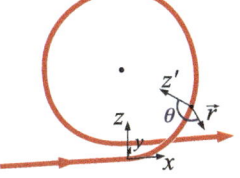

After these two introductory subchapters, we now turn to the actual topic, kinematics. This refers to the description of movements, i.e., the specification of where a body is located at certain times. We will discuss the cause of the motion (influence of forces) as dynamics in the next chapter.

5.3 Uniform Motion

In uniform motion, a body moves in a straight line through space at a constant velocity. Neither the direction nor the magnitude of the velocity changes. The distance covered by the body is proportional to the elapsed time. If you plot the distance covered against the elapsed time, you get a straight line (see ◘ Fig. 5.4).

To further describe the motion, we define the velocity:

$$v = \frac{s}{t} \tag{5.8}$$

Here, v is the velocity, s is the distance covered, and t is the time required for it. The unit of speed is m/s or derived units such as e.g. km/h. Some typical velocities can be found in ◘ Table 5.1.

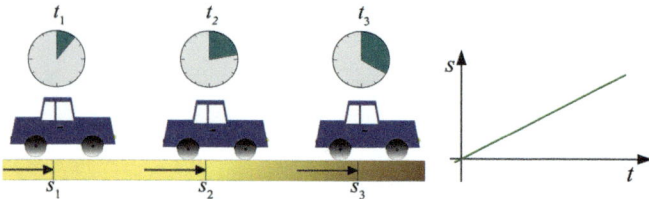

◘ **Fig. 5.4** A car travels at a constant velocity. It covers fixed distances in fixed times

◘ Table 5.1 Approximate velocities of some processes or movements		
Growth of stalactites	0.1 mm/a	3.2 pm/s
Growth of grass	5 cm/week	83 nm/s
Slugs	5 cm/min	0.8 mm/s
Pedestrian	5 km/h	1.4 m/s
Fastest Sprint (Usain Bolt 2019)	200 m/19.19 s	10.4 m/s
Cheetah (fastest land animal)	110 km/h	30 m/s
Long-distance flight	860 km/h	240 m/s
Speed of sound in air	1236 km/h	343.2 m/s
Orbital speed of the Earth around the Sun	29.8 km/s	$8.3 \cdot 10^3$ m/s
α-particles in radioactive decay	0.05 % c	$1.5 \cdot 10^7$ m/s
Speed of light	300,000 km/s	$3 \cdot 10^8$ m/s

5.3 · Uniform Motion

Example 5.4: Uniform Motion

A car travels on an empty, but speed-limited highway covering 120 km in one hour. Its velocity is:

$$v = \frac{120\,\text{km}}{1\,\text{h}} = 120\,\frac{\text{km}}{\text{h}} = \frac{120{,}000\,\text{m}}{60 \cdot 60\,\text{s}} = 33.3\,\frac{\text{m}}{\text{s}}.$$

Experiment 5.1: Projectile Speed

On an axis driven by a drill, two cardboard discs are mounted one after the other at a distance of 50 cm. The rifle bullet (air rifle) penetrates the cardboard discs at the outer edge. Angle markings are attached to the cardboard discs, aligned between the two discs. After the projectile has penetrated the first disc, the axis continues to rotate, so that it penetrates the second disc offset at an angle α. The projectile is stopped in a stable wooden wall behind the second disc.

© Photo: Hendrik Brixius

The drill rotates at 3000 revolutions per minute. The bullet hole on the second disc is offset from the first by $\alpha = 27°$.

We determine the projectile's speed: 3000 rev/min corresponds to 50 revolutions per second or 20 ms time for a full rotation. From cross-multiplication, the flight time from the first to the second disc is then

$$t = \frac{27°}{360°}\,20\,\text{ms} = 1.5\,\text{ms}$$

and from this, the projectile speed is determined to be

$$v = \frac{0.5\,\text{m}}{1.5\,\text{ms}} = 333\,\frac{\text{m}}{\text{s}}$$

© Photo: Hendrik Brixius

The uniform motion considered here is a special case. Generally, the velocity can change over time, e.g., when a car accelerates or brakes.

The velocity was introduced for simplicity as a scalar quantity. Strictly speaking, it is a vector quantity with a direction (see Appendix A3.2). The velocity vector points in the direction of motion (in our example with the car in the direction of travel). The correct definition is

$$\vec{v} = \frac{\vec{s}}{t} \tag{5.9}$$

Now we have also written the distance as a vector ($\vec{s}$). This vector points from the starting point of the motion to its end point. If the direction of the speed does not change, we speak of a linear motion. Both linear and uniform motion are special cases of a general motion.

Finally, a note on nomenclature: The symbols introduced here v, s and t are the usual symbols for these quantities, which are derived from their Latin names. This is a convention that you are allowed to break. For example, in a problem with several different velocities, other velocities are often denominated as u. However, if you deviate from the usual conventions, you should clearly define your symbols to the reader.

Example 5.5: Curling

In curling, the friction is low. The movement of the curling stone can be approximated as uniform linear motion (until it encounters obstacles).

Example 5.6: Swimming in flowing water

An average swimmer can reach a speed of $1\,\text{m/s}$ in standing water. Now imagine he wants to swim across a $50\,\text{m}$ wide river with a flow of $0.5\,\text{m/s}$. How far will he drift downstream if he always swims perpendicular to the direction of flow?

Let's choose a coordinate system (a two-dimensional one is sufficient) so that the river flows in the x-direction and the swimmer starts in y-direction. Then its velocity in relation to the shore is given by the vector sum of the flow velocity $\vec{v}_{\text{flow}}$ and the velocity of the swimmer $\vec{v}_S$:

$$\vec{v}_{\text{flow}} = \left(0.5\,\frac{\text{m}}{\text{s}}, 0\right)$$
$$\vec{v}_S = \left(0, 1\,\frac{\text{m}}{\text{s}}\right)$$
$$\vec{v} = \vec{v}_{\text{flow}} + \vec{v}_S = \left(0.5\,\frac{\text{m}}{\text{s}}, 1\,\frac{\text{m}}{\text{s}}\right)$$

Now we can use Thales's theorem of intersecting lines to determine the drift Δs

$$\frac{0.5\,\text{m/s}}{1\,\text{m/s}} = \frac{\Delta s}{50\,\text{m}} \Rightarrow \Delta s = 25\,\text{m}$$

Perhaps more interesting is the question at which angle the swimmer must swim against the current to avoid any drift. Now, $\vec{v}_S = (\cos\phi, \sin\phi) \cdot 1\,\frac{m}{s}$, where ϕ is the angle to the x-axis (direction of flow). The swimmer does not drift, if the x-component of $\vec{v} = \vec{v}_{\text{flow}} + \vec{v}_S$ disappears. So

$$0.5\,\frac{m}{s} + \cos\phi \cdot 1\,\frac{m}{s} = 0 \Rightarrow \cos\phi = -0.5 \Rightarrow \phi = 120°$$

He must swim with 120° to the x-axis or with 30° against the current.

Finally, let's determine how long the crossing takes in both cases. In the first case (with drift), the distance to be covered is $s = \sqrt{(50\,m)^2 + (25\,m)^2} = 55.9\,m$, in the second case it is 50 m. The speeds are (magnitudes of $\vec{v}$) 1.12 m/s or 0.87 m/s. From this, the times are calculated to be ($t = s/|\vec{v}|$) 50 s or 57.7 s. If the swimmer accepts to drift, he crosses the river faster.

5.4 Non-uniform Motion

The Example 5.4 seems contrived. A car rarely drives at constant speed. In ◘ Fig. 5.5 a completely different example is shown, which a motorsport enthusiast recorded with the GPS in his car on a German racetrack called Nürburgring. If one were to insert the time used for the total distance of 21.5 km into the definition of the previous section (Eq. 5.8), his average speed would result. But if we want to describe the actual course of the drive, we need to rethink the definition of speed.

We first turn to a simpler example, the free fall (Experiment 5.2).

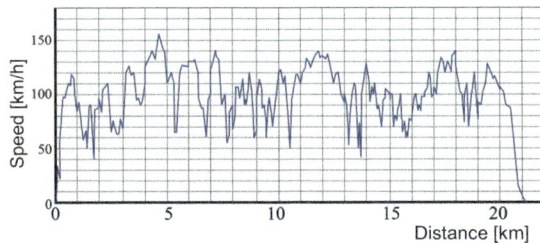

◘ **Fig. 5.5** Speed profile of a sports car on the northern loop of the Nürburgring

> **Experiment 5.2: Stroboscopic Recording of a Falling Ball**
>
> We observe the free fall of a steel ball. The steel ball is held at the top of a drop tower approximately 1.5 m high by an electromagnet. The droptower is lined with black cloth on the inside. Light barriers are attached to the side to record the movement. First, we want to capture the movement photographically. We darken the room and illuminate the drop tower with a stroboscope (frequency approx. 20 Hz). When the electromagnet is opened, the camera is triggered for the duration of the fall (exposure time 2 s). The two images show the apparatus and the stroboscopic recording. One can see a semicircular reflection of the stroboscope light from the ball a total of 13 times. The distance between the reflections increases downwards. The speed of the ball increases as it falls.

© Photos: Hendrik Brixius

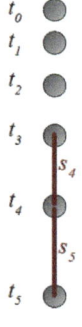

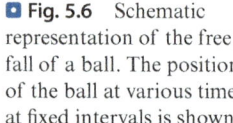

Fig. 5.6 Schematic representation of the free fall of a ball. The position of the ball at various times at fixed intervals is shown

Figure 5.6 shows snapshots of the falling ball at fixed time intervals Δt. We want to determine the velocity at the time t_4. There are two ways to apply $v = s/t$:

$$v_4 = \frac{s_4}{\Delta t} \qquad v_5 = \frac{s_5}{\Delta t} \qquad (5.10)$$

The velocity v_4 is smaller than v_5, the sought-after velocity at time t_4 is somewhere in between. It cannot be determined exactly from the figure. If you decrease Δt, the difference between v_4 and v_5 decreases. You are approaching the sought-after velocity.

You can narrow it down with any desired accuracy by making Δt correspondingly small. Mathematically, this corresponds to a limit transition from the difference quotient to the derivative. The velocity at the time $t = t_4$ is:

$$v(t_4) = \lim_{\Delta t \to 0} \frac{s(t_4 + \Delta t) - s(t_4)}{\Delta t} = \lim_{\Delta t \to 0} \frac{\Delta s}{\Delta t} \qquad (5.11)$$

where one could have just as well started from v_5 (see Appendix A3.2 to Appendix A3.5).

5.5 Acceleration

Here too, we initially suppressed the direction of the velocity. It can be obtained from

$$\vec{v}(t) = \lim_{\Delta t \to 0} \frac{\vec{s}(t + \Delta t) - \vec{s}(t)}{\Delta t} = \frac{d\vec{s}}{dt} \quad (5.12)$$

The velocity vector points in the direction of the tangent to the path.

5.5 Acceleration

In colloquial terms, acceleration is understood as an increase in speed, as can be seen in ◘ Fig. 5.7 with a motorcycle.

In physics, the term is used a bit more generally. It refers to any change in velocity. This can also be a decrease of velocity (a braking) then the value of the acceleration is negative. In physics, a change in the direction of velocity (a turn), even with a constant magnitude of velocity, is also referred to as "acceleration". The symbol for acceleration is usually an a.

The acceleration can change over time. Therefore, it is defined as a derivative in time, just as the velocity:

$$\vec{v}(t) = \lim_{\Delta t \to 0} \frac{\vec{s}(t + \Delta t) - \vec{s}(t)}{\Delta t} = \frac{d\vec{s}}{dt} \quad (5.13)$$

◘ **Fig. 5.7** An accelerating motorcycle. The acceleration vector points horizontally forward. © MOTORCYCLE/jkünstle

$$\vec{a}(t) = \lim_{\Delta t \to 0} \frac{\vec{v}(t+\Delta t) - \vec{v}(t)}{\Delta t} = \frac{d\vec{v}}{dt} = \frac{d}{dt}\left(\frac{d\vec{s}}{dt}\right) = \frac{d^2\vec{s}}{dt^2}$$
(5.14)

Like velocity, acceleration is a vector quantity. It has a direction. It points in the direction of the change of the velocity vector. Two special cases are distinguished in relation to the direction of velocity:
- Longitudinal acceleration: Acceleration points in the direction of the velocity,
- Centripetal or normal acceleration: Acceleration is perpendicular to the velocity.

Any acceleration can be decomposed into these components (there are two normal components, as there are two independent directions perpendicular to the motion).

The unit of acceleration can be read from the definition. It is m/s².

If the acceleration does not change over time, it is referred to as uniform acceleration. In this case, the acceleration can be calculated from

$$\vec{a} = \frac{\Delta \vec{v}}{\Delta t}$$
(5.15)

where $\Delta \vec{v}$ is the change in velocity over the time Δt. Typical accelerations for some processes can be found as examples in ▪ Table 5.2.

Example 5.7: Acceleration of a sports car

The Porsche 911 Carrera accelerates from zero to one hundred kilometers per hour in 4.3 s. The average acceleration is (100 km/h equals 27.8 m/s):

$$a = \frac{27.8\,\text{m/s}}{4.3\,\text{s}} = 6.5\,\frac{\text{m}}{\text{s}^2}.$$

© Porsche AG

5.5 · Acceleration

Table 5.2 Typical accelerations in some processes

Subway trains	1 m/s²
Car entering the highway	2 m/s²
Ariane 5 rocket at launch	5 m/s²
Gravity acceleration	9.81 m/s²
Car at full brake	–10 m/s²
Extreme values in a roller coaster	40 m/s²
Car: frontal collision with wall at 50 km/h	200 m/s²
Human load limit for very short stresses (~ ms)	1000 m/s²
Ultracentrifuges	10^6 m/s²
Electron in a television tube	10^{17} m/s²

If the acceleration is constant over time, the velocity and distance covered can be easily calculated. Then the following applies (see Mathematical Appendix A3.7):

$$a(t) = a_0 = \text{const.}$$
$$v(t) = \int a_0 dt = \int \frac{dv(t)}{dt} dt = a_0 t + v_0 \quad \text{with } v_0 = v(t=0)$$
$$s(t) = \int v(t) dt = \int (a_0 t + v_0) dt$$
$$= \frac{1}{2} a_0 t^2 + v_0 t + s_0 \quad \text{with } s_0 = s(t=0).$$
(5.16)

As you can see, the velocity increases linearly with time. The distance covered increases quadratically with time.

Our result for $s(t)$ is called the position function. It indicates where the body is at a time t.

The trajectory describes the path of a body through space. An example can be seen in ◘ Fig. 5.8. The green curve indicates the locations along which the body has moved. The direction of acceleration can be graphically determined from the trajectory. In the figure, the position of the body at the times t_0 and t_1 is indicated (black vectors). The velocity vector is the tangent to the trajectory at these positions (red vectors). From the velocity vectors, one can determine $\Delta \vec{v} = \vec{v}_1 - \vec{v}_0$ (blue vector). The acceleration vector points in this direction.

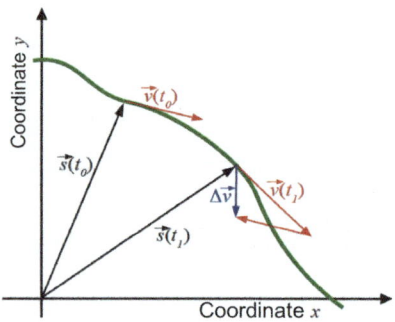

Fig. 5.8 The trajectory of a fictional body (green line). The position and velocity vectors are given at the times t_0 and t_1 (black and red respectively)

Note that the lengths of the position, velocity, and acceleration vectors cannot be directly compared with each other, as they carry different units.

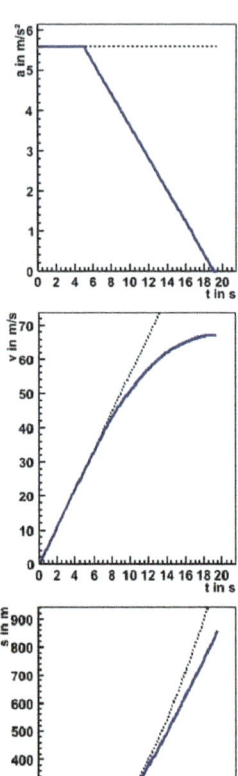

Example 5.8: Acceleration of a car

We return once again to the acceleration of a sports car. Let's assume that it accelerates at $5.6\,\text{m/s}^2$. However, this cannot go on indefinitely. To get a rough impression of the car's true behavior during acceleration, let's assume that it accelerates uniformly up to a speed of $100\,\text{km/h}$ ($27.8\,\text{m/s}$) and then the acceleration decreases linearly, so that in the end it reaches $240\,\text{km/h}$ ($66.7\,\text{m/s}$).

The calculation up to $100\,\text{km/h}$ gives the already known result and for speeds above it:

$$a(t) = a_0 - ct_2$$

$$v(t) = a_0 t - \frac{1}{2} c t_2^2$$

$$s(t) = \frac{1}{2} a_0 t^2 - \frac{1}{6} c t_2^3$$

Here, t_2 is the time after reaching $100\,\text{km/h}$ and $c = 0.4\,\text{m/s}^3$ a constant. The result is shown in the figures (solid curves) compared to the case without reduction of acceleration (dashed curves).

5.6 Free fall

Free fall is a special case of an accelerated motion. It is well known to us from everyday life. It refers to the falling of bodies without air drag. Gravity causes this motion. The Galileo's law of motion, which mathematically describes the free fall, is one of the oldest laws of modern times. It states that all bodies fall equally fast.

> **Galileo's Law of Motion**
> All bodies fall at the same speed.

The law is surprising as it contradicts our everyday experience. In everyday life, we find that heavy bodies fall faster to the ground than light ones. If we accidentally pull a napkin off the table on which a glass stood, the glass will hit the ground before the napkin. The reason is aerodynamic drag. In the case of the falling glass, it is negligible, but it significantly slows down the napkin. In an experiment, the influence of air drag can be eliminated. Then Galileo's law shows itself in its full breadth. In Experiment 5.3 you can see a feather and a lead ball falling in a vacuum tube. They fall at the same speed.

© RWTH Aachen, Physics Experiment Collection

Experiment 5.3: Free Fall in Vacuum

In a glass tube, there is a lead disc and a feather. The air in the glass tube is removed with a vacuum-pump. A pressure below 1 mbar no longer affects the fall of the bodies. The tube is sealed with a valve and held vertically. If you now quickly turn the lower end upwards, the lead piece and the feather are taken upwards. Afterwards, you can observe their fall through the tube. The fall lasts just under half a second. You may have to repeat the experiment a few times until you can clearly see the fall.

Experiment 5.4: Free Fall in the Slipstream

As we have seen, the free fall is not directly accessible from every-day experience due to air drag. However, there is a simple trick to eliminate it, namely by falling in slipstream. You can easily try this yourself. You need a sheet of paper and a book[2]. You take the sheet in one

---

2 Be kind to this book and find another, less valuable one.

hand and the book in the other, hold them at the same height and let them fall simultaneously. Due to the air drag, the sheet hits the ground after the book. Now repeat the experiment by placing the sheet on the book. When you let go of the book, the sheet and the book fall at the same speed. The book displaces the air and without air drag, the sheet falls as fast as the book.

The law of the free fall goes back to Galileo Galilei (◘ Fig. 5.9). The manuscript dates from the year 1604, as an analysis of the ink with a proton beam could show, (*New Scientist* No. 2343, p. 17). How could Galilei prove the law with the means of his time?

The story goes that he conducted free fall experiments at the Leaning Tower of Pisa (◘ Fig. 5.10). He claimed to have thrown stone and wooden balls from the tower and observed a hundred times that they arrive at the bottom simultaneously. One can easily verify that this is not true. Due to air drag, the wooden ball hits the ground after the stone ball. At the height of the tower, the difference is large enough to be noticeable.

Even if this story probably did not happen this way, it shows Galileo's most important achievement. He in-

◘ **Fig. 5.10** The Leaning Tower of Pisa. Allegedly, Galileo Galilei conducted his free fall experiments from the tower

◘ **Fig. 5.9** Galileo Galilei (1564–1642) introduced the experiment into natural sciences. © CPA Media Co. Ltd/picture alliance

troduced the experiment into physics. His predecessors sought confirmation of their thoughts (models) in the Aristotelian writings. Galileo sought it in nature itself, by conducting experiments.

But how did he go from observing that the stone ball hits the ground before the wooden ball, to the law of free fall? He had to realize that air drag interferes with the experiment. To successfully conduct the experiment, he had to abstract from it and other environmental influences. A task, that every experimenter still faces today. In every experiment, there are influences from the environment that affect the measurement and distort the actual effect one wants to measure. It is part of the "art of experimenting" to design an experiment in such a way that environmental influences are negligible, or to find a way to estimate and correct them in the result.

Galileo took the following path. He shifted the free fall from the vertical to the inclined plane. There, bodies "fall" slower. Since the aerodynamic drag depends quadratically on speed, it is negligible at slower speeds and one can observe the "pure" law of free fall. ◘ Figure 5.11 shows Galileo's inclined plane. The track is designed for different metal and wooden balls. Above the running groove, movable arches are attached, each with a bell that is struck by the passing balls. Can you imagine how the motion is measured with the bells?

◘ **Fig. 5.11** Reconstruction of the inclined plane with which Galileo tested the law of free fall. © Museo Galileo—Instituto e Museo di Storia della Scienza in Florence

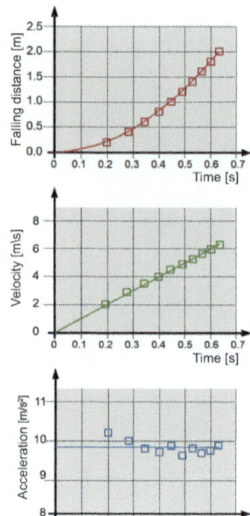

> **Experiment 5.5: Free Fall with the Drop Tower**
>
> We use the drop tower again, which you already know from Experiment 5.2. Light barriers are installed at regular intervals along the side of the falling distance, which register the passage of the ball. The signals are recorded with a computer. It displays the measurement of the recorded times graphically and determines velocity and acceleration from it. The three illustrations show the results. In the graph of the acceleration, you may notice the limited accuracy of the time measurements.

The law of free fall states that all bodies fall downwards with the same acceleration. It is called the gravitational acceleration g. On Earth, it is approximately 9.81 m/s^2. In fact, it is not the same value everywhere. It varies slightly from place to place. An effect that Galileo could not yet measure.

Several effects are responsible for the variation. In addition to gravity, the centrifugal force caused by the Earth's rotation must be taken into account. Since this is maximum at the equator, g is smallest there. With increasing altitude, g also decreases, as one moves further from the center of the Earth and thus the gravitational force decreases and also the centrifugal force increases. The flattening of the Earth due to rotation also plays a role. In addition, there are local effects due to variations in the density of the surrounding rock and the topology (whether one is standing on a mountain peak or on a plain).

As an example, some places are listed[3]:

$g = 9.83219 \, \text{m/s}^2$ at the poles (sea level)

$g = 9.81101 \, \text{m/s}^2$ in Aachen, home of the authors

$g = 9.78033 \, \text{m/s}^2$ at the equator (sea level)

There are formulas which allow to approximate g, for example, the WELMEC formula

---

[3] The information is taken from the Gravity Information System SIS of the Physikalisch-Technische Bundesanstalt Braunschweig: ▶ http://www.ptb.de/cartoweb3/SISproject.php.

5.6 · Free fall

$$g = 9.780318(1 + 0.0053024 \sin^2 \varphi - 0.0000058 \sin^2 2\varphi) \frac{\text{m}}{\text{s}^2}$$

$$- 0.000003085 \frac{1}{\text{s}^2} h \qquad (5.17)$$

Here, φ is the geographical latitude and h is the height above sea level in meters. For Aachen (geographical latitude 50.7667°; height above sea level 173 m) the WELMEC formula yields 9.81084 m/s². The correction due to the geographical latitude is +0.318 %, the correction due to the sea level is −0.005 %. The remaining difference to the above number can be explained by the local circumstances.

Example 5.9: High Diving

Galileo's law states that all bodies fall at the same speed. From the constant acceleration g, the motion of a falling body is given by

$$h(t) = h_0 + v_0 t - \frac{1}{2} g t^2$$

$$v(t) = v_0 - gt$$

with the initial height h_0 and the initial velocity v_0.
Let's consider as an example an athlete diving into the pool from different heights. The diving board or the diving tower has the height h_0. The diver jumps with a speed of 4.5 m/s upwards, which takes her to a height of just over a meter before she falls into the water. We determine some interesting quantities:
Duration of the jump (from $h(t = t_{\text{end}}) = 0$):

$$t_{\text{end}} = \frac{v_0 + \sqrt{2 g h_0 + v_0^2}}{g}.$$

Speed upon hitting the water $v(t = t_{\text{end}})$:

$$v_{\text{end}} = -\sqrt{2 g h_0 + v_0^2}.$$

The duration of the jump and the impact speed increase with the square root of the height of the diving board. Here are some values

h_0	1 m	3 m	10 m
t_{end}	1.10 s	1.36 s	1.96 s
v_{end}	22.7 km/h	32.0 km/h	53.0 km/h

5.7 Projectile Motion

Somebody throws a ball in the gravitational field of the Earth. What does the trajectory of the body look like? The uniform motion due to the initial velocity is superimposed by the free fall. The resulting motion is the sum of the two individual motions, which do not influence each other. We demonstrate this in Experiment 5.6.

Experiment 5.6: Superposition with the Train

© RWTH Aachen, Physics Experiment Collection

We use an electric train. A pitching machine is loaded onto an open freight wagon. It can fire a plastic ball vertically upwards. The mechanism is mechanically triggered at a certain point on the track. The ball rises half to one meter high and then falls back down. If the train is moving at a constant speed, the ball falls back into the funnel of the pitching machine.

We now want to try to describe a projectile motion mathematically. We choose a right-handed coordinate system with the z-direction upwards and the x-direction in the direction of travel of the train. We start the clock with the firing of the ball ($t = 0$). Before the launch, the ball has a horizontal velocity that corresponds to that of the train $v_x = v_{\text{Train}}$. The launch creates a motion in the vertical with the initial velocity $v_z = v_0$.

We assume that the horizontal and vertical motion do not influence each other. There is no acceleration in the horizontal direction. Consequently, the horizontal motion is uniform:

$$x(t) = v_{\text{Train}} t + x_0 \tag{5.18}$$

In y direction, there is no movement ($y(t) = y_0$) and in the vertical direction, the acceleration due to gravity acts

5.7 · Projectile Motion

downwards, hence negative. This results in a free-fall motion:

$$z(t) = -\frac{1}{2}gt^2 + v_0 t + z_0 \quad (5.19)$$

The coordinates of the ball at the time $t = 0$ are x_0, y_0 and z_0 are given. The actual movement arises as a superposition of the horizontal and vertical movements. The position is

$$\vec{r}(t) = \begin{pmatrix} v_{\text{train}} t + x_0 \\ y_0 \\ -\frac{1}{2}gt^2 + v_0 t + z_0 \end{pmatrix} \quad (5.20)$$

Horizontal and vertical movements remain unchanged.

Here, you have learned an example of the principle of superposition. We calculated the horizontal and vertical motion independently of each other and superimposed the results at the end. The application of the principle of superposition greatly simplifies the calculation of this motion and other motions. However, note that you cannot apply a principle of superposition in all cases. A crucial prerequisite is that the two motions (here the horizontal and vertical) do not influence each other. This must be checked in each individual case before the principle of superposition is applied.

The parameters in Experiment 5.6 are chosen so that air drag becomes negligible. If we shoot the ball even higher, this is no longer the case. As we will learn later, the air drag depends on all three components of the velocity. You can no longer determine one motion without the other. They say that the motions are coupled by air drag. The principle of superposition no longer applies.

In kinematics, two different representations are used to describe motion. You already saw the first representation—the position as a function of time. It gives the position of the body in the coordinate system at any time t. It contains all information about the motion, but sometimes it is difficult to imagine the path the body takes from the position function. Then we use the trajectory instead.

We want to determine the trajectory in our example. For simplicity, we choose $(x_0, y_0, z_0) = (0, 0, 0)$. From the horizontal motion (Eq. 5.18), it follows that

$$x(t) = v_{\text{train}} t \rightarrow t = \frac{x}{v_{\text{train}}}, \quad (5.21)$$

which we substitute into the vertical motion

$$z(x) = -\frac{1}{2}g\left(\frac{x}{v_{\text{train}}}\right)^2 + v_0 \frac{x}{v_{\text{train}}} = -\frac{1}{2}\frac{g}{v_{\text{train}}^2}x^2 + \frac{v_0}{v_{\text{train}}}x.$$
(5.22)

As you can see, a parabola results (z depends quadratically on x). It is called the projectile motion. It can be beautifully represented with water (see Experiment 5.7).

Mathematically speaking, the position function is nothing more than a specific parametric representation of the trajectory, namely the one that uses time as a parameter.

Experiment 5.7: Projectile trajectory with a water jet

A thin stream of water comes out of a hose. Each individual water molecule follows a projectile trajectory. If you attach graph paper behind the jet, you can directly read its shape.

© RWTH Aachen, Physics Experiment Collection

Experiment 5.8: Superposition with the crossbow

A poacher hunts monkeys. He has spotted a victim in a tree and aims his crossbow at the monkey. But the monkey recognizes the danger and drops from the tree at the moment the poacher pulls the trigger. What happens? The experiment recreates this story.

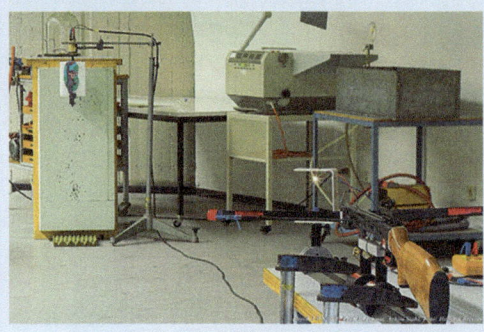

© Photo: Hendrik Brixius

© Photo: Hendrik Brixius

The principle of superposition applies here, too. The arrow from the crossbow is shot horizontally. The horizontal movement is overlaid by a vertical falling mo-

5.7 · Projectile Motion

tion. This is not affected by the horizontal movement. Vertically, the arrow makes a free fall, just like the monkey. They fall at the same speed. The arrow will hit the monkey, regardless of how sharp the shot is. Even if you only pull the crossbow very lightly and the arrow falls by a whole meter on the nearly 10-meter flight path to the monkey, it will hit the monkey precisely. Dropping was a fataly wrong decision by the monkey.

Experiment 5.9: Superposition with the Ski Jump

This is another experiment that impressively demonstrates the superposition of horizontal and vertical motion. A steel ball rolls down a ski jump and leaves the track at the bottom with a horizontal velocity. A second ball is held by an electromagnet at the same height as the bottom of the ski jump. At the end of the ski jump, it triggers a light barrier that interrupts the current to the electromagnet. The second ball falls at the same moment (delays due to the electronics are negligible) as the first ball leaves the ski jump horizontally. After a short flight, which is hardly noticeable to the eye, one hears the impact of the balls against each other. For the experiment to succeed, the track must be aligned with the stationary ball so that the balls do not miss each other laterally and the distance should not be chosen too large that the balls reach the ground before they collide.

Light barrier

(©) RWTH Aachen, Physics Experiment Collection

Example 5.10: Optimization of a Projectile Motion

A football player kicks a ball from a balcony into the garden. He tries to kick the ball as far as possible. In doing so, he will not kick the ball horizontally, but upwards. With what launch angle β relative to the horizontal does he achieve the maximum distance?

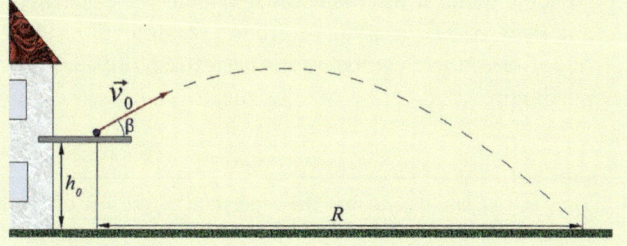

The calculation of the flight distance (without air drag) results in:

$$R = \frac{v_0^2}{2g} \sin 2\beta \left(1 + \sqrt{1 + \frac{2gh_0}{v_0^2 \sin^2 \beta}}\right).$$

Here, g is the standard gravity, the other quantities are explained in the sketch. If you want to check the formula, use superposition!

At which angle does R become maximum?

We consider, as indicated in the formula, the range as a function of the launch angle $R(\beta)$. At the maximum of this function, its slope must vanish. We calculate the derivative with respect to β using the product rule and multiple applications of the chain rule. It results in:

$$\frac{d}{d\beta} R(\beta) = \frac{v_0^2}{g} \cos 2\beta \left(1 + \sqrt{1 + \frac{2gh_0}{v_0^2 \sin^2 \beta}}\right)$$

$$+ \frac{v_0^2}{g} \sin 2\beta \; \frac{1}{\sqrt{1 + \frac{2gh_0}{v_0^2 \sin^2 \beta}}} \; \frac{gh_0}{v_0^2 \sin^3 \beta} \cos \beta.$$

Now we require $\frac{d}{d\beta} R(\beta = \beta_{max}) = 0$ and obtain (see Mathematical Appendix A3.8):

$$\beta_{max} = \arccos\left(\frac{\sqrt{2gh_0 + v_0^2}}{\sqrt{2gh_0 + 2v_0^2}}\right).$$

Strictly speaking, we should still check whether it is actually a maximum by showing that[4]

$$\left. \frac{d^2}{d\beta^2} R(\beta) \right|_{\beta = \beta_{max}} < 0.$$

In this case, however, physical intuition helps. For $\beta = 0$ the range is initially low and increases with increasing β. At some point, it will reach the maximum we calculated and then reduces, as a range of 0 will result for $\beta = 90°$. In between, there can be no further extrema besides the maximum.

---

[4] The vertical line means that the expression to the left of the line must be evaluated and then, as indicated at the bottom right, β is to be replaced by β_{max}.

5.8 Circular Motion

At the end of this chapter, we turn to a special type of accelerated motion, circular motion. Although a body does not get faster during circular motion, it is an accelerated motion. The body continuously changes its direction of motion and thus its vector of velocity, and as we have seen, any change in velocity is associated with acceleration.

The acceleration that forces a body into a circular motion is called "centripetal acceleration". It always points towards the center of the circular path. The speed on the circular path is called "orbital speed". It runs tangentially, i.e., in the direction of the tangent to the circular path. Acceleration and velocity are perpendicular to each other. There is a fixed relationship between the orbital speed and the centripetal acceleration. This is shown in ◘ Fig. 5.12.

A body reaches point A. The orbital speed is v_O. If we were to ignore the change in the direction of velocity on the circular path, the body would cover the distance AB in time t. This is given by AB = $v_O t$. To follow the circle, the body's path must be supplemented by the distance h. If we choose t very small (limit $t \to 0$), then h is perpendicular to AB and the initial speed in this direction is zero. To cover the distance h in time t, a centripetal acceleration a_{CP} must act, so that:

$$h = \frac{1}{2} a_z t^2. \tag{5.23}$$

Now we just need to calculate h. To do this, we apply the Pythagorean theorem to the triangle OAB:

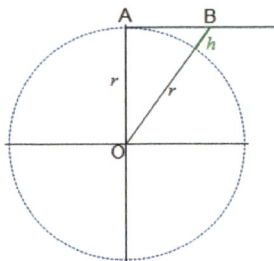

◘ **Fig. 5.12** Sketch for determining the centripetal acceleration on a circular path

$$r^2 + (v_O t)^2 = (r+h)^2$$

$$v_O^2 t^2 = 2rh + h^2 \approx 2rh \to h = \frac{v_O^2 t^2}{2r}. \qquad (5.24)$$

Substituting into (Eq. 5.23) results in:

$$a_{CP} = \frac{v_O^2}{r}. \qquad (5.25)$$

As you can see, the centripetal acceleration increases quadratically with increasing orbital speed. The faster the body moves, the more it must be accelerated towards the center to maintain the circular path, and this is even more the case the smaller the radius of the path. What would you have expected from your everyday experience?

? Problems

1. Provide the following velocities in m/s and km/h:
   (a) The average growth rate of a tree that has reached a height of 30 m in 100 years.
   (b) The speed of the Earth as it orbits the Sun (Earth-Sun distance approx. 149 600 000 km).
   (c) The speed of the tip of the 0.7 m long hour hand of a tower clock.
   (d) The speed of sound, which is inferred from the observation that the echo from a 0.5 km distant rock face is received 3 s after calling out.
2. A bike race consists of a mountain stage and subsequently a downhill ride on the same way back to the start. Uphill, the winner's speed averages $v_1 = 15 \frac{km}{h}$ and downhill $v_2 = 50 \frac{km}{h}$. What is the average speed on the entire loop? Is it greater or smaller than $\frac{1}{2}(v_1 + v_2)$?
3. A glow ball is shot vertically upwards. How fast must it be so that a height of 100 m is reached? How long must the delay for lighting up be selected so that this happens at the highest point?
4. A suburban train accelerates at $0.8 \frac{m}{s^2}$ leaving the platform and decelerates at $-1.2 \frac{m}{s^2}$ when approaching. What is the minimum travel time between two stations that are 1 km apart? What is the maximum speed the train can reach? By how much does the travel time extend with a speed limit of 100 $\frac{km}{h}$?
5. We consider one-dimensional motion with constant acceleration a, where at the time $t = 0$ the location

with s_0 and the velocity with v_0 are given. Establish the relationships between the variable quantities t, s and v and for each of the three pairs $(t, v), (t, s)$ and (s, v)!

6. Two drivers are driving in town, one at the speed limit of $v_1 = 50\,\frac{km}{h}$, the other at an excessive speed of $v_2 = 70\,\frac{km}{h}$. Both cars have the same deceleration. What is the ratio of the braking distances of the two cars? Both cars are approaching a pedestrian crossing side by side, which is just being entered by a pedestrian. Both brake simultaneously with the same deceleration. The first car just manages to stop before the pedestrian. What speed v_z does the second car have when reaching the pedestrian crossing? The reaction time of both drivers should be so short that you can neglect it.

7. Usain Bolt completed the 100-meter race in 9.58 s during his world record run in 2009. Assume that he accelerated constantly up to the 25-meter mark and then ran at a constant speed to the finish line. What were his acceleration and final speed?

8. In clay pigeon shooting, the shooters must hit a clay pigeon that flies at a speed of $19\,m/s$ on a path perpendicular to the line of fire, at a distance of $19\,m$. How far ahead must the shooter aim, i.e., what is the distance between the clay pigeon and the target point that the shooter must aim at in order to hit? The bullet speed is $360\,m/s$.

9. An airship moves at a constant speed v relative to the surrounding air. Calculate generally the time t, that the airship needs for the round trip between two places at a distance s for the following cases:
   (a) No wind,
   (b) Side wind with the wind speed w,
   (c) Tailwind on the outward journey and headwind on the return journey with wind speed w.
   Arrange the three cases along their travel times.

10. Water is sprayed horizontally from a nozzle at a speed of $12\,m/s$. At what distance and at what angle to the horizontal does the jet hit the ground, if the nozzle is $0.5\,m$ above the ground?

11. A trick ski artist jumps over a horizontal ramp and lands $10\,m$ behind the take-off edge, measured on the slope inclined at $45°$. What was the speed at take-off?

12. In a parabolic flight, an airplane follows a projectile motion, so that there is zero gravity inside the air-

plane. An airplane starts such a parabolic flight with a speed of 600 km/h from a climb at a climb angle of 45° and ends it at a descent angle of 45°. How long does the parabolic flight last, i.e., how long is there zero gravity in the airplane?

13. (a) Show that for a projectile motion where the release and impact points are at the same height, the range of the thrown object is given by:

$$R = 2h \sin 2\beta$$

where β is the release angle relative to the horizontal and $h = v^2/(2g)$ would be the maximal height if thrown vertically upwards.

(b) At which release angle β_{max} does the maximum range occur?

(c) Show that for a projectile motion, where the release point is H above the point of impact, the range of the object is given by:

$$R = 2h \cos \beta \left(\sin \beta + \sqrt{\sin^2 \beta + \frac{H}{h}} \right)$$

(d) Show that for the release angle β_{max}, at which the range R becomes maximum, it results in:

$$\sin \beta_{max} = \frac{1}{\sqrt{2 + \frac{H}{h}}}$$

14. Calculate from the parabolic trajectory, neglecting air drag, the optimal launch angle for a shot putter who releases the shot at a height of 2.1 m with a speed of 13 m/s. In reality the athletes choose shallower angles. Why?

15. A self-propelled howitzer hits a practice target at a distance of 22 km with a barrel inclination of 60°. This is called "high-angle fire," as the firing angle is greater than 45°. Afterwards, the same target should be hit with the same propellant charge and the same projectile at a barrel inclination less than 45° ("low-angle fire"). How large must this barrel inclination be chosen for the second shot? What time interval must the shots be fired so that the two projectiles arrive at the target simultaneously? Neglect air drag, projectile spin, and the influence of the Earth's rotation.

5.8 · Circular Motion

16. For training purposes, an astronaut is to be subjected to four times standard gravity in a centrifuge. At the end of a 6 m long arm rotates a cabin in which the astronaut's seat is suspended rotatably against the horizontal. What rotation frequency must the centrifuge be operated at? At what angle to the horizontal does the astronaut then sit?
17. By how much is one lighter at the equator due to the Earth's rotation than at the poles? Assume that the Earth is an ideal sphere with the radius 6400 km.

Dynamics of a Point Mass

Contents

6.1 The Inertia Principle – 82

6.2 The Basic Law of Mechanics – 87

6.3 The Principle of Reaction – 97

6.4 Superposition of Forces – 101

6.5 Measurement of Forces – 104

© The Author(s), under exclusive license to Springer-Verlag GmbH, DE, part of Springer Nature 2025
S. Roth and A. Stahl, *Mechanics*,
https://doi.org/10.1007/978-3-662-68079-7_6

6.1 The Inertia Principle

Throughout the entire Middle Ages, the teachings of Aristotle (384–322 BC, ◘ Fig. 6.1) from antiquity were the decisive source of knowledge about nature. We want to start this chapter with a review of his theory of motion:

Aristotle defines the concept of motion much more generally than we use it today. He refers to any kind of change in an object as motion. He uses the ancient Greek word *kínesis*, from which the word "kinematics" as the study of motion is derived. We discussed it in the previous chapter. Aristotle defines motion as change of any kind. He assumes that such a change must already be inherent in the object as a possibility. He speaks about "motion", when this possibility is finally realized.

Aristotle distinguishes four types of motion. He calls them motion
- according to essence → coming to be and perishing,
- according to quantity → growth and decay,
- according to quality → change of a property,
- according to place → local motion, i.e. motion in the modern sense.

◘ **Fig. 6.1** Aristotle (384–322 BC). © CPA Media Co. Ltd/picture alliance

The last of these four types of motion corresponds to what we exclusively understand as motion today. We will focus on it in the following. Aristotle further divides the local motions into (◘ Fig. 6.2):
- Natural local motion, which results in light bodies moving upwards and heavy bodies moving downwards. Thus, smoke (a light body) rises, while a stone (a heavy body) falls to the ground.
- Forced local motion, which requires a drive. Today we would speak of a force. Without this drive, the forced local motion comes to a standstill. If the drive is strong, a fast motion results, if it decreases, it becomes slower.

This concept seems foreign to us. It may also contradict your understanding of physics. But do not dismiss it lightly. It directly stems from our everyday experience. Heavy bodies seemingly fall to the ground without propulsion, light bodies rise. If you want to move a body against its natural tendency, it requires a force. For example, when you try to pick up the fallen stone again, or when you pull a heavy cart across a street. If you pull

6.1 · The Inertia Principle

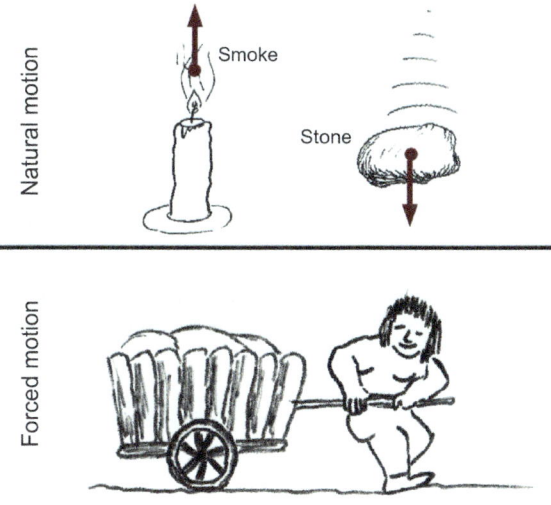

◘ **Fig. 6.2** Natural and forced motion according to Aristotle

hard, the cart rolls faster. If you stop pulling, the cart stops, the movement is over.

So much for Aristotle. With the beginning of modern science, physicists finally break with Aristotle's ideas about motion. We turn to Newton's ideas (◘ Fig. 6.3):

Sir Isaac Newton (1643–1727) laid the foundations of modern mechanics in 1687 with his *Philosophiae Naturalis Principia Mathematica* (Natural Philosophy Based on Mathematics, ◘ Fig. 6.4). It contains a different concept of motion. Newton builds his theory of mechanics on three axioms. The first of the three axioms is the so-called principle of inertia.

◘ **Fig. 6.3** Portrait of Sir Isaac Newton. Painted by Godfrey Kneller (National Portrait Gallery, London). © CPA Media Co. Ltd/ picture alliance

> **Newton's First Axiom (lex prima), Principle of Inertia**
> Every object perseveres in its state of rest, or of uniform motion in a right line, except insofar as it is compelled to change that state by forces impressed thereon.

This contradicts Aristotle. According to Newton, no force (no propulsion) is necessary to maintain motion. Once a body is in motion, it continues to move. On the contrary, a force is necessary to stop the motion once it has started.

But Newton also contradicts everyday experience. If you want to pull a heavily loaded cart, you need a force to maintain the motion. Without force, the cart comes to a standstill. To explain everyday experience with

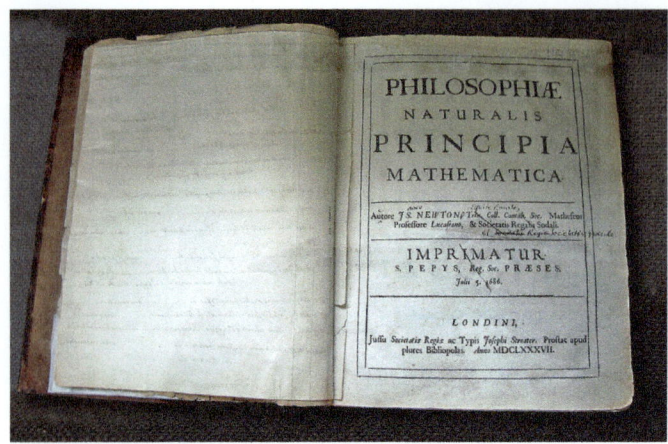

◨ **Fig. 6.4** Newton's Principa © Wikimedia: Andrew Dunn, November 5, 2004

Newton's axiom, friction must be introduced everywhere. Newton's explanation of everyday motion is much more complicated than the Aristotelian one, even if it may seem more familiar to you due to your education.

Axioms

At this point, some thoughts on axioms are appropriate. Axioms are the starting points of physical theories. They are general propositions on which the theory is based. They should be immediately understandable. They cannot be derived from the theory or from experiments. Such axioms are found in both Aristotle's explanation of motion and Newton's principle of inertia.

Axioms cannot be right or wrong, but they can be more or less meaningful, useful, or suitable. This becomes apparent with the success of the theory that builds on them. Aristotle's theory can accurately represent a range of natural phenomena. But with the development of new experimental techniques in Newton's time, new areas became experimentally accesible that Newton's, but not Aristotle's, theory of motion can capture. Here, the movements of the planets are particularly noteworthy. Newton's theory can explain these, but Aristotle's theory cannot. Also, frictionless movements, which could be approximated in Newton's time, can only be explained with Newton's theory. Because of the larger scope, New-

6.1 · The Inertia Principle

> ton's axioms are more useful and are almost exclusively used today.
>
> Only with the experiments on the speed of light in the late nineteenth century were new areas discovered that required an extension of Newton's axioms. As long as the speeds are small compared to the speed of light, they may still be used today.

We saw in ▶ Sect. 5.2 that the description of a motion always requires the specification of a reference system. Newton assumed that there is a global reference system in which the entire world rests. This reference system is anchored with the fixed stars. It defined the universe in his time. Newton's statements refer to this reference system. For him, this was the natural choice. But modern cosmology has demonstrated that such a reference system does not exist. To which reference system must we then refer the principle of inertia? Or does this mean that we can use any reference system?

Certainly not. Imagine you are sitting in a train that is accelerating out of the station. You see an empty beer bottle, left behind by a passenger, on the floor. It should remain at rest according to the principle of inertia, but it rolls to the end of the carriage, although, there is no force causing the movement. The cause is not a force, it lies in the reference system. We understand the principle of inertia today in the following sense: it asserts that there are reference systems in which the principle of inertia can be applied. These are called "inertial frames".

The axiom also provides a rule for identifying inertial frames: you must check the principle of inertia. If it applies, you found an inertial frame. For example, you could hang a ball on a string. If it hangs straight downwards without moving further, the principle of inertia is fulfilled. You are in an inertial frame. But if the ball moves out of its resting position, your reference frame cannot be an inertial frame. In practice, the distinction between inertial and non-inertial frames is rarely a problem. As a rule, one can estimate effects due to the acceleration of the frame and decide whether they are relevant or not, but logically the formulation of the principle of inertia appears like a circular argument: The principle of inertia applies in inertial frames, which we define as those in which the principle of inertia applies.

Finally, a remark about Newton's global reference system. It is associated with an absolute space, which exists independently of everything else. The universe is located in this space, the fixed stars rest in this space. It was accessible through the fixed stars. Today we know that there is no absolute space, in which our universe is located, and there are also no resting fixed stars. Nevertheless, this system of fixed stars is sometimes helpful, for example, when you want to estimate deviations from the principle of inertia that arise from the movement of the earth around the sun. You can use it as long as you do not want to address cosmological questions.

Experiment 6.1: Inertia of a heavy ball

A ball is hung on a thin thread. At the bottom, a handle is attached to the ball by means of another thin thread. Both threads have the same tensile strength. Now you pull on the handle until one of the two threads breaks. Which one will it be[1]?

The question cannot be answered like this. It depends on how you pull. If you pull slowly on the handle, the upper thread will break. It additionally carries the weight of the ball and therefore breaks before the lower thread. However, if you pull abruptly, the lower thread will break. Now the inertia of the ball comes into play. The short jerk puts a strain on the lower thread. This pulls the ball downwards and should tear the upper thread, but the ball resists the movement due to its inertia, so that the tensile force is only gradually passed upwards. If the pull is short enough, the lower thread breaks before the upper one is further strained, so that it "survives". It is recommended to use a heavy metal ball so that the effect becomes as clear as possible.

Experiment 6.2: Inertia at the set table

Imagine you are home alone in the evening and have prepared a cozy dinner. You put a tablecloth on and set

---

1 If you want to demonstrate the experiment, the ball must be secured against falling.

the table festively. But then the tablecloth seems a bit exaggerated for a dinner without guests. But how do you remove the tablecloth without having to clear everything again?—Right!—You remember the law of inertia, grab the tablecloth at one end and pull it out from under the porcelain and glasses with a strong jerk. Due to the inertia, the dishes remain in place and you have achieved the desired goal[2].

Experiment 6.3: Inertia in Rotation

This experiment is particularly simple. All you need is a raw egg. You place the egg on a smooth surface. This could be the ceramic hob in your kitchen. It has an edge that prevents the egg from falling off. Now you set the egg in as rapid a rotation as possible. Then you briefly stop the egg until it has completely come to rest and let it go again. The egg will restart to rotate all by itself. How can this be?
Inside the hard shell is a liquid (egg white and yolk). When you spin the egg, this is gradually set in rotation by friction. Now you stop the egg. The shell is stopped, but the liquid continues to rotate. It is only slowly decelerated by friction. If you let go of the egg in time, the liquid is still rotating. It now takes the shell with it again via friction. The egg begins to rotate again.
You don't believe that the restarting has something to do with the liquid inside the egg? Test a hard-boiled egg!

6.2 The Basic Law of Mechanics

Newton's first axiom deals with free fall. It states that without the influence of force, a uniform motion results. About a motion with force, it merely says that it is non-uniform. The second Newtonian axiom now quantifies the case with force:

---

[2] If you want to perform the experiment yourself (it's not as difficult as it seems at first), the table and tablecloth should not be too large. Make sure to use a smooth tablecloth (e.g., a paper tablecloth) and don't use your most expensive dishes for the first attempt. Authors and publisher assume no liability.

> **Newton's Second Axiom (lex secunda), Principle of Action**
> The change of motion of an object is proportional to the force impressed; and is made in the direction of the straight line in which the force is impressed.

It can be understood as the fundamental law of mechanics. If you know the force acting on a body, you can calculate its motion from it. Today we usually express it as an equation:

$$\vec{F} = m\vec{a}. \tag{6.1}$$

Here, Newton introduced mass as a property of physical bodies that describes their inertia, i.e., their resistance to changes in motion. Sometimes we also explicitly speak of inertial mass. There are bodies in a wide range of different masses. In ◘ Table 6.1 we have compiled some examples for you.

Eq. 6.1 assumes that the mass m of the moving body is constant. You will learn about a generalization for changing masses further below.

◘ **Table 6.1** The mass of some objects

Electron	$9.109 \cdot 10^{-31}$ kg
Hydrogen atom	$1.67 \cdot 10^{-27}$ kg
Protein molecule (Actin)	$6.97 \cdot 10^{-23}$ kg
Bacterium	$\approx 10^{-15}$ kg
Egg cell (Human)	4 µg
Hair (5 cm)	100 µg
Fly	100 mg
Human	50 … 100 kg
Car	1 t
Airbus A380 (max. takeoff weight)	578 t
Moon	$7.348 \cdot 10^{22}$ kg
Earth	$5.97 \cdot 10^{24}$ kg
Sun	$1.989 \cdot 10^{30}$ kg

6.2 · The Basic Law of Mechanics

Example 6.1: Acceleration of a Sports Car

In Example 5.7 we described that a Porsche 911 Carrera can accelerate with up to $6.5\,\mathrm{m/s^2}$. What force must the engine transmit to the road through the tires to achieve this? To answer this question, we need to know the weight of the car. We estimate it, including fuel and driver, at $1.5\,\mathrm{t}$. Using $F = ma$ this results in a force of almost $10\,\mathrm{kN}$. Each of the two rear tires must therefore transmit $5\,\mathrm{kN}$ to the road (rear-wheel drive).

Example 6.2: Braking Force of a Super Oil Tanker

The Jahre Viking, later renamed Knock Nevis, was an oil tanker and the largest vessel ever built. It had a maximum displacement of $647{,}955\,\mathrm{t}$, which corresponded to its weight when fully loaded, with a maximum load of $564{,}000\,\mathrm{t}$ of crude oil. The maximum speed was 15.8 knots or $8.13\,\mathrm{m/s}$. After a conversion, it was $458.45\,\mathrm{m}$ long. The ship was scrapped in 2010.

Like every tanker, it had to prove during the test drive that it could be brought to a standstill within 15 ship lengths by setting the engines to "Full speed back". What braking force is necessary for this?

With constant acceleration, the ship's speed is reduced to zero in the time $t = v/|a|$. The acceleration is negative here. During this time, it covers the distance $s = \frac{1}{2}at^2 = \frac{1}{2}\frac{v^2}{a}$, from which the acceleration can be calculated:

$$|a| = \frac{1}{2}\frac{v^2}{s} = \frac{1}{2}\frac{(8.13\frac{\mathrm{m}}{\mathrm{s}})^2}{15 \cdot 485\,\mathrm{m}} \approx 0.0048\frac{\mathrm{m}}{\mathrm{s^2}}.$$

The braking force is:

$$F = m\,|a| \approx 3.11 \cdot 10^6\,\mathrm{N}.$$

However, some remarks should be made:
- The power of the ship's engines was even higher, so the Jahr Viking could be brought to a standstill within the maximum allowed braking distance.
- During braking, the drag from the displaced water also plays an important role, which further shortens the braking distance.
- The radius of the turning circle is, as with most ships of this type, less than the braking distance. In the

> event of a sudden obstacle, the captain is likely to turn and maneuver rather than stop.

Newton's second axiom is a differential equation (see Mathematical Appendix A3.10). It does not directly indicate the position $\vec{r}(t)$ of a body, but the acceleration acting on it. The motion must be calculated from the acceleration. Let's first consider the simplest case in which the accelerating force vanishes, $\vec{F} = 0$. Then:

$$\vec{a}(t) = \frac{d\vec{v}(t)}{dt} = 0$$
$$\vec{v}(t) = \int \frac{d\vec{v}(t)}{dt} dt = \int \vec{0} \, dt = \vec{c},$$
(6.2)

where $\vec{c}$ is a constant vector (the constant of integration). This simple case results in a constant velocity, which we already knew from the principle of inertia.

Let's assume that a non-vanishing force is acting. We assume it to be constant over time, $\vec{F}(t) = \vec{F}_0$. Then the acceleration is also constant over time, with

$$\vec{a} = \frac{1}{m} \vec{F}_0.$$
(6.3)

This is the case of uniform acceleration (see ▶ Sect. 5.5, Eq. 5.16).

The solution of the differential equation becomes more difficult when the force varies with position or time. In courses on theoretical physics you will be taught how to solve the differential equation. Here, only a few examples are given.

> **Example 6.3: Motion in the Gravitational Field**
>
> In ▶ Sect. 5.7 we discussed the projectile motion and assumed that the acceleration due to gravity is constant during the motion. At very large heights, the body moves out of the Earth's gravitational field, causing the acceleration due to gravity to decrease. We want to discuss this case as an example of motion under the influence of a position-dependent force. We will learn later that the gravitational force acting on a body in the Earth's gravitational field decreases with the distance from the center of the Earth as:

6.2 · The Basic Law of Mechanics

$$\vec{F}_G = -GM\frac{m}{r^2}\hat{e}_r,$$

where m is the mass of the body, M is that of the Earth, G is a constant (the gravitational constant) and $\hat{e}_r$ is a vector pointing outward from the center of the Earth. The gravitational force acts in the opposite direction (towards the center). We launch a body vertically upwards at high speed. Then the problem can be reduced to a one-dimensional calculation in the direction of r. From Newton we get:

$$ma = m\frac{dv_r}{dt} = -GM\frac{m}{r^2},$$

with the radial velocity v_r. Instead of expressing it through its temporal change, it can be expressed through its spatial change:

$$\frac{dv_r}{dt} = \frac{dv_r}{dr}\frac{dr}{dt} = \frac{dv_r}{dr}v_r.$$

We substitute and separate the variables r and v_r, i.e., we bring all quantities containing r to one side of the equation and those containing v_r to the other:

$$mv_r\frac{dv_r}{dr} = -GM\frac{m}{r^2}$$
$$mv_r\,dv_r = -GM\frac{m}{r^2}\,dr$$
$$v_r\,dv_r = -GM\frac{1}{r^2}\,dr.$$

We now integrate and obtain:

$$\int v_r\,dv_r = -GM\int\frac{1}{r^2}\,dr$$
$$\frac{1}{2}v_r^2 = \frac{GM}{r} + c$$

where c is a constant of integration that we need to determine from the boundary conditions. We choose c such that at launch, i.e., at a distance $r = r_E$, which corresponds to the surface of the Earth, v_r matches the initial velocity v_0:

$$\frac{1}{2}v_0^2 = \frac{GM}{r_E} + c \rightarrow c = \frac{1}{2}v_0^2 - \frac{GM}{r_E}$$

and thus

$$\frac{1}{2}v_r^2 = \frac{GM}{r} + \frac{1}{2}v_0^2 - \frac{GM}{r_E} = \frac{1}{2}v_0^2 + GM\left(\frac{1}{r} - \frac{1}{r_E}\right)$$

or

$$v_r = \sqrt{v_0^2 + 2GM\left(\frac{1}{r} - \frac{1}{r_E}\right)}.$$

Now, for example, the maximal altitude r_{max} can be calculated as the distance r, for which the radial velocity becomes zero. After that, the body falls back to Earth.

$$0 = \sqrt{v_0^2 + 2GM\left(\frac{1}{r_{max}} - \frac{1}{r_E}\right)} \Rightarrow r_{max} = \left(\frac{1}{r_E} - \frac{v_0^2}{2GM}\right)^{-1}$$

The result is shown in the figure as a function of the initial velocity. For small velocities, the flight altitude is hardly distinguishable from the Earth's radius, but with increasing v_0 it increases more and more until it tends towards infinity above 10,000 m/s. The escape velocity is reached. The body escapes from the Earth's gravitational field. You can certainly determine the exact value yourself.

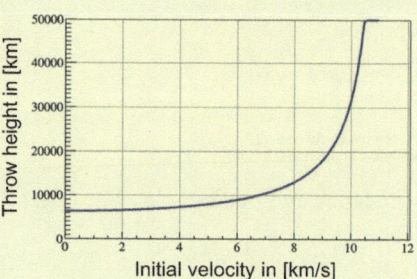

Example 6.4: Motion with a Time-Dependent Force

A soccer player kicks a ball. The force on the ball grows from the first touch to a maximum and then decreases again when the ball takes off. We want to approximate the temporal course of the force by a Gaussian curve. We also assume that the ball is hit centrally and thus flies straight away. Then it is sufficient to consider the magnitude of the force:

$$F(t) = F_0 e^{-t^2/t_0^2}.$$

6.2 · The Basic Law of Mechanics

How large is the velocity after the impact? We substitute into Newton's second law:

$$ma = m\frac{dv}{dt} = F_0\, e^{-\frac{t^2}{t_0^2}}$$

$$dv = \frac{F_0}{m} e^{-\frac{t^2}{t_0^2}}\, dt$$

$$\int_0^{v_e} dv = \int_{-\infty}^{\infty} \frac{F_0}{m} e^{-\frac{t^2}{t_0^2}}\, dt .$$

In the first step, we have again separated the variables. Since we are only interested in the final velocity, we have integrated over the entire impact in the second step. The impact occurs at $t = 0$. At $t = -\infty$ the speed is still zero, at $t = +\infty$ the final speed v_e is reached. The integral can be looked up in a table. It results in:

$$v_e = \frac{F_0}{m}\sqrt{\pi}\, t_0 .$$

Example 6.5: Numerical Integration of the Equation of Motion

In some cases, the dependence of the accelerating force on position and time can become so complicated that no analytical solution can be found. The differential equation can still be solved numerically. A particularly simple method (which you may be able to program yourself) is presented here.

We start at the time t. At this time, the body is located at the position $\vec{r}$ and has the velocity $\vec{v}$. The force acting on it is $\vec{F}(\vec{r}, t)$. This force leads to a change in velocity, which can be calculated from Newton's second law:

$$m\vec{a} = m\frac{d\vec{v}}{dt} = \vec{F}(\vec{r}, t)$$

$$d\vec{v} = \frac{1}{m}\vec{F}(\vec{r}, t)\, dt$$

$$\Delta\vec{v} = \int_{\vec{v}(t)}^{\vec{v}(t+\Delta t)} d\vec{v}$$

$$= \frac{1}{m}\int_t^{t+\Delta t} \vec{F}(\vec{r}, t')\, dt' \approx \frac{1}{m}\vec{F}(\vec{r}, t)\int_t^{t+\Delta t} dt' = \frac{1}{m}\vec{F}(\vec{r}, t)\Delta t .$$

In the last step, we have assumed that the time interval Δt is so small that the force can be assumed to be constant over this interval. Then the force can be pulled out of the integral and the integral only yields Δt. We have determined the speed at the end of the time interval Δt. Now the position can be obtained from

$$\vec{r}(t + \Delta t) = \vec{r}(t) + \vec{v}(t) \cdot \Delta t.$$

We determined position and speed at the time $t + \Delta t$. We iterate this calculation for each further time step Δt. Since we have to choose small time steps, many iterations will be necessary. This is best done with a computer.

We made a few choices which are not obvious. We used the force at the beginning of the time interval $\vec{F}(\vec{r}, t)$ for the calculation of the change in velocity. We could just as well have used the force at the end $\vec{F}(\vec{r}, t + \Delta t)$ or in the middle. For the calculation of the change in position, we used the velocity at the beginning of the time interval. Here too, we could have used the velocity at the end or in the middle of the interval. Similar considerations apply to the position. If the time intervals are small enough, all choices lead to the same result. Comparing different options is a good method to check whether the time intervals are small enough.

At first glance, such a procedure may seem unsatisfactory. In the end, you get tables of numbers about the motion, but no analytic functions. In practice this makes no difference. You can determine the trajectory with such a procedure with an accuracy that is limited only by computing time and numerical inaccuracies of the computer. An accuracy that is sufficient in practice.

Experiment 6.4: Acceleration on the Air Track

Newton's secon axiom gives two proportionalities that can be checked experimentally. The acceleration should be proportional to the accelerating force and inversely proportional to the accelerated mass.

6.2 · The Basic Law of Mechanics

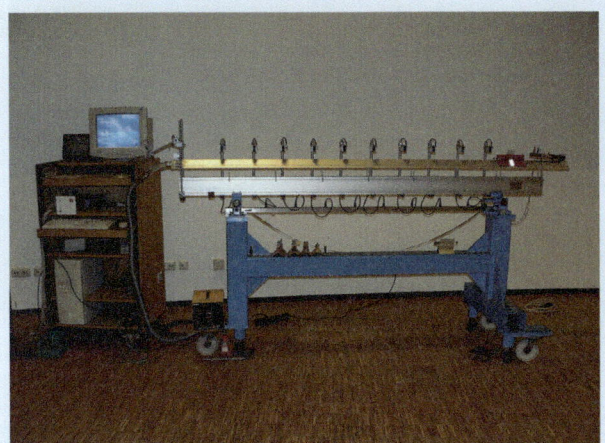

© RWTH Aachen, Physics Experiment Collection

To conduct the experiment, we use an air track to eliminate any distortions due to friction. Pressurized air escapes through small holes in the track, on which the carts float. The cart is accelerated by a thread and a pulley with weights. There are 10 light barriers installed on the rail, which record the movement and determine the acceleration. The picture shows the apparatus. Three measurements were carried out, which are presented in the three sketches and the table. Predictions on the second and third measurement were derived from the first. Note that both the mass of the cart and the accelerating weight itself must be accelerated.

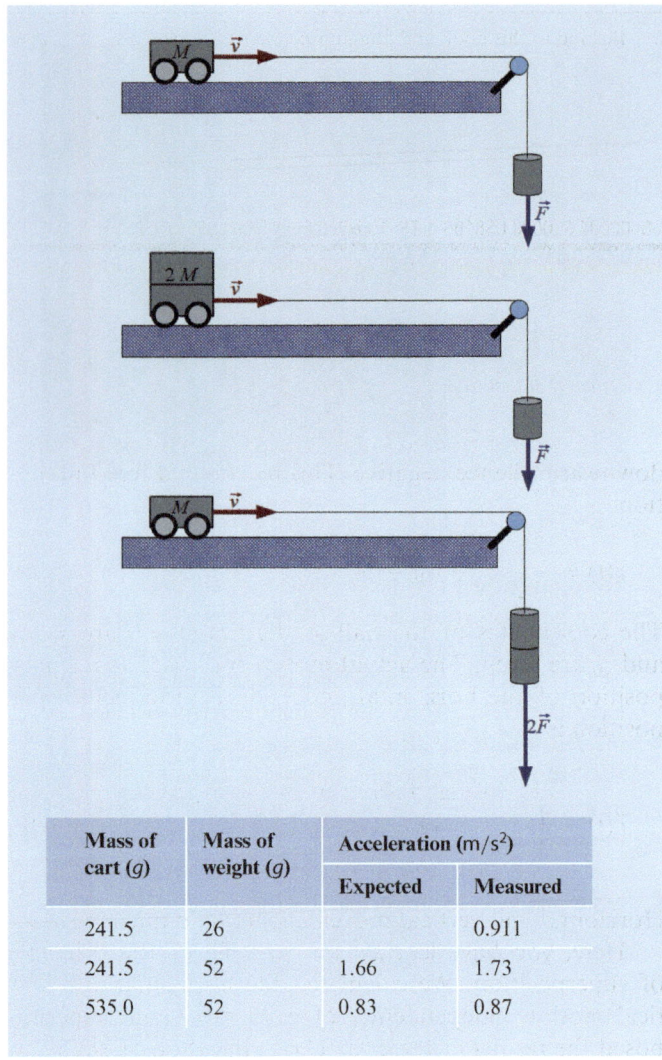

Mass of cart (g)	Mass of weight (g)	Acceleration (m/s²)	
		Expected	Measured
241.5	26		0.911
241.5	52	1.66	1.73
535.0	52	0.83	0.87

In honor of Newton, the unit of force is referred to as 1 Newton, with the symbol N. Through Newton's second law, the unit can be related to the base units m, s, kg:

$$1\,\text{N} = 1\,\frac{\text{kg m}}{\text{s}^2}.$$

1 N is the force that must be applied to accelerate a body with a mass of 1 kg with 1 m/s². To get an estimate for the less familiar unit of Newton, one can remember that 1 N approximately corresponds to the weight of a mass of 100 g.

6.3 The Principle of Reaction

The third Newtonian axiom deals with the cause of forces. Every force must have a material cause. There must be a body from which the force originates. Newton's third axiom states that an equally large force reacts back on the body that exerts the force.

> **Newton's Third Axiom (lex tertia), Principle of Reaction**
> To every action, there is always opposed an equal reaction.

This axiom is often abbreviated with the words "*actio = reactio*". This axiom can also be expressed as an equation:

$$\vec{F}_{A \to B} = -\vec{F}_{B \to A},$$

with the force exerted by body *A* on body *B* on the left and on the right is the reactive force of *B* on *A*.

Reactive forces are familiar to us from everyday life. You lift a heavy bag by pulling it upwards. You clearly feel the reaction as the force with which the bag pulls your hand downwards. Or you pull a sled, on which a child is sitting, up a hill. You pull upwards (*actio*), but the sled pulls you downwards (*reactio*). That's why you have difficulty moving forward when the snow is frozen and slippery.

Example 6.6: Reaction on the Hammer

You hammer a nail into a wall. The hammer transfers a force to the nail, driving it into the wall. But consider the movement of the hammer. It approaches the nail at high speed and is abruptly slowed down. Its speed suddenly becomes zero. According to Newton's second axiom, a force is always responsible for a change in speed. This force is the reactive force of the nail. The nail exerts a force back on the hammer, slowing it down.

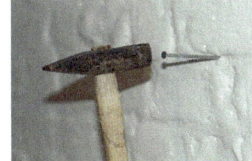

Experiment 6.5: Reactive Forces on Skateboards

This experiment is a sporting competition. Two approximately equally heavy opponents each stand on a skateboard. A mark is placed in the middle between the two skateboards (finish line). Now the they are given a rope, with which they try to pull the opponent over the finish line. Whoever's skateboard reaches the finish line first, loses. Who will win?

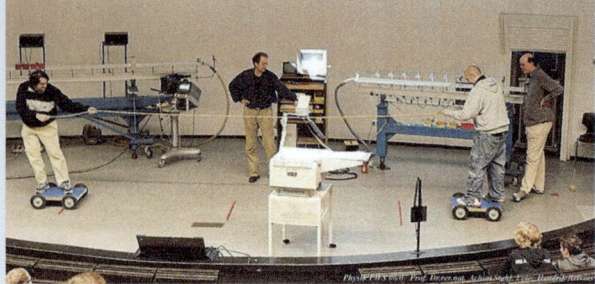

© Photo: Hendrik Brixius

No one will win! Both skateboards arrive at the finish line at the same time. When one of them pulls on the rope, he pulls his oponent towards the finish line, but a reaction from the opponent will pull him towards the finish line to the same extent. Since the forces are equal, they arrive at the same time[3].

Experiment 6.6: Recoil from the Medicine Ball

Do you know the heavy medicine balls found in sports halls? Have you ever tried to throw such a ball? Stand firmly on both legs and push the ball away from your chest. You transfer a force to the ball, which accelerates it away from you, but you will clearly feel that a force from the ball acts back on you. You are pushed backwards.

---

[3] A referee may need to ensure that the opponents do not gain advantages through friction effects.

6.3 · The Principle of Reaction

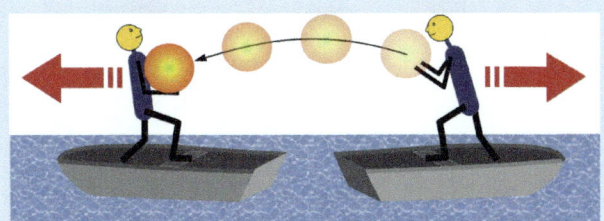

This recoil can be made visible by standing on a movable surface, e.g. in a boat on a lake. The boat moves backwards after the throw. The same happens when you catch the heavy ball standing in a boat. When catching, you slow down the ball with a force directed against the direction of the motion of the ball. The reaction to this moves the boat backwards.

The person in one boat seems to exert a force on the person in the other boat without touching him. This is how forces are imagined to act at a distance in modern physics.

Experiment 6.7: Water Rocket

Rockets are accelerated by the reaction. The rocket is filled with a fuel that is expelled at high speed. A force acts inside the rocket, accelerating the fuel backwards. As a reaction, the accelerated fuel exerts a force on the rocket that points forward. This force accelerates the rocket.

In the picture, you see a simple, yet impressive example, a cold water rocket. The rocket is a simple plastic container with a volume of about half a liter. It is filled with 40 ml of water. After that, the rocket is inflated with compressed air to 5 bar. When the hose is removed, the rocket launches. The compressed air pushes the water out of the opening at the back and accelerates the rocket. It easily flies across the lecture hall.

© Photo: Hendrik Brixius

Example 6.7: Tension in Strings

The sketch shows two masses connected by a string over a pulley. We assume that the masses of the string and friction can be neglected. If the masses are not equal, the heavier one will pull the lighter one upwards. We want to determine the acceleration. For simplicity, we assume $m_1 > m_2$. The accelerating force is

$$F = m_1 g - m_2 g \,.$$

However, the inertia of both masses must be overcome when accelerating. Consequently, .

$$a = \frac{F}{m_1 + m_2} = \frac{m_1 - m_2}{m_1 + m_2} g$$

Now we want to determine the forces acting on the string and on the suspension of the pulley. Let's start with the right string. The weight m_2 is suspended by this string. Its gravitational force pulls the rope downwards. In addition, the weight is accelerated upwards with the acceleration a, for which a force $m_2 a$ is required. The reaction to this force adds to the gravitational force, resulting in the tension:

$$F_{\text{string}} = m_2 g + m_2 \frac{m_1 - m_2}{m_1 + m_2} g = \frac{2 m_1 m_2}{m_1 + m_2} g \,.$$

The same force must act on the left string. In the suspension, the two forces add up.

6.4 Superposition of Forces

Newton's second law is about a single force acting on a body. The generalization to multiple forces is simple. Newton describes it in a corollary. One must vectorially add the forces to a total force $\vec{F}_{tot}$ which is then entered into the calculation of the motion. So

$$\vec{F}_{tot} = \sum_i \vec{F}_i. \tag{6.4}$$

Here we encounter another superposition principle. It is closely linked with the superposition of velocities, which we discussed in ▶ Sect. 5.7. If we have multiple motions that do not influence each other, we can calculate these individually from Newton's axioms. We can then assume superposition for the velocities:

$$\vec{v}_{tot} = \sum_i \vec{v}_i. \tag{6.5}$$

Superposition also applies to the accelerations, because

$$\vec{a}_{tot} = \frac{d\vec{v}_{tot}}{dt} = \frac{d}{dt}\sum_i \vec{v}_i = \sum_i \frac{d\vec{v}_i}{dt} = \sum_i \vec{a}_i \tag{6.6}$$

and from this follows the superposition for the forces:

$$m\vec{a}_{tot} = m\sum_i \vec{a}_i = \sum_i m\vec{a}_i = \sum_i \vec{F}_i = \vec{F}_{tot}. \tag{6.7}$$

However, the converse does not hold. The superposition of forces is universally valid. It is part of the axiomatic foundation of classical mechanics. The velocities $\vec{v}_i$ in Eq. 5.7 can only be calculated individually from $\vec{F}_i = m\vec{a}_i$ if none of the forces $\vec{F}_i$ is influenced by the movements generated by the other forces $\vec{F}_j (j \neq i)$. Please reconsider experiment 5.6 and the result of our calculation of the motion (Eq. 5.20). What would change if we took air drag into account? We will see later that drag represents a force $\vec{F}_L$ that is opposed to the movement, i.e., it points in the direction $-\vec{v}$, and its magnitude depends quadratically on the magnitude of the velocity, i.e., $|\vec{F}_L| \propto v_x^2 + v_y^2 + v_z^2$. If we take this force into account, the ascent and descent of the plastic ball is influenced by the velocity v_x of the train. The magnitude and direction of the drag depend on the vertical and horizontal components of the motion. One cannot be calculated without knowing the

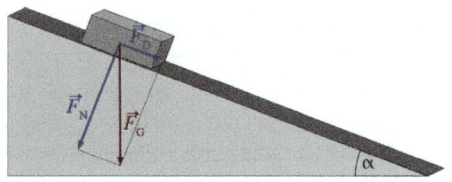

Fig. 6.5 Decomposition of the weight of a body on an inclined plane

other, and vice versa. Superposition still applies to the forces, but can no longer be transferred to the velocities. If you want to use the superposition of velocities to simplify the calculation of a motion, you must check in each individual case whether there is actually no mutual influence between the components of the motion.

Experiment 6.8: Addition of Forces

We can demonstrate the vectorial addition of forces with spring scales. In the figure, a weight is shown hanging on two inclined spring scales. In the sketch next to it, the two forces $\vec{F}_1$ and $\vec{F}_2$ are shown in dark red. They add up vectorially to the net force $\vec{F}_R$ (green). This is represented by the parallelogram of forces. The net force is exactly opposite to the gravitational force of the weight.

 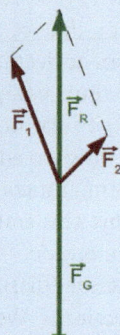

© Photo: Hendrik Brixius

The principle of superposition is also used in reverse when we decompose forces into components. For example, **Fig. 6.5** shows a body on an inclined plane. The weight $\vec{F}_G$ acts on the body. It is decomposed into the

6.4 · Superposition of Forces

downhill-slope force $\vec{F}_D$ parallel to the plane and the normal force $\vec{F}_N$ perpendicular to it. The vector sum of $\vec{F}_D$ and $\vec{F}_N$ reproduces $\vec{F}_G$. Of course, other decompositions are possible. One could decompose $\vec{F}_G$ into more than two components, or into two components that have different directions than $\vec{F}_D$ and $\vec{F}_N$. These decompositions are possible, but less helpful. The benefit of the decomposition shown in the figure is that $\vec{F}_N$ is perpendicular to the direction of motion and therefore has no influence on the motion on the plane. It is solely determined by $\vec{F}_D$, which points exactly in the direction of motion.

Example 6.8: Superposition in Hammer Throw

In hammer throw, several forces act on the hammer, which is placed today by a steel ball weighing 7.26 kg. The essential forces are $\vec{F}_1$, with which the athlete pulls on the approximately 1.2 m long steel cable with outstretched arms, the centrifugal force $\vec{F}_2$, which acts radially outward from the pivot point, and maybe the air drag $\vec{F}_3$, which points against the motion. The sum of the forces defines the acceleration of the steel ball. We have constructed it geometrically in the figure. We attached the vector of the force $\vec{F}_2$ to the end point of $\vec{F}_1$ (green dashed arrows) and the vector $\vec{F}_3$ to its end point. The end point of $\vec{F}_3$ then marks the endpoint of the resulting total force $\vec{F}_{tot}$ (red).

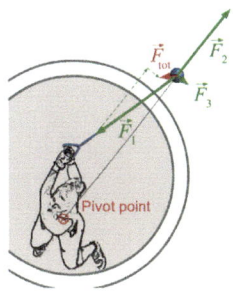

Example 6.9: Force Decomposition on Ropes

You can decompose forces geometrically with a force parallelogram. In our example, a weight hangs on two strings of different lengths attached to two eyes in the walls at different heights. The weight $\vec{F}_G$ pulls it downwards. The reaction that holds the weight is distributed between the two strings: $\vec{F}_1$ and $\vec{F}_2$. You can determine those forces with a force parallelogram defined by the suspension point and the tip of the reaction force $-\vec{F}_G$. The sides of the parallelogram are aligned with the strings, as can be seen in the sketch. Note that the two forces $\vec{F}_1$ and $\vec{F}_2$ are not equal in magnitude and that configurations can be found (which ones?), in which the magnitude of one or both rope forces exceeds the magnitude of the weight. Try it out!

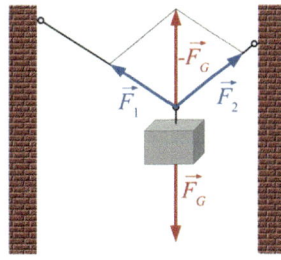

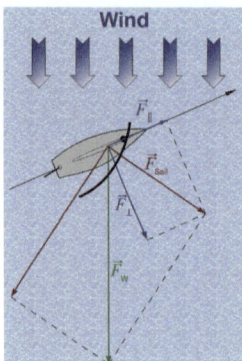

Example 6.10: Forces on a Sailing Boat

The illustration shows a rough representation of the force on a sailing boat. The force of the wind points straight down. Due to the Bernoulli effect (see ▶ Chap. 16), a force component perpendicular to the sail is created, which drives the boat. It is marked with $\vec{F}_{\text{Sail}}$. A centerboard (or keel) and the rudder prevent the boat from drifting perpendicular to the course. We decompose the force $\vec{F}_{\text{Sail}}$ into a component perpendicular to the centerboard $\vec{F}_\perp$ and one along its direction $\vec{F}_\parallel$. The first component has no influence on the motion. It is absorbed by the centerboard and the rudder. Since $\vec{F}_{\text{Sail}}$ pushes approximately in the middle of the sail, but the counterforce of the centerboard acts much lower (in the middle of the sword), the boat tends to lean away from the wind (heeling). The second component $\vec{F}_\parallel$ moves the boat against the drag of the water and against the air drag on the chosen course. Note that the decomposition of forces on the sail creates a force component that makes a course against the wind possible.

6.5 Measurement of Forces

Forces are usually measured with spring scales. In ◉ Fig. 6.6 you can see one in use. It uses a spring that is stretched by the force. The elongation of the spring is proportional to the strength of the force. The spring is enclosed by a shell on which a scale is printed on alternating red and white background. The scale is directly calibrated in N. The scale has a movable envelope with which the scale can be zeroed before the force is applied.

The spring scale exerts a force on the body via the hook at the front and the string. This force is what the spring scale is supposed to measure. The force causes a reaction of the same size, with which the rope pulls on the spring and stretches it. We determine the elongation via Hooke's law[4]: The elongation s of the spring is proportional to the applied force $F = ks$, where the stiffness k characterizes the strength of the spring. Hooke's

---

[4] We will deal with Hooke's law in more detail in ▶ Sect. 14.1.

6.5 · Measurement of Forces

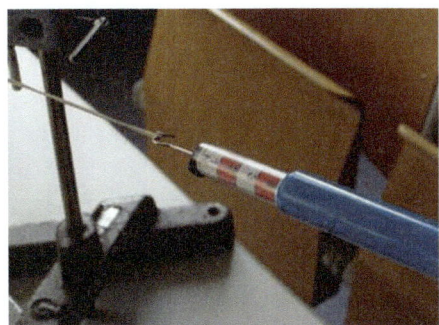

Fig. 6.6 A spring scale measures the force in a string © RWTH Aachen, Physics Experiment Collection

law is an approximation that only applies for small elongations of the spring. Therefore, spring scales should not be overloaded. From the known stiffness, the scale is divided into Newtons. For forces of different magnitudes, there are spring scales with springs of corresponding stiffness.

Alternatively, forces can be determined by counterweights (Fig. 6.7). The force is redirected vertically with a string and pulleys. A weight is chosen whose weight force exactly compensates for the force being sought. The weight then remains at rest. The force can be calculated from the mass of the weight. However, one must know the standard gravity g exactly, which depends on the geographical latitude, the sea level, and other parameters.

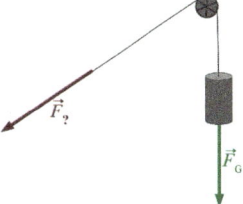

Fig. 6.7 Determination of an unknown force by a counterweight

❓ Problems

1. A railroad engine of the 189 series of Deutsche Bahn (German railways) with a mass of 87 t has an initial tractive force of 300 kN. How strongly can the engine accelerate on a flat track with an attached container car of mass 100 t or with an attached ore car of mass 4000 t? Calculate for both cases the force on the hitch, with which the freight cars are attached to the engine. Ignore friction.
2. A braking distance of 2800 m is listed for an emergency braking of an ICE 3 train from a speed of 300 km/h. What force must the braking system be able to apply? The ICE 3 has a mass of 409 t.

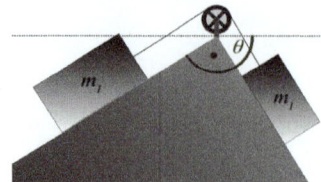

3. In an elevator, the cabin of mass 1100 kg is held by a rope. What force acts on the rope when the cabin is accelerated upwards at 0.9 $\frac{m}{s^2}$? What is the force with the same acceleration downwards?

4. A vehicle is standing on a slope with a 12% gradient. Suddenly the brakes fail. What acceleration does the vehicle experience? Ignore friction effects.

5. Consider two masses connected by a string over a pulley. They slide frictionlessly on the two planes, which are perpendicular to each other (see sketch). What must the ratio of the masses be for a given angle θ to remain at rest? What is the acceleration a when the masses are equal ($m_1 = m_2$)?

6. A rope of mass m and total length L hangs with a section of length l over the edge of a table and is initially held firmly. At time $t = 0$ it is released and slides frictionlessly over the edge of the table. Calculate the position $s(t)$ of the lower end of the rope, with respect to the edge of the table. After what time has the rope completely slid off the table?

7. The cars of roller coasters are braked after the ride with eddy current brakes, whose braking force is proportional to the speed, $F_{break} = k v$. Calculate the speed $v(t)$ of a car of mass m and initial velocity v_0 when it is decelerated from time $t = 0$. How long does it take to decelerate a 400 kg car from 50 km/h to 5 km/h, if the braking force at $v = 10 \frac{m}{s}$ is exactly 5 kN?

8. As a passenger in a car, you observe during a braking process that a thread with an object at the end, which otherwise hangs vertically down from the rearview mirror, is deflected forward by 8°. What is the deceleration?

9. An airplane performs a climb with an acceleration of 2.5 $\frac{m}{s^2}$ at an angle of 20° to the horizontal. What force does a passenger of mass 80 kg exert on his seat?

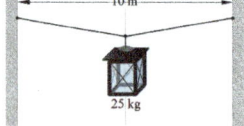

10. A street lamp of mass 25 kg is hung in the middle between two houses 10 m apart. How much must the rope sag at the location of the lamp at minimum so that the load on the rope does not exceed 10 kN?

Work and Energy

Contents

7.1　Work and Power – 108

7.2　Energy – 116

7.3　Energy Conservation – 119

7.4　Symmetries – 131

© The Author(s), under exclusive license to Springer-Verlag GmbH, DE, part of Springer Nature 2025
S. Roth and A. Stahl, *Mechanics*,
https://doi.org/10.1007/978-3-662-68079-7_7

7.1 Work and Power

The term "work" has many different meanings in common language. We talk about work as a profession with the words "going to work" or about the work of an artist in the sense of an artwork. Etymologically, the word is related to the concept of effort. These everyday terms are quite vague and therefore not suitable for immediate use in physics. We need to define the physical terms more precisely. We want to approach the physical concept of work through a simple machine, the tackle.

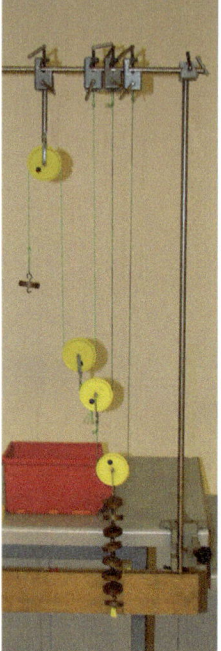

© RWTH Aachen, Physics Experiment Collection

Experiment 7.1: Lifting with a Tackle

A load consisting of eight weights is to be lifted with a pulley tackle. The tackle is assembled from three loose pulleys and one fixed pulley (the pulley on the far left), which only serves to deflect the pulling force downwards. As can be seen in the photo, a single weight on the rope on the left is sufficient to keep the load in balance. Each of the loose pulleys reduces the weight of the load by a factor of two, so that in total only 1/8 of the force is needed to lift the load.

This arrangement is called a gun tackle. It reduces the weight by a factor 2^r, where r is the number of loose pulleys.

When we pull down on the rope, the load moves upwards. In doing so, we can make another important observation: We have to pull a much greater distance on the rope than the load lifts. To lift the load by 10 cm, we have to pull the pull rope 80 cm.

Example 7.1: Tackle Systems

The illustrations show three tackles with increasing force reduction. The upper pulleys of blocks are fixed, the lower ones are loose. The first tackle reduces the force by a factor of 2, the second by a factor of 3, and the last one by a factor of 4. The distance that the pull rope must be pulled increases accordingly by a factor of 2, 3, or 4.

7.1 · Work and Power

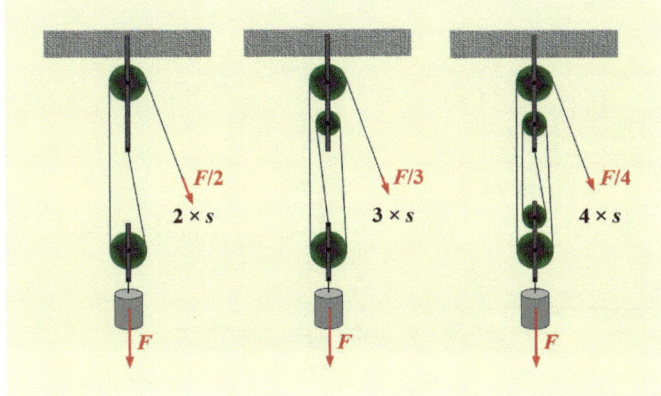

As shown in Experiment 7.1 and Example 7.1, you can reduce the force to lift a weight using a tackle. Unfortunately, it increases the distance that you must pull. The force to lift the load is reduced to $1/n$, the distance extends by a factor of n. The product of distance and force is constant. This observation is not only valid for tackles. It is much more general.

The product of the force and the distance covered under the force - the displacement - is called work in physics $W = F \cdot s$.

In this simple definition, we have implicitly assumed that the displacement occurs along the direction of the force. If this is not the case, only the component of the force in the direction of the displacement is taken into account. This can be expressed by a scalar product:

$$W = \vec{F} \cdot \vec{s}. \tag{7.1}$$

The unit of work is a Joule (Symbol J), named after James Prescott Joule. In terms of the base units, it is

$$1\,\text{J} = 1\,\text{N}\,\text{m} = 1\,\frac{\text{kg}\,\text{m}^2}{\text{s}^2}. \tag{7.2}$$

> **Example 7.2: Lifting Weights with a Crane**
>
> A crane lifts a load of $m = 200\,\text{kg}$ from the street to the upper floor of a building shell at a height of $h = 3\,\text{m}$. The work performed is
>
> $$W = \vec{F}_G \cdot \vec{h} = F_G h = m g h$$
>
> $$W = 200\,\text{kg} \cdot 9.81\,\frac{\text{m}}{\text{s}^2} \cdot 3\,\text{m} = 5886\,\frac{\text{kg}\,\text{m}^2}{\text{s}^2} \approx 5.9\,\text{kJ}$$

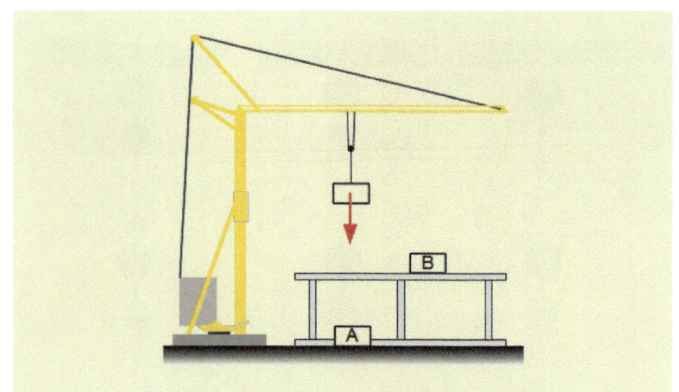

Example 7.3: Lever

The invariance of the work performed with different tackles can be found in all mechanical machines. In this example we consider the lever. The figures show a one-sided and a two-sided lever. The force needed to lift the load is reduced, depending on the distance of the impact point of the force from the pivot point. However, the height to which the load is lifted is reduced by the same factor.

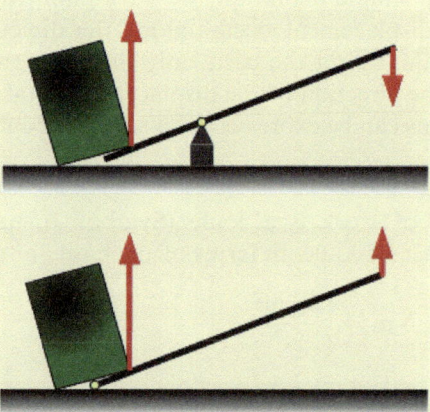

Levers are widely used in everyday life. Can you recognize their function in the objects depicted?

7.1 · Work and Power

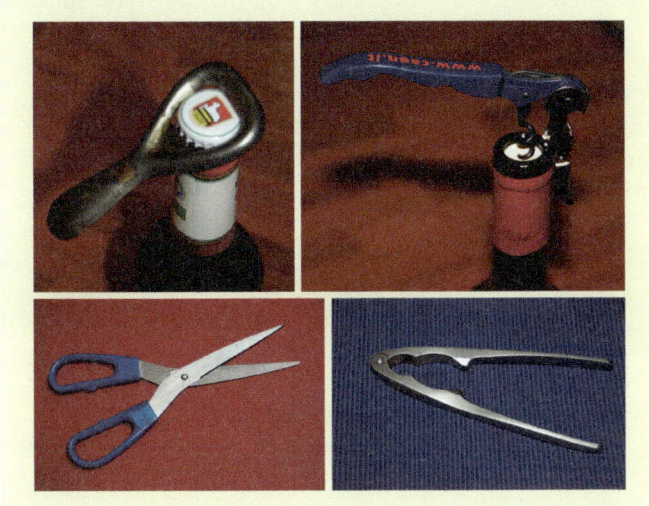

Let's now consider the situation in ◘ Fig. 7.1. A cart is being pulled up a slope. Depending on the gradient, a different amount of force must be applied. The force changes along the path. We apply a horizontal force, which results in a changing angle between the force and the displacement. We therefore use the scalar product $\vec{F} \cdot \vec{s}$. We account for the variable force by dividing the path into small segments $\Delta \vec{s}_i$ and calculate the work for each segment. Finally, we sum the work over the entire path:

$$\Delta W_i = \vec{F}_i \cdot \Delta \vec{s}_i, \quad W = \sum_i \Delta W_i. \quad (7.3)$$

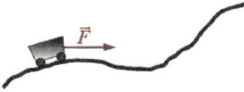

◘ **Fig. 7.1** Work along a variable path

If we make the segments $\Delta \vec{s}_i$ increasingly smaller, the sum describes the actual work performed more and more accurately. If we finally make the segments infinitesimally small, we get an exact result. Mathematically, this corresponds to the transition to the integral:

$$W = \int_A^B \vec{F}(\vec{s}) \, d\vec{s}, \quad (7.4)$$

where A and B are the starting and ending points of the path.

Please note that work can take both positive and negative values. The work is positive when the applied force points in the direction of the displacement. This is the case when you pull the cart up the slope as just outlined.

It is said that you perform work on the cart. Now imagine, you pull the cart with too little force and despite your effort, the cart slowly rolls backwards down the slope. Now the work is negative. The cart performs work on you.

Finally, in this section, we want to introduce power as another physical quantity. It is defined as the work performed per unit of time:

$$P = \frac{W}{t}. \tag{7.5}$$

However, this definition is only valid as long as the work is constant within the time interval t. If the work changes over time, we must proceed to the derivative, similar to what we discussed when defining velocity. ▶ Sect. 5.4):

$$P(t) = \frac{dW(t)}{dt}. \tag{7.6}$$

The unit of power is the Watt (symbol W), named after James Watt, the inventor of the steam engine. With Eq. 7.5, this unit can be related to the base units

$$1\,\text{W} = 1\,\frac{\text{J}}{\text{s}} = 1\,\frac{\text{kg}\,\text{m}^2}{\text{s}^3}. \tag{7.7}$$

We have compiled some typical powers in ◘ Table 7.1.

◘ **Table 7.1** Typical power of some processes

Minimum power at the antenna for radio reception	2.5 fW
Power of sound at the hearing threshold	2.5 pW
Light power of a laser pointer	1 mW
Human heart	1.5 W
Energy saving light bulb	10 W
Cyclist, peak performance	400 W
Solar radiation per square meter	1.37 kW
Water kettle	2 kW
Car	70 kW
Rated power of large wind turbines	5 MW
Power of a high-speed train	20 MW
Nuclear reactor, thermal power	5 GW
World energy consumption	20 TW

7.1 · Work and Power

Example 7.4: Units in Everyday Life

In everyday life, in addition to the watt, several other units of work and power are in use. The power of engines is expressed in comparison to the power of horses by horsepower (PS). 4 PS is approximately equivalent to a power of 3 kW. More precisely: 1 PS = 0.7354 kW. Historically, one PS corresponded to the continuous power of a working horse. It is defined as the power necessary to lift a weight of 75 kg to a height of 1 m in one second. However, a horse can certainly perform more than one PS. It can achieve up to 20 PS at a gallop, and even a strong human can exceed the power of one PS.

We have to pay the power companies for the electrical work of the current they provide. The SI unit would be Joule. However, most power companies bill the electricity in kilowatt-hours (kWh). This is a quite illustrative unit. It means that with one kilowatt-hour you can operate a consumer of 1 kW power for 1 hour. 1 kWh corresponds to 3.6 MJ.

Example 7.5: Work on a Pendulum

The sketch shows a weight on a string, i. e. a pendulum. The weight is deflected from the rest position to the angle Φ_E. How much work needs to be done?

The weight is $F_G = mg$. We decompose it into the components along the string and perpendicular to it. The first component is compensated by the string. Work must be performed against the second component. Its amount is $F_{G\perp} = F_W = mg \sin \Phi$. We could replace the sine of the angle Φ with $\sin \Phi = x/l$, which shows that the restoring force F_W is proportional to the displacement of the pendulum from its equilibrium position, but we will stick with the expression using the angle Φ. The work is calculated according to

$$W = \int_0^{\Phi_E} \vec{F}_W \cdot d\vec{s}$$

Since force and displacement point in the same direction, we may ignore the scalar product and obtain

$$W = \int_0^{\Phi_E} F_W \cdot ds = \int_0^{\Phi_E} mg \sin \Phi \, ds$$

The displacement can be expressed by the angle. It is $ds = l \, d\Phi$ and thus

$$W = \int_0^{\Phi_E} mgl \sin \Phi \, d\Phi = mgl \int_0^{\Phi_E} \sin \Phi \, d\Phi$$

An antiderivative of the sin is $-\cos$. This results in

$$\begin{aligned} W &= mgl(-\cos \Phi)\Big|_0^{\Phi_E} \\ &= mgl(-\cos \Phi_E - (-\cos 0)) \\ &= mgl(1 - \cos \Phi_E) \end{aligned}$$

This is the work that must be done to excite the pendulum to Φ_E.

Example 7.6: Work against the Gravitational Force

If a body wants to escape from the surface of a planet (e.g., with a rocket), work has to be performed against the gravitational force. We want to calculate the work to completely escape from the gravitational field of the planet. To do this, one has to move it infinitely far away from the planet.

We will learn later that gravity decreases inversely proportional to the square of the distance from the center of the planet r. So we can write

$$F_G = k \frac{1}{r^2},$$

where the weight of the body is included in the constant k. To calculate the work, we assume that the rocket launches vertically upwards, so the body follows a radial path, meaning that force and motion point in the same direction. The work is then calculated as the path integral from the surface of the planet with radius R to infinity (see Mathematical Appendix A3.11):

$$W = \int_R^{\infty} k \frac{1}{r^2} \, dr.$$

An antiderivative of $1/r^2$ is $-1/r$, so that we obtain

$$W = -k\left.\frac{1}{r}\right|_R^\infty = -k\left(\frac{1}{\infty} - \frac{1}{R}\right) = \frac{k}{R}$$

Example 7.7: Acceleration at Constant Power

A car (mass 1.4 t) is driving at a speed of $v_0 = 50\,\text{km/h}$. It approaches a slow-moving truck. The driver decides to overtake. He increases the engine's power to $P = 100\,\text{PS}$. How does the speed change?

To answer the question, we first need to determine the force with which the car accelerates. It is:

$$P = \frac{dW}{dt} = \frac{F\,ds}{dt} = F(t) \cdot v(t).$$

Since the power is supposed to be constant, the accelerating force decreases with increasing speed. We insert this force into the basic law. We get:

$$\frac{dv}{dt} = a = \frac{F(t)}{m} = \frac{1}{m}\frac{P}{v(t)}.$$

This differential equation can be solved by separating the variables v and t i.e., we bring all terms in which t occurs on one side of the equation and all with v to the other. In a somewhat casual notation, we get:

$$\frac{m}{P} v(t)\,dv = dt.$$

We integrate, taking care to match the integration limits on both sides of the equation. We start the integrations at time 0 with the speed v_0 and end at time t and the speed v:

$$\int_{v_0}^{v} \frac{m}{P} v'(t)\,dv' = \frac{m}{P}\int_{v_0}^{v} v'\,dv' = \int_0^t dt'.$$

The integration yields:

$$\frac{1}{2}\frac{m}{P}\left(v^2 - v_0^2\right) = t,$$

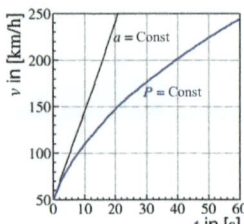

from which

$$v(t) = \sqrt{\frac{2P}{m}t + v_0^2}$$

follows. The sketch shows the increase in speed during the acceleration process in blue. For comparison, we have depicted an overtaking process with constant acceleration $a = P/(m\,v_0)$ in black. The initial acceleration is the same in both cases, but in the more realistic case of constant engine power, the acceleration decreases with increasing speed. You may wonder why a car with 100 PS reaches such high speeds. This is because we have neglected air drag, which becomes noticeable from about 80 km/h on and limits the maximum speed that the car can reach given the power[1].

7.2 Energy

In the previous ▶ Sect. 7.1, we defined work as a physical quantity. We now want to turn to the closely related concept of energy. Energy is a more abstract concept. It refers to the ability of a physical system to perform work. The physical system can be many things, a simple body or a complicated machine.

For example, a body that is moving has the ability to perform work. One could attach a rope to a moving cart and connect it via a pulley to a weight (◘ Fig. 7.2). The cart would lift the weight, which according to ▶ Sect. 7.1 is a performance of work ($W = F_G h$). In doing so, the cart would be slowed down until it eventually comes to rest. Thus, its energy is consumed. It cannot perform any further work. However, the lifted mass now has the ability to perform work. It could lift another mass via another pulley. The energy of the cart has not been lost. It has been transferred to the lifted weight.

We want to work out this example quantitatively. How much work can the cart perform if it initially moves at the velocity v_0 and is completely decelerated on the dis-

---

[1] The case with air resistance $F_{air} = k\,v^2$ can be solved using the same method, although the solution is somewhat lengthy.

7.2 · Energy

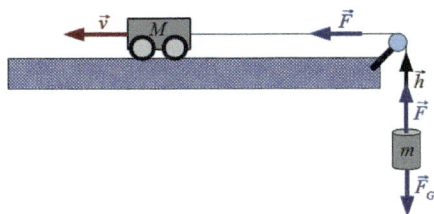

Fig. 7.2 A cart that is moving at a velocity $\vec{v}$ performs the work $W = \vec{F} \cdot \vec{h}$, by lifting a weight

tance of s_1 to s_2? The mass of the cart is M, that of the weight m. The cart is decelerated with a force $F = mg$, which according to Newton via the relationship:

$$F = Ma = M\frac{dv}{dt}. \tag{7.8}$$

leads to a deceleration of magnitude a. The work performed is:

$$W = \int_{s_1}^{s_2} F\,ds = \int_{s_1}^{s_2} M\frac{dv}{dt}ds = \int_{v_0}^{0} M\frac{ds}{dt}dv$$
$$= M\int_{v_0}^{0} v\,dv = \frac{1}{2}Mv^2\Big|_{v_0}^{0} = -\frac{1}{2}Mv_0^2. \tag{7.9}$$

It was lost by the cart. We have determined the initial energy content of the cart. Since the energy was contained in its motion, we also call it kinetic energy E_{kin}. It applies to any body:

$$E_{kin} = \frac{1}{2}mv^2. \tag{7.10}$$

To perform work, it is not necessarily required to completely decelerate the body. If you decelerate it to the velocity v_1, the work performed is:

$$W = \frac{1}{2}mv_1^2 - \frac{1}{2}mv_0^2 = \Delta E_{kin} < 0. \tag{7.11}$$

Another gedanken experiment is intended to illustrate a different form of energy. Consider the setup in **Fig. 7.3**. A load is lifted with a pulley from the height h_A at A to the height h_B at B. Due to to Newton's third axiom we have $F = -F_G$. The work performed is:

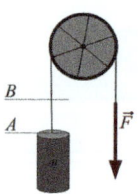

Fig. 7.3 A force lifts a load

$$W = \int_A^B F\,ds = \int_A^B m g\,ds = m g(h_B - h_A). \tag{7.12}$$

Through the force F we performed work on the load. Now, it has the ability to perform work itself. We have supplied it with energy. The energy is stored in the position of the load in the Earth's gravitational field (e.g., relative to sea level). We refer to it as gravitational or potential energy:

$$E_{\text{pot}} = m g h. \tag{7.13}$$

Analogous to kinetic energy, we find:

$$W = m g h_0 - m g h_1 = \Delta E_{\text{pot}}. \tag{7.14}$$

A stretched or compressed spring can perform work. When the spring relaxes, it exerts a force of magnitude $F_{\text{spring}} = k s$ (Hooke's Law), which can be used, for example, to accelerate a cart. We calculate the work performed by the spring during relaxation:

$$W = \int_{-s_0}^{0} k s\,ds = \frac{1}{2} k s^2 \Big|_{-s_0}^{0} = -\frac{1}{2} k s_0^2 \tag{7.15}$$

Here, s_0 is the displacement from the equilibrium position. The energy was stored in the tension in the material of the spring. It is another type of potential energy. The work done and the energy are related by

$$W = \frac{1}{2} k s_0^2 - \frac{1}{2} k s_1^2 = \Delta E_{\text{pot}} \tag{7.16}$$

There are many other types of energy that you will get to know in the following. In a system, different forms of

Table 7.2 Typical energies of some objects

Brownian molecular motion of an air molecule at room temperature	Kinetic energy	$4.0 \cdot 10^{-21}$ J
Binding energy of a water molecule	Chemical energy	$8.3 \cdot 10^{-19}$ J
Binding energy of the electron in a hydrogen atom	Electrical and kinetic energy	$2.2 \cdot 10^{-18}$ J
Flight of a mosquito	Kinetic energy	10^{-7} J
Picking up a bar of chocolate from the ground	Potential energy	1 J
Heating one liter of water by 10°	Thermal energy	42 kJ
Drive of a car at 120 km/h	Kinetic energy	0.5 … 1 MJ
Ride of a person with an elevator to the 10th floor	Potential energy	2 MJ
Calorific value of a liter of gasoline	Chemical energy	33 MJ
Fission of one kilogram of ^{235}U	Nuclear energy	86 TJ
A mass of 1 kg ($E = mc^2$)	Rest energy	$9 \cdot 10^{16}$ J
Formation of the sun from a gas cloud	Gravitational energy	10^{41} J

energy may occur simultaneously. They are divided into kinetic energy and potential energy. The latter includes all forms of energy that are not of a kinetic nature. Like work, they all carry the Joule as a unit. Some typical energies are listed in ▶ Table 7.2.

7.3 Energy Conservation

We have introduced energy as an abstract quantity. One might wonder what use this definition has. The concept of energy only makes sense through the law of conservation of energy. Initially it is an empirical fact, i.e., it has been deduced from experience with experiments. It states:

> **Energy Law**
> In a closed system, the sum of all types of energy does not change.

We must first clarify the concept of a closed system. It is a system that is separated from its environment in such a way that neither energy nor matter can escape into the environment, nor can energy or matter be transferred from the environment to the system.

Back to the conservation of energy: It is an important law in all of physics. To date, no violation has been observed. We assume that it is valid without restriction in all areas of physics.

It states that the sum of all energies in a closed system is constant over time, it is conserved. This does not imply that each type of energy is constant. Energy can indeed be converted from one type to another, only the sum of all types of energy always remains the same.

Example 7.8: Downhill Skiing

A skier is skiing frictionless down a slope. At the beginning of the slope, his speed and thus his kinetic energy are zero. But he has potential energy of the magnitude mgh_0, where h_0 is the height from which he starts. During the ride, the potential energy transforms into kinetic energy, until the potential energy at the foot of the mountain is finally used up (final speed v_E). At any point in time, the following applies:

$$m g h_0 = m g h(t) + \frac{1}{2} m v(t)^2 = \frac{1}{2} m v_E^2$$

Experiment 7.2: Energy Conversion with a Dynamo

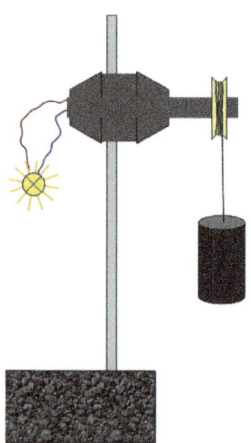

This experiment demonstrates the conversion of energy with a dynamo. The dynamo is attached to a stand. The friction wheel is replaced by a larger wheel, on which a string is wound. A weight hangs on this string. If you let go of the weight, it will fall to the ground. It will drive the dynamo and the connected incandescent light bulb lights up.

At the beginning, the weight has a certain potential energy. This is converted into other types of energy. A small part is transformed into the kinetic energy of the weight and the rotation of the wheel and its axle (rotational energy). However, this is only a small part, as the weight sinks slowly. The larger part of the potential energy is converted into electrical energy in the dynamo,

7.3 · Energy Conservation

which in turn is converted into heat in the light bulb. The filament radiates this heat, part of it as visible light.

Experiment 7.3: Conservation of Energy on the Pendulum

This is an exciting experiment. You can check how much trust you have in the laws of physics.

You stand with your back against a wall (or a screen). In front of you hangs a heavy pendulum on a long steel wire. You hold the pendulum's body in front of your face, so that it just touches the tip of your nose. You may need to adjust the length of the wire. Then you let go of the pendulum and remain motionless. The pendulum gradually converts its potential energy into kinetic energy. At the lowest point, it has maximum kinetic energy. Then, on the opposite side, the kinetic energy is converted back into potential energy as the pendulum swings through. The pendulum has reached the opposite turning point and is now coming towards you. It converts the potential energy back into kinetic energy (zero crossing) and then back into potential energy. If the conservation of energy works, the pendulum should have the same potential energy at the end as at the beginning. Then it should arrive back at exactly the same point where you let it go. It will turn right in front of your nose. Do you trust it to stop before it hits your nose? However, what should you avoid?[2]

We regularly conduct the experiment in lectures with a volunteer. We use a 5 kg steel ball on an approximately 8 m long steel cable. We project the face of the volunteer in close-up onto a screen. The facial expression of the volunteer usually leads to great amusement in the lecture hall.

P.S.: So far, the conservation of energy has always worked. No nose has been damaged, yet.

Example 7.9: Conservation of Energy in Pole Vaulting

A very beautiful sequence of energy transformations can be observed in pole vaulting:

---

2 Do not push the pendulum body when letting go!

The athlete starts running. With powerful strides, he accelerates. His kinetic energy increases until it reaches its maximum just before the pole is planted.

The vaulter plants the pole. The pole bends. Kinetic energy is converted into elastic energy in the pole until the pole is maximally bent and the athlete jumps.

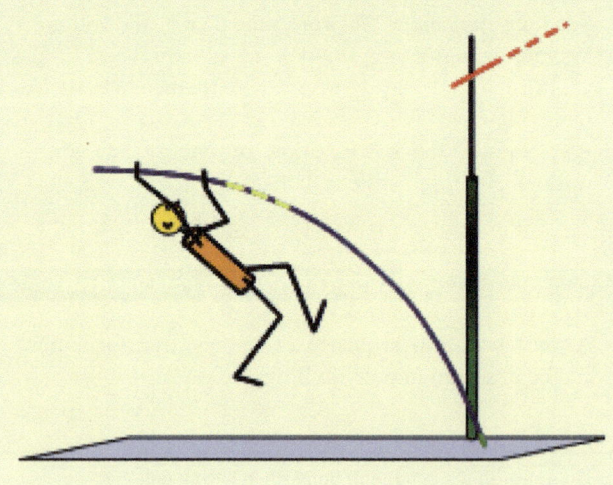

7.3 · Energy Conservation

Now the athlete is ascending. The energy stored in the pole is converted into the kinetic energy of the upward motion and into potential energy due to the gained height. In addition, there is the energy from the jump (acceleration upwards).

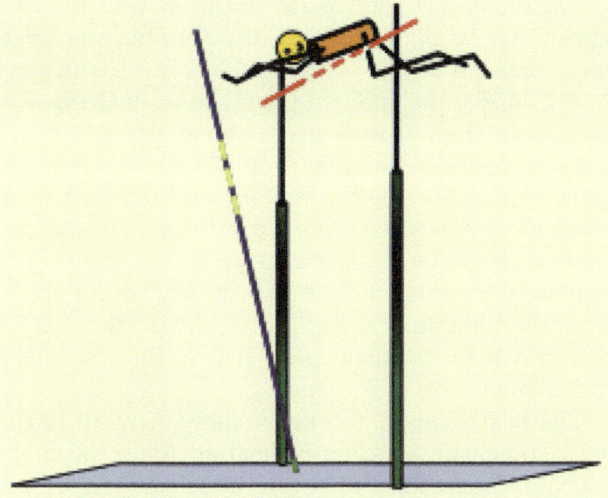

With the take-off, the athlete gains height. If the jump is successful, he converts all the kinetic energy and the elastic energy stored in the pole into potential energy. At the highest point, the pole is relaxed again and the athlete has almost no kinetic energy left.

> Finally, he falls down. The potential energy is converted back into kinetic energy, which in the end is absorbed by friction in the mat.

We turn to the amounts of energy. How much energy is needed to lift a weight to the height h? Can the energy expenditure be optimized, i.e., can we find ways or machines that can lift the weight to the height h with less energy? Consider the different paths shown in ◘ Fig. 7.4. Is the same amount of work needed on all paths?

First, consider example A in ◘ Fig. 7.4. This example is simple. The force is $F = mg$ and thus the work is $W = mgh$. We must take this work from an energy reservoir (e.g., from the power socket) to lift the weight.

If we use a ramp as in B to bring the weight up, it reduces the force to $F = mg \sin \alpha$. However, the path over the ramp is longer than in A. It is $h/\sin \alpha$, so that the work again results in $W = mgh$.

The next example (C) again shows a ramp (inclined plane), now with a greater inclination. Since the angle of inclination was cancelled out in the calculation for example B, the result is again $W = mgh$.

In Example D, a tackle is used to reduce the force. With similar techniques (e.g., gears), modern machines reduce the effort required to lift the weight. Here it is a 2-fold tackle. The force is only $F = mg/2$, but the displacement is $2h$, so in the end, it still results in $W = mgh$.

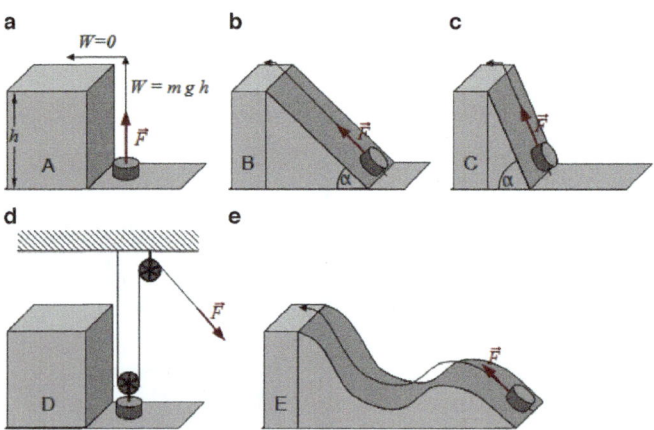

◘ **Fig. 7.4** Possible ways to lift a weight to the height h

7.3 · Energy Conservation

What will result in the general case, symbolized in Example E by a road with variable inclination? Unfortunately, this case is difficult to calculate. One must break down the path into infinitesimal ramps. But the result will be the same again: $W = mgh$.

The result is always the same regardless of the path:

$$W = mgh. \tag{7.17}$$

Did that surprise you? If so, consider what you could have achieved with a path where the work to lift the weight would be reduced. You could manage to lift the weight on this path and then bring it down again on path A. With the energy released on A, you could manage to lift the next weight on your path upwards. There would still be some energy left over. With this surplus, you could power a generator and generate electricity. You would have created a perpetual motion machine.

We initially introduced the conservation of energy as an empirical fact. Upon closer inspection, however, one realizes that it is already contained in Newton's axioms. If you calculate a motion from the axioms, it will automatically respect conservation of energy, provided that you have considered only energy-conserving forces (no friction).

As an example, we calculate the skier, which you already know from Example 7.8, from Newton's second axiom. The downhill force is (◘ Fig. 7.5):

$$\left|\vec{F}_{\text{slope}}\right| = mg \sin \alpha. \tag{7.18}$$

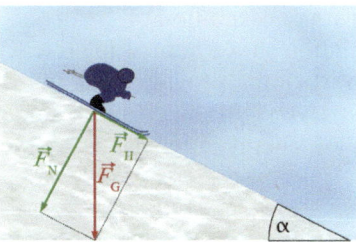

◘ **Fig. 7.5** Forces on a skier

We obtain from

$$F = ma = mg \sin \alpha$$
$$a(t) = g \sin \alpha = \text{const.}$$
$$v(t) = \sin \alpha \, g \, t \tag{7.19}$$
$$s(t) = \frac{1}{2} \sin \alpha \, g \, t^2$$

We want to determine the final velocity v_E. The distance to the end of the slope is $h_0/\sin \alpha$, where h_0 is the height from which the skier started. From the equation of motion, the duration t_E of the entire descent is

$$\frac{h_0}{\sin \alpha} = \frac{1}{2} \sin \alpha \, g \, t_E^2 \rightarrow t_E = \sqrt{\frac{2 h_0}{\sin^2 \alpha \, g}} \tag{7.20}$$

This results in a final velocity of $v_E = v(t_E)$:

$$v_E = \sin \alpha \, g \, t_E = \sqrt{2 g h_0} \tag{7.21}$$

We use the conservation of energy for a cross-check. From

$$m g h_0 = \frac{1}{2} m v_E^2 \tag{7.22}$$

follows

$$v_E = \sqrt{2 g h_0} \tag{7.23}$$

in agreement with the initial calculation. We derived a motion from Newton's axioms and obtained a result that is consistent with conservation of energy.

This calculation shows something else: Our derivation from Newton's axioms provided full information about the motion ($s(t), v(t), \ldots$), but is often somewhat tedious. Conservation of energy often leads to the result much faster.

Example 7.10: Roller Coaster with Looping

How fast must a roller coaster run so that in a loop nobody falls out of the car? Consider the sketch of a looping track with radius r, which we want to assume is frictionless. The car starts from rest at point A at a height h.

7.3 · Energy Conservation

First, we determine the velocities at points B and C from conservation of energy ($v_A = 0$ and $h_B = 0$)

$$mgh = \frac{1}{2}m v_B^2 = mg(2r) + \frac{1}{2}m v_C^2$$

We obtain

$$v_A = 0 \qquad v_B = \sqrt{2gh} \qquad v_C = \sqrt{2g(h-2r)}$$

and from $a = v^2/r$ we determine the accelerations in radial direction

$$a_A = 0 \qquad a_B = 2g\frac{h}{r} \qquad a_C = 2g\left(\frac{h}{r} - 2\right).$$

Now we can establish the condition for safe travel through the loop. It must hold

$$a_C \geq g$$

$$2g\left(\frac{h}{r} - 2\right) \geq g$$

$$2\left(\frac{h}{r} - 2\right) \geq 1$$

$$\frac{h}{r} - 2 \geq \frac{1}{2}$$

$$\frac{h}{r} \geq \frac{5}{2}$$

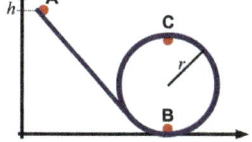

The starting height must be at least 2.5 times the radius of the loop.

In the amusement park Geiselwind, you may find the Boomerang, a roller coaster with a loop of the Shuttle Coaster type. In such a roller coaster, the cars are pulled up a ramp backwards. After releasing the cars, they coast freely, i.e., without propulsion through the circuit. The ramp has a slope of 45° and a maximum height of 35 m. The train is 15 m long. The loop is assumed to be a perfect circle with a radius of 8 m. We want to calculate the accelerations affecting the passengers for this roller coaster. We have to consider that the center of gravity of the train, when it has reached the top end of the ramp, is still lower, namely at $h = 35\,\text{m} - \frac{1}{2}15\,\text{m}\sin 45° \approx 30\,\text{m}$. Inserted into the formulas from above, this results in the highest and lowest point of the loop

$$v_B = 24\,\frac{\text{m}}{\text{s}}; \quad a_B = 74\,\frac{\text{m}}{\text{s}^2} \approx 7g$$
$$v_C = 17\,\frac{\text{m}}{\text{s}}; \quad a_C = 34\,\frac{\text{m}}{\text{s}^2} \approx 3g$$

At the deepest point, the passengers experience the acceleration $a_B + g = 8g$ downwards and at the highest point the acceleration $a_C - g = 2g$ upwards. In fact, the loop does not have the shape of a circle, but that of a clothoid. Can you imagine why?

We go back once again to the examples in ◘ Fig. 7.4. The work we have to perform to lift the weight always resulted in the same value, regardless of the chosen path. This does not always have to be the case. As soon as forces occur that drain energy from the system (so-called dissipative forces), this will no longer apply.

Our example changes as soon as we consider frictional forces. Compare B and C in ◘ Fig. 7.4 and F in ◘ Fig. 7.6. In a frictionless case, you do not need any force for the horizontal distance CF. It does not contribute to the work balance. With friction, it requires an additional contribution to the work, which is greater the longer the horizontal distance is. Thus, F can no longer yield the same work as C, if friction is included.

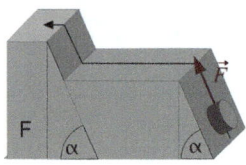

◘ **Fig. 7.6** Another possible way to lift a weight to the height h

There is no general rule that determines whether the work in a process is independent of the path or not. This must be examined on a case-by-case basis. You must check whether dissipative effects are negligible or not. The work is independent of the path and energy is conserved only if dissipative effects are negligible. Then the following three equivalent statements apply. Here, equivalent means that each statement can be derived from any of the other two and vice versa.

— The work to move a body from any point A to another arbitrary point B is independent of the path.
— The work to move a body along any closed path is always zero.
— A potential energy can be uniquely assigned to each location.

If these conditions are met, we speak of conservative forces as opposed to dissipative ones. We want to show that the second and third statement follow from the first.

7.3 · Energy Conservation

Consider ◘ Fig. 7.7. The work from A to B along the two arbitrary paths $S1$ and $S2$ is

$$W_{S1} = \int_{S1} \vec{F}(\vec{r})\, d\vec{s},$$

$$W_{S2} = \int_{S2} \vec{F}(\vec{r})\, d\vec{s}, \qquad (7.24)$$

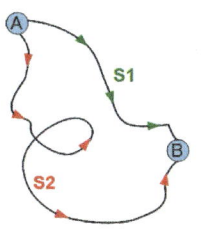

◘ **Fig. 7.7** Two paths from A to B

where, according to the first statement

$$W_{S1} = W_{S2}. \qquad (7.25)$$

We first show that the second statement follows from this assumption. To do this, we follow the path $S2$ backwards, i.e. we determine the work from B to A on path $S2$. We denote this path as $-S2$. The direction of motion is reverted and thus the work changes sign. It is $W_{-S2} = -W_{S2}$. Then we have

$$\begin{aligned} W_{S1} + W_{-S2} &= \int_{S1} \vec{F}(\vec{r})\, d\vec{s} + \int_{-S2} \vec{F}(\vec{r})\, d\vec{s} \\ &= W_{S1} - W_{S2} = 0. \end{aligned} \qquad (7.26)$$

But the sum of these two integrals is nothing other than the integral along the closed path from A over $S1$ to B and back again through $S2$. This is a closed path. We started at A and end at the same position. Therefore, we get (the circle on the integral sign indicates that the path is closed):

$$\oint \vec{F}(\vec{r})\, d\vec{s} = 0. \qquad (7.27)$$

Since both A and B as well as $S1$ and $S2$ are arbitrary, it must apply to every closed path. We have shown, that the second statement follows from the first.

To derive the third statement from the first, we need to provide a clear definition of potential energy. We need a formula that allows us to calculate the potential energy at any point in space. We choose a reference point A and define the potential energy at any location $\vec{r}$ as:

$$E_{\text{pot}}(\vec{r}) := E_{\text{pot}}(A) + \int_{A}^{\vec{r}} \vec{F}\, d\vec{s}. \qquad (7.28)$$

Due to the first statement, the integral does not depend on the path. It provides a unique value. The value of the potential energy $E_{pot}(A)$ at point A can be chosen arbitrarily. It cancels in all calculations. Usually, it defines the zero point of the energy scale.

It is instructive to consider the reversal of the last proof. For simplicity, we consider the one-dimensional case only. We express the potential energy as an integral

$$E_{pot}(x_0) = \int_A^{x_0} dE_{pot} = \int_A^{x_0} \frac{dE_{pot}}{dx} dx \qquad (7.29)$$

at the same time, according to the definition of work

$$E_{pot}(x_0) = \int_A^{x_0} F\, dx \qquad (7.30)$$

where F is the force that must be applied to move the body against the system's forces from A to x_0. It is opposed to the force generated by the system. We obtain a relation between the potential energy and the force imposed by the system.

$$F(x) = -\frac{dE_{pot}(x)}{dx} \qquad (7.31)$$

In three dimensions, we obtain correspondingly

$$\vec{F}(\vec{r}) = -\left(\frac{\partial E_{pot}(\vec{r})}{\partial x}, \frac{\partial E_{pot}(\vec{r})}{\partial y}, \frac{\partial E_{pot}(\vec{r})}{\partial z}\right) = -\vec{\nabla} E_{pot}(\vec{r}) \qquad (7.32)$$

with the gradient operator

$$\vec{\nabla} = \left(\frac{\partial}{\partial x}, \frac{\partial}{\partial y}, \frac{\partial}{\partial z}\right) \qquad (7.33)$$

In some cases, the potential energy depends not only on the location, but also on the properties of the moving body. In such cases, we may normalize the potential energy to the properties of the body. A good example is the gravitational field. It is a conservative force field. For a given body, one can define a potential energy in the gravitational field. However, this depends on the mass of the body and is therefore not transferable to other bodies. To circumvent this shortcoming, one defines a potential as

7.4 · Symmetries

$$\varphi(\vec{r}) = \frac{E_{\text{pot}}(\vec{r})}{m} \qquad (7.34)$$

It is the potential energy of a unit mass. Similar definitions of a potential can be found in many other areas of physics.

At the end of this important chapter, we want to turn to a special machine, the perpetual motion machine. A perpetual motion machine (1st kind) is a machine that runs forever without external energy input and performs work in the process. There exist plenty of ideas. One is shown in Example 7.11. However, after everything we have discussed in this chapter, it should be clear to you that none of these machines can function.

> **Example 7.11: Perpetual Motion Machine of 1st Kind**
>
> The illustration shows a proposal by Leonardo da Vinci with weights that fold out or in depending on their position. Will the disc rotate?

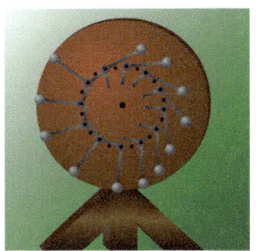

> **Experiment 7.4: The Drinking Duck**
>
> Here is our version of a perpetual motion machine: The Drinking Duck. You place a bowl of water in front of it and push its head forward once so that the beak dips into the water. The duck will right itself and from time to time bend down to sip water. It does this for days on end despite significant friction. Where does the energy come from?[3]

© RWTH Aachen, Physics Experiment Collection

7.4 Symmetries

Symmetries play an important role in modern physics. Usually, objects are examined for their symmetry. ◘ Figure 7.8 shows an example, a snowflake. The snowflake is symmetric under a series of symmetry transformations, e.g., under a rotation by 60°. This means that the same

---

[3] We will return to the drinking duck in the volume on thermodynamics.

◘ **Fig. 7.8** The image of a snowflake. It shows a high symmetry

image results when the flake is rotated by 60° degrees. The rotation by 60° is called a symmetry transformation of the snowflake. There are others. For example, reflections about several axes or point reflection at the center.

In physics, symmetry is considered as the invariance of a physical theory under various symmetry transformations. As with the snowflake, this means that the laws of the theory do not change when the symmetry transformation is applied to nature. For example, one could ask whether Newton's axioms are invariant under spatial reflection, that is, whether they would equally correctly describe the mirror image of nature. Indeed, they do.

The mathematician and physicist Emmy Noether (◘ Fig. 7.9) has shown that such invariance of a theory under a symmetry transformation always leads to a conservation law. This is now called the Noether theorem. It belongs in the field of theoretical physics and is only briefly mentioned here. Noether also indicated how to determine the conserved quantity.

In that sense, the law of conservation of energy follows from the invariance of the laws of nature under a symmetry transformation. It is the invariance under time translation, which is the shift of the zero point of the time axis. All natural laws only involve temporal distances between events, never absolute times in relation to a global $t = 0$. This implies that the laws of nature do not change over time.

7.4 · Symmetries

Fig. 7.9 Emmy Noether. In an obituary in the New York Times, Albert Einstein wrote about her: "*In the judgment of the most competent living mathematicians, Fräulein Noether was the most significant creative mathematical genius thus far produced since the higher education of women began.*" © Science Source/mauritius images

Example 7.12: Constancy of the Laws of Nature on the Pendulum

We cannot derive conservation of energy from the symmetry transformation here, but we can understand the connection with simple examples. We want to consider a pendulum. The temporal constancy of the laws of nature implies, among other things, that the frequency with which the pendulum oscillates is the same, no matter when you set the pendulum swinging. Now try to imagine that energy was not conserved. Then there could be no constancy of the laws of nature, because if you deflect the pendulum and hold it and its energy content would change over time, then it would have to oscillate at different speeds, depending on the energy content at the time you let go.

❓ Problems

1. Calculate
   - the work to carry a bicycle (mass 15 kg) up the basement stairs (height 3 m).
   - the kinetic energy that the bicycle has at a speed of 15 km/h.
   - the potential energy related to the Earth's surface of an Airbus 380 (mass 430 t) at cruising altitude (10,000 m).
   - the kinetic energy of an Airbus 380 at cruising speed (920 km/h).
   - the thermal power of the sun from the heat radiation that arrives from the sun on Earth (solar constant $E_0 = 1367 \, \text{W/m}^2$, distance Earth–Sun $1.5 \cdot 10^{11}$ m).
   - Potential energy of the water of a fully filled pumped storage power station with an upper basin of volume $7 \cdot 10^6 \, \text{m}^3$ at a height of 250 m.
   - Power of this pumped storage power station, if the water flows out at a rate of $350 \, \text{m}^3/\text{s}$.

2. In a spindle press, the pressure required to squeeze out the fruit juice is generated by a spindle unscrewing from a fixed thread and pressing on a plunger. The thread has a pitch of 8 mm and the spindle is turned by a handle, which is 1 m away from the axis, with a force of 200 N. What is the force exerted by the plunger?

3. A bicycle has a chain shift with three chainrings at the front on the pedals and seven sprockets at the back on the wheel. The largest chainring at the front has 45, the smallest 29 teeth. The largest sprocket at the back has 34, the smallest 13 teeth. What is the transmission range of the gear shift, i.e., the ratio of the force to be applied between the largest and smallest adjustable gear?

4. A car is accelerated from the speed $v_1 = 100 \, \text{km/h}$ to $v_2 = 130 \, \text{km/h}$, with the gain in kinetic energy being supplied by the motor. Calculate the speed v_3 that the car would have achieved if the same energy had been used for acceleration from a standstill.

5. In an oblique throw (projectile motion), the thrown projectile has kinetic energy at the apex of the trajectory that is half as large as its potential energy relative to the launch height. At what angle to the horizontal was the projectile thrown?

7.4 · Symmetries

6. A child swings on a swing until the reversal point of the swing's angle of deflection remains evenly perpendicular to the vertical. The child is now considering at which deflection angle β it must jump off to land as far as possible from the swing's rest point. Calculate this distance S in relation to the swing length L and the angle β, assuming that the child's center of gravity is at the same height when landing as it was previously on the swing at the lowest point of motion. Determine the optimal angle β numerically. You can find corresponding programs on the internet as well.

7. A boat is powered by a 5 kW engine. It reaches a speed of 40 m/s. How large is the drag that opposes the movement?

8. The Drachenfelsbahn is the oldest of the four still operating cog railways in Germany. It uses electric railcars with a motor power of 175 kW. The railcars have an unladen mass of 12,400 kg and can accommodate a maximum of 80 passengers. Estimate whether the stated motor power can be sufficient to reach the speed of 14 km/h at a gradient of 20% as specified in the data sheet.

9. A small van with a mass of 2 t is driven on the highway at full throttle (maximum motor power). Uphill, with a road gradient of 4%, a speed of 130 km/h is achieved, while on the corresponding downhill drive with a gradient of 4% a speed of 150 km/h is reached. Calculate the maximum motor power of the van under the simplifying assumption that the frictional resistance (air drag and rolling friction) is directly proportional to the square of the speed. Assume that the maximum motor power is constant and in particular independent of the vehicle's speed, which is only a good approximation if it is an electric drive.

Momentum

Contents

8.1　Conservation of Momentum – 138

8.2　Center of Mass – 140

8.3　Collision Processes – 146

8.4　Systems with Variable Mass – 155

8.5　The Impulse – 158

8.1 Conservation of Momentum

We define the momentum p of a body as the product of its mass and its velocity

$$\vec{p} = m\vec{v}. \tag{8.1}$$

Like velocity, momentum is also a vector. The unit of momentum can be derived from its definition. It is

$$[p] = 1\frac{\text{kg m}}{\text{s}}. \tag{8.2}$$

The unit has no specific name or symbol.

With the definition of momentum, Newton's second law can be reformulated

$$\vec{F} = \frac{d\vec{p}}{dt} = \frac{d}{dt}(m\vec{v}). \tag{8.3}$$

Assuming that the mass of the body does not change, we obtain the well-known form

$$\vec{F} = \frac{d}{dt}(m\vec{v}) = m\frac{d\vec{v}}{dt} = m\vec{a} \tag{8.4}$$

In fact, the form given by $\vec{F} = d\vec{p}/dt$ is more generally valid. Newton's formulation is a special case for constant masses.

Example 8.1: The Force of a Water Jet

A jet from a water cannon hits a billboard. What force does the jet exert on the board? The cannon can spray water at a rate of $1200\,\text{l/min}$ and, even at a distance of several meters, the water has a speed of up to $v = 30\,\text{m/s}$. One liter of water corresponds to a mass of one kilogram, i.e., in one minute, $m = 1200\,\text{kg}$ mass is expelled. This carries a momentum $p = m \cdot v = 36{,}000\,\text{kg m/s}$. To estimate the change in momentum Δp, we assume that the water is stopped at the billboard and then drops down. We neglect any possible back-spray of the water. Then $\Delta p = p$ and the force results in

$$F = \frac{\Delta p}{\Delta t} = \frac{36{,}000\,\text{kg m/s}}{60\,\text{s}} = 600\,\text{N},$$

which corresponds to the weight of a mass of approximately 60 kg. If the water were to bounce elastically off the billboard, we would double the force.

8.1 · Conservation of Momentum

We want to consider a system of bodies in which only internal forces act, i.e., forces exist between the bodies contained within the system, but there are no forces exerted between the bodies within the system and any bodies outside the system. For simplicity, we assume that the system consists of only two bodies. Then the equations of motion are

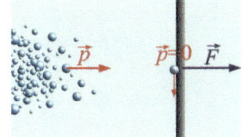

$$\vec{F}_{12} = \frac{d}{dt}\vec{p}_1, \quad \vec{F}_{21} = \frac{d}{dt}\vec{p}_2, \quad (8.5)$$

where $\vec{F}_{12}$ is the force that body 2 exerts on body 1, and $\vec{F}_{21}$ vice versa. The two forces are linked via *actio = reactio*:

$$\vec{F}_{12} = -\vec{F}_{21} \quad \text{or} \quad \vec{F}_{12} + \vec{F}_{21} = 0. \quad (8.6)$$

We get

$$\vec{F}_{12} + \vec{F}_{21} = \frac{d}{dt}\vec{p}_1 + \frac{d}{dt}\vec{p}_2 = \frac{d}{dt}(\vec{p}_1 + \vec{p}_2) = 0, \quad (8.7)$$

from which it follows that

$$\vec{p}_1 + \vec{p}_2 = \text{konst}. \quad (8.8)$$

We have just derived from Newton's laws of motion that for a system of two bodies, momentum is a conserved quantity. This applies generally to systems with any number of bodies. The derivation is analogous, just requiring more writing. Thus, we formulate in general

> **Momentum Theorem**
> In a system where only internal forces act, the total momentum is conserved.

Like the conservation of energy, the law of conservation of momentum originates from an invariance within the laws of nature under a symmetry transformation, as per Noether's theorem. The momentum theorem is based on the translational symmetry of space. This symmetry states that a physical process (e.g., an experiment) yields the same result at any location in space.

Experiment 8.1: Conservation of Momentum on an Air Cushion Track

We demonstrate the conservation of momentum with an air track. In the middle of the track, there are two carts. In ◻ Fig. 8.1, you can see two carts of the same mass. The experiment can be repeated with carts of different

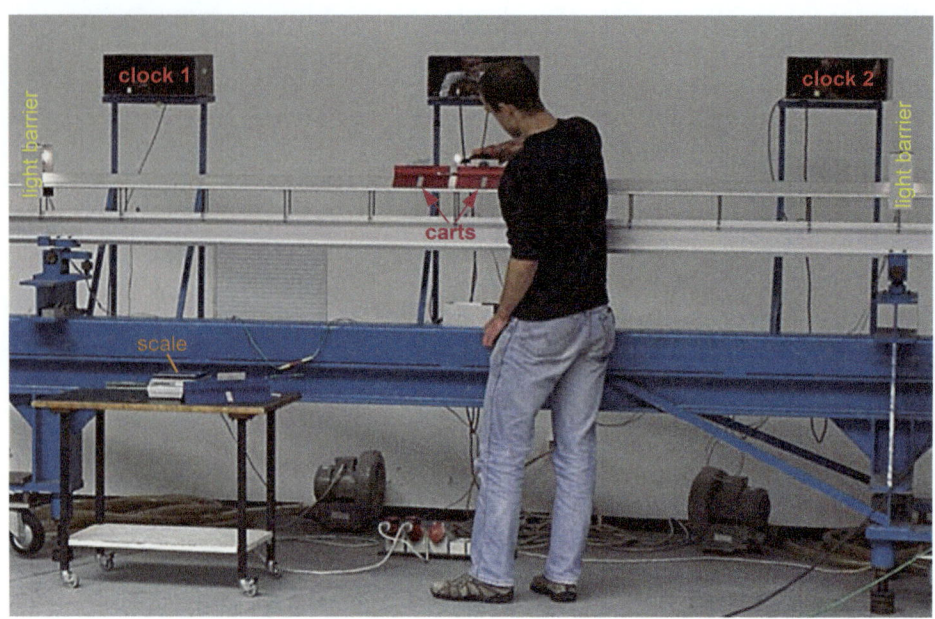

☐ **Fig. 8.1** © Photo: Hendrik Brixius

masses. The carts are weighed before the start of the experiment. The two carts are connected by a compressed spring and a thread. The spring pushes the carts apart, the thread prevents this.

At the beginning, the carts rest in the middle of the track. The initial total momentum is therefore zero. Then, the thread is fused with a lighter. The two carts push each other away and then move uniformly away from each other. Their speed is measured using a light barrier. Since the spring represents an internal force, the total momentum must be conserved, this means $\vec{p}_1 = -\vec{p}_2$. Accordingly, the speeds must behave inversely to the masses, which can be checked with the measurements.

8.2 Center of Mass

In the previous chapter, we discussed systems of bodies on which no external forces act. We found that the total momentum of such a system does not change. We now want to investigate their dynamics. To do this, we introduce the concept of the center of mass. It is calculated

8.2 · Center of Mass

from the positions at which the individual mass points of the system are located:

$$\vec{r}_{cm} = \frac{\sum_i m_i \vec{r}_i}{\sum_i m_i} \qquad (8.9)$$

An example is shown in ◘ Fig. 8.2. We want to determine the center of mass of a system consisting of three mass points with the masses $2m$, $3m$ and $4m$. They are located at the positions

$$\begin{aligned} \vec{r}_1 &= (4, 0, 2.5) & m_1 &= 2\,m \\ \vec{r}_2 &= (2, 4, 3) & m_2 &= 3\,m \\ \vec{r}_3 &= (0, 0, 3) & m_3 &= 4\,m \end{aligned} \qquad (8.10)$$

Inserting into Eq. 8.9 yields

$$\vec{r}_{cm} = (1.56, 1.33, 2.89). \qquad (8.11)$$

The center of mass (cm) is shown in ◘ Fig. 8.2. It lies in the plane that is spanned by the three mass points of the system. In the case with three mass points, this is always true.

For an extended body with a continuous mass distribution, the sums in the definition of the center of mass must be replaced by integrals over the volume of the body (see Mathematical Appendix A3.12 and A3.13):

$$\vec{r}_{cm} = \frac{\int_V \vec{r}\,dm}{\int_V dm} = \frac{1}{M} \int_V \rho(\vec{r})\,\vec{r}\,dV, \qquad (8.12)$$

where M is the total mass of the body and $\rho(\vec{r})$ its local density.

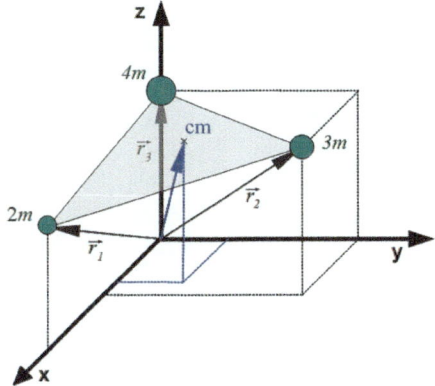

◘ **Fig. 8.2** The center of mass of a system of 3 mass points

We first want to consider the motion of the center of mass without the influence of external forces. We learned in the previous chapter that the total momentum is conserved in this case. We determine the velocity of the center of mass:

$$\frac{d}{dt}\vec{r}_{cm} = \frac{d}{dt}\left(\frac{\sum_i m_i \vec{r}_i}{\sum_i m_i}\right) = \frac{\sum_i m_i \frac{d}{dt}\vec{r}_i}{\sum_i m_i}$$
$$= \frac{\sum_i m_i \vec{v}_i}{\sum_i m_i} = \frac{\vec{p}_{ges}}{M} = \text{konst.} \quad (8.13)$$

The velocity of the center of mass is constant over time in this case.

Let's summarize this:

> **Center of Mass Theorem**
> In the absence of external forces, the center of mass moves in a straight line at a constant velocity.

Example 8.2: Center of Mass Theorem: Two-Stage Rocket

Consider the trajectory of this two-stage rocket. The rocket starts with the first stage. At the highest point of the trajectory, the first stage is burned out and is detached. When detaching, a force acts between the first and second stage. This is an internal force. The center of mass continues to follow its parabolic trajectory (dotted line). The first stage falls—slowed down by the detachment—to the ground. The second stage flies far beyond the trajectory of the center of mass.

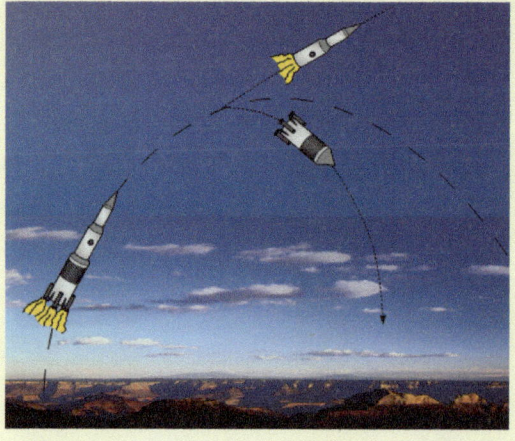

8.2 · Center of Mass

Experiment 8.2: Pendulum Cart

The pendulum cart is actually a children's toy. A pendulum is mounted on a chassis with four wheels. If you deflect the pendulum on the stationary cart, the cart performs a counter-movement to the swinging pendulum. Whenever the pendulum moves to the right, the cart rolls to the left, and vice versa. The center of mass remains stationary.

If you push the cart, the pendulum also starts to swing. The individual parts perform complicated, jerky-looking movements. Now you can only guess that the center of mass is moving evenly.

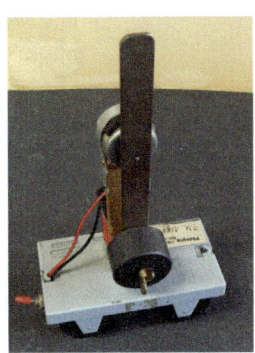

© Photo: Hendrik Brixius

So far, we have dealt with the case where no external forces act on the system. Now we want to consider the general case with external forces and calculate the acceleration of the center of mass:

$$\frac{d^2}{dt^2}\vec{r}_{cm} = \frac{d}{dt}\vec{v}_{cm} = \frac{d}{dt}\frac{\vec{p}_{cm}}{M} = \frac{d}{dt}\frac{\sum_i \vec{p}_i}{\sum_i m_i}. \tag{8.14}$$

We multiply the equation by the total mass $M = \sum_i m_i$:

$$M\vec{a}_{cm} = \frac{d}{dt}\sum_i \vec{p}_i = \sum_i \frac{d}{dt}\vec{p}_i = \sum_i \vec{F}_i, \tag{8.15}$$

in which we used Newton's second law for the individual mass points in the last step. Due to Newton's third law of motion, the sum of forces includes, with each internal force, the corresponding counterforce, such that the internal forces cancel one another and the sum is reduced to only the external forces:

$$M\vec{a}_{cm} = \sum_i \vec{F}_{i,\text{ext}}. \tag{8.16}$$

The result has the form of Newton's second law, applied to the center of mass. We see that the center of mass moves as if the total mass were combined in it and the external forces were acting on it.

Example 8.3: The Two-Body Problem

We want to consider a system consisting of two mass points on which no external forces act. Can we generally determine the motion of the bodies?

We associate the position vectors $\vec{r}_1$ and $\vec{r}_2$ of the two point masses with the center of mass $\vec{r}_{cm}$, i.e.:

$$\vec{r}_1 = \vec{r}_{cm} + \vec{r}_{1cm}$$
$$\vec{r}_2 = \vec{r}_{cm} + \vec{r}_{2cm}$$

Then we move to a coordinate system in which the center of mass is at rest, and first calculate the motion with respect to the center of mass. The equations of motion represent a system of two coupled differential equations. The coupling is in the force, because each force depends on the position of both point masses, more precisely on the relative position. $\Delta\vec{r} = \vec{r}_{2cm} - \vec{r}_{1cm}$ of the two point masses to each other. The differential equations are:

$$m_1 \frac{d^2\vec{r}_{1cm}}{dt^2} = \vec{F}_{21}(\Delta\vec{r})$$

$$m_2 \frac{d^2\vec{r}_{2cm}}{dt^2} = \vec{F}_{12}(\Delta\vec{r}).$$

We multiply the first equation with m_2 and the second with m_1 and use *actio = reactio* in the form of $\vec{F}_{21}(\Delta\vec{r}) = -\vec{F}_{12}(\Delta\vec{r})$:

$$m_1 m_2 \frac{d^2\vec{r}_{1cm}}{dt^2} = -m_2 \vec{F}_{12}(\Delta\vec{r})$$

$$m_1 m_2 \frac{d^2\vec{r}_{2cm}}{dt^2} = +m_1 \vec{F}_{12}(\Delta\vec{r}).$$

We subtract the two equations from each other and obtain:

$$\frac{m_1 m_2}{m_1 + m_2} \frac{d^2\Delta\vec{r}}{dt^2} = \vec{F}_{12}(\Delta\vec{r}).$$

This is the equation of motion for a single point mass of size $m' = m_1 m_2/(m_1 + m_2)$ called the reduced mass of the system. We must first solve this single differential equation, with the solution method and solution depending on the exact form of the force. Then we need to calculate the motion of the center of mass. Since no external forces are acting, this is a simple, uniform motion. Once we have solved both equations, we can derive the motion of the two individual point masses from $\Delta\vec{r}(t)$ and $\vec{r}_s(t)$. According to the definition of the center of mass

$$(m_1 + m_2)\,\vec{r}_{cm} = m_1\vec{r}_1 + m_2\vec{r}_2.$$

We add $m_2\Delta\vec{r} = m_2\vec{r}_2 - m_2\vec{r}_1$ and then solve for $\vec{r}_1$. We obtain:

$$\vec{r}_1(t) = \vec{r}_{cm}(t) + \frac{m_2}{m_1 + m_2}\,\Delta\vec{r}(t)$$

and correspondingly:

$$\vec{r}_2(t) = \vec{r}_{cm}(t) + \frac{m_1}{m_1 + m_2}\,\Delta\vec{r}(t).$$

With this, we have reduced the two-body problem to the solution of a one-body problem with the reduced mass, m'.

Example 8.4: The Moon's Orbit

The Sun, the Earth, and the Moon are huge bodies. Nevertheless, their motion can be described by considering them as point masses. One might wonder, considering the deviations of the shape of the bodies from a perfect sphere, where exactly to place the point masses. We have found the answer: The point masses must be applied at the center of mass of the celestial bodies. The bodies move as if the mutual forces were acting on the centers of mass. However, this is a 3-body problem, the solution of which goes beyond this book.

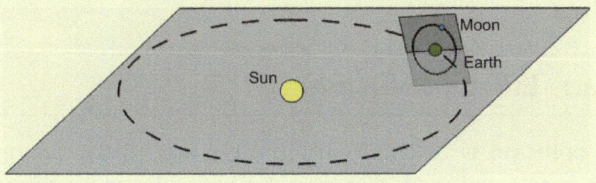

8.3 Collision Processes

An important application of the conservation laws to systems with internal forces are so-called collision processes. In these processes, individual point masses collide. Usually, we are interested in the velocities of the collision partners after the collision. In mechanics, collisions occur, for example, in billiards or in the motion of molecules in gases. They also occur in other areas of physics, such as Rayleigh scattering of light in the atmosphere. We will discuss these processes in the following.

When two or more bodies collide, strong internal forces act, against which any external forces can be neglected. Therefore, we can utilize conservation of momentum. Although the momenta of the individual bodies will change, the total momentum must be the same before and after the collision. Consequently, it must apply

$$\left(\sum_i \vec{p}_i\right)_{\text{before}} = \left(\sum_i \vec{p}_i\right)_{\text{after}} \tag{8.17}$$

where $\vec{p}_{i,\text{before}}$ and $\vec{p}_{i,\text{after}}$ are the momenta of the individual collision partners before and after the collision.

The kinetic energy is not necessarily preserved in a collision process. There are collisions in which the collision partners bounce off each other, so that the sum of the kinetic energies before and after the collision remains the same. However, this is a special case. Generally, bodies will deform and heat up during a collision process, leading to the loss of kinetic energy. Therefore, we distinguish between elastic and inelastic collisions. Some configurations are demonstrated in Experiment 8.3.

In principle, any number of bodies can collide. However, in reality, it is very unlikely that more than two bodies will collide at exactly the same moment. Therefore, we want to mostly limit ourselves to the collision of two bodies.

8.3.1 Elastic Collision

A collision is called "elastic" when no kinetic energy is lost. The collision partners bounce off each other. A good example is the collision of balls in a game of billiards. The sum of the kinetic energies after the collision is equal to the sum before:

$$\left(\sum_i E_{\text{kin},i}\right)_{\text{before}} = \left(\sum_i E_{\text{kin},i}\right)_{\text{after}} \qquad (8.18)$$

8.3.2 Perfectly Inelastic Collision

This is the other extreme. In a perfectly inelastic collision, as much kinetic energy is converted as it is possible, according to the conservation of momentum. This results in the colliding partners moving together after the collision. An example is traffic accidents. When two cars collide, they deform and can become wedged together. They then slide together after the collision, a motion that is quickly slowed down by the high friction on the road after the collision:

$$(\vec{v}_1)_{\text{after}} = (\vec{v}_2)_{\text{after}} = \ldots = (\vec{v})_{\text{after}} \qquad (8.19)$$

It is most easily seen that the maximum amount of kinetic energy is indeed lost in such a collision when one shifts the reference frame to that of the centre of mass of the two colliding bodies. This system moves with the center of mass, the total momentum is zero. Before the collision, the bodies move towards each other with opposite momentum. After the collision, they both rest. The kinetic energy of the relative motion has disappeared, only the kinetic energy of the center-of-mass motion remains. This cannot be converted due to the conservation of momentum.

8.3.3 Inelastic Collision

Elastic and perfectly inelastic collisions are limiting cases. In the first, no kinetic energy is lost, while in the latter, the maximum possible is lost. These limiting cases can only be approximated in reality. Real collisions always lie between the limiting cases. Such collisions are called "inelastic". A portion of the kinetic energy of the colliding partners is converted into other forms of energy. However, note that in all cases, the total momentum must be conserved. There are various ways to quantify the loss of kinetic energy in a collision. One possibility is the so-called collision number k, which is also known as the coefficient of restitution. In the collision of two point masses, it is:

$$k = \frac{\left|(\vec{v}_2)_{\text{after}} - (\vec{v}_1)_{\text{after}}\right|}{\left|(\vec{v}_2)_{\text{before}} - (\vec{v}_1)_{\text{before}}\right|}. \tag{8.20}$$

It is the ratio of the relative velocities before and after the collision. For an elastic collision, $k = 1$ and for a perfectly inelastic collision $k = 0$. Real collisions have a value between 0 and 1.

We now want to discuss some special cases in more detail. First, we assume that the collision partners can only move in one dimension, such as wagons on a rail. Then we move on to collisions in two (balls on a billiard table) and three dimensions (molecules in a gas).

8.3.4 Elastic Collision in One Dimension

Two bodies collide elastically in one direction (see Fig. 8.3). To simplify the representation, we denote the velocities before the collision with $\vec{u}_1$ and $\vec{u}_2$ and the velocities after the collision with $\vec{v}_1$ and $\vec{v}_2$. Positive velocities always point to the right. The masses of the bodies are m_1 and m_2. We may omit the vector arrows in one dimension. How large are the velocities after the collision?

In an elastic collision, the kinetic energy must be conserved. In addition, as with all collisions, the total momentum is conserved. Therefore, the following applies:

$$\frac{1}{2}m_1 u_1^2 + \frac{1}{2}m_2 u_2^2 = \frac{1}{2}m_1 v_1^2 + \frac{1}{2}m_2 v_2^2,$$
$$m_1 u_1 + m_2 u_2 = m_1 v_1 + m_2 v_2. \tag{8.21}$$

We multiply the first equation by 2 and rearrange the terms:

$$m_1 u_1^2 - m_1 v_1^2 = m_2 v_2^2 - m_2 u_2^2,$$
$$m_1 u_1 - m_1 v_1 = m_2 v_2 - m_2 u_2. \tag{8.22}$$

Now we isolate the masses and factorize the squares:

$$m_1(u_1 - v_1)(u_1 + v_1) = m_2(v_2 - u_2)(v_2 + u_2),$$
$$m_1(u_1 - v_1) = m_2(v_2 - u_2). \tag{8.23}$$

We divide the first equation by the second and obtain:

$$(u_1 + v_1) = (v_2 + u_2) \Rightarrow v_2 = u_1 + v_1 - u_2, \tag{8.24}$$

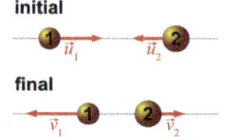

Fig. 8.3 Elastic collision in one dimension

8.3 · Collision Processes

which we then substitute into the first relation of Eq. 8.23:

$$m_1(u_1 - v_1)(u_1 + v_1) = m_2(u_1 + v_1 - 2u_2)(u_1 + v_1)$$
$$m_1(u_1 - v_1) = m_2(u_1 + v_1 - 2u_2)$$
$$m_1 u_1 - m_1 v_1 = m_2 u_1 + m_2 v_1 - 2 m_2 u_2$$
$$m_1 v_1 + m_2 v_1 = m_1 u_1 - m_2 u_1 + 2 m_2 u_2$$
$$(m_1 + m_2) v_1 = 2 m_1 u_1 + 2 m_2 u_2 - m_2 u_1 - m_1 u_1$$
$$v_1 = 2 \frac{m_1 u_1 + m_2 u_2}{m_1 + m_2} - u_1 \tag{8.25}$$

and analogously, we obtain:

$$v_2 = 2 \frac{m_1 u_1 + m_2 u_2}{m_1 + m_2} - u_2. \tag{8.26}$$

The first term of this result indicates the velocity of the center of mass. It is

$$v_{\text{cm}} = \frac{m_1 u_1 + m_2 u_2}{m_1 + m_2} \tag{8.27}$$

The calculation could have been significantly simplified if we had used the center of mass system from the beginning, in which the center of mass is at rest.

It is interesting to consider the special case in which the two masses are equal $m_1 = m_2 = m$. Then the result simplifies to

$$\begin{aligned} v_1 &= u_2 \\ v_2 &= u_1 \end{aligned} \tag{8.28}$$

The two bodies exchange velocities. The initially faster body is decelerated and the slower one is accelerated. For example, if a railway wagon runs onto an equally heavy, stationary wagon, the first wagon will be stopped and the second will start rolling at the speed of the first, provided the speed is not too large and the buffers' repulsion is elastic.

8.3.5 Perfectly Inelastic Collision in one Dimension

In the case of a perfectly inelastic collision, both bodies move together after the collision (◘ Fig. 8.4). It is $v_1 = v_2 = v$. We have conservation of momentum as the sole condition

$$m_1 u_1 + m_2 u_2 = m_1 v + m_2 v. \tag{8.29}$$

From this follows:

$$v = \frac{m_1 u_1 + m_2 u_2}{m_1 + m_2} \tag{8.30}$$

or for bodies of equal weight:

$$v = \frac{1}{2}(u_1 + u_2). \tag{8.31}$$

If two bodies of equal weight collide perfectly inelastically with the same speed ($u_1 = -u_2$), they remain at the location of the collision. In a real collision, the result is:

$$\begin{aligned} v_1 &= \frac{m_1 u_1 + m_2 u_2 - m_2 (u_1 - u_2) k}{m_1 + m_2} \\ v_2 &= \frac{m_1 u_1 + m_2 u_2 - m_1 (u_2 - u_1) k}{m_1 + m_2}, \end{aligned} \tag{8.32}$$

with the collision number k.

The result of the perfectly inelastic collision (Eq. 8.30) also applies to collisions of more than two bodies. The same result is obtained in more than one dimension, but it must then be written in vector form

$$\vec{v} = \frac{m_1 \vec{u}_1 + m_2 \vec{u}_2}{m_1 + m_2}. \tag{8.33}$$

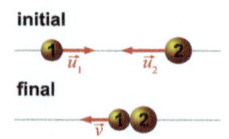

Fig. 8.4 Perfectly inelastic collision in one dimension

Experiment 8.3: Collisions on an Air Cushion Track

Both elastic and inelastic collisions in one dimension can be demonstrated on the air track. Carts with the same or different masses can be used. Elastic collisions are realized by attaching a spring to one of the carts. Perfectly inelastic collisions can be achieved by placing a ball of modeling clay between the carts.

For example, if we use two carts of equal weight, one of which is initially at rest, then only two speed measurements are sufficient to fully understand the kinematics. The speed of the incoming cart is first determined. In the elastic case, it will come to a stop after the collision, and the struck cart will move away at the same speed. In the inelastic case, both carts continue together after the collision at half the initial speed.

8.3 · Collision Processes

8.3.6 Elastic Collision in Two Dimensions

In elastic collisions in more than one dimension, no unique result can be given. Even with a clearly defined initial configuration, many configurations are possible after the collision. This is due to the approximation of the bodies as point masses. If one were to consider extended balls, the point at which the balls touch would uniquely determine the velocities and directions after the collision. Good billiard players take advantage of this.

In two dimensions, we have four unknown quantities that we want to calculate, namely the components of the final velocities of the two bodies in the two considered spatial directions (Fig. 8.5). In the case of point masses, however, we only have three constraints available. These are the conservation of momentum in the two spatial directions and the conservation of energy. Consequently, we cannot find a unique solution. We can freely assign a value to one of the four unknown quantities. We choose the deflection of the first point mass from its original direction of motion, i.e., the angle between $\vec{u}_1$ and $\vec{v}_1$. We call it θ_1.

To simplify the calculation, we want to choose a special coordinate system (Fig. 8.6). It should initially move with the second point mass, so that it remains at rest before the collision in this coordinate system, i.e. $\vec{u}_2 = \vec{0}$. We also have the freedom to choose the orientation of the axes. We rotate the coordinate system such that the first point mass initially moves along the x-axis.

The following conservation laws apply (see Mathematical Appendix A3.15):

$$\frac{1}{2}m_1 u_1^2 = \frac{1}{2}m_1 v_1^2 + \frac{1}{2}m_2 v_2^2$$
$$m_1 u_1 = m_1 v_{1,x} + m_2 v_{2,x} \qquad (8.34)$$
$$0 = m_1 v_{1,y} + m_2 v_{2,y}.$$

We switch to planar polar coordinates.

$$\frac{1}{2}m_1 u_1^2 = \frac{1}{2}m_1 v_1^2 + \frac{1}{2}m_2 v_2^2$$
$$m_1 u_1 = m_1 v_1 \cos\theta_1 + m_2 v_2 \cos\theta_2 \qquad (8.35)$$
$$0 = m_1 v_1 \sin\theta_1 + m_2 v_2 \sin\theta_2.$$

Now, if we input a value for θ_1, we can determine v_1, v_2 and θ_2 from this system of equations. The solution is

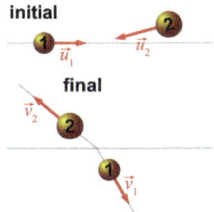

Fig. 8.5 Elastic collision in two dimensions

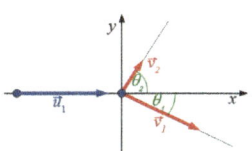

Fig. 8.6 Elastic collision in two dimensions in the coordinate system

lengthy. We leave it to the reader as an exercise (see Mathematical Appendix A3.15). Instead, we want to derive an interesting result for the special case of equal masses. So we set $m_1 = m_2 = m$ and rewrite the conservation of momentum in vector notation again:

$$m\vec{u}_1 = m\vec{v}_1 + m\vec{v}_2 \quad \text{or} \quad \vec{u}_1 = \vec{v}_1 + \vec{v}_2. \tag{8.36}$$

The right equation states that the vectors $\vec{u}_1$, $\vec{v}_1$ and $\vec{v}_2$ form a triangle. The conservation of energy now takes the following form:

$$\frac{1}{2}m\vec{u}_1^2 = \frac{1}{2}m\vec{v}_1^2 + \frac{1}{2}m\vec{v}_2^2 \quad \text{or} \quad u_1^2 = v_1^2 + v_2^2. \tag{8.37}$$

This is nothing other than the Pythagorean theorem for this triangle. The three vectors thus form a right-angled triangle with the two vectors $\vec{v}_1$ and $\vec{v}_2$ as catheti. They enclose a right angle, i.e., the velocities of the two bodies after the collision are perpendicular to each other. However, this simple relation only applies in case of equal masses.

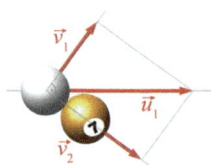

> **Example 8.5: A Collision in Billiards**
>
> While the direction of the velocities after the collision of two point masses in a plane is not uniquely determined by the initial conditions, this is indeed the case for the collision of extended spheres. Our sketch shows the example of a billiard collision. The center of the white ball is not exactly aimed at the center of the yellow ball. It hits the yellow ball at a certain offset. A force is created during the collision, which points from the center of the white ball through the point of contact of the two balls to the center of the yellow ball. The yellow ball is accelerated along this line. The direction of the white ball after the collision then follows from the conservation of momentum, as we had calculated with point masses. If the two balls are of equal weight, a right angle forms between their directions.

8.3.7 Elastic Collision in Three Dimensions

The elastic collision in three dimensions does not really bring anything new. We now have three unknown components for each of the unknown velocities, a total of six unknowns. There are four determining equations: the conservation of energy and the three components of the momentum equation.

We want to consider the special case again that the second point mass is at rest before the collision. We choose a similar coordinate system as in the 2-dimensional case. The x-axis is again given by the velocity of the first point mass before the collision. We rotate the coordinate system around the x-axis until the first point mass moves in the x-y-plane after the collision. The z-axis is perpendicular to it. Let's discuss the z-component of the total momentum first. It is zero before the collision. After the collision, due to the special choice of the y-axis, the z-component of the first point mass is still zero. Then, due to the conservation of momentum, the z-component of the momentum of the second point mass must vanish, too. This means that none of the point masses ever move in the z-direction. The collision actually takes place in a plane. We have reduced the collision in three dimensions to the 2-dimensional case.

Experiment 8.4: Newton's Cradle

Newton's Cradle is a physics gadget. A series of equally heavy balls are suspended on strings as pendulums. If you deflect the first ball and let it fall onto the resting balls, the last ball on the opposite side jumps away. The others remain at rest. A periodic back and forth is created, in which only the two outer balls are involved.

The balls perform nearly elastic collisions in a direction predetermined by the suspension. The first ball hits the second. During the collision, it is stopped. Its momentum is completely transferred to the second ball. This now hits the third one and transfers its momentum, and so on, until the last ball is reached, which then swings out. The movement of the inner balls from the impact to the transfer of momentum is so small that it is not noticeable.

It is interesting to observe what happens when you deflect two balls instead of one. One might naively expect that a ball with double the speed would be repelled on the opposite side. However, this is not the case. Two balls are repelled, both with the simple speed. Why must this be the case? The reason is that in addition to momentum, kinetic energy must also be fully transferred

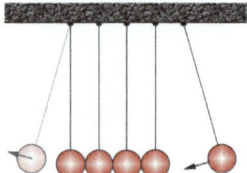

during elastic collisions. This is only possible if the repelled object has the same mass as the impacting one, so two balls are repelled as well. Calculate it yourself!

Experiment 8.5: The Bouncy Ball Pyramid

The Bouncy Ball Pyramid is a simple but quite impressive experiment. You will need two to four bouncy balls of different sizes. Hold them one above the other as shown in the sketch and then let them fall. With several balls, it takes a bit of practice to let them fall straight. The balls fall at the same speed. The bottom one will hit the ground first, bounce up and hit the one above. It kicks this one upwards and transfers some of its energy to it. The second then kicks the third one upwards and so on. The top one is repelled upwards with a surprisingly high speed. Please do not perform this experiment in your apartment!

Using Eq. 8.26 you can determine the velocity of a upwards-kicked bouncy ball. Let's assume that the bouncy balls differ in mass by a factor of 2. Then, from Eq. 8.26 with $m_1 = 2m$, $m_2 = m$, $u_1 = u$, $u_2 = -u$, it is seen that after the collision with the upper ball, a velocity of $v_2 = 5/3\,u$ results. In a pyramid with four bouncy balls, three collisions occur, so that the final velocity of the upper bouncy ball amounts to $v_{\text{end}} = \left(\frac{5}{3}\right)^3 u \approx 4.6\,u$.

Experiment 8.6: Ballistic Pendulum

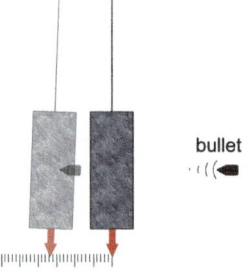

© RWTH Aachen, Physics Experiment Collection

The ballistic pendulum is used to determine the velocity of projectiles. A projectile catcher, e.g., a thick wooden board, hangs from a long string. The projectile, whose mass must be known, is fired at the board. It gets stuck in the board. The pendulum swings a little in recoil. The deflection is measured. The length of the pendulum, as well as the weight and thickness of the board, must be adapted to the weapon. For the air rifle that we use in the lecture, a small board of a few centimeters thickness on a 2-meter long pendulum is already sufficient.

The impact of the projectile in the board represents a perfectly inelastic collision. The entire kinetic energy of the projectile is transferred, the projectile moves together with the board after the collision. We first determine the speed u of the pendulum body (board) after the collision. The mass of the pendulum body is $M = 1800$ g, the length of the pendulum $l = 220$ cm and $g = 9.81$ m/s^2 the standard gravity. The horizontal deflection was measured to be $a = 4.5$ cm. The mass of the projectile is $m = 0.75$ g. Then, the conservation of energy for the pendulum after the impact results in

$$\frac{1}{2}(m+M)u^2 = (m+M)gh = (m+M)gl\left(1 - \cos\frac{a}{l}\right).$$

From this results

$$u = \sqrt{2gl\left(1 - \cos\frac{a}{l}\right)} = 2\sqrt{gl\sin^2\frac{a}{2l}} \approx 0.095\,\frac{\text{m}}{\text{s}}.$$

Now we apply momentum conservation for the inelastic collision and obtain the desired velocity of the projectile v_0:

$$m\,v_0 = (m+M)\,u \Rightarrow v_0 = \frac{m+M}{m}\,u \approx 228\,\frac{\text{m}}{\text{s}}.$$

We get a projectile velocity of 228 m/s.

We can now numerically verify the elasticity of the collision. The energies before and after the collision are

$$\frac{1}{2}m\,v_0^2 \approx 19.5\,\text{J}$$

$$\frac{1}{2}(m+M)\,u^2 \approx 0.008\,\text{J}.$$

Indeed, almost all of the kinetic energy is lost in the collision.

8.4 Systems with Variable Mass

Up to this point, we have assumed that the mass of the system remains constant. There are some applications where this is not the case. We will now examine these.

We consider a system of mass M, which experiences a mass increase dM in an infinitesimal step. The system initially moves at the velocity $\vec{v}$. The mass that is transferred to the system can also have an initial velocity. We denote it with $\vec{u}$. To calculate the motion, we start from the general form of Newton's equation of motion:

$$\vec{F} = \frac{d\vec{p}}{dt} = \frac{d}{dt}(m\vec{v}). \tag{8.38}$$

We determine the momentum $\vec{p}$ of the system before and after the absorption of dM:

$$\begin{aligned} \vec{p}_{\text{before}} &= M\vec{v} + dM\,\vec{u} \\ \vec{p}_{\text{after}} &= (M + dM)(\vec{v} + d\vec{v}) \\ &= M\vec{v} + dM\,\vec{v} + M\,d\vec{v} + dM\,d\vec{v}, \end{aligned} \tag{8.39}$$

where the last term can be neglected as it represents a product of two infinitesimally small quantities. We calculate the change in momentum

$$\begin{aligned} d\vec{p} = \vec{p}_{\text{after}} - \vec{p}_{\text{before}} &= dM\,\vec{v} + M\,d\vec{v} - dM\,\vec{u} \\ &= M\,d\vec{v} - dM\,\vec{v}_{\text{rel}}. \end{aligned} \tag{8.40}$$

In the last step, we introduced the relative velocity $\vec{v}_{\text{rel}} = \vec{u} - \vec{v}$, which is the speed at which dM moves from the perspective of the system. We differentiate with respect to time:

$$\frac{d\vec{p}}{dt} = M\frac{d\vec{v}}{dt} - \frac{dM}{dt}\vec{v}_{\text{rel}} = \vec{F}_{\text{ext}} \tag{8.41}$$

or

$$M\vec{a} = \vec{F}_{\text{ext}} + \frac{dM}{dt}\vec{v}_{\text{rel}}. \tag{8.42}$$

We had already obtained the first part of this result in ▶ Sect. 8.2 for the center of mass of a system with constant mass. If the mass of the system changes, the equation must be supplemented by the second term. An increase or decrease in mass acts like an external force of the magnitude $\frac{dM}{dt}\vec{v}_{\text{rel}}$.

Example 8.6: Conveyor Belt

We consider a horizontal conveyor belt. At one end, bulk material is dropped onto the belt. This increases the mass of the belt by the mass of the bulk material. How much force is needed to drive the belt?

We assume that the belt runs without friction, i.e., no force is necessary to move the belt without the cargo.

8.4 · Systems with Variable Mass

The mass increase dM/dt is positive. The bulk material falls vertically downwards. The velocity component in the direction of the belt's motion is zero. The relative velocity between the bulk material and the belt is $v_{\text{rel}} = u - v = -v$. It points against the direction of the belt's velocity. To ensure a smooth run of the belt, the motor must exert a force on the belt, even in the frictionless case, so that

$$F_{\text{motor}} - \frac{dM}{dt} v = 0.$$

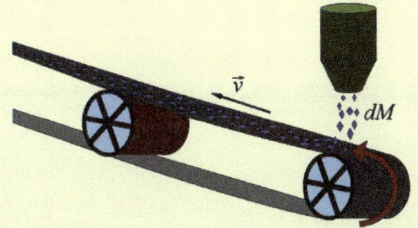

Example 8.7: Rocket Equation

A rocket accelerates by expelling gases at high speed $v_{\text{rel}} < 0$ against the direction of flight. The mass of gases that are expelled per second is dM/dt or $\dot{M}$. The value is also negative. The rocket develops a thrust

$$F_{\text{thrust}} = \dot{M} v_{\text{rel}}.$$

The mass of the rocket decreases linearly due to the expulsion

$$M(t) = M_0 + \dot{M} t.$$

Since no external force acts on the rocket, it follows from Eq. 8.41,

$$(M_0 + \dot{M} t) \frac{dv(t)}{dt} = \dot{M} v_{\text{rel}}.$$

We separate the variables and integrate:

$$dv = \dot{M} v_{\text{rel}} \frac{1}{M_0 + \dot{M} t} dt$$

$$\int dv = v_{\text{rel}} \int \frac{1}{\frac{M_0}{\dot{M}} + t} dt$$

$$v(t) = v_{\text{rel}} \ln \left(\frac{M_0}{\dot{M}} + t \right) + c = v_{\text{rel}} \ln \frac{M(t)}{\dot{M}} + c.$$

© NASA

The integration constant is to be determined in such a way that the velocity is zero at the start ($t = 0$). We obtain $c = -v_{\text{rel}} \ln M_0/\dot{M}$ and thus

$$v(t) = v_{\text{rel}} \ln \frac{M(t)}{M_0}.$$

If you substitute for $M(t)$ the weight of the burned-out rocket (which will essentially be the payload), you get the final velocity of the rocket. Note that we have not taken Earth's gravity into account in our calculation. The result is only valid in the weightlessness of space.

Experiment 8.7: Rocket Car

With a gas bottle, we constructed a simple rocket engine. A CO_2 bottle is mounted horizontally under the sheet metal hood of the cart. A pipe directs the escaping gas to the rear. When the bottle is opened, the cart accelerates slowly but steadily with a loud noise. In the photo, a colleague is driving into the lecture hall with the rocket cart for the Christmas lecture.

8.5 The Impulse

Collision processes cause a change in momentum. During the usually short collisions between the collision partners, a force acts. The temporal course of the force is difficult to determine and often irrelevant. What matters is the change in momentum. We integrate Newton's basic law of mechanics over the duration of the collision:

8.5 · The Impulse

$$\vec{F} = \frac{d\vec{p}}{dt}$$

$$\int_{t_1}^{t_2} \vec{F} dt = \int_{t_1}^{t_2} \frac{d\vec{p}}{dt} dt = \vec{p}(t_2) - \vec{p}(t_1) = \Delta \vec{p}. \quad (8.43)$$

The integral over the force is called the impulse $\vec{T}$:

$$\vec{T} = \int_{t_1}^{t_2} \vec{F} dt. \quad (8.44)$$

Example 8.8: Sport Climbing

In sport climbing, elastic ropes are used to reduce the impact load on the climber. If the climber falls from a certain height, he has a downward momentum that the rope must reduce to zero when catching. This creates an impulse, the size of which is given by the height of the fall. An elastic rope extends the duration of the impact (slow deceleration) and thereby reduces the force acting on the climber during the catch.

© Wikimedia: Stefanos Nikologianis

Example 8.9: Car Crash Test

In a car accident, the occupants are decelerated by an impulse. The size of the impulse is given by external conditions (initial speed, deformation of the car, impact on wall or oncoming vehicle, etc.). Without a seatbelt, the body initially flies freely forward. Dangerous force peaks occur upon impact on the steering wheel or dashboard. As you can see in the diagram below, the seatbelt smooths the force curve and thereby reduces the maximum force to which an occupant is exposed. The impulse, which is the integral over the curves, is the same in both cases.

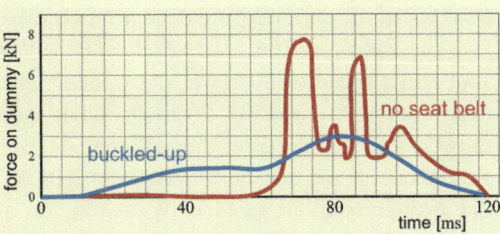

Experiment 8.8: Impulse with the Skateboard

An impulse can be demonstrated with a skateboard. Two people should stand on the skateboard. If the person at the back gently jumps off the skateboard, an impulse is created that accelerates the skateboard with the other person forward.

❓ Problems

1. Determine the position of the center of mass of a homogeneous cone with a height of H and whose base has the radius R.
2. Determine the location of the center of mass of the flat shape in the sketch, assuming a homogeneous density.

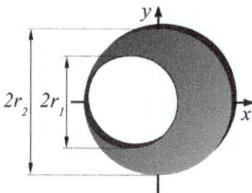

3. In a variation of the "Ballistic Pendulum" experiment, a projectile with a mass of 0.75 g and a speed of 300 m/s is shot through a pendulum body with a mass of 1800 g, which hangs on a 2.2 m long string. The pendulum body swings out horizontally by 2.5 cm. What is the speed of the projectile after passing through?
4. A rifle with a mass of 4.4 kg is fired, using 2900 J of energy to shoot a bullet with a mass of 11.5 g. What speed does the bullet reach and what is the recoil speed of the rifle if it is not held? Why should rifles be pressed firmly against the shoulder when firing?
5. At an intersection, two vehicles collide on black ice and then slide further, locked together. The directions of travel were perpendicular to each other before the collision. The first vehicle with a mass of 1.4 t had a speed of 30 km/h, the second vehicle of mass 1.8 t had a speed of 20 km/h. What speed do the two vehicles have after the collision? At what angle relative to the original direction of vehicle 1 do the vehicles continue to slide? What fraction of the original kinetic energy is lost to other forms of energy?

8.5 · The Impulse

6. In billiards, the laws of collision resulting from the conservation of momentum are manifested:
   (a) Consider the collision of a billiard ball (white) with a stationary billiard ball (blue) of the same mass in the approximation of point masses. Specify the possible velocities and directions of the balls after the collision.
   (b) Now take into account the expansion of the balls (diameter $d = 57.2$ mm). Calculate the direction of the two balls depending on the impact parameter b. This is the perpendicular distance between the path of the center of the white ball and the center of the blue ball.

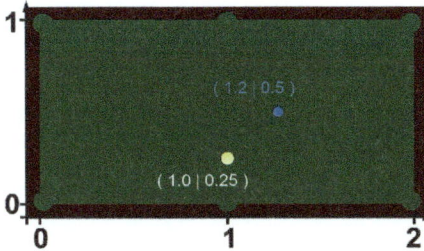

   (c) The image shows a billiard table with a playing field of length 2540 mm and width 1270 mm. The white ball is located in at $(1.0|0.25)$ in the indicated coordinate system, the blue ball at $(1.2|0.5)$. The player wants to sink the blue ball directly into the hole in the bottom right, which is at the coordinates $(2.0|0.0)$. Calculate the scattering angle of the blue ball in the approximation of point masses. Calculate the scattering angle of the blue ball considering the ball's expansion. Can the center of the hole be reached by the blue ball? Now also consider the expansion of the hole (diameter $D = 135$ mm) and calculate where the blue ball must be hit by the white ball to sink it in the hole.
   (d) The direct hit proved difficult in part c. Therefore, the blue ball should be sunk into the bottom right hole via the cushion. Calculate the scattering angle of the blue ball in the point mass approximation. Calculate the scattering angle of the blue ball considering the ball's expansion. Consider the expansion of the hole and calculate where and with what precision the blue ball must be hit by the white ball.

7. With what force must a firefighter hold the nozzle if 10 l of water are ejected per second at a speed of 15 m/s?

8. A conveyor belt runs at a speed v and transports bulk material with a throughput of dm/dt to a height h. What is the required power? Compare this with the gain in potential and kinetic energy per unit of time and explain where the energy "disappears".

9. A student enthusiastic about space technology builds a rocket. The rocket has a mass of 4 kg, which includes the mass of the fuel of 3.5 kg. The fuel burns evenly within 7 s, with the combustion gases being expelled relative to the rocket at a speed of 80 m/s.
   (a) Derive the equation of motion of the rocket and show that the following applies:

   $$m\frac{dv}{dt} = -\frac{dm}{dt}v_{rel} - mg,$$

   where $m = m(t)$ represents the current mass, $v = v(t)$ its current speed, and $v_{rel} = $ const. denotes the velocity of the gas relative to the rocket.
   (b) The student loads the rocket with an additional "payload" of 3.5 kg and ignites the propellant. When will the rocket lift off? How large can the payload of the rocket be so that it lifts off immediately?
   (c) Now solve the above equation of motion from the time $t_0 = 0$, at which the rocket lifts off from the ground $x_0 = 0$ with the mass m_0. Assume the acceleration due to gravity g to be constant.
   (d) The student has learned from part b and therefore removes the payload of the rocket in the following. Calculate the maximum speed of the rocket!
   (e) Calculate the flight altitude of the rocket $s(t)$ by integrating $v(t)$!
   (f) Calculate the maximum altitude of the rocket!

10. A frog with a mass of 0.1 kg sits on a scale. When it jumps vertically upwards from the scale, it briefly shows a force of 2.5 N. Assume a duration of the jump of 0.2 s. How high does the frog jump?

Friction

Contents

9.1 Overview – 164

9.2 Static Friction – 166

9.3 Kinetic Friction – 172

9.4 Rolling Friction – 174

© The Author(s), under exclusive license to Springer-Verlag GmbH, DE, part of Springer Nature 2025
S. Roth and A. Stahl, *Mechanics*,
https://doi.org/10.1007/978-3-662-68079-7_9

9.1 Overview

Friction is a force that resists the relative motion of bodies, slowing them down. It arises upon contact of real bodies sliding along each other. It strongly depends on the characteristics of the surfaces in contact. On rough surfaces, the friction is stronger than on smooth ones. Imagine that the two surfaces interlock. A force is necessary to break this lock and to move the bodies, i.e., to overcome the friction. Without this force, the bodies gradually come to a standstill. Friction also acts between stationary bodies. It can prevent the bodies from moving against each other at all, even if other forces are acting on them.

Friction is often undesirable. It can be reduced, but in the real world, it can never be completely eliminated. However, there are also situations where friction is desired and helpful. Just imagine how difficult it is to walk on an ice surface where friction is greatly reduced. Friction is crucial for our movement. It can be reduced by smoothing the contact surfaces, for example, it is easier to slide a body over a polished parquet floor than over rough asphalt.

In this chapter, we want to discuss the friction between solid bodies. Friction also occurs with liquids and gases, even among these. It is usually low with liquids and even lower with gases. Thus, a liquid film (lubricating film) between solids helps to reduce friction, and friction with air is often negligible at low speeds (air drag).

Wherever friction occurs, the motion is deprived of its energy. It is mostly converted into heat (frictional heating). This effect is particularly pronounced where strong friction occurs, e.g., in brakes. These slow down motion through friction and become hot if they are not sufficiently cooled. Friction changes the surfaces. Microscopic peaks on the surface are worn away and either deposited in the valleys or completely removed. This results in wear and tear.

9.1 · Overview

Friction is a dissipative force. It is not conservative in the sense discussed in ▶ Sect. 7.3. In processes with friction, conservation of energy appears to be violated. Energy is "lost".

> **Example 9.1: Friction in Everyday Life**
>
> Friction has a great significance in everyday life. Many simple things rely on friction. But friction also has negative aspects. It wears out machines and consumes energy. Here are some examples:
>
> No knot holds without friction
>
>
>
> Without friction, the nail falls out of the wall
>
>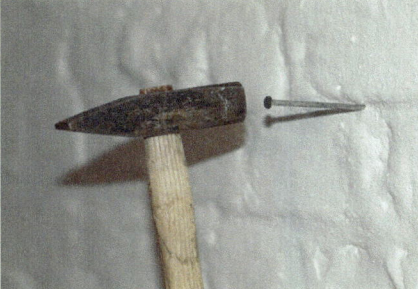
>
> Optimal friction for sports

But frictional heating and wear can become a problem

© Wikimedia: CZmarlin

9.2 Static Friction

We begin the treatment of friction with a simple experiment (Experiment 9.1):

The body on the board is subjected to gravity. If the board is horizontal, gravity acts perpendicular to the board and is completely absorbed by it. Already, if the board is tilted by a small angle, we need to decompose gravity into a normal force perpendicular to the board and the downhill-slope force parallel to the board (▶ Sect. 6.4 and Eq. 7.18). Only the normal force is compensated by the reactio of the board. According to Newton's law, the body should move under the influence of the

9.2 · Static Friction

non-vanishing downhill-slope force, but as we have seen in the experiment, it does not. Obviously, there must be another force that compensates the downhill-slope force. This is the frictional force. Since the body in this experiment adheres to the plane and does not yet slide, we speak of static friction. The maximum static friction is the force that must be overcome to set the body in motion.

> **Experiment 9.1: Static Friction on an Inclined Plane**
>
> We place a body on a board. In our example (see photo), it is a block of black plastics. Even if we tilt the board a little, the body remains in place. With a set screw and a protractor, we determine the angle at which the body starts to slide. We repeat the measurement several times, as the readings vary.
>
>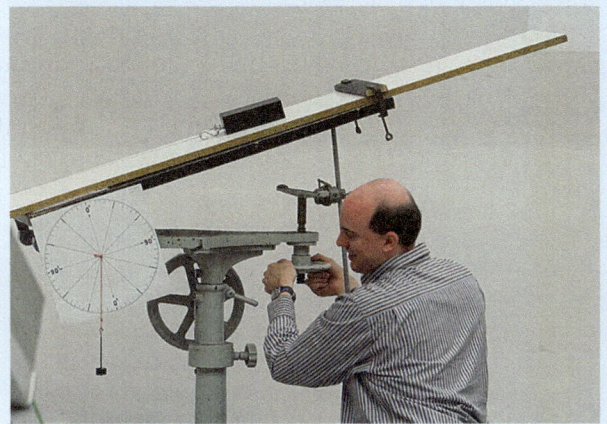
>
> © Photo: Hendrik Brixius

The individual forces are shown in ◘ Fig. 9.1. The weight $\vec{F}_G$ is decomposed into the normal force $\vec{F}_N$ perpendicular to the inclinded plane and the downhill-slope force $\vec{F}_{\text{slope}}$ parallel to the plane. The body presses on the plane with the normal force. This is balanced by the reaction of the plane $\vec{F}_{\text{reac}}$, which points in the opposite direction. The static friction force $\vec{F}_{\text{fric}}$ must compensate for the downhill-slope force, so that $\vec{F}_G$, $\vec{F}_{\text{reac}}$ and $\vec{F}_{\text{fric}}$ add up to zero and the body remains at rest.

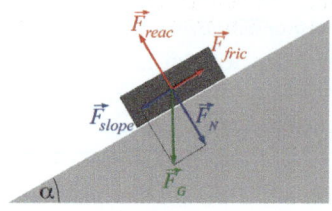

Fig. 9.1 Decomposition of forces on the inclined plane

If we decrease the inclination of the plane, the downhill-slope force decreases. The body remains at rest. The frictional force has decreased correspondingly. Regardless of the inclination of the plane and the size of the downhill-slope force, we always have

$$\vec{F}_{\text{fric}} = -\vec{F}_{\text{slope}} \tag{9.1}$$

If we increase the inclination of the plane, we find a limiting angle at which the body eventually starts to slide (see Experiment 9.1). The frictional force reaches a maximum value at this angle, which it cannot exceed. We want to determine this force.

Phenomenological Approach

Before we start determining the limiting angle, a remark is appropriate. The first chapters of this book were axiomatically structured. You have learned the axioms of Newtonian mechanics, and we have applied these in different situations. In doing so, we derived new laws, such as the conservation of momentum. The phenomena we have discussed can all be explained from the Newtonian axioms.

This should seem to you the best way and in principle we could proceed along this approach with friction. The frictional forces go back to the forces between the atoms in the surfaces in contact. When the surfaces approach each other, forces form between their atoms, which are the cause of friction. They must be overcome if the body is to move. The frictional force is the superposition of these atomic forces and can in principle be calculated from them.

But unfortunately, this is impractical. The forces depend on the detailed structure of the surface and even if one knew the structure in all of its details, the calcu-

9.2 · Static Friction

lation of the forces would still be too difficult. Overall, this problem is too complex for a strict derivation from the axioms. It would still not succeed if we were to use quantum mechanics to describe the atoms. This makes the problem even more complicated.

Such problems often arise in physics. In some areas, one encounters the limits of what can be derived from the axioms earlier, in others later. Then one must take a different approach. We try to systematically investigate the phenomenon in question with experiments and to deduce the essential relationships from these experiments. This is called an empirical approach, i.e., an approach based on experience. In this way, new physical laws are found, which we may use to describe the phenomena. They are therefore called "phenomenological laws". As with the axioms, the larger the scope of its application, the more valuable the law.

In many cases, we will not be able to fully clarify all relationships. For example, in the case of friction, the dependence of the frictional force on the material of the surfaces is difficult to quantify. We are forced to introduce constants, which depend on the material. Such material constants are only calculable in principle; in practice they must be measured. They represent our inability to actually carry out these calculations.

However, one should not valuate phenomenological laws as inferior to axiomatically derived laws.

We reuse the setup from Experiment 9.1 to systematically investigate which parameters the maximum static friction depends on (Experiment 9.2).

Experiment 9.2: Static Friction on the Inclined Plane 2

We set the inclined plane from Experiment 9.1 horizontally and place a body on the plane. We determine the maximum static friction by pulling on the body with a spring scale until the body starts to slide.
- We investigate the dependence on the weight of the body by placing additional weights on the body. As the weight increases, the frictional force at the sliding limit becomes larger. A more detailed investigation shows that the frictional force is proportional to the weight.

- We investigate the dependence on the size of the contact area by placing the cuboid body on its three different sides. We find that this has no influence on the frictional force.
- We change the inclination of the inclined plane and find that the maximum frictional force changes with the cosine of the inclination angle of the plane.
- Finally, we use bodies made of different materials and find that different materials lead to different frictional forces. However, this relationship is difficult to quantify, as the texture of a surface is hard to quantify.

The table summarizes the results:

Size	Dependence
Weight	Proportional
Contact Area	None
Inclination of the plane	Proportional to $\cos\alpha$
Surface texture	Yes, but complicated

From the first three observations in Experiment 9.2, we conclude:

$$|\vec{F}_{\text{fric,max}}| \propto \cos\alpha \cdot |\vec{F}_G| = |\vec{F}_N| \tag{9.2}$$

We hide the complicated dependency on the nature of the surface by inserting a material-dependent constant. This constant is called the coefficient of static friction μ_s. It depends on the material and the texture of the surfaces in contact. Some values can be found in ◘ Table 9.1. Overall, we have

$$|\vec{F}_{\text{fric}}| \leq \mu_s |\vec{F}_N| \tag{9.3}$$

The actual amount and direction of the static friction are determined by the applied force. The static friction opposes it and is equal in magnitude. If the applied force is increased, the static friction increases, too, until the limit of static friction (Eq. 9.3) is reached and the body starts to move.

We have created a macroscopic picture of friction. Now, we want to return to the microscopic picture and attempt a simple explanation of static friction. We can explain friction qualitatively by looking at the texture of the surface (◘ Figs. 9.2 and 9.3).

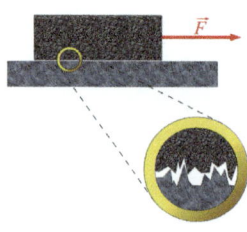

◘ **Fig. 9.2** Microscopic image of a surface during friction

9.2 · Static Friction

◘ **Table 9.1** Coefficients of friction of some material pairs

Materials	Static friction μ_s	Kinetic friction μ_k
Steel–steel	0.3	0.2
Steel with oil film	0.1	0.05
Teflon–steel	0.04	0.04
Leather–steel	0.6	0.3
Ice skate–ice	0.03	0.01
Wood–wood	0.5	0.3
Rubber–asphalt (dry)	1.0	0.8
Rubber–asphalt (wet)	0.6	0.4

◘ **Fig. 9.3** Image of a zinc surface taken with a scanning electron microscope

The surface of the body is locked with the surface of the base. Microscopic irregularities interlock. These prevent movement. If we want to overcome static friction, we must first lift the body slightly. This requires an additional force, a force we recognize as friction. This is a simple image, but it can explain some of the relationships we have found. The friction is proportional to the weight of the body. This can be explained from the microscopic image, as we need more force to lift the body, the heavier it is. Similarly, the image can explain that the force on the inclined plane is reduced. Finally, the image can also explain that the friction does not depend on the contact area. We have to lift the body a certain height, regardless of how large the area is. It is also apparent that the frictional force for rough surfaces is greater than for smooth ones. But unfortunately, this simple image does not provide a deeper insight into the material dependency. Reality is complicated, after all.

9.3 Kinetic Friction

If enough force is applied, static friction can be overcome and movement begins. With motion friction is reduced, but it does not disappear. We now speak of kinetic friction. We define the kinetic friction force as the force necessary to maintain the relative motion between the body and its base. That this is indeed the case can be demonstrated with a very simple experiment (Experiment 9.3).

> **Experiment 9.3: Static and Kinetic Friction on a Rod**
>
> The first experiment demonstrates the quantitative difference between static friction and kinetic friction. The experimental setup is sketched in the figure. A rotatable rod is mounted at an inclination. A weight is hanging from the rod with a string. The string is not tied to the rod. It hangs in a loop over the rod. Now we adjust the inclination of the rod so that the weight just does not slide down. It is held by static friction. If we now turn the rod, we switch from static friction to kinetic friction between the rod and the string. Since kinetic friction is less than static friction, the string slides down the rod.

The experimental investigation of kinetic friction leads to a similar law as already with static friction:

$$|\vec{F}_{\text{fric,kin}}| = \mu_k |\vec{F}_N| \tag{9.4}$$

The kinetic friction always acts against the movement of the body. Within reasonable limits, it depends only slightly on the velocity of the body. The proportionality constant is the material-dependent coefficient of kinetic friction. You may find some values in Table 9.1.

9.3 · Kinetic Friction

Experiment 9.4: Measurement of the Coefficient of Kinetic Friction

With the apparatus outlined in the figure, we can measure the coefficients of kinetic friction. A small electric motor drives a winch, which slowly but steadily pulls the body over the corresponding surface. With a spring scale in the rope, we can read the friction force. Some examples of coefficients of kinetic friction are listed in ▪ Table 9.1.

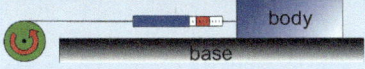

Experiment 9.5: Sliding Rod

This is a simple experiment with a surprising result. You may try yourself. You take a rod and place it on your outstretched index fingers. You initially support the rod at both ends. Then you slowly move the two fingers towards each other. The rod will sometimes slide on the right finger, sometimes on the left. Whatever you do, the two fingers meet almost exactly in the middle of the rod. The rod remains in balance. Naively, one might have expected that once the rod starts sliding on one side, it will continue to slide on that side due to the reduced friction, until the rod eventually loses its balance. Can you explain why you always end up in the middle? Friction depends on weight. But how much is the weight on a single finger?

© Photo: Hendrik Brixius

Experiment 9.6: Heat through Friction

Overcoming kinetic friction requires work. This work is performed by the force that moves the body against the kinetic friction. The work is

$$W = -\vec{F}_{\text{fric,kin}} \cdot \vec{s}$$

This work generates heat and causes wear. An effect that can be easily observed. Have you ever tried to drill into a board with a blunt drill, possibly at high speed? The drill barely penetrates the wood, but the drill and board become so hot that the wood turns black and may start to smoke. This is the effect of frictional heat.

9.4 Rolling Friction

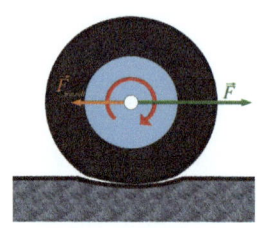

■ **Fig. 9.4** Rolling friction on a wheel

When a body rolls on a surface, both the body and the surface deform due to the weight of the body (normal force, see ■ Fig. 9.4). This usually elastic deformation consumes energy. The body sinks a little into the surface. The surface rises in front of the contact point. A force is necessary for it to continue rolling. This creates a frictional force, the rolling friction.

Like static and kinetic friction, rolling friction also depends on the material. In addition, it depends on the shape of the rolling body, namely its radius r. The following applies

$$|\vec{F}_{\text{fric,roll}}| = \mu_r \frac{|\vec{F}_N|}{r} \tag{9.5}$$

Example 9.2: Rolling Friction on a Bicycle

A bicycle tire deforms when it contacts the road. This creates rolling friction. The air in the tire heats up. High tire pressure reduces deformation and thus rolling resistance. An effect that you have probably noticed with underinflated tires.

Example 9.3: The Bearing of an Axle

The illustrations show some technical solutions for mounting an axle. In a we show the simplest solution, a plain bearing. The axle rotates in the bearing. This

9.4 · Rolling Friction

causes static or kinetic friction, which can be reduced by an oil film. In b, a ball bearing is shown. The axle rolls on balls that are housed in an outer shell of the bearing. The friction of such a bearing is significantly lower than in a, as only rolling friction occurs here. The roller bearing in figure c works similarly. Here, the balls are replaced by cylindrical rollers.

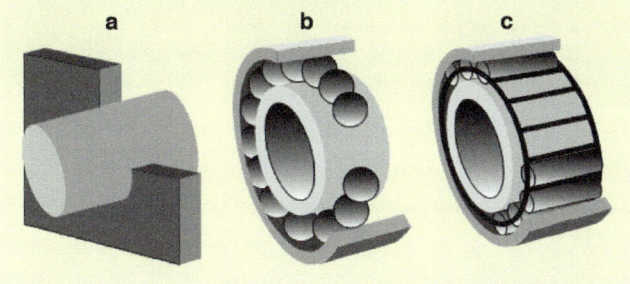

a   b   c

You have now learned about static, kinetic, and rolling friction. As you have seen in Experiment 9.3, kinetic friction is less than static friction. Usually, rolling friction is even significantly less than these two.

❓ Problems

1. How far does a railway wagon roll on a flat track if it is slowed down from a speed of 40 km/h only by frictional losses of rolling friction (Effective coefficient of friction $\mu = \frac{F_{fric}}{F_N} = 0.004$)?
2. A car with a mass of 1.5 t is uniformly decelerated from a speed of 130 km/h and brought to a standstill without the wheels locking. What is the minimum braking time if the coefficient of static friction between rubber and asphalt is assumed to be $\mu_s = 0.6$? What heat power is then generated at the beginning of the braking process at the brakes?
3. In curling sport, a curling stone is pushed and slides a total of 6 s for a distance of 7 m. What is the coefficient of friction μ_k?
4. A cart rolls down a ramp of length $L = 10$ m, which is inclined at an angle $\alpha = 5°$ to the horizontal, and immediately afterwards rolls up a ramp with the same inclination. How far up does it make it on the second ramp, if the coefficient of friction is $\mu = \frac{F_{fric}}{F_N} = 0.02$?

5. A boy is dragging a board with a mass of 20 kg behind him using a rope attached at the board's center of gravity. The board has a friction coefficient of $\mu = 0.3$ with the ground. What is the smallest force with which the board can be pulled, and at what angle α between the rope and the ground must the boy pull the board in this case?

Fictitious Forces

Contents

10.1   Overview – 178

10.2   Uniformly Accelerated Reference Systems – 180

10.3   Centrifugal Force – 183

10.4   Coriolis Force – 189

10.5   Absolute Motion? – 205

© The Author(s), under exclusive license to Springer-Verlag GmbH, DE, part of Springer Nature 2025
S. Roth and A. Stahl, *Mechanics*,
https://doi.org/10.1007/978-3-662-68079-7_10

10.1 Overview

We have based our previous considerations on Newton's axioms. From these, we have learned (▶ Chap. 6) that they only apply in inertial systems. Newton's first axiom can be understood as a definition of an inertial system. It defines inertial systems as those in which a force-free body remains in its motion or at rest. Now we want to show how Newton's axioms can also be used in non-inertial systems.

Once an inertial system has been found, others arise from it. Any system that moves uniformly against an inertial system or rests against it is itself an inertial system (◻ Fig. 10.1). A system that, on the other hand, accelerates against an inertial system is not an inertial system itself.

The calculation of movements is particularly simple in inertial systems. However, there are also situations where one wants to describe and calculate a motion from a non-inertial system. Strictly speaking, the widely used laboratory system is not an inertial system. It is anchored to the Earth, which rotates around its own axis and orbits the Sun. Nevertheless, we frequently use this system. In order to apply the basic law of mechanics, additional inertia- or fictitious forces are introduced. Then

$$m\vec{a} = \sum \vec{F}_{\text{real}} + \sum \vec{F}_{\text{fic}} \tag{10.1}$$

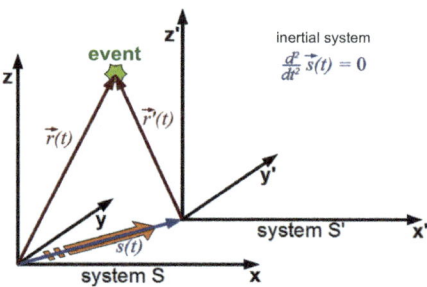

◻ **Fig. 10.1** If S is an inertial system, then S′ is an inertial system, if $\frac{d^2\vec{s}(t)}{dt^2} = 0$.

10.1 · Overview

applies, where the first sum contains the forces that also occur in inertial systems and the second sum the additional fictitious forces. The fictitious forces have no material cause. They are not accompanied by a Reactio. They perform no work. Their purpose is to explain the deviation of the behavior of bodies in an accelerated reference system compared to an inertial system.

> **Example 10.1: The Train Departs**
>
> In subway trains, you often find the notices "Please hold on tightly". The reason for this is the fictious forces that arise when the train accelerates and decelerates. For a passenger standing in a train that accelerates at $1.0\,\text{m/s}^2$, the backward-directed inertial force is added to the downward-directed gravitational force, which corresponds in terms of forces to the situation on an inclined plane that rises at a gradient of about $10\,\%$.
>
> $$\frac{|\vec{F}_\text{fic}|}{|\vec{F}_G|} = \frac{|\vec{a}_\text{train}|}{|\vec{g}|} = 0.1$$
>
> To avoid tipping over, the passenger must hold on (or shift his weight). Once the train reaches its constant travel speed, the inertial force disappears and the floor appears flat again. Again, the passenger must compensate for this.

> **Example 10.2: Ball on Carousel**
>
> A girl is standing on a carousel. Imagine she places a ball on the floor of the carousel and lets it go. How will she see the ball move from her perspective?
> The ball, which was originally at rest, begins to roll outward. It is accelerated outward. From this, you as an observer on the carousel would conclude that a force must be acting on the ball. This is the fictitious force.

Chapter 10 · Fictitious Forces

10.2 Uniformly Accelerated Reference Systems

We want to start with an example. As an accelerated reference system we consider a glass elevator that is accelerating downwards with constant acceleration a_0 at the start of a downward journey (◘ Fig. 10.2). A person in the elevator drops an object. We observe the fall from the perspective of this person and from the perspective of an observer at the bottom of the building.

As reference systems, we choose the elevator shaft for the observer on the ground. We call this system S. This is an inertial system[1]. The observer observes a free fall of the object, which falls with standard gravity acceleration $\vec{g}$ downwards. The person in the elevator chooses the elevator cabin (system S′) as the reference system. This is an accelerated reference system. In this reference system, the person will also observe a falling motion, but with a reduced acceleration $\vec{a}'$.

The z-axis points upwards, with the body initially at the origin of the coordinate system. The fall begins at $t = 0$:

$$\text{System S:} \quad z(t) = -\frac{1}{2}g t^2$$
$$\text{System S′:} \quad z'(t) = -\frac{1}{2}a' t^2$$
(10.2)

◘ **Fig. 10.2** An elevator accelerating downwards

---

1 Here we neglect the motion and rotation of the Earth.

10.2 · Uniformly Accelerated Reference …

The motion of the elevator cabin from the perspective of S is

$$z_0(t) = -\frac{1}{2}a_0 t^2 \tag{10.3}$$

A coordinate transformation from S′ to S results in:

$$z(t) = z_0(t) + z'(t)$$
$$-\frac{1}{2}gt^2 = -\frac{1}{2}a_0 t^2 - \frac{1}{2}a' t^2 \tag{10.4}$$

We divide both sides by $-\frac{1}{2}t^2$ and obtain:

$$g = a_0 + a'$$
$$\Rightarrow a' = g - a_0. \tag{10.5}$$

We want to determine the associated fictitious force. In S′ we have:

$$\vec{F}_G + \vec{F}_{\text{fic}} = m\vec{a}' = m(\vec{g} - \vec{a}_0), \tag{10.6}$$

with $\vec{F}_G = m\vec{g}$

$$\vec{F}_{\text{fic}} = -m\vec{a}_0. \tag{10.7}$$

The fictious force points in the direction opposite to the acceleration of the reference system S′. The magnitude of the apparent force is given by the acceleration of S′. This relation holds for all linearly, uniformly accelerated reference systems. If the acceleration is not uniform, then $\vec{a}_0$ must be replaced by $\vec{a}_0(t)$.

> **Example 10.3: Linearly Accelerated Reference Systems**
>
> An elevator represents an accelerated reference system, independent of the height of the elevator. The elevator is accelerated when starting and stopping. The acceleration thereby changes its direction. In between, the elevator behaves like an inertial system. Other everyday examples of linearly accelerating systems are starting or braking cars or trains.

Experiment 10.1: Fictious Force in the Accelerated Reference System

The result of Eq. 10.7 can be easily demonstrated. You don't necessarily need an elevator. The figure shows the apparatus. On the right side, the accelerated reference system S' is hanging on two pulleys, with a counterweight on the left side. If we choose the counterweight to be slightly lighter than the right side, our system S' will fall downwards with constant acceleration. In S' a simple experiment is set up. A weight is hanging on a spring scale. The spring scale is observed via a wireless camera. According to Eq. 10.7, the weight should decrease, which can indeed be observed. If you let S' fall, the deflection of the spring scale decreases until S' is stopped by a foam rubber plate on the floor underneath the apparatus. The experiment can also be reversed by accelerating S' upwards with a too large counterweight. Then the deflection of the spring scale increases.

10.3 · Centrifugal Force

© RWTH Aachen, Physics Experiment Collection

10.3 Centrifugal Force

We now want to deal with rotating reference systems, which are accelerated non-inertial reference systems, too. We start with a simple experiment (◘ Fig. 10.3). On a rotating disc, a ball is attached to the axis of rotation via a spring scale. The ball and spring scale rotate with the disc. An observer watches the experiment from outside from a reference system S, which is at rest (inertial system). A second observer S′ stands on the disc and rotates with it. What do they observe?

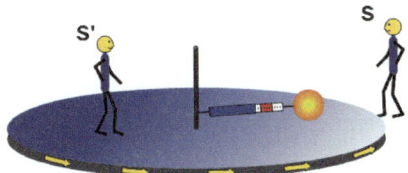

◘ Fig. 10.3 Thought experiment on the centrifugal force

From the perspective of the stationary observer S, the spring exerts a force $\vec{F}_{\text{Hook}}$ on the ball, which forces it onto a circular path. This force is called the centripetal force $\vec{F}_{\text{CP}}$. It is directed towards the center of the circular path and always occurs in circular motion. It causes the centripetal acceleration, which you have already learned about in ▶ Sect. 5.8. It is

$$\vec{a}_{\text{CP}} = -\frac{v_B^2}{r}\frac{\vec{r}}{r},$$
$$\vec{F}_{\text{CP}} = m\,\vec{a}_{\text{CP}}.$$
(10.8)

The vector $\vec{v}_B$ gives the orbital velocity of the ball with magnitude v_B. The centripetal acceleration can also be expressed by the angular velocity $\vec{\omega}$, where $\vec{v}_B = \vec{\omega} \times \vec{r}$ (for the vector product or cross product see mathematical appendix A3.16). The angular velocity indicates the angle by which an object rotates per time interval. The unit is 1/s (more precisely rad/s). Then

$$\vec{a}_{\text{CP}} = -\omega^2\, r\, \frac{\vec{r}}{r} = -\omega^2\, \vec{r}.$$
(10.9)

We now consider the motion from the perspective of the rotating observer (S′). For him, the ball is at rest, but he recognizes that the spring scale exerts a force on the ball. He concludes that there must be another force that "neutralizes" the spring force, so that the sum of the forces acting on the ball vanishes and it remains at rest. This is the centrifugal force $\vec{F}_{\text{CF}}$. It is a fictitious force that only appears in the accelerated reference system. From the condition $\vec{F}_{\text{Hook}} + \vec{F}_{\text{CF}} = 0$ follows:

$$\vec{F}_{\text{CF}} = -\vec{F}_{\text{Hook}} = m\,\frac{v_B^2}{r}\,\frac{\vec{r}}{r} = m\,\omega^2\,\vec{r}.$$
(10.10)

Please note: The centrifugal force is <u>not</u> the reaction force (Reactio) to the centripetal force. The centripetal force is a real force, the reaction force of which acts on the spring mount. The centrifugal force, on the other hand, is a fictitious force. It has no reaction force.

Experiment 10.2: Measurement of the Centrifugal Force

With the experiment shown in the sketch, we can measure the centrifugal force. A toy wagon is placed on a rotatable rail. The rail can be driven by a rubber band

10.3 · Centrifugal Force

from an external motor. The wagon is attached to a string that is redirected upwards along the axis of rotation via a pulley. It is attached to a spring scale outside the photo. When the rail is set in rotation, the wagon feels a centrifugal force, which is transferred to the spring scale via the string. With faster rotation, the centrifugal force increases quadratically, the spring scale is stretched accordingly and the wagon moves outward on the rail. We can read the centrifugal force from the scale on the rail.

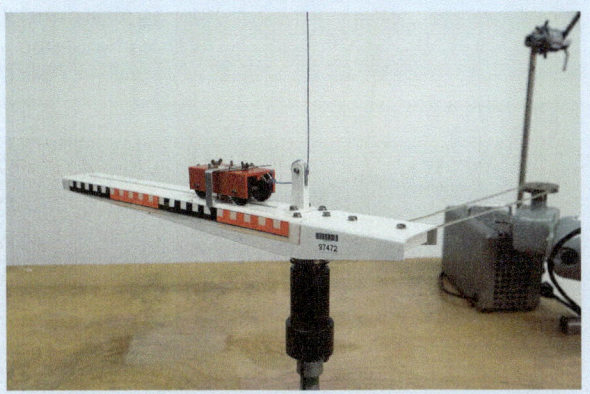

Example 10.4: Everyday Examples of Centrifugal Force

You are probably familiar with the feeling of flying in a carousel. The feeling is generated by the centrifugal force. In the carousel, you are in a rotating reference system. Another everyday example of rotating reference systems are cars driving through a sharp turn. At high speed, the centrifugal force is clearly noticeable. The car tilts outwards and the passengers are pushed to the outside.

© Photo: Hans@pixabay.com

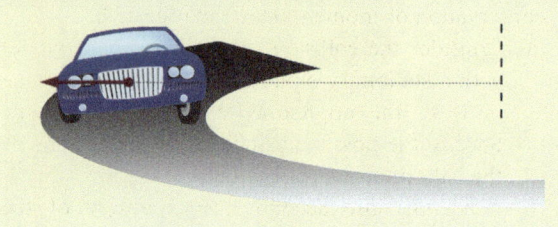

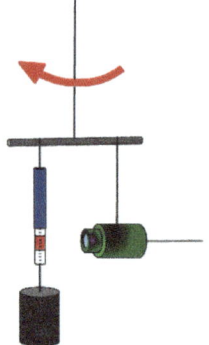

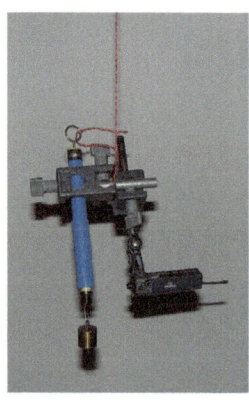

© RWTH Aachen, Physics Experiment Collection

Experiment 10.3: Centrifugal Force

We demonstrate the centrifugal force with a modification of the apparatus from Experiment 10.1. This time, the spring scale and weight are attached to a string and swung overhead like a lasso by the experimenter. With some skill, a sufficiently uniform rotation is achieved. The spring scale is again observed via the wireless camera. It displays the magnitude of the centrifugal force.

Experiment 10.4: Earth Ellipsoid

The Earth is not an exact sphere. Due to centrifugal force, it is slightly flattened at the poles. This model can illustrate the flattening. The model consists of two crossed steel bands, which are attached to a rod with a fixed (top) and a sliding (bottom) bracket. When the model is at rest, the bands form circles (indicated by the yellow line in the left picture). The model

10.3 · Centrifugal Force

is clamped into an experimental motor and set into rotation. The rotation frequency can be adjusted continuously. The centrifugal force $\vec{F}_{CF} = m\omega^2 \vec{r}$ is greatest at the furthest distance r from the axis of rotation, i.e. at the equator. This results in a deformation that depends on the rotation frequency ω. When rotating, the bands form the image of a transparent, rotating sphere, which takes the shape of a flattened spheroid at high rotation. For comparison, we copied the shape from the left picture to the right (dotted yellow line). The flattening is clearly visible.

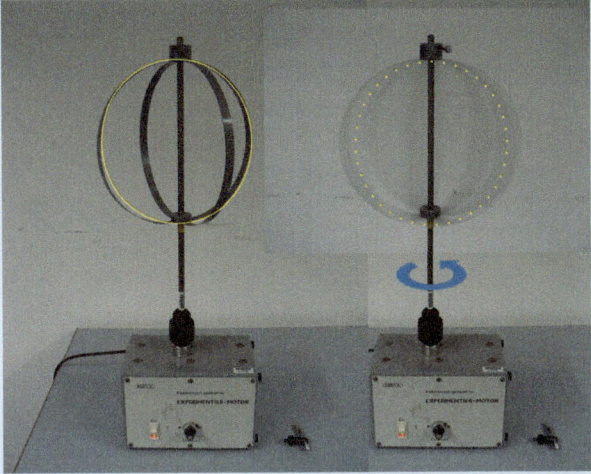

The effect on Earth is small. The equatorial diameter is 12 756 km compared to 12 716 km from pole to pole, a difference of 0.3%.

Experiment 10.5: The Surface of Rotating Liquids

A stationary liquid forms a flat surface. This changes when the liquid is set in rotation. The jar is placed on an electrically driven turntable, and after a short start-up phase, the rotation of the jar is transferred to the liquid through friction. Then a curved surface forms, as can be seen in the photo.
You may wonder what exact shape the surface takes. It can be calculated. After the start-up phase, the shape of the surface no longer changes. It is static. We use this as a condition to determine the shape. To do this, we

© RWTH Aachen,
Physics Experiment
Collection

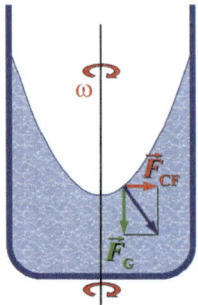

consider the force on a molecule of the liquid at the surface (see sketch). We decompose this force into a component normal to the surface and one parallel to it. The normal component is compensated by the pressure of the underlying molecules, but not the parallel component. This leads to molecules moving along the surface. This would cause the surface to change its shape. Since this no longer happens after the start-up phase, we must require that the force component parallel to the surface vanishes as soon as a static situation is reached. We base the calculation on this condition.

For the calculation, we use cylindrical coordinates with the z-axis upwards and the r-coordinate radially outward from the axis of rotation. Let's call $h(r)$ the mathematical form of the surface, i.e. the height of the surface as a function of the radial position. From the triangle of forces in the sketch, we read off the slope of the surface. It is:

$$\frac{dh(r)}{dr} = \frac{|\vec{F}_{CF}|}{|\vec{F}_G|} = \frac{m\omega^2 r}{mg} = \frac{\omega^2}{g} r.$$

Integrating this differential equation with respect to r results in

$$h(r) = \frac{\omega^2}{2g} r^2 + k.$$

This is a parabola with respect to r with a minimum at zero, whose curvature is determined by the angular velocity. Such a shape will develop in all liquids regardless of the shape of the jar.

Experiment 10.6: Rotating Bucket

This experiment can be easily performed at home. It demonstrates the influence of the centrifugal force on water in a bucket. However, it requires confidence in the physical laws we have discovered. If you hesitate, you will have the laughter of the audience on your side.

Fill a bucket about halfway with water and rotate it quickly by the long arm as indicated in the sketch. The centrifugal force always acts outward and keeps the water in the bucket even in passing through the upside-down position. The first rotation is the only difficult part.

10.4 Coriolis Force

In the previous chapter, we studied the forces acting on a body at rest in a rotating reference system and arrived at the centrifugal force. We now want to deal with the general case of a moving body in a rotating reference system.

10.4.1 Derivation of the Coriolis Force

We will treat this problem generally, without referring to a specific example. We start from an inertial system S. In this system, another system S′ rotates with constant angular velocity around an axis $\vec{\omega}$ (◘ Fig. 10.4). The origins of both systems coincide and lie on the axis of rotation. A body moves through the systems. Since both systems have a common origin, the two position vectors $\vec{r}(t)$ and $\vec{r}'(t)$, which describe the path, coincide. The trajectory of the body is:

$$\begin{aligned} \text{System S:} \quad & \vec{r}(t) = x(t)\hat{e}_x + y(t)\hat{e}_y + z(t)\hat{e}_z, \\ \text{System S′:} \quad & \vec{r}'(t) = x'(t)\hat{e}'_x + y'(t)\hat{e}'_y + z'(t)\hat{e}'_z. \end{aligned} \quad (10.11)$$

Here, $\hat{e}_x$, $\hat{e}'_x$, etc. are unit vectors in the direction of the coordinate axes of the two systems. Due to the rotation of S′, its unit vectors are constantly changing. More correctly, we should write $\hat{e}'_x(t)$, but we suppress the time-dependence in the notation. We now consider the velocities on the trajectory:

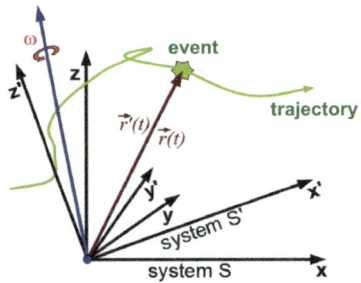

◘ Fig. 10.4 Representation of a trajectory in a rotating reference system

$$\text{System S:} \quad \vec{v}(t) = \frac{dx(t)}{dt}\hat{e}_x + \frac{dy(t)}{dt}\hat{e}_y + \frac{dz(t)}{dt}\hat{e}_z,$$

$$\text{System S':} \quad \vec{v}'(t) = \frac{dx'(t)}{dt}\hat{e}'_x + \frac{dy'(t)}{dt}\hat{e}'_y + \frac{dz'(t)}{dt}\hat{e}'_z.$$
(10.12)

So far so good. Now we want to represent the velocity that the observer sees in system S, not by his own coordinates, but by the coordinates of system S'. The observer in S recognizes that the object is moving relative to the dashed coordinates and that the coordinate axes of the system S' are also moving. It is:

$$\begin{aligned}\text{System S:} \quad \vec{v}(t) &= \frac{dx'(t)}{dt}\hat{e}'_x + \frac{dy'(t)}{dt}\hat{e}'_y + \frac{dz'(t)}{dt}\hat{e}'_z \\ &+ x'(t)\frac{d\hat{e}'_x}{dt} + y'(t)\frac{d\hat{e}'_y}{dt} + z'(t)\frac{d\hat{e}'_z}{dt} \\ &= \vec{v}'(t) + \vec{u}(t).\end{aligned}$$
(10.13)

As expected, the velocity from the perspective of S is composed of the velocity from the perspective of S' and a velocity vector, which describes the movement of the coordinate vectors of the system S' relative to S, which we have called $\vec{u}(t)$.

Now the unit vectors of the system S' rotate around the rotation axis with the angular velocity $\vec{\omega}$. Hence, we get:

$$\frac{d\hat{e}'_x}{dt} = \vec{\omega} \times \hat{e}'_x, \quad \frac{d\hat{e}'_y}{dt} = \vec{\omega} \times \hat{e}'_y, \quad \frac{d\hat{e}'_z}{dt} = \vec{\omega} \times \hat{e}'_z$$
(10.14)

and thus:

$$\begin{aligned}\vec{u}(t) &= (\vec{\omega} \times \hat{e}'_x)x'(t) + (\vec{\omega} \times \hat{e}'_y)y'(t) + (\vec{\omega} \times \hat{e}'_z)z'(t) \\ &= \vec{\omega} \times (\hat{e}'_x x'(t) + \hat{e}'_y y'(t) + \hat{e}'_z z'(t)) \\ &= \vec{\omega} \times \vec{r}'(t) \\ &= \vec{\omega} \times \vec{r}(t).\end{aligned}$$
(10.15)

In the last step, we took advantage of the fact that the position vectors coincide in both systems.

As an intermediate result, we obtain for the transformation of velocities:

$$\vec{v}(t) = \vec{v}'(t) + \vec{\omega} \times \vec{r}(t).$$
(10.16)

10.4 · Coriolis Force

We now want to derive a corresponding relation for the accelerations:

$$\vec{a}(t) = \frac{d\vec{v}(t)}{dt} = \frac{d\vec{v}'(t)}{dt} + \vec{\omega} \times \frac{d\vec{r}(t)}{dt} \quad (10.17)$$

We make an intermediate step and first calculate $\frac{d\vec{v}'(t)}{dt}$:

$$\begin{aligned}\frac{d\vec{v}'(t)}{dt} &= \frac{dv'_x(t)}{dt}\hat{e}'_x + \frac{dv'_y(t)}{dt}\hat{e}'_y + \frac{dv'_z(t)}{dt}\hat{e}'_z \\ &\quad + v'_x(t)\frac{d\hat{e}'_x}{dt} + v'_y(t)\frac{d\hat{e}'_y}{dt} + v'_z(t)\frac{d\hat{e}'_z}{dt} \quad (10.18)\\ &= \vec{a}'(t) + (\vec{\omega} \times \hat{e}'_x)v'_x(t) + (\vec{\omega} \times \hat{e}'_y)v'_y(t) \\ &\quad + (\vec{\omega} \times \hat{e}'_z)v'_z(t) \\ &= \vec{a}'(t) + \vec{\omega} \times \vec{v}'(t).\end{aligned}$$

We obtain with $\vec{v}(t) = \vec{v}'(t) + \vec{\omega} \times \vec{r}(t)$

$$\begin{aligned}\vec{a}(t) &= \vec{a}'(t) + \vec{\omega} \times \vec{v}'(t) + \vec{\omega} \times \vec{v}(t) \\ &= \vec{a}'(t) + \vec{\omega} \times \vec{v}'(t) + \vec{\omega} \times (\vec{v}'(t) + \vec{\omega} \times \vec{r}(t)) \quad (10.19)\\ &= \vec{a}'(t) + 2\vec{\omega} \times \vec{v}'(t) + \vec{\omega} \times (\vec{\omega} \times \vec{r}(t))\end{aligned}$$

and finally solved for $\vec{a}'(t)$ ($\vec{a} \times \vec{b} = -\vec{b} \times \vec{a}$):

$$\vec{a}'(t) = \vec{a}(t) + 2\vec{v}'(t) \times \vec{\omega} + \vec{\omega} \times (\vec{r}'(t) \times \vec{\omega}) \quad (10.20)$$

We first examine the last term. We consider the special case that $\vec{\omega}$ and $\vec{r}$ are perpendicular to each other. With the fingers of the right hand, it can be easily determined that the two cross products generate a vector in the direction of $\vec{r}$. The magnitude of the vector is $\omega^2 r$. It is the centrifugal acceleration already known from the previous chapter.

The first term is also known. This is the actual acceleration of the body along its path. However, the middle term is new. It only occurs when the body moves in the rotating reference system ($\vec{v}' \neq 0$). Therefore, it was not noticed in ▶ Sect. 10.3. It is called the Coriolis acceleration $\vec{a}_C$ and the associated fictitious force the Coriolis force, named after the French mathematician and physicist Gaspard Gustave de Coriolis (◘ Fig. 10.5), who first derived this force in the early 19th century:

$$\vec{a}'(t) = \vec{a}(t) + \vec{a}_C + \vec{a}_{CF}. \quad (10.21)$$

Fig. 10.5 Gaspard Gustave de Coriolis. © MP/Leemage/picture-alliance

First, we want to consider special cases of the Coriolis force for certain directions of $\vec{v}'$ (see Fig. 10.6):

1. Radial motion ($\vec{v}' \parallel \vec{r}'$)

   The body moves radially outward (or inward) on the disc. It feels a Coriolis force perpendicular to its motion, against (or in) the direction of the orbital velocity. A stationary observer sees that the body is moving away from the axis of rotation. This should increase the orbital velocity. However, since there is no accelerating force, it lags behind the rotation of the disc. The moving observer perceives this as a Coriolis force.

2. Tangential motion ($\vec{v}' \perp \vec{r}'$)

   The body moves along a circle around the axis of rotation at a constant radius. If the motion is directed in the direction of rotation, it experiences a Coriolis force directed radially outward. If it moves against the direction of rotation, the Coriolis force is directed in-

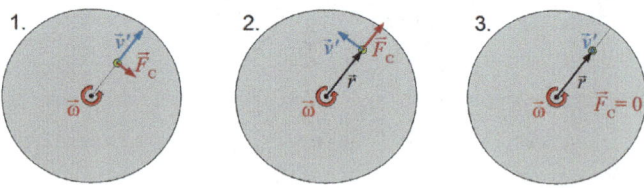

Fig. 10.6 Coriolis force with different movements

10.4 · Coriolis Force

ward. From the perspective of a stationary observer, the following picture emerges. Due to the additional motion in the direction of rotation, the body rotates faster than the disc. This increases the centrifugal force and it feels a force outward.

3. Axial motion ($\vec{v}' \parallel \vec{\omega}$)
   The body moves parallel to the axis of rotation. No Coriolis force occurs.

Any motion in an arbitrary direction can be represented by superposition of these three special cases.

10.4.2 Examples

Coriolis forces occur on Earth due to the its rotation. ◘ Fig. 10.7 shows an example. If a body moves north on the northern hemisphere of the Earth, the Coriolis force points east. The velocity $\vec{v}'$ has a radial component inward, which generates the Coriolis force. On the southern hemisphere, the Coriolis force points west when moving in a northern direction.

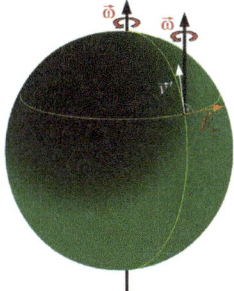

◘ **Fig. 10.7** Coriolis force on the northern hemisphere of the Earth

Experiment 10.7: Polar Bear Hunting with Difficulties

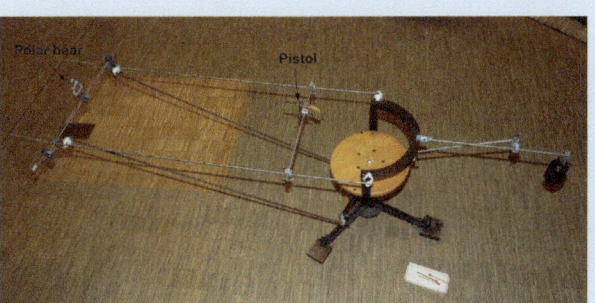

© RWTH Aachen, Physics Experiment Collection

This experiment beautifully demonstrates the influence of the Coriolis force on movements in a rotating system. A person (the hunter) sits on a swivel chair. On the chair, about 2 m in front of the person, the polar bear is painted as a target on a plexiglass panel. The person has a toy gun in front of them with arrows that stick to the panel. Initially, when the chair is not rotating yet, the hunter aims the gun, which is clamped to the frame, at the target and check for a perfect hit. Then the chair is

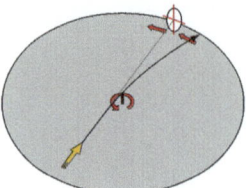

set in rotation. The hunter fires the gun again with the same adjustment, but now the shot misses. While the arrow flies from the gun to the target, the polar bear continues to move and escapes the shot. The arrow sticks horizontally next to the target.

A wireless camera on the hunter's chair records the shot from his perspective. He sees that the arrow drifts sideways in flight to the target. This is the influence of the Coriolis force.

Example 10.5: Coriolis Force on an Air Plane

An airplane flies eastward along the equator at a speed of $v = 900$ km/h. This corresponds to the special case of pure tangential motion. The Coriolis force points away from the earth's axis (upwards). It is

$$\omega_{\text{Earth}} = \frac{2\pi}{\text{day}} = 7.27 \cdot 10^{-5} \, \frac{\text{rad}}{\text{s}}$$

$$v_{\text{plane}} = 250 \, \frac{\text{m}}{\text{s}}$$

$$a_C = 0.04 \, \frac{\text{m}}{\text{s}^2} \approx 0.5 \, \% \, g \, .$$

© Wikimedia: u278

Example 10.6: Pirouette in Ice Skating

The ice skater, Isabeau Levito, performs a spin during the warm-up for the women's short program at 2022 Skate America. If she pulls her arms, legs, and body towards the axis of rotation, a Coriolis force acts on them

10.4 · Coriolis Force

in the direction of rotation. This accelerates the pirouette. The ice skater spins faster. This is also called the pirouette effect.

© Wikimedia: Flowering Dagwood

Example 10.7: Cyclones and Anticyclones

The Coriolis force is responsible for the fact that air masses around large-scale central regions of high atmospheric pressure, known as anticyclones in the northern hemisphere move clockwise, while they move counter-clockwise around regions of low atmospheric pressure (cyclones). In a low-pressure area, the air flows inward due to the pressure gradient. This flow is deflected to the right in the northern hemisphere by the Coriolis force. This results in a counter-clockwise rotation, as can be seen in the image of an Icelandic low.

In large-scale air currents, the Coriolis force can have a significant influence despite its small magnitude, which can go so far that the air masses move perpendicular to the pressure gradient. In the sketch, the direction of the pressure gradient is marked by blue arrows. The velocity of the air masses is to be broken down into an axial movement parallel to the rotation axis of the earth and a radial movement perpendicular to it. The latter part causes the Coriolis force, the direction of which is indicated in the sketch by red arrows. It generates the large-scale vortex.

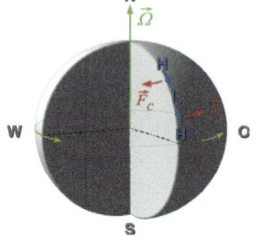

Example 10.8: Drain Vortex in the Sink

When water drains from the bathtub, a Coriolis force acts on the water. This results in a preferred direction of rotation of the drain vortex. Although this example is often cited, it is not true. The Coriolis force is too small to leave a noticeable effect. The direction of rotation is determined by other disturbing influences (air currents, asymmetrical tub shape, etc.).

Example 10.9: Proof of Earth's rotation

A body falling from a high tower is deflected eastward from its vertical fall due to the rotation of the Earth, an effect predicted by Galilei and Newton. The experimental proof of this deflection would provide evidence for the Earth's rotation. Giovanni Battista Guglielmini first succeeded in doing this in 1791 with fall experiments

10.4 · Coriolis Force

from the 120 meter high Torre degli Asinelli in Bologna (picture).
We want to calculate the deflection. We neglect air drag and treat the fall as free. Then we can calculate the duration of the fall t_0 from the height h_0:

$$s(t) = h_0 - \frac{1}{2} g t^2 \Rightarrow t_0 = \sqrt{\frac{2 h_0}{g}}$$

The speed increases linearly during the fall $v(t) = gt$. The component perpendicular to the Earth's axis (radial motion) is $v_\perp(t) = gt \cos\varphi$, with the geographical latitude $\varphi = 44.5°$ of Bologna. The component parallel to the Earth's axis (axial motion) is irrelevant here. The Coriolis acceleration is:

$$a_C(t) = 2 \omega_{\text{Earth}} \, g \, t \cos\varphi$$

From this, we can integrate the motion. For the speed and distance covered in the eastern direction, we get

$$v_C(t) = \omega_{\text{Earth}} \, g \, t^2 \cos\varphi$$
$$s_C(t) = \frac{1}{3} \omega_{\text{Earth}} \, g \, t^3 \cos\varphi$$

Now we just need to insert the fall time and we get the deflection

$$\Delta s_C = \frac{1}{3} \omega_{\text{Earth}} \, g \cos\varphi \left(\frac{2 h_0}{g}\right)^{\frac{3}{2}} = \frac{2}{3} \omega_{\text{Earth}} \cos\varphi \sqrt{\frac{2 h_0^3}{g}} \quad (10.22)$$

which results in approximately 2 cm for $h_0 = 120$ cm.

Example 10.10: Meander

The flowing water of a river is affected by a Coriolis force due to the Earth's rotation, which causes the right river bank (in the direction of flow) to erode more than the left. This leads to the formation of meanders as can be seen in the photo taken at the Innoko National Wildlife Refuge. Even Einstein addressed this phenomenon: A. Einstein, "The cause of the meander formation of river courses and the so-called Baer's law", in: *The Natural Sciences* 14, No. 11, 1926, pp. 223–224.

© U.S. Fish and Wildlife Service

Example 10.11: Coriolis Scale

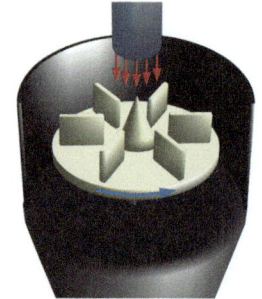

The Coriolis scale is used to weigh the throughput of bulk material, such as grain, in a conveyor system. Our sketch shows the basic structure. The bulk material falls from above through the inlet pipe onto a rotating disc. Over a conical elevation in the middle of the disc, the material slides outward, where vertical plates set it in rotation at the speed of the disc. The centrifugal force transports it further outward, increasing the orbital speed and thus the centrifugal force until it is finally thrown against the walls of the scale, slides down into a funnel, and leaves the scale through the funnel.

As the bulk material moves from the center of the plate outward, it must be accelerated. The motor that drives the plate has to perform work. This work is determined, for example, by the power consumption of the motor. From the power consumption, the mass flow of the bulk material, which is the mass transported per unit of time through the scale, can be determined.

Example 10.12: Rotating Linkage

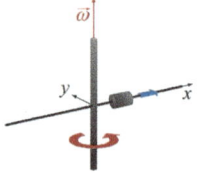

A rod rotates with an angular velocity $\vec{\omega}$ around an axis perpendicular to it. A cylinder is placed on the rod, which can move outward without friction. We want to determine the force $\vec{F}_S$ that the rod exerts on the cylinder. We choose a rotating reference system with the x-axis along the rod. If the cylinder slides outward, the rod must exert a force on it to accelerate the cylinder in the direction of rotation. It points in y-direction. We start from Eq. 10.20.

$$m\vec{a} = \vec{F}_D + 2m\vec{v} \times \vec{\omega} + \vec{\omega} \times (\vec{r} \times \vec{\omega}).$$

We determine the directions of these components. Due to the restriction of the cylinder to the rod, the acceleration must point in the x-direction (This is called a constraint).

$$\begin{pmatrix} m\frac{d^2r}{dt^2} \\ 0 \\ 0 \end{pmatrix} = \begin{pmatrix} 0 \\ F_D \\ 0 \end{pmatrix} + \begin{pmatrix} 0 \\ -2m\omega\frac{dr}{dt} \\ 0 \end{pmatrix} + \begin{pmatrix} m\omega^2 r \\ 0 \\ 0 \end{pmatrix}.$$

As you can see, the centrifugal force determines the acceleration in the radial direction. It results in $r(t)$ from the differential equation

$$m\frac{d^2r(t)}{dt^2} = m\omega^2 r(t)$$

with the solution: $r(t) = r_0 e^{\omega t}$.

Since the centrifugal force is a fictitious force, it cannot perform any work. However, the rod exerts the force $\vec{F}_D$ against the cylinder and performs work on it. As the calculation shows, this force is equal in magnitude to the Coriolis force

$$\vec{F}_D = 2m\omega^2 r \hat{e}_y = -\vec{F}_C.$$

10.4.3 The Foucault Pendulum

Due to the Earth's rotation, a pendulum on Earth changes its direction of oscillation. This is easy to understand for a pendulum at the North Pole (see Experiment 10.9). An observer watching the Earth from space sees a pendulum with a fixed direction of oscillation, under which the Earth rotates. However, the observer at the North Pole will claim that the pendulum continuously changes its direction of oscillationat a rate of 360° per day. The effect also occurs when the pendulum is located in temperate latitudes, although, at a daily rotation of less than 360°.

■ **Fig. 10.8** The Foucault's pendulum in the Panthéon in Paris. © wikimedia: Arnaud 25

The pendulum, known today as the Foucault Pendulum, became famous when the French physicist Jean Bernard Léon Foucault installed a 67 m long and 28 kg heavy pendulum in the Panthéon in Paris in 1851 to demonstrate the Earth's rotation to the public. Today, a reconstruction of his pendulum swings in the Panthéon (■ Fig. 10.8).

> **Experiment 10.8: Model Experiment on Foucault's Pendulum**
>
> The experiment demonstrates a Foucault pendulum at the North Pole. A pendulum is mounted on the axis of a rotating disc slowly turned by a motor. If you view the pendulum from above with a camera, you can clearly see how the 'Earth' rotates under the pendulum. The pendulum remains in its original direction of oscillation.

© RWTH Aachen, Physics Experiment Collection

10.4 · Coriolis Force

Experiment 10.9: Foucault's Pendulum

© Photo: Hendrik Brixius

To demonstrate Earth's rotation wih the Foucault Pendulum, one needs a long pendulum with a sufficiently heavy pendulum body, so that it swings at least for a few hours. The picture shows our pendulum before the start. It is held by an electromagnet, so that no disturbances are transmitted to the pendulum at the start. A point is attached to the bottom of the pendulum. Foucault himself scattered sand on the floor of the Panthéon, in which the point recorded traces of the movement. We have set up screws on their heads in a circle around the rest position of the pendulum, which the pendulum knocks over one after the other, to the extent that the plane of oscillation rotates.

Example 10.13: Equation of Motion of the Foucault Pendulum

As already mentioned, the plane of the Foucault Pendulum rotates due to the rotation of the Earth. We want to calculate how fast the pendulum plane changes.

We describe the pendulum in a coordinate system that is anchored at the rest point of the pendulum with the Earth and therefore rotates with the Earth (figure). The x-axis points horizontally in the direction of the Earth's rotation, the y-axis points horizontally towards the North. The z-axis indicates the vertical. It points against thedirection of gravity.

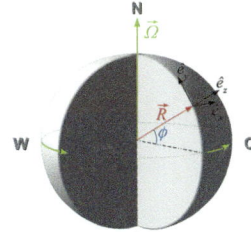

We begin our discussion with an approximation: We assume that the pendulum length l, that is, the length of the thread on which the pendulum body hangs, is much greater than the displacement of the pendulum body from its rest position. Among other things, this means that the pendulum body moves in the x-y-plane and we can set its z-coordinate to approximately zero during the entire movement. Its position vector is therefore $\vec{r} = (x, y, 0)$.

Now we begin the calculation with the equation of motion. Three forces act on the pendulum body:

- The gravitational force $\vec{F}_G = -mg\,\hat{e}_z$: It results in a restoring force that always points to the rest point and is proportional to the displacement (see example 7.5). It has the amount $m g r/l$, where r is the deflection, m is the mass of the pendulum body and g is the standard gravity. We write the force as

$$\vec{F}_R = -m\omega_0^2 \begin{pmatrix} x \\ y \\ 0 \end{pmatrix},$$

with the constant $\omega_0^2 = g/l$.

- The centrifugal force $\vec{F}_Z = -m\vec{\Omega} \times (\vec{\Omega} \times (\vec{R} + \vec{r}))$: Here, $\vec{\Omega}$ denotes the angular velocity of the Earth and $\vec{R}$ the vector from the center of the Earth to the origin of our coordinate system. The magnitude of $\vec{\Omega}$ corresponds to a rotation of 2π per day[2]. This centrifugal force is perpendicular to the Earth's axis. It changes the gravitational force. This change is already included in the value of g at $\vec{r} = \vec{0}$ and in the definition of the vertical of our coordinate system. The variation of the centrifugal force due to the fact that the pendulum body moves out of the rest position is so small that we can neglect it. We therefore do not need to consider the centrifugal force further.

- The Coriolis force $\vec{F}_C = -2m\vec{\Omega} \times \vec{v}$: From the illustration, the components of the vector can be determined. They amount to $\vec{\Omega} = \Omega\,(0, \cos\phi, \sin\phi)$.

---

[2] Here, a sidereal day should be used (86 164.099 s). In a year, the Earth rotates 366 times on its own axis relative to the fixed stars, which appears to us as 365 solar days due to the additional orbit of the Earth around the Sun.

10.4 · Coriolis Force

The force exerted by the y-component points in the z-direction and thus in the direction of the gravitational force. It can be easily verified that it can be neglected compared to the gravitational force. The component that originates from Ω_z, on the other hand, falls in the orbital plane. It causes the rotation of the pendulum, which we want to calculate.

To simplify the notation, we will abbreviate derivatives with respect to time by dots on the symbols. We write $v_x = \frac{d}{dt}x(t) = \dot{x}, a_x = \frac{d^2}{dt^2}x(t) = \ddot{x}$, etc. The equation of motion is therefore:

$$m\ddot{\vec{r}} = -m\omega_0^2 \begin{pmatrix} x \\ y \\ 0 \end{pmatrix} - m\vec{\Omega} \times \left(\vec{\Omega} \times \left(\vec{R} + \vec{r}\right)\right) - 2m\vec{\Omega} \times \dot{\vec{r}}.$$

With the approximations described above, the first two components of this differential equation—only these are relevant here—are:

$$\ddot{x} = -\omega_0^2 x + 2\sin\phi\, \Omega\, \dot{y}$$
$$\ddot{y} = -\omega_0^2 y - 2\sin\phi\, \Omega\, \dot{x}.$$

Here we have introduced a constant $\omega_0 = \sqrt{g/l}$ which gives the oscillation frequency of the pendulum, assuming it oscillates in an inertial system. To solve these two coupled differential equations, we transform to plane polar coordinates:

$$x = r\cos\varphi$$
$$y = r\sin\varphi,$$

with the two time-dependent polar coordinates $r(t)$ and $\varphi(t)$. We differentiate the two equations twice, taking into account the product rule, and obtain:

$$\dot{x} = \dot{r}\cos\varphi - r\sin\varphi\, \dot{\varphi}$$
$$\dot{y} = \dot{r}\sin\varphi + r\cos\varphi\, \dot{\varphi}$$

as well as

$$\ddot{x} = \ddot{r}\cos\varphi - 2\dot{r}\sin\varphi\, \dot{\varphi} - r\cos\varphi\, \dot{\varphi}^2 - r\sin\varphi\, \ddot{\varphi}$$
$$\ddot{y} = \ddot{r}\sin\varphi + 2\dot{r}\cos\varphi\, \dot{\varphi} - r\sin\varphi\, \dot{\varphi}^2 + r\cos\varphi\, \ddot{\varphi}.$$

With this, we could transform the two differential equations into polar coordinates. But before we do that, we

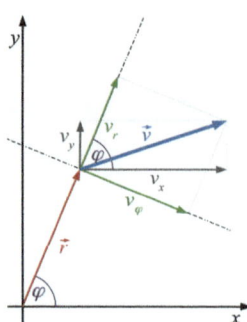

insert another step. We want to express the accelerations in x- and y-direction through the radial acceleration a_r and the angular acceleration a_φ. From the sketch we read off:

$$v_r = \cos\varphi\,\dot{x} + \sin\varphi\,\dot{y} = \dot{r}$$
$$v_\varphi = -\sin\varphi\,\dot{x} + \cos\varphi\,\dot{y} = r\dot{\varphi}$$

or

$$\dot{x} = \cos\varphi\,v_r - \sin\varphi\,v_\varphi$$
$$\dot{y} = \sin\varphi\,v_r + \cos\varphi\,v_\varphi$$

and

$$a_r = \cos\varphi\,\ddot{x} + \sin\varphi\,\ddot{y} = \ddot{r} - r\dot{\varphi}^2$$
$$a_\varphi = -\sin\varphi\,\ddot{x} + \cos\varphi\,\ddot{y} = r\ddot{\varphi} + 2\dot{r}\dot{\varphi}.$$

Now we insert into the last two equations $\ddot{x}$ and $\ddot{y}$ from the differential equation and we get:

$$\ddot{r} - r\dot{\varphi}^2 = \cos\varphi\left(-\omega_0^2 x + 2\sin\phi\Omega\,\dot{y}\right)$$
$$+ \sin\varphi\left(-\omega_0^2 y - 2\sin\phi\Omega\,\dot{x}\right)$$
$$r\ddot{\varphi} + 2\dot{r}\dot{\varphi} = -\sin\varphi\left(-\omega_0^2 x + 2\sin\phi\Omega\,\dot{y}\right)$$
$$+ \cos\varphi\left(-\omega_0^2 y - 2\sin\phi\Omega\,\dot{x}\right),$$

where we finally express the right side through polar coordinates, too. We obtain:

$$\ddot{r} - r\dot{\varphi}^2 = 2\sin\phi\,\Omega\,r\dot{\varphi} - \omega_0^2 r$$
$$r\ddot{\varphi} + 2\dot{r}\dot{\varphi} = -2\sin\phi\,\Omega\,\dot{r}.$$

The first differential equation describes the radial acceleration, the second the angular acceleration of the pendulum mass. We will now convert the equations into normal form:

$$\ddot{r} + r\left(\omega_0^2 - \dot{\varphi}^2 - 2\sin\phi\Omega\,\dot{\varphi}\right) = 0$$
$$r\ddot{\varphi} + 2\dot{r}(\dot{\varphi} + \sin\phi\Omega) = 0.$$

We are primarily interested in the angular velocity $\dot{\varphi}$ which describes the rotation of the pendulum plane. In fact, one can read a special solution for $\dot{\varphi}$ directly from the second differential equation. Let's assume that the angular velocity has a constant value $\dot{\varphi} = -\sin\phi\,\Omega$, then $\ddot{\varphi} = 0$. This solves the differential equation, which you can easily verify by substitution. Therefore, the rotation of the pendulum plane on the northern hemisphere is opposite to the rotation $\vec{\Omega}$ of the Earth's axis and is slowed down by a factor $\sin\phi$ relative to it, where ϕ represents the geographical latitude.

If we substitute this angular velocity into the differential equation of radial acceleration, we obtain

$$\ddot{r} + \omega^2 r = 0,$$

with $\omega = \sqrt{\omega_0^2 + \sin^2\phi\,\Omega^2}$. As we will see in ▶ Sect. 17.1, this is the differential equation of a harmonic oscillation with a frequency ω, which is slightly increased compared to the frequency in an inertial system. However, it should be noted that this is only a specific solution of the differential equation, the general solution[3] leads to an anharmonic oscillation, the treatment of which would go beyond the scope of this discussion.

In conclusion of these sections on fictitious forces in rotating reference systems, it should be noted that in addition to the centrifugal force and the Coriolis force, another fictitious force can occur when the angular velocity $\vec{\omega}$, with which the reference system rotates, is not constant over time. This additional fictitious force is called the Euler force. It has the magnitude

$$\vec{F}_E = -m\frac{d\vec{\omega}}{dt} \times \vec{r}. \qquad (10.23)$$

10.5 Absolute Motion?

Newton assumed that our solar system is embedded in an immovable and unchangeable universe. A reference system was connected to it that he considered to be absolutely at rest. It was experimentally accessible via the fixed stars (◘ Fig. 10.9). In this reference system, he formulated his axioms.

This simple picture has been destroyed by modern cosmology. The universe is constantly changing, on small scales under the influence of stars, on large scales it is ex-

---

3 The general solution for the angular velocity is $\dot{\varphi} = c/r^2 - \sin\phi\,\Omega$.

Fig. 10.9 The starry sky, Newton's absolute reference system. © NASA, ESA, and The Hubble Heritage Team (STScI/AURA)

panding. There is no way to define an absolute reference system throughout the universe.

Today we start from the principle of relativity. It states that there is no distinguished, absolutely resting reference system. In Einstein's words, it is:

> **Principle of Relativity**
> The laws of physics are the same for all observers in any inertial frame of reference relative to one another.

It goes back to Galileo and is therefore also called Galilean invariance of the laws of nature. Newton also knew about it, but it only gained significance in the 20th century with the theory of relativity.

But what about accelerations? Does a principle of relativity also apply here? Can we define reference systems that are not accelerated?

The Earth moves with acceleration relative to the fixed stars. It orbits the sun and circles with it around the center of the Milky Way. This acceleration can be measured (e.g., from the 3-K background radiation). But does this mean that the fixed stars are not accelerated? Or would it be equivalent in the sense of a principle of relativity to say that the Earth moves uniformly and the fixed stars are accelerated relative to the Earth? We leave this

10.5 · Absolute Motion?

question open here. It belongs in the realm of the theory of general relativity.

In conclusion, it should be emphasized once again that there is no absolutely stationary reference system, but the fixed stars can still be helpful in some considerations as an approximately stationary reference system.

> **Example 10.14: Principle of Relativity**
>
> If we imagine two rockets passing each other in otherwise empty space, statements about velocities are obviously relative:
> From the perspective of observer A, who perceives his rocket as being at rest, rocket B passes at considerable speed. From the perspective of observer B, on the other hand, his rocket is at rest while A is moving.

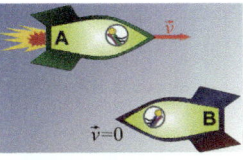
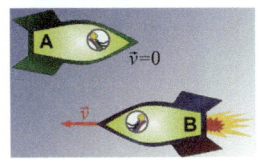

❓ Problems

1. A hammer thrower rotates around his own axis, holding the ball with the taut rope of length 1 m at the extended arm of length 0.5 m, so that it moves on a circle whose axis is generally inclined to the horizontal. How many revolutions per second must the hammer thrower make at least to be able to throw at an angle of 45°?
2. At what speed does the gravitational pull on a body flying parallel to the equator around the Earth get balanced by the centrifugal force? How short would an Earth day have to be to cancel out the Earth's gravity at the equator?
3. Calculate for the depicted centrifugal governor, the opening angle α as a function of the rotation frequency f given the length of the rod l and sketch the solution. You can neglect the mass of the rod compared to the weights. At which rotation frequencies can the centrifugal governor effectively regulate?
4. A motorcyclist leans into a curve. What is the proper angle of inclination ϕ against the vertical depending on his speed v and the radius r of the curve? What is the maximum speed he can travel through the curve if the tires on the asphalt have a coefficient of friction μ? What is the maximum angle of inclination ϕ the motorcyclist can ride without slipping, assuming a coefficient of friction of $\mu = 1$?

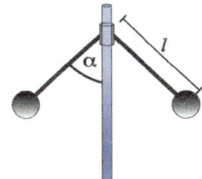

5. A body falling freely from a point vertically above a location on the Northern Hemisphere with the geographical latitude φ experiences a deflection due to the Coriolis and centrifugal forces. Calculate both the magnitude and direction of the resulting accelerations. Determine the deviation from the vertical that a body experiences when dropped from a height of 100 m in Aachen (51 ° northern latitude).

Celestial Mechanics

Contents

11.1 Kepler's First Law – 210

11.2 Kepler's Second and Third Laws – 214

11.3 Newton's Law of Universal Gravitation – 222

11.4 Gravitational and Inertial Mass – 239

11.5 Potential and Potential Energy – 241

© The Author(s), under exclusive license to Springer-Verlag GmbH, DE, part of Springer Nature 2025
S. Roth and A. Stahl, *Mechanics*,
https://doi.org/10.1007/978-3-662-68079-7_11

11.1 Kepler's First Law

In 1600, Johannes Kepler (Fig. 11.1) becomes the assistant of Tycho Brahe in Prague. A year later, after Brahe's death, he advances to the position of imperial court mathematician. This gives him access to Brahe's lifetime achievements, the most extensive and accurate observations of the positions and movements of stars of his time.

Kepler is a proponent of the Coperinican, i.e. heliocentric system. He had recognized that describing the motion of the planets in this system is much simpler than in the Ptolemaic system. The observation of a planet's position with respect to the stars fixed to the celestial dome results in a complicated pattern of loops when the planet's motion is related to the Earth (Fig. 11.2). As we will see below, the movements are much simpler when the Sun is chosen as the reference point. The orbits become approximate circles.

Today it is a matter of course that we try to describe the orbits of the planets mathematically. But modern science only emerged in Kepler's time. And even with him, one can feel this transition. In his early work *Mysterium Cosmographicum*, he still tries, firmly rooted in the tradition of the time, to describe the motion of the planets as a harmony of geometric shapes. Fig. 11.3 shows our solar system with the spheres of the six then-known planets Mercury, Venus, Earth, Mars, Jupiter, and Saturn, which revolve around the Sun at well-defined dis-

Fig. 11.1 Johannes Kepler 1610. © akg-images/picture-alliance

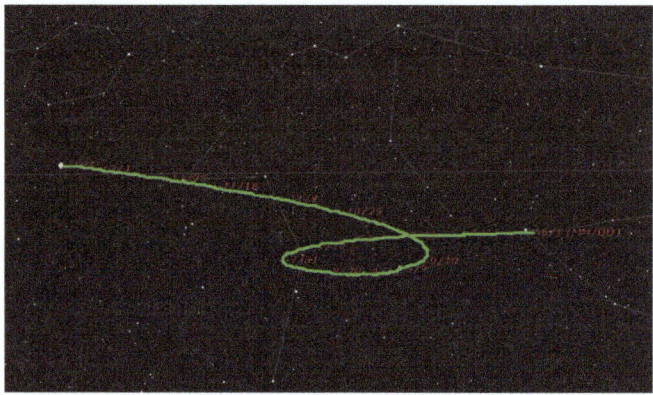

Fig. 11.2 The positions of Mars in the fixed star sky in the summer of 2003. Such loops are called epicycles. © Graphic: Robert J. Ballantyne (► http://www.ballantyne.com)

11.1 · Kepler's First Law

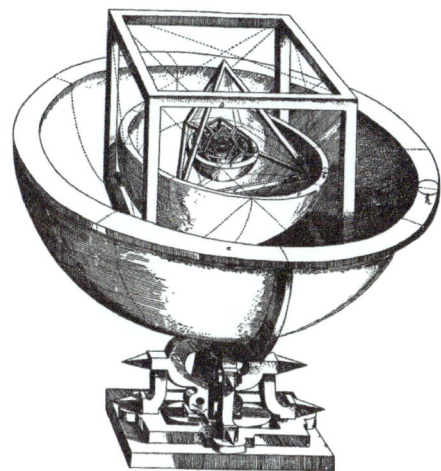

Fig. 11.3 Kepler's model of planetary orbits from his early work Mysterium Cosmographicum (1596)

tances. Kepler tried to explain the distance ratios by the dimensions of the five Platonic solids. According to a divine plan, there were exactly six planets and exactly five Platonic solids that should fit into the interstices. An approach that seems highly strange to us today.

Initially, Kepler, like all proponents of the heliocentric system, assumed that the planets orbit the Sun in circles. Eventually, Kepler had to recognize that Brahe's precise observations could not be described by circles. He had to abandon the "divine" form of the circle and allow ellipses. This led him to the following realization

1. Kepler's Law

The orbit of every planet is an ellipse with the Sun at one of the two foci.

The properties of the planetary orbits are compiled in Table 11.1 (see also Fig. 11.4). The numerical eccentricity indicates the deviation of the orbit from a perfect circle. The values are very small. The orbits resemble a circular path.

The elliptical shape of the orbits is not a special feature of our solar system, but a consequence of a central force binding the planets to the Sun. The movement of two bodies—here Sun and planet—always leads to a conic section as a trajectory when the force between the bodies is central, i.e. it points along the direct connection be-

Fig. 11.4 The planets of our solar system in realistic ratios of their size

■ Table 11.1 Orbital data of the planets (Pluto is no longer officially a planet)

		Semi-major axis a	Numerical eccentricity ϵ	Orbital period T	Orbital inclination	T^2/a^3
		10^6 km	–		Degree	$s^2/10^9$ km^3
Mercury	☿	57.9	0.206	87.97 d	7.0	0.297
Venus	♀	108.2	0.0068	224.7 d	3.4	0.300
Earth	♁	149.6	0.0167	365.256 d	0.0	0.299
Mars	♂	228.0	0.0935	686.98 d	1.9	0.305
Jupiter	♃	778.4	0.0484	11.86 a	1.3	0.300
Saturn	♄	1433.4	0.0542	29.46 a	2.5	0.299
Uranus	♅	2872.4	0.0472	84.01 a	0.8	0.298
Neptune	♆	4495.0	0.0113	164.79 a	1.8	0.300
Pluto	♇	5906.4	0.2488	247.68 a	17.2	0.299

tween the two bodies.[1] This is the case here. Another example can be found in classical models of the atoms. The electric attraction of the nucleus exerted on the electrons also points along their connecting line. Consequently, the electrons orbit the nucleus on elliptical paths[2].

> **Example 11.1: Perihelion Precession**
>
> We explained that the planets orbit the Sun on a Kepler ellipse, whose position relative to the fixed stars is unchangeable. In fact, there are disturbances to the orbits, which result in a rotation of the perihelion of the

---

[1] However, they are only conic sections in 3-dimensional space if the force decreases like $1/r^2$.
[2] This applies to Bohr's model of the atom, a semi-classical approach. In quantum physics, it is no longer possible to define an orbit.

planetary orbit around the Sun. The planets actually move on rosettes, as indicated in the illustration. The effect is most pronounced with Mercury. The perihelion of its orbit rotates per century by 571.91″ or 0.1589°. Towards the outer planets, the rotation of the perihelion decreases. For Neptune, it is only 104″.

The majority of the motion of the perihelion can be explained by the disturbances of the orbit by the other planets. For Mercury, about half of the effect can be explained by the influence of Venus on its orbit, 25% of the effect is due to Jupiter, and 18% to the other planets. The discovery of the planet Neptune was triggered by the observation of the rotation of the perihelion of Uranus, which could only be explained by another planet—namely Neptune. Since all planets rotate around the Sun in the same direction, their mutual influence on the orbits adds up to a perihelion rotation in this direction.

However, this is not the only effect that influences the position of the perihelion. Deviations from the rotational symmetry of the central star can also cause a perihelion rotation. For example, the equatorial bulge of the Earth rotates the perigee of satellites. In the case of planetary orbits, the influence of deviations of the sun from a spherical shape is very small.

If you have added up the predictions for the perihelion rotation of Mercury in your mind, you have noticed that these do not fully explain the observed value of 571.91″. According to current measurements, 43.11″ remain unexplained. This was a mystery for a long time and led to speculations about the deviation of the gravitational force from a strict $1/r^2$ dependence, which ultimately resulted in a triumph of Einstein's General Theory of Relativity, which explains such a deviation by distortions of space-time near large masses and correctly predicts the perihelion rotations of the planets.

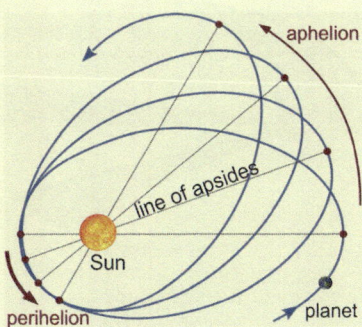

11.2 Kepler's Second and Third Laws

Kepler's second law deals with the speed at which the planets orbit on the ellipse. It is illustrated in ◘ Fig. 11.5.

> **Kepler's Second Law**
> A line segment joining a planet and the Sun sweeps out equal areas during equal intervals of time.

The planet orbits the Sun on the depicted ellipse. A line connects the Sun to the planet. The depicted distances s_1, s_2 and s_3 are chosen such that the planet takes the same amount of time to traverse each. The law then states that the grey shaded areas must be equal in size. Because of the reference to the areas, this law is also called the area law. Note that the speed at which the planet moves along the orbit changes continuously. As the planet approaches the Sun, the Sun's gravitational pull accelerates the planet. At the perihelion, the point closest to the Sun, it reaches its maximum speed. After the perihelion, it is decelerated again by Sun's gravity. It reaches its lowest speed at the aphelion, the point furthest from the Sun in its orbit.

Kepler's second law is a consequence of conservation of angular momentum. We will only get to know the angular momentum in ▶ Chap. 13. For the sake of completeness, the connection is already shown here. We consider the motion of the planet in an infinitesimally small time interval dt (◘ Fig. 11.6). During this time, the planet covers the distance $\vec{v}(t)\,dt$. The area dA, which the line segment sweeps out during this time, is

$$dA = \frac{1}{2m}|\vec{r}|\,|\vec{v}\,dt|\sin\alpha\,, \tag{11.1}$$

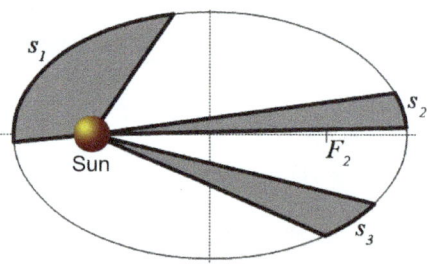

◘ **Fig. 11.5** Illustration of Kepler's second law

11.2 · Kepler's Second and Third Laws

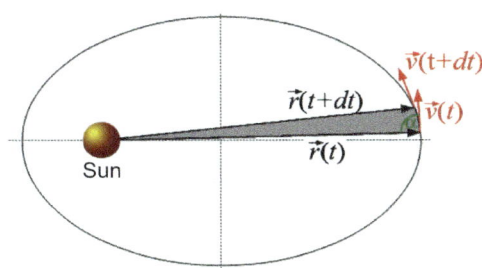

Fig. 11.6 Determination of angular momentum from the area law

where α is the angle between the position vector and the velocity vector. Then it is:

$$\begin{aligned}\frac{dA}{dt} &= \frac{1}{2}|\vec{r}|\,|\vec{v}|\sin\alpha = \frac{1}{2m}|\vec{r}|\,|\vec{p}|\sin\alpha = \frac{1}{2m}|\vec{r}\times\vec{p}|\\ &= \frac{1}{2m}|\vec{L}| = \text{const.}\end{aligned} \quad (11.2)$$

Here, $\vec{L}$ is the already mentioned angular momentum. Kepler's second law is therefore a direct consequence of the conservation of angular momentum. It applies as long as we can neglect forces on the planet that do not originate from the Sun. Such disturbances could be caused for example by the other planets in the solar system.

Example 11.2: Planetary Orbits Part 1: Centrifugal Potential

The motion of a planet around the Sun takes place in a plane, called the orbital plane. To calculate the orbit, we must capture the revolution mathematically. The energy of the planet is an important parameter in the calculation. We want to determine it here. We use polar coordinates with the origin of the coordinate system in the Sun. We define the zero point of the polar angle φ as the direction from the sun to the perihelion. The planet orbits counterclockwise.

The Sun is so heavy that we can assume it to be stationary and we want to neglect the influence of other planets on the orbit. Then the energy of the planet (mass m) is given by

$$E_{\text{tot}} = \frac{1}{2}mv^2 + E_{\text{pot}}(r).$$

The potential energy E_{pot} results from the gravitational force of the Sun on the planet. It depends solely on the distance of the planet from the Sun, but not on the polar angle. In Cartesian coordinates, it would be $v^2 = v_x^2 + v_y^2$, but we want to use polar coordinates here. Then $v^2 = v_r^2 + v_\varphi^2$, with the radial velocity $v_r = \frac{dr}{dt}$ and the velocity component perpendicular to it $v_\varphi = r\omega = r\frac{d\varphi}{dt}$. Hence, we get:

$$E_{tot} = \frac{1}{2}m(v_r^2 + v_\varphi^2) + E_{pot}(r)$$

$$= \frac{1}{2}m\left(\left(\frac{dr}{dt}\right)^2 + r^2\left(\frac{d\varphi}{dt}\right)^2\right) + E_{pot}(r).$$

In Eq. 11.2 we introduced the angular momentum as $\vec{L} = \vec{r} \times m\vec{v}$. Due to the cross product, only the velocity component perpendicular to $\vec{r}$ contributes, which is v_φ. In terms of magnitude:

$$L = mrv_\varphi = mr^2\frac{d\varphi}{dt} \Rightarrow \frac{d\varphi}{dt} = \frac{L}{mr^2}.$$

This results in:

$$E_{tot} = \frac{1}{2}m\left(\frac{dr}{dt}\right)^2 + \frac{L^2}{2mr^2} + E_{pot}(r).$$

According to Kepler's second law, the angular momentum L is a constant. The middle term thus only depends on the distance r from the Sun. It resembles the term of potential energy. Therefore, this term is treated as a potential, even though this is not strictly correct. It is called "centrifugal potential". It is combined with the potential energy to form an effective potential:

$$E_{tot} = \frac{1}{2}m\left(\frac{dr}{dt}\right)^2 + E_{eff}(r) \quad \text{with} \quad E_{eff}(r) = \frac{L^2}{2mr^2} + E_{pot}(r).$$

Now look at the equation again carefully. What have we achieved? The planetary motion is two-dimensional, but the energy balance only contains the coordinate r. We have reduced the problem to one dimension. This will simplify the calculation, which we can only deal with once we can specify $E_{pot}(r)$.

How did we achieve this? With a fixed angular momentum, the orbital speed is determined by the radius. If the planet moves closer to the Sun, it must rotate faster to maintain the orbital angular momentum, and if it moves away from the Sun, correspondingly slower. We have exploited this to express the kinetic energy, which is in the rotation, through the distance r and thereby eliminated the polar angle.

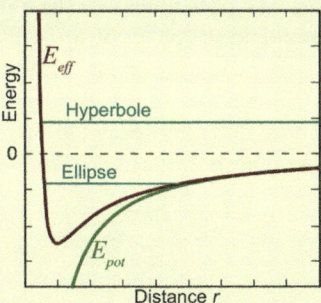

The illustration shows the course of the effective potential. It rises sharply toward small distances, as the rotational speed increases strongly there. Also plotted is the total energy of a planet on an elliptical orbit, which must be constant due to energy conservation. The line is limited by the points where the effective potential reaches the value of the total energy. Within this range, the planet has a positive kinetic energy in the radial movement, outside these points it would have to be negative, which is not possible. The planet must stay between these points. The two points mark the distance of the planet to the Sun in the perihelion and in the aphelion. If the total energy is less than zero, elliptical orbits result, with positive energies we get hyperbolas. At the border at $E_{tot} = 0$ the path is parabolic. Hyperbolic (and parabolic) paths occur, for example, with aperiodic comets. They are not bound. The celestial body can move infinitely far away from the Sun.

With the bound orbits (ellipses), another tendency can be seen: The lower (more negative) the total energy, the more the orbital radii in perihelion and aphelion approach each other. The eccentricity of the ellipses decreases until they are finally equal in the minimum of the effective potential. Then the planet orbits on a circular path.

Example 11.3: Planetary Orbits Part 2: Co-Movement of the Sun

Kepler's first law states that the Sun rests in one of the two foci of the orbital ellipses. Strictly speaking, this is not correct. If one considers the Sun and one of its planets and neglects the forces of the other planets, no external forces act on the Sun and the planet. Their center of mass rests, or moves uniformly through the universe (center of mass theorem). The planet, but also the Sun, move on elliptical orbits around the center of mass resting in the focus. However, the semi-axes of the Sun's orbit are much smaller than the semi-axes of the planetary orbits, so Kepler did not notice the co-movement of the Sun.

We want to show how it can be taken into account.

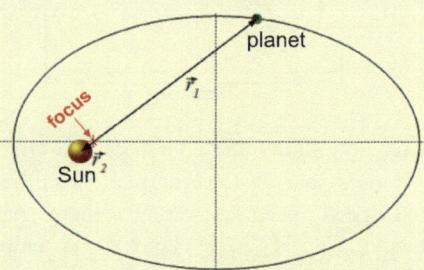

We had denoted with r the distance between planet and Sun. Let $\vec{r}_1$ be the vector from the focus to the planet and $\vec{r}_2$ the corresponding vector of the Sun. Then $r = r_1 + r_2$. In relation to the center of mass, it must also apply $m\,r_1 = M\,r_2$, where m is the mass of the planet and M that of the Sun. From the two relationships, it follows that

$$r_1 = \frac{M}{m+M}r \qquad r_2 = \frac{m}{m+M}r.$$

Now let's consider the total energy from example 11.2, where we now have to take into account the energy of the planet and the Sun:

$$\begin{aligned}E_{\text{tot}} &= \frac{1}{2}mv_1^2 + \frac{1}{2}Mv_2^2 + E_{\text{pot}}(r) \\ &= \frac{1}{2}m\left(\left(\frac{dr_1}{dt}\right)^2 + r_1^2\left(\frac{d\varphi}{dt}\right)^2\right) \\ &\quad + \frac{1}{2}M\left(\left(\frac{dr_2}{dt}\right)^2 + r_2^2\left(\frac{d\varphi}{dt}\right)^2\right) + E_{\text{pot}}(r).\end{aligned}$$

11.2 · Kepler's Second and Third Laws

Please note that the angular velocities $\frac{d\varphi}{dt}$ for the planet and the sun are the same, as they always face each other. We examine the terms individually and start with the radial velocities:

$$\frac{1}{2}m\left(\frac{dr_1}{dt}\right)^2 + \frac{1}{2}M\left(\frac{dr_2}{dt}\right)^2$$

$$= \frac{1}{2}m\frac{M^2}{(m+M)^2}\left(\frac{dr}{dt}\right)^2 + \frac{1}{2}M\frac{m^2}{(m+M)^2}\left(\frac{dr}{dt}\right)^2$$

$$= \frac{1}{2}\frac{mM^2 + Mm^2}{(m+M)^2}\left(\frac{dr}{dt}\right)^2 = \frac{1}{2}\frac{mM(m+M)}{(m+M)^2}\left(\frac{dr}{dt}\right)^2$$

$$= \frac{1}{2}\frac{mM}{m+M}\left(\frac{dr}{dt}\right)^2.$$

For the angular velocities, the result is:

$$\frac{1}{2}mr_1^2\left(\frac{d\varphi}{dt}\right)^2 + \frac{1}{2}Mr_2^2\left(\frac{d\varphi}{dt}\right)^2$$

$$= \frac{1}{2}\left(m\frac{M^2}{(m+M)^2} + M\frac{m^2}{(m+M)^2}\right)r^2\left(\frac{d\varphi}{dt}\right)^2$$

$$= \frac{1}{2}\frac{mM}{m+M}r^2\left(\frac{d\varphi}{dt}\right)^2.$$

The potential energy remains unchanged, as $E_{\text{pot}}(r)$ depends solely on the distance Sun-planet, but not on the center of gravity. With this, we have eliminated r_1 and r_2 and transformed all terms to r. We introduce another quantity m', which is called the "reduced mass":

$$m' = \frac{mM}{m+M}.$$

We insert everything into the energy balance and obtain:

$$E_{\text{tot}} = \frac{1}{2}mv_1^2 + \frac{1}{2}Mv_2^2 + E_{\text{pot}}(r)$$

$$= \frac{1}{2}m'\left(\left(\frac{dr}{dt}\right)^2 + r^2\left(\frac{d\varphi}{dt}\right)^2\right) + E_{\text{pot}}(r).$$

This is exactly the same relation that we started with in example 11.2, only that the mass of the planet has been replaced by the reduced mass. We have reduced the two-body problem to a one-body problem. With the reduced mass, we first calculate the motion in r and then determine the motion of the Sun and the planet from $r_1 = \frac{M}{m+M} r$ and $r_2 = \frac{m}{m+M} r$.

Due to the large mass of the Sun, the correction is small. For the motion of the Earth around the Sun, it is approximately

$$m' = (1 - 3 \cdot 10^{-6}) m_\oplus .$$

and the semi-axes of the Sun's orbit are smaller than those of the Earth by the same factor $3 \cdot 10^{-6}$. However, the Sun's co-movement is significant if one wants to consider, for example, the influence of the other planets on the Earth's orbit. The other planets also move the Sun and thus indirectly the center of gravity of the Earth's orbit.

Kepler's third law finally makes a statement about the orbital period of the planets. As can be seen in Table 11.1, the orbital period increases with increasing distance from the Sun. Kepler's third law quantifies this:

> **3. Kepler's Law**
> The square of a planet's orbital period is proportional to the cube of the length of the semi-major axis of its orbit.

Expressed in an equation, this means:

$$\frac{T_{\text{planet}}^2}{a_{\text{planet}}^3} = \text{const.} \tag{11.3}$$

A look at Table 11.1 (last column) shows that this relationship is indeed very well fulfilled in our solar system.

Example 11.4: Trojans

On top of the planets, our solar system includes a large number of smaller celestial bodies, the asteroids. The Trojans are a group of such asteroids that orbit the Sun at the same distance as Jupiter. According to Kepler's third law, the orbital period is independent of the mass of the celestial body. In fact, the Trojans orbit at the same speed as Jupiter, which is many orders of magnitude heavier. Therefore, they never get in each other's way.

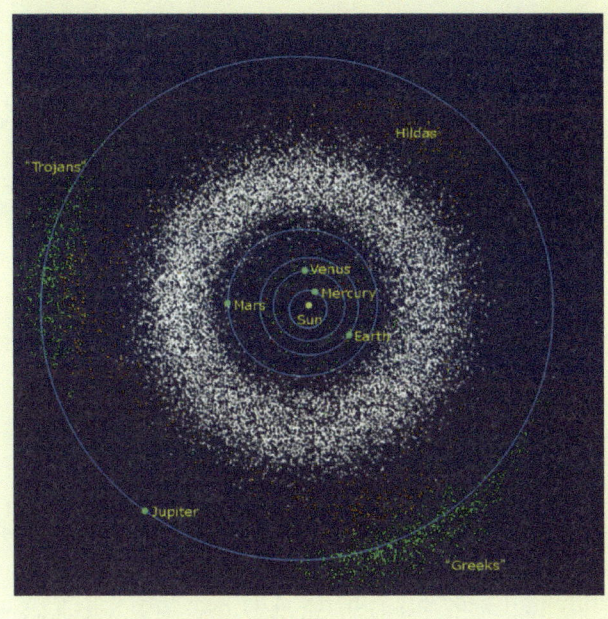

In conclusion, it should be noted that planetary orbits are actually more complicated. This is because the Sun does not rest at the focus, but strictly speaking, the common center of gravity of the Sun and the planets rests at the focus. The Sun orbits the focus like the planet on an ellipse, but with much smaller semi-axis. However, since there is more than one planet, this leads to a complicated motion of the Sun, which in turn affects the planetary orbits. In addition, the gravitational forces between the planets are not entirely negligible.

Example 11.5: Determination of the Mass of the Central Star

Kepler's third law provides a way to determine the mass of a central star from the data of the orbiting celestial bodies. We are jumping ahead a bit by stating the value of the constant in Kepler's third law. We will determine it further below:

$$\frac{T^2}{a^3} = \frac{4\pi^2}{MG},$$

where M is the mass of the central star and G is the gravitational constant. If one has captured the motion of a body around a central star through astronomical observations (e.g., a moon orbiting a planet or a planet orbiting its sun), one can calculate the mass of the central star using Kepler's third law.

As an example, we determine the solar mass $M_\odot$ from the distance of the Earth from the Sun and the length of a year:

$$M_\odot = \frac{4\pi^2}{G} \frac{a^3}{T^2} = \frac{4\pi^2}{6.674 \cdot 10^{-11} \, \frac{\mathrm{m}^3}{\mathrm{kg\,s}}} \frac{(1.496 \cdot 10^{11}\,\mathrm{m})^3}{(365.62 \cdot 24 \cdot 3600\,\mathrm{s})^2}$$

$$= 1.984 \cdot 10^{30}\,\mathrm{kg}.$$

11.3 Newton's Law of Universal Gravitation

From antiquity to Newton, the world was divided into earthly and heavenly spheres. The heavenly spheres were considered divinely perfect. They were not subject to the laws of the earthly world. One of Newton's greatest achievements was to recognize that both earthly and heavenly physics rest on the same basis. Newton realized that the same laws of nature that cause an apple to fall from a tree, also determine the motion of the moon around the Earth and the planets around the Sun.

He derived his conclusion with a simple thought experiment, which he illustrated in his main work *Philosophiae Naturalis Principia Mathematica* with the sketch reproduced in ◘ Fig. 11.7. He imagined someone standing on a high mountain (V). The person throws a stone from the mountain that will land at position (D). If the person throws the stone faster, it will fly further, perhaps to position (E). Newton writes: "A stone thrown from a mountain top with ever increasing speed will eventually not land on the ground, but 'fall around' the Earth. In this sense, the moon also 'falls around' the Earth." According to an anecdote, this idea came to him as he sat under an apple tree and watched an apple fall to the ground.

Once one has accepted that Newton's axioms also apply to the motion of the planets, one can calculate the

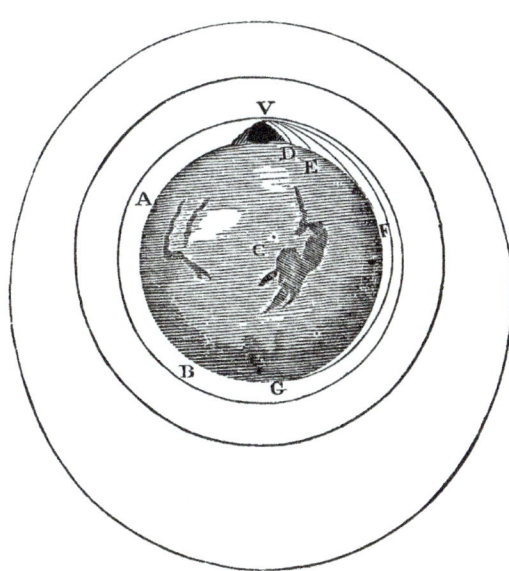

Fig. 11.7 Illustration from Newton's *Philosophiae Naturalis Principia Mathematica*

motion from them. However, we must know the force that binds the planets to the Sun, the gravitational force. It should not be surprising that the law of gravity is already contained in Kepler's laws. We want to derive it from them just as Newton did.

From Kepler's first law (elliptical orbits), it follows that gravity must be a central force. This is immediately obvious for the special case of a circular orbit. For this special case, we have already seen several times that the centripetal force points to the center of the circle, i.e., towards the Sun. Expressed in an equation, we can write:

$$\vec{F}_G(r) = f(r)\,\hat{e}_r \,. \tag{11.4}$$

where $\hat{e}_r$ is a unit vector from the planet to the Sun. From Galilei's law of motion, we know that the force must be proportional to the mass of the falling body. The reaction to the Earth must be of the same size. However, this only makes sense if we assume that the gravitational force is proportional to the mass of both involved objects. It therefore has the form:

$$\vec{F}_G(r) = G\,m_1\,m_2\,g(r)\,\hat{e}_r \,. \tag{11.5}$$

We have introduced a proportionality constant G called the gravitational constant. It is a natural constant that determines the strength of the gravitational force. The value of the gravitational constant must be determined by experiments. We will discuss this further below. The measured value is:

$$G = (6.67384 \pm 0.00080)\, 10^{-11} \frac{\text{m}^3}{\text{kg s}^2}. \tag{11.6}$$

The function $g(r)$ describes the dependence of the gravitational force on the distance between the two bodies. It can be determined for planetary motion from Kepler's third law. The gravitational force acts as a centripetal force in planetary motion. This has—as we already know—the magnitude $m\,\omega^2\, r$. Therefore, it must hold:

$$G\, m_{\text{planet}}\, m_{\text{Sun}}\, g(r) = m_{\text{planet}}\, \omega^2\, r. \tag{11.7}$$

The angular velocity is inversely proportional to the orbital period $\omega \sim 1/T$ and this is linked to the semi-major axis according to Kepler's third law $\omega^2 \sim T^{-2} \sim r^{-3}$. It follows from Eq. 11.7 $g(r) \sim r^{-2}$. Thus, we obtain for the law of gravity:

> **Law of Gravity**

$$\vec{F}_{\text{G}}(\vec{r}) = \pm G \frac{m_1\, m_2}{r^2}\, \hat{e}_r.$$

Example 11.6: Gravitation of a Rotationally Symmetric Star

Usually, we approximate the gravitational effect of an extended star by that of a point mass. We want to show why this is a good approximation even in its immediate vicinity. It will turn out that the star's gravitation is identical to that of a point mass, provided the star is built rotationally symmetric.

We calculate the gravitational force on a mass m_1, which is at a distance r from the star with radius R and mass m_2 ($r > R$). We choose a coordinate system with the origin at the center of the star and rotate the axes so that the mass m_1 is on the z-axis (see sketch).

11.3 · Newton's Law of Universal Gravitation

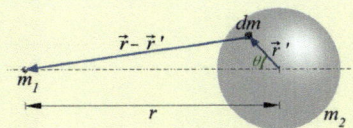

We denote the mass distribution of the star with $\rho(r')$ and incorporate the rotational symmetry of the star in that the mass distribution depends only on the magnitude of the vector $\vec{r}'$.

For reasons of symmetry, the gravitational force must point along the z-axis. It is sufficient to consider for each mass element dm of the star the projection of the gravitational force on the z-axis:

$$\vec{F}_G = -F_r \hat{e}_z = \int \left| d\vec{F}_G \right| \cos \gamma \, \hat{e}_z.$$

We can read the angle γ from the figure. It is:

$$\cos \gamma = \frac{r - r' \cos \theta}{|\vec{r} - \vec{r}'|}.$$

The distance $|\vec{r} - \vec{r}'| = \sqrt{(\vec{r} - \vec{r}')^2}$ between m_1 and dm can be expressed by

$$\left| \vec{r} - \vec{r}' \right|^2 = r^2 + r'^2 - 2rr' \cos \theta.$$

All of this inserted results in:

$$F_r = G m_1 \int_0^R \int_0^\pi \int_0^{2\pi} \frac{\rho(r') \cos \gamma}{|\vec{r} - \vec{r}'|^2} \, d\varphi \sin \theta \, d\theta \, r'^2 dr'$$

$$= 2\pi \, G m_1 \int_0^R \int_0^\pi \frac{\rho(r') (r - r' \cos \theta)}{(r^2 + r'^2 - 2rr' \cos \theta)^{3/2}} \sin \theta \, d\theta \, r'^2 dr',$$

where in the second step we have already performed the integration over $d\varphi$. It results in a factor 2π. To solve the next integral, we substitute $x = \cos \theta$, $dx = -\sin \theta \, d\theta$ and obtain:

$$F_r = 2\pi \, G m_1 \int_0^R \int_{-1}^{+1} \frac{\rho(r') (r - r'x)}{(r^2 + r'^2 - 2rr'x)^{3/2}} \, dx \, r'^2 dr'.$$

This integral can be further simplified by a second substitution. We set $y = r - r'x$, $dy = -r' dx$ and obtain:

$$F_r = 2\pi\, G m_1 \int_0^R \int_{r-r'}^{r+r'} \frac{\rho(r')\, y}{(r'^2 - r^2 + 2ry)^{3/2}}\, dy\, r'\, dr'.$$

The solution to this integral can be found in tables. It is:

$$F_r = 2\pi\, G m_1 \int_0^R \rho(r')\frac{2}{4r^2}$$

$$\cdot \left[(r'^2 - r^2 + 2ry)^{\frac{1}{2}} + \frac{r'^2 - r^2}{(r'^2 - r^2 + 2ry)^{\frac{1}{2}}} \right]_{r-r'}^{r+r'} r'\, dr'$$

$$= 2\pi\, G m_1 \int_0^R \rho(r')\frac{2}{4r^2}[2r' + 2r']\, r'\, dr'$$

$$= G\frac{m_1}{r^2}\left(4\pi \int_0^R \rho(r') r'^2\, dr'\right).$$

The term in the last bracket, however, yields nothing other than the mass m_2 of the star. Thus, we obtain as a result:

$$\vec{F}_G = -G\frac{m_1 m_2}{r^2}\hat{e}_z,$$

which, despite the star's expansion, corresponds to the force of a point mass m_2.

Example 11.7: The Gravitational Field Inside a Star

After we have determined the gravitational force of a rotationally symmetric star on an external mass in Example 11.6, we now want to ask ourselves how the force changes when the mass m_0 is located inside the star. The star has the mass M and the radius R. The mass m_0 is located at a distance r from the center of the star. Now let $r < R$. We divide the star into concentric mass shells and first consider the outer shells, i.e., those with a radius $r' > r$. From each mass shell, forces are acting in different directions on the mass m_0. For each force, there is a corresponding force from the opposite direction. A pair of $\vec{F}_1, \vec{F}_2$ of such forces is plotted in the sketch. We want to show that the two forces are equal in magnitude and thus add up to zero. Since we have made no assumption about the direction of the pair, this would apply to all pairs, and we would come to the conclusion

11.3 · Newton's Law of Universal Gravitation

that an external mass shell exerts no resultant force on the mass m_0.

We consider the surface elements dA_1 and dA_2, which cover the same solid angle on the spherical shell when viewed from m_0. In our example, dA_1 is larger than dA_2, which increases $\vec{F}_1$, but the distance r_1 from dA_1 to the mass m_1 is also larger than the distance r_2 from dA_2, which reduces $\vec{F}_1$. It is:

$$dm_1 = \rho(r') \, dr' \, dA_1$$
$$dm_2 = \rho(r') \, dr' \, dA_2,$$

where dr' is the thickness of the spherical shell. The two forces are:

$$\left|\vec{F}_1\right| = G \frac{m_0 \, dm_1}{r_1^2} = G m_0 \, \rho(r') \, dr' \, \frac{dA_1}{r_1^2}$$
$$\left|\vec{F}_2\right| = G \frac{m_0 \, dm_2}{r_2^2} = G m_0 \, \rho(r') \, dr' \, \frac{dA_2}{r_2^2}$$

However, according to the intercept theorem,

$$\frac{dA_1}{r_1^2} = \frac{dA_2}{r_2^2},$$

which shows that the forces are equal in size and thus the outer spherical shells do not exert any force on the mass m_0.

Now let's consider the inner spherical shells, i.e., the spherical shells with $r' < r$. Here we can refer to the arguments from example 11.6, because the mass m_0 is outside these spherical shells. The force of the inner spherical shells corresponds to the force of a point mass of the same mass, which is located at the center C of the star. The gravitational force $\vec{F}_0$ of the star on the mass m_0:

$$\vec{F}_0 = -G \frac{m_0 \, m_r}{r^2} \, \hat{e}_r,$$

where m_r indicates the mass that is within the radius r:

$$m_r = 4\pi \int_0^r \rho(r') \, r'^2 \, dr'.$$

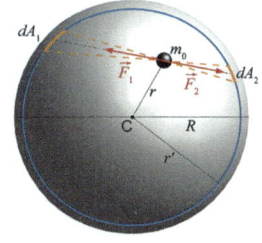

> In the case of a star with homogeneous density, the result is $m_r = \frac{4}{3}\pi \rho\, r^3$, which leads to a force $\vec{F}_0$ that increases linearly with the distance r from the center of the star.

11.3.1 Measurement of the Gravitational Constant

Gravity is capable of keeping the Earth with its massive weight in its orbit, but for this, it requires a celestial body with even more mass: the Sun. For masses that are available in the laboratory, gravity is very weak. Between two bodies of mass $m = 1\,\text{kg}$, which are at a distance of $r = 10\,\text{cm}$ from each other, the force is only $6.67 \cdot 10^{-9}\,\text{N}$. This makes the measurement of the gravitational constant so difficult. The slightest disturbances from the environment affect the measurement. Although it was the first natural constant to be discovered, it is still one of the least accurately known today.

We want to discuss the measurement of the gravitational constant with the gravitational balance, which goes back to the first measurements by Henry Cavendish (1797) (○ Fig. 11.8). The method was later perfected by Loránd Eötvös.

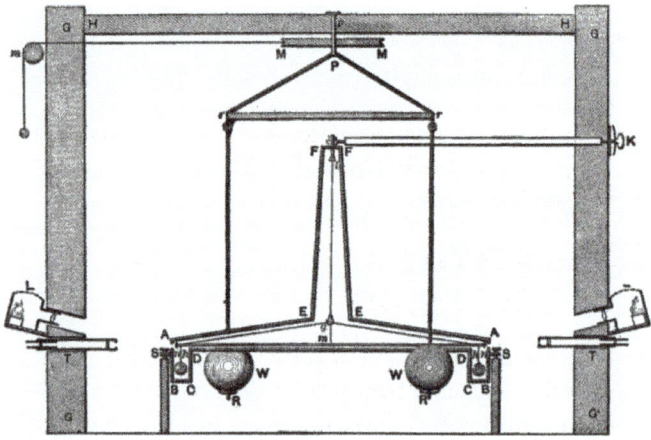

○ **Fig. 11.8** Sketch of Cavendish's apparatus from his publication "Experiments to determine the Density of the Earth" from 1798

11.3 · Newton's Law of Universal Gravitation

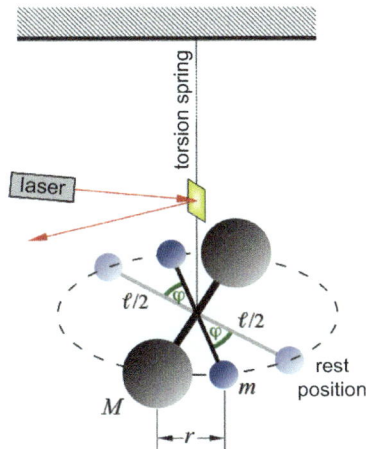

Fig. 11.9 Schematic setup of a gravitational balance

The gravitational balance is used to measure the gravitational force between metal balls. The balls are positioned at a small distance to each other to generate the largest possible force. The big balls with mass M (see ■ Fig. 11.9) are stationary during the measurement. The small balls (mass m) hang from a thin metal wire, which creates a restoring force by twisting at the angle φ. The gravitational force between the balls is

$$F_G = 2\,G\frac{m\,M}{r^2}, \tag{11.8}$$

where the factor 2 arises because there are two pairs of balls. With this force, the small balls are attracted to the big balls. The force exerted by the opposite ball is neglected. A measurement is shown in Experiment 11.1. We will return to another evaluation method in ▶ Sect. 14.4.

Experiment 11.1: Gravitational Balance

The illustration shows a gravitational balance that is firmly anchored to the concrete wall of the lecture hall. The torsion spring, a thin fiber, is located in a tube, at the lower end of which there is a housing closed with two glass plates. Inside are the mirror and the beam with the two small balls. The housing shields the small balls from air currents and other environmental influences.

First, the big balls are brought as close as possible to the small balls as shown in the illustration. The apparatus is given time to find its equilibrium. This takes several hours. Then the big balls are rotated to the opposite position near the other small ball and their motion is observed with a laser and a mirror. The small balls "fall" initially accelerated towards the big ones, until a damped oscillation is created by the restoring force of the torsion spring, around a zero point, which is shifted by the gravitational force compared to the original equilibrium position.

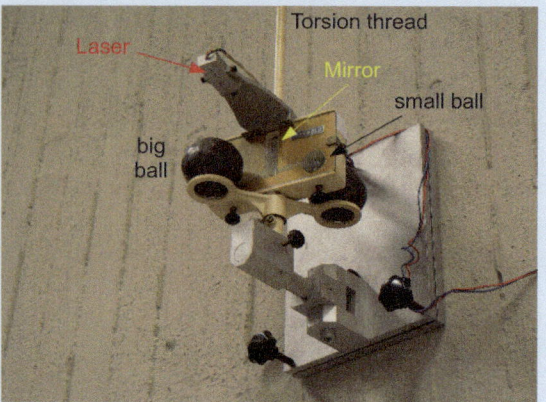

© RWTH Aachen, Physics Experiment Collection

There are several ways to evaluate the measurement. Here, the relatively simple acceleration method, which only uses the initial motion, is presented. At the beginning of the motion, the speed is still small, and the change of the forces due to the motion can be neglected. In the original rest position, the torsional force of the spring compensates for the gravitational force of the balls. It is:

$$F_{\text{tors}} = 2G\frac{mM}{r^2}.$$

After the big balls are moved to the opposite position, the gravitational force acts in the reverse direction. Torsional force and gravitational force add up. Now the small balls "fall" towards the big ones with initially constant acceleration. We get:

11.3 · Newton's Law of Universal Gravitation

$$2ma = F_G + F_{\text{tors}} = 4G\frac{mM}{r^2}.$$

We record the motion and determine the gravitational constant G from the acceleration. The figure shows a measurement over a longer period of time. A parabola (blue curve), as we would expect from a constant acceleration, is fitted to the initial data points. The fit matches the first data points quite well. In addition, a harmonic, damped oscillation is fitted to the data points of the entire measurement (red curve). Please, note that the balls really move in an anharmonic oscillation, which leads to systematic deviations of the fit from the data points.

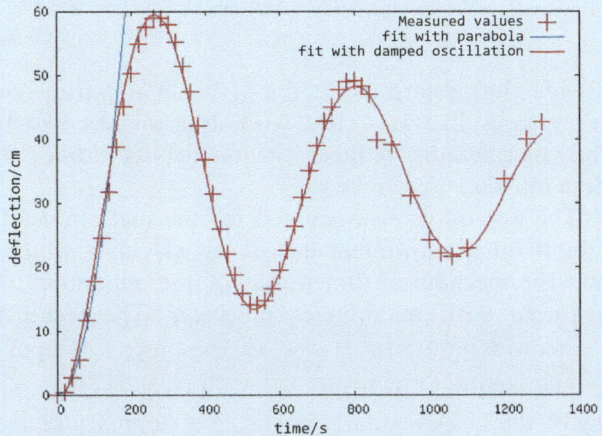

By the way, it is quite possible to build a gravitational balance with simple means yourself. Give it a try! Instructions can be found on the web, for example, here:
▶ https://www.fourmilab.ch/gravitation/foobar/.

The experiments to measure the gravitational constant were successively improved. ◘ Fig. 11.10 shows an apparatus with which an international research group at the Bureau International des Poids et Mesures published a precise measurement in 2001. The experiment is named "Big G". In this experiment, instead of pairs of balls, four cylinders are used. The outer large cylinders can be clearly seen in the photo. The torque on the inner small bodies exerted by disturbing masses

Fig. 11.10 The gravitational balance of the 'Big G' experiment at the Bureau International des Poids et Mesures in Sèvres near Paris. © Bureau International des Poids et Mésures

outside the apparatus decreases in an apparatus with two masses like $1/r^3$, but with four masses like $1/r^5$. This significantly reduces the disturbing influence of these masses.

The apparatus is evacuated for the measurement. It is built on a coordinate measuring machine, which allows the mechanical dimensions of the apparatus to be measured with micrometer precision. The result was $G = 6.667559(67) \cdot 10^{-11} \, \frac{\text{m}^3}{\text{kg}\,\text{s}^2}$, an accuracy of 41 ppm.

The National Institute of Standards and Technology of the USA regularly publishes a summary of measurements and recommends a value for G. The recommended value is given in Eq. 11.6. It is based on 11 measurements, which are summarized in Table 11.2. As can be seen, significantly improved accuracies are possible today, but the table also shows discrepancies between the measurements. For example, the two most accurate measurements differ by $0.0019 \cdot 10^{-11} \, \frac{\text{m}^3}{\text{kg}\,\text{s}^2}$ with errors smaller than $0.0002 \cdot 10^{-11} \, \frac{\text{m}^3}{\text{kg}\,\text{s}^2}$, an indication of how difficult these measurements are.

11.3.2 Gravitation at the Earth's Surface

The weight that bodies experience at the Earth's surface is a consequence of the Earth's gravitational force. We

11.3 · Newton's Law of Universal Gravitation

Table 11.2 Measurements of the gravitational constant G

	G in $10^{-11} \frac{m^3}{kg\,s^2}$	Uncertainty in ppm
Luther and Towler (1982)	6.67248(43)	64
Karagioz and Izmailov (1996)	6.6729(5)	75
Baglay and Luther (1997)	6.67398(70)	100
Gundlach and Merkowitz (2000)	6.674255(92)	14
Quinn et al. (2001)	6.67559(27)	40
Kleinvoß et al. (2002)	6.67422(98)	150
Armstrong and Fitzgerald (2003)	6.67387(27)	40
Hu, Guo and Luo (2005)	6.67228(87)	130
Schlamminger et al. (2006)	6.67425(12)	19
Luo et al. (2010)	6.67349(18)	27
Parks and Faller (2010) [a]	6.67260(25)	37

[a] The authors of this measurement found an error in the evaluation of their measurement after the printing of our first edition and had to correct their result: See H. Parks, J. Faller, Physical Review Letters, 122 (2019) 199901.

want to calculate the weight of a body of mass m ($m_\oplus$ is the mass of the Earth, $r_\oplus$ its radius):

$$F_G = m g = G \frac{m\, m_\oplus}{r_\oplus^2} = \frac{G m_\oplus}{r_\oplus^2} m \Rightarrow g = \frac{G m_\oplus}{r_\oplus^2}. \quad (11.9)$$

If we insert the mass and diameter (equator) of the Earth, we get

$$g = \frac{6.674 \cdot 10^{-11} \frac{m^3}{kg\,s^2} \times 5.974 \cdot 10^{24}\, kg}{(12\,756\, km/2)^2} = 9.801 \frac{m}{s^2} \quad (11.10)$$

As can be seen from the law of gravitation, the weight on the Earth's surface is not constant. It decreases with increasing height. To what extent is the simple formula $F = mg$ justified? We investigate this question by approximating the law of gravitation for radii close to the Earth's radius. We express the distance of a body from the center of the Earth as $r = r_\oplus + h$, where h indicates the height above the Earth's radius, and we expand the law of gravitation in a first-order Taylor series:

$$\begin{aligned}
F_G(r_\oplus + h) &= F_G(r_\oplus) + \frac{d}{dr}\left[G\frac{m\,m_\oplus}{r^2}\right]_{r=r_\oplus} \cdot h + \ldots \\
&= F_G(r_\oplus) + \left[-2G\frac{m\,m_\oplus}{r^3}\right]_{r=r_\oplus} \cdot h + \ldots \\
&= F_G(r_\oplus) - 2\frac{h}{r_\oplus}\left(G\frac{m\,m_\oplus}{r_\oplus^2}\right) + \ldots \\
&= F_G(r_\oplus) \cdot \left(1 - 2\frac{h}{r_\oplus} + \ldots\right).
\end{aligned}$$

(11.11)

For example, at a height of 1 km, the weight is reduced by 0.03%. As long as such deviations are negligible, one can use $F = m\,g$.

Approximations

In the calculation just performed, we determined the weight of a body at the Earth's surface approximately. Such approximation methods are important in physics. They simplify the calculation and often make it possible to get a result in the first place. Approximate results are not less valuable than exact ones. Note that in every calculation of a real process, approximations must already be included in the approach. There are always some effects that you neglect. Before you even start calculating, you need to be clear about the accuracy you want to achieve. If you then make approximations during the calculation, this is not a disadvantage, as long as the approximations do not japordize the desired accuracy.

Example 11.8: Low Tide and High Tide

The Moon also exerts a gravitational force on objects on Earth, even on the water in the oceans. Moon's gravitational force is $F_M = G\frac{mM_M}{r_M^2}$, where M_M is the mass of our Moon and r_M is the distance between Earth and Moon (distance of the centers). The quantity m denotes the mass of the object on Earth. However, the water is not located at the center of the Earth, but on its surface (radius of the Earth R_E, mass M_E). Depending on whether the water is on the side facing or opposing the Moon, the gravitational acceleration generated by the Moon differs by

$$\pm a_M = G \frac{M_M}{(r_M \mp R_E)^2} - G \frac{M_M}{r_M^2}$$

$$= G \frac{M_M}{r_M^2} \left(\frac{1}{\left(1 \mp \frac{R_E}{r_M}\right)^2} - 1 \right) \approx \pm G \frac{M_M}{r_M^2} \cdot 2 \frac{R_E}{r_M}$$

$$= \pm 2g \frac{M_M}{M_E} \frac{R_E^3}{r_M^3}.$$

If you plug in values, you will find that the acceleration due to gravity caused by the Moon is approximately $10^{-7}g$. Or in other words, the weight of a person with 100 kg changes by 10 mg due to the influence of the Moon.

The surface of the oceans adjusts so that the total acceleration at the surface of the water is the same everywhere. From this, the tidal range d can be determined.

$$G \frac{M_E}{(R_E \pm d)^2} \pm a_G = g$$

$$g \frac{1}{(1 \pm \frac{d}{R_E})^2} = g \mp a_G$$

$$1 \pm \frac{d}{R_E} = \sqrt{\frac{1}{1 \mp \frac{a_G}{g}}} \approx 1 \pm \frac{1}{2} \frac{a_G}{g}$$

$$d = \pm \frac{1}{2} \frac{a_G}{g} R_E.$$

The estimate results in a tidal range of ± 30 cm. This is the tidal range in the open ocean. The tidal range on the coast is usually significantly higher due to current effects.

Example 11.9: Trojans, Part 2

In example 11.4 we introduced the Trojans on Jupiter's orbit. We argued that they are stable there because they have the same orbital periods as Jupiter itself due to

Kepler's third law. But this is not the whole story. Have you noticed that there are asteroids only in certain positions on Jupiter's orbit? Kepler's laws assume that only the gravitational force of the Sun acts on the orbiting celestial bodies. But the asteroids on Jupiter's orbit certainly feel the gravitational pull of Jupiter in addition to that of the Sun. And indeed, this force has caused the asteroids to disappear from some positions. They have crashed into Jupiter.

But why, given Jupiter's gravitational pull, are there stable positions for asteroids on Jupiter's orbit at all? This is largely due to the co-movement of the Sun. Neither Jupiter nor the asteroids orbit the Sun. They orbit the common center of gravity of the Sun and Jupiter (we can safely neglect the contribution of the asteroids to the center of gravity). The balance of the forces is shown in the sketch. It shows the positions in a coordinate system that rotates with the Sun and Jupiter around their center of gravity S. In this system, all objects are stationary. The Sun is to the left of the center of gravity at a distance r_S, Jupiter is to the right at a distance r_J. An asteroid is placed at the bottom of the sketch. The gravitational forces $\vec{F}_S$ and $\vec{F}_J$, which the Sun and Jupiter exert on the asteroid, are shown. In addition, we must consider the centrifugal force $\vec{F}_{CF}$ because it is an accelerated (rotating) reference system. The gravitational forces point to the Sun or Jupiter, the centrifugal force points from the center of gravity outwards.

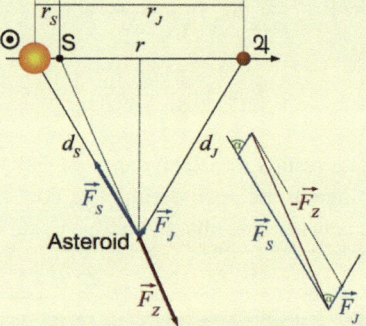

The forces are shown again as a parallelogram in the insert on the right. As a condition for static positions we have:

11.3 · Newton's Law of Universal Gravitation

$$\vec{F}_S + \vec{F}_J + \vec{F}_Z = 0 \quad \text{with} \quad \begin{cases} |\vec{F}_S| = G\frac{mM_S}{d_S^2} \\ |\vec{F}_J| = G\frac{mM_J}{d_J^2} \\ |\vec{F}_Z| = m\omega^2 d. \end{cases}$$

The rotation frequency of the three celestial bodies can be obtained from the balance of centrifugal force and gravity. We consider the rotation of Jupiter, neglecting the forces exerted by the asteroid. The gravitational force is determined by the distance r between the Sun and Jupiter, while the centrifugal force is determined by the distance of Jupiter from the center of mass r_J. It is $r_J = M_S/(M_S + M_J)\, r$ (Example 11.3).

$$G\frac{M_S M_J}{r^2} = M_J \omega^2 \frac{M_S}{M_S + M_J} r \Rightarrow \omega^2 = \frac{G(M_S + M_J)}{r^3}.$$

Leonhard Euler and Joseph-Louis Lagrange have worked on this problem and came to the surprising result that the three involved celestial bodies form an equilateral triangle, regardless of their masses. We want to show that this is indeed the case. We apply the cosine theorem to the force parallelogram and obtain

$$F_Z^2 = \left(G\frac{mM_S}{d_S^2}\right)^2 + \left(G\frac{mM_J}{d_J^2}\right)^2$$
$$- 2\cos(\pi - \alpha)\left(G\frac{mM_S}{d_S^2}\right)\left(G\frac{mM_J}{d_J^2}\right)$$

$$F_Z^2 = \left(G\frac{mM_S}{d_S^2}\right)^2 + \left(G\frac{mM_J}{d_J^2}\right)^2$$
$$+ 2\cos\alpha\left(G\frac{mM_S}{d_S^2}\right)\left(G\frac{mM_J}{d_J^2}\right)$$

$$F_Z^2 = G^2 m^2 \left[\frac{M_S^2}{d_S^4} + \frac{M_J^2}{d_J^4} + 2\cos\alpha\,\frac{M_S M_J}{d_S^2 d_J^2}\right] \quad \text{(result A)},$$

where α refers to the angle enclosed by $\vec{F}_S$ and $\vec{F}_J$.

On the other hand, the centrifugal force can be determined from the geometry of the arrangement of the celestial bodies. We consider the triangle that is spanned by the center of gravity, the center of the connecting line between the Sun and Jupiter, and the asteroid. The lengths of the sides are $r_J - r/2$ (upper side), the height $h = \frac{\sqrt{3}}{2}r$ of the equilateral triangle Sun-Jupiter-Asteroid and d. From this, we can determine d. We obtain:

$$d^2 = \left(\frac{\sqrt{3}}{2}r\right)^2 + \left(\frac{r}{2} - \frac{M_S}{M_S + M_J}r\right)^2$$

$$= \left(1 - \frac{M_S}{M_S + M_J} + \left(\frac{M_S}{M_S + M_J}\right)^2\right)r^2.$$

Now we can calculate the centrifugal force. We get:

$$F_Z^2 = m^2 \omega^4 d^2$$

$$= \frac{G^2 m^2}{r^4}(M_S + M_J)^2 \left(1 - \frac{M_S}{M_S + M_J} + \left(\frac{M_S}{M_S + M_J}\right)^2\right)$$

$$= \frac{G^2 m^2}{r^4}(M_S^2 + M_J^2 + M_S M_J) \quad \text{(result B)}.$$

We obtained two results for F_Z^2, where we had assumed for result B that the Sun, Jupiter, and the asteroid form an equilateral triangle. Then we insert $\alpha = 60°$ and $d_S = d_J = r$ in result A and we see that the two results match. At this special point on Jupiter's orbit, the centrifugal force compensates for the joint attraction of the Sun and Jupiter. An asteroid located at this point will remain there. At other points on Jupiter's orbit, this balance of forces is not given. Asteroids that may have been there have left these points. They have crashed into Jupiter or the Sun or have been thrown out of orbit.

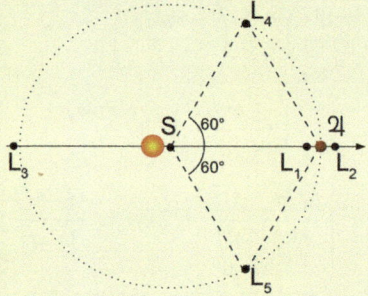

Overall, there are five points in such a three-body system where gravity and centrifugal force compensate each other. They are called the Lagrange points of the system. The sketch shows the location of the Lagrange points. We had determined the location of L_5 above. L_4 is symmetric to L_5 on the other side of Jupiter. The points L_1 to L_3 can be determined from the balance of gravity and centrifugal force, too. At the points L_4 and L_5 are the asteroids mentioned in example 11.4 (Trojans and Greeks). The points L_1 to L_3 are empty. They are not stable. If the asteroid moves slightly from its position along the line connecting the Sun and Jupiter, e.g., due to a force exerted by another planet like Saturn, the asteroid will plunge into the Sun or Jupiter. This is different at the points L_4 and L_5. If an asteroid is slightly deflected from these points, a force is created that points back to the Lagrange point.

Lagrange points do not only occur in the Sun-Jupiter system, but in the orbit of all planets and also in the Earth-Moon system. In fact, some asteroids have been found in the Lagrange points of Mars' orbit, and there are indications that there is also a (small) companion in Earth's orbit. However, there are thousands of Asteroids with Jupiter. This is because Jupiter is by far the heaviest planet in the solar system. The displacement of the Sun from the center of gravity r_S is greatest in the Sun-Jupiter system. At the same time, the disturbance of the movements by the other planets is the least. Due to the influence of the other planets, the Sun moves. With the lighter planets, this movement can become so strong that the Lagrange point becomes temporarily unstable and the asteroids are lost.

11.4 Gravitational and Inertial Mass

In ▶ Sect. 6.2 we discussed Newton's second law and learned about the mass of a body as a quantification of its inertia. In $\vec{F} = m\vec{a}$ the mass describes the resistance that a body opposes to a change in its motion, i.e., its inertia. Therefore, it is also referred to as "inertial mass". It indicates how much force must be applied to accelerate a body. Have you noticed that in the previous ▶ Sect. 11.3 we saw another definition of mass, namely in the law of gravity $F_G = G\frac{mM}{r^2}$. Here, the mass determines the strength with which the gravitational force acts on the body. We call it the "gravitational mass" (from gravity). The two definitions initially have nothing to do with each other. They define the two masses as independent physical quantities. It should be surprising that we use the same name and the same symbol for the two quantities. They could indeed be different! We have learned that this is not the case from observations, for example, from Galileo's law of motion(▶ Sect. 5.6).

Let us look again at Galileo's law and pay attention to the difference between gravitational and inertial mass. We determine the acceleration due to gravity from Newton's second axiom with the inertial mass m_t:

$$\vec{F} = m_t \vec{a}. \tag{11.12}$$

In free fall, the acting force stems from gravity $\vec{F} = m_s \vec{g}$. Since this is derived from the law of gravity, we must use the gravitational mass m_s here. So,

$$m_s \vec{g} = m_t \vec{a} \quad \text{or} \quad \vec{a} = \frac{m_s}{m_t} \vec{g}. \tag{11.13}$$

Now, Galileo's law states that $\vec{a}$ is the same for all bodies. According to Eq. 11.13, this can only be the case if m_s and m_t are proportional to each other, so that the ratio of masses can be reduced to a constant, thus $m_s = c\, m_t$. The constant c can be absorbed by a redefinition of the gravitational constant:

$$F_G = G \frac{m_s M_s}{r^2} = G \frac{c\, m_t\, c\, M_t}{r^2} = G c^2 \frac{m_t M_t}{r^2} = G' \frac{m_t M_t}{r^2}. \tag{11.14}$$

This formula expresses the law of gravity through inertial masses by introducing a new gravitational constant $G' = G c^2$. We had already carried out this redefinition, the gravitational constant used in ▶ Sect. 11.3 already relates to inertial masses with $c = 1$, so we can drop the prime on G. Therefore, $m_s = m_t$ applies, and we can abandon the distinction between heavy and inertial mass again.

This equality between the gravitational force acting on a body and its inertia is a special property of gravitation. Compare, for example, with the Coulomb force $F_{el} = k \frac{qQ}{r^2}$. Here, a new physical quantity appears in the numerator—the electric charge—which is not proportional to the inertia of the body. The connection between inertial and gravitational mass was long known, but its significance was not recognized. Only Albert Einstein elevated the equality to a postulate and built his theory of gravitation, the general theory of relativity, on it.

11.5 Potential and Potential Energy

11.5.1 The Potential Energy

Gravitation is a conservative force. We can therefore follow the arguments of ▶ Sect. 7.3 and define a potential energy. Let's consider the gravitational field of the Earth as an example. As a zero point, we initially want to choose the Earth's surface, which is approximately given by the Earth's radius $r_\oplus$.

$$E_{\text{pot}}(r) = -\int_{r_\oplus}^{r} F_G(r')\, dr' = -\int_{r_\oplus}^{r} -G\frac{m\, m_\oplus}{r'^2}\, dr'$$

$$[2mm] = -G\frac{m\, m_\oplus}{r} + G\frac{m\, m_\oplus}{r_\oplus}.$$

(11.15)

In many cases, the changes in altitude that occur are small compared to the Earth's radius itself ($r_\oplus \approx 6370\,\text{km}$). In these cases, an approximation is useful, which we derive through a Taylor expansion around $r = r_\oplus$. We set $r = r_\oplus + h$ and define $E_{\text{pot}}(r_\oplus) = 0$:

$$E_{\text{pot}}(r_\oplus + h) = E_{\text{pot}}(r_\oplus)$$
$$+ \frac{d}{dr}\left[-G\frac{m\, m_\oplus}{r} + G\frac{m\, m_\oplus}{r_\oplus}\right]_{r=r_\oplus} \cdot h + \ldots$$
$$= \left[G\frac{m\, m_\oplus}{r^2}\right]_{r=r_\oplus} \cdot h + \ldots$$
$$= G\frac{m\, m_\oplus}{r_\oplus^2} h + \ldots$$
$$= mgh + \ldots .$$

(11.16)

With the standard gravity g from Eq. 11.9 we obtain the well-known equation for potential energy:

$$E_{\text{pot}} = mgh.$$

(11.17)

If we do not want to make this approximation of small altitudes, it is usually more practical to choose a point infinitely far away from the mass M, which generates the gravitational field, as the zero point of potential energy. Then the potential energy is given by

$$E_{\text{pot}}(r) = -G\frac{m M}{r}.$$

(11.18)

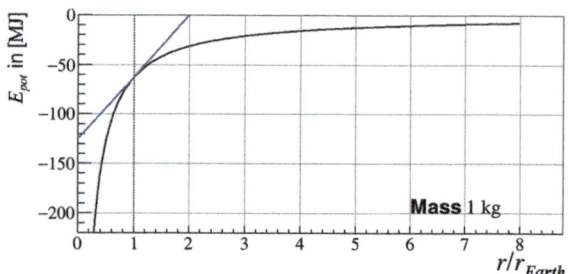

Fig. 11.11 Potential energy of a body with a mass of 1 kg in the gravitational field of the Earth. *Black*: exact curve; *blue*: approximation (C) NASA

As an example, Fig. 11.11 shows the potential energy of a body of mass 1 kg in the gravitational field of the Earth. The black curve (hyperbola) shows the exact shape. In comparison, the approximation $m\,g\,h$ is plotted in blue (straight line). They coincide at the height of the Earth's radius. The potential energy inside the Earth was determined under the assumption that its entire mass is concentrated at its center. In reality, the curve flattens below 1.

© NASA

Example 11.10: Escape Velocity

From conservation of energy, we can calculate the escape velocity from Earth, which is the speed at which a rocket must be launched so that it leaves the Earth's gravitational field without further propulsion.

At infinity, the rocket then has neither potential nor kinetic energy. Therefore, at the beginning of the journey, potential and kinetic energy must also add up to zero:

$$-G\frac{m\,m_\oplus}{r_\oplus} + \frac{1}{2}m v^2 = 0$$

$$\Rightarrow v = \sqrt{\frac{2\,G\,m_\oplus}{r_\oplus}} = \sqrt{2\,g\,r_\oplus} \approx 40\,000\,\text{km/h}\,.$$

The escape velocity is independent of the mass of the rocket (see also example 6.3).

11.5.2 The Potential and the Field Strength

The gravitational force that a body (test body) experiences in the sphere of influence of a planet, a sun, or another object, depends on the test body's mass. If we want to generally characterize the gravitational effect of a certain body, for example, a planet, the dependence on the mass of the test body is distracting. We introduce a new physical quantity that abstracts from the mass of the test body. This quantity is called "gravitational field" or "gravitational field strength" $\vec{G}$. It gives the gravitational force on a body with a unit mass of 1 kg as follows:

$$\vec{G} = \frac{\vec{F}}{m}. \tag{11.19}$$

In ◨ Fig. 11.12, a two-dimensional cut through the Earth's gravitational field can be seen. The field lines (green) indicate the direction of the gravitational field. The line density is a measure of the magnitude of the field at the respective location.

We can also abstract the potential energy from the mass of the body located in the gravitational field. The new quantity is called the "gravitational potential" Φ_G

$$\Phi_G = \frac{E_{\text{pot}}}{m}. \tag{11.20}$$

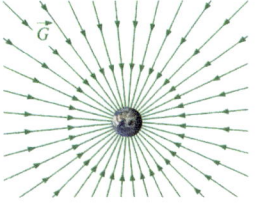

◨ **Fig. 11.12** The Earth's gravitational field

◨ Fig. 11.13 shows the gravitational potential of the Earth, which we have assumed to be a homogeneous sphere. Again, a two-dimensional cut through the center

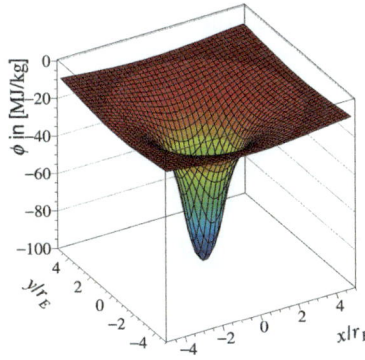

◨ **Fig. 11.13** The gravitational potential of the Earth. Shown is a cut through the center of the Earth. The distances are given in units of the Earth's radius

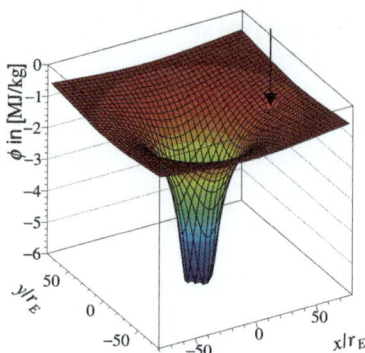

Fig. 11.14 The gravitational potential of the Earth and the Moon. The Moon is located at the point marked by the arrow. To make the depression of the potential by the Moon visible, the potential was cut off downwards at −6 MJ/kg

of the sphere is shown. The height of the surface (*z*-coordinate) indicates the value of the potential at the respective location. The colors refer to the value of the potential, too. The transition from one color to the next occurs at a fixed value of the potential. The boundaries represent locations of a constant potential. They are called "equipotential lines". As we have seen in ▶ Sect. 4.3, the force always points in the direction of the gradient of the surface, i.e., in the direction of the greatest slope. The steeper the inclination of the surface at a certain position, the larger the gravitational force at that location.

Just as potential energies can be added, the potential of several bodies is calculated as the sum of the individual potentials. In ◘ Fig. 11.14, the combined gravitational potential of the Earth and the Moon can be seen. The Earth is located at the center at $x, y = 0$. The coordinate system is chosen so that the Moon lies on the *x*-axis, at a distance of about 60 Earth radii from the Earth. The potential of the Moon is recognizable at this location as a small indentation. Note the change of scale of the *z*-axis compared to ◘ Fig. 11.13.

> **Example 11.11: Planetary Orbits Part 3: Equation of Motion**
>
> We now want to continue with Example 11.2 and try to calculate a planetary orbit. We start from the energy balance, which consists of kinetic energy and effective potential, and insert the gravitational energy:

11.5 · Potential and Potential Energy

$$E_{\text{tot}} = \frac{m}{2}\left(\frac{dr}{dt}\right)^2 + \frac{L^2}{2mr^2} - G\frac{mM}{r}.$$

It follows:

$$\frac{dr}{dt} = \pm\sqrt{\frac{2}{m}\left(E_{\text{tot}} + G\frac{mM}{r}\right) - \frac{L^2}{m^2 r^2}}.$$

Now it is:

$$d\varphi = \frac{d\varphi}{dt}dt = \frac{d\varphi}{dt}\frac{dt}{dr}dr = \frac{\frac{d\varphi}{dt}}{\frac{dr}{dt}}dr.$$

From example 11.2 we know that $\frac{d\varphi}{dt} = \frac{L}{mr^2}$, and for $\frac{dr}{dt}$ we insert the relation from above:

$$d\varphi = \frac{\frac{L}{mr^2}}{\pm\sqrt{\frac{2}{m}\left(E_{\text{tot}} + G\frac{mM}{r}\right) - \frac{L^2}{m^2 r^2}}}dr$$

and integrate:

$$\varphi(r) = \pm\int \frac{\frac{L}{mr^2}}{\sqrt{\frac{2}{m}\left(E_{\text{tot}} + G\frac{mM}{r}\right) - \frac{L^2}{m^2 r^2}}}dr$$

$$= \pm L \int \frac{1}{\sqrt{2mE_{\text{tot}} + 2G\frac{m^2 M}{r} - \frac{L^2}{r^2}}}\frac{dr}{r^2}.$$

We substitute $u = 1/r$ with $du = -1/r^2\, dr$ and we obtain:

$$\varphi(u) = \mp L \int \frac{1}{\sqrt{2mE_{\text{tot}} + 2Gm^2 M u - L^2 u^2}}du.$$

This integral can be found in tables. It is:

$$\int \frac{1}{\sqrt{au^2 + bu + c}}du$$

with
$$a = -L^2 < 0$$
$$b = 2Gm^2 M$$
$$c = 2mE_{\text{tot}}$$
$$\Delta = 4ac - b^2 < 0$$

solution: $-\frac{1}{\sqrt{-a}}\arcsin\left(\frac{2au + b}{\sqrt{-\Delta}}\right)$

We substitute and obtain:

$$\varphi(u) = \mp L\left(-\frac{1}{\sqrt{L^2}} \arccos\left(\frac{-2L^2 u + 2Gm^2 M}{\sqrt{8L^2 m E_{tot} + 4G^2 m^4 M^2}}\right)\right) + \varphi_0$$

$$= \pm \arccos\left(\frac{-L^2 u + Gm^2 M}{\sqrt{2L^2 m E_{tot} + G^2 m^4 M^2}}\right) + \varphi_0,$$

which we rearrange in trying to solve for u:

$$\cos(\varphi - \varphi_0) = \frac{-L^2 u + Gm^2 M}{\pm\sqrt{2L^2 m E_{tot} + G^2 m^4 M^2}}$$

$$= \frac{-\frac{L^2}{m} u + GmM}{\pm\sqrt{\frac{2L^2 E_{tot}}{m} + G^2 m^2 M^2}} = \frac{-\frac{L^2}{Gm^2 M} u + 1}{\pm\sqrt{\frac{2L^2 E_{tot}}{G^2 m^3 M^2} + 1}}.$$

We introduce two abbreviations, the meaning of which we will see later:

$$p = \frac{L^2}{Gm^2 M}$$

$$\epsilon = \sqrt{\frac{2L^2 E_{tot}}{G^2 m^3 M^2} + 1}.$$

Then it is:

$$\cos(\varphi - \varphi_0) = \frac{-pu + 1}{\pm \epsilon}$$

$$\pm \epsilon \cos(\varphi - \varphi_0) = -pu + 1$$

$$r(\varphi) = \frac{1}{u} = \frac{p}{1 \mp \epsilon \cos(\varphi - \varphi_0)}.$$

This is the equation of an ellipse, as we already know from the discussion at the end of ▶ Sect. 11.1. The quantities $p = b^2/a$ and $\epsilon = \sqrt{a^2 - b^2}/a$ are the semi-latus rectum and the numerical eccentricity of the ellipse (a, b: major and minor semi-axis). As a rule, we choose the negative sign. The positive sign describes a 90° rotated ellipse, whose major semi-axis points along the y-axis.

11.5 · Potential and Potential Energy

Example 11.12: Conic Sections

Already in ▶ Sect. 11.1 we mentioned that movements under the influence of a central force lead to conic sections as paths. In example 11.5 we implicitly proved this. We calculated that the path can be represented by

$$r(\varphi) = \frac{p}{1 - \epsilon \cos \varphi}.$$

The parameter ϵ determines the type of conic section. We obtain

- for $\epsilon = 0$ a circular orbit,
- for $0 < \epsilon < 1$ the Kepler ellipses,
- for $\epsilon = 1$ a parabolic path,
- and finally for $\epsilon > 0$ hyperbolic paths.

Only the first two cases describe closed orbits, where the celestial body is bound to the central star. In the case of $\epsilon \geq 1$ the object approaches the central star from infinity, is deflected from its straight path by it, and moves away again. Which of the four possible paths is established depends on the initial conditions, which we define by the total energy E_{tot} and the orbital angular momentum L.

We had determined the parameter ϵ as

$$\epsilon = \sqrt{\frac{2 L^2 E_{\text{tot}}}{G^2 m^3 M^2} + 1}.$$

A parabolic path results in case of $E_{\text{tot}} = 0$. The celestial body starts its path with vanishing speed at infinity, passes through the parabolic path, and approaches zero speed again at infinity.

A circular path results for $\epsilon = 0$, i.e. for

$$\frac{2 L^2 E_{\text{tot}}}{G^2 m^3 M^2} = -1 \quad \Rightarrow \quad E_{\text{tot}} = -\frac{1}{2} \frac{G^2 m^3 M^2}{L^2}.$$

This corresponds to the minimum of the effective potential (see example 11.1), which you can easily verify by calculating the derivative of the curve. At this energy, $r = \text{const.}$, so that $dr/dt = 0$ and the effective potential already makes up the entire energy. Kepler ellipses result for values of the total energy between the minimum (circular orbit) and zero (parabola). Hyperbolic paths belong to movements with $E_{\text{tot}} > 0$.

Problems

1. To appear stationary above the reception area, communication satellites fly on geostationary orbits, where they orbit the Earth at the angular velocity of the Earth's rotation. Calculate the height above the Earth's surface at which such a satellite is located. A satellite has lost altitude and thus its position in the sky shifts by $0.3°$ within a year. By how much must it be raised again using the control nozzles?

2. From Kepler's third law, calculate the length of the major semi-axis of the orbit of Halley's comet in astronomical units (AU). The comet reaches its perihelion every 76 years and has a minimum distance to the Sun of $0.6\,\text{AU}$ (The astronomical unit AU denotes the average distance between the Sun and the Earth). What maximum distance to the Sun will it reach on its orbit?

3. An astronaut explores an asteroid on a path from pole to pole and notices that he becomes lighter with increasing speed. Starting at a speed of $1.5\,\text{m/s}$, he begins to float and can orbit the asteroid like a satellite. What is the diameter of the asteroid, assuming it has an average density like that of Earth ($\rho = 5.5 \cdot 10^3 \,\text{kg/m}^3$)? The asteroid also rotates about its own axis. At which rotation frequency would the astronaut feel weightless when standing at the equator? How long is an asteroid day then? Can celestial bodies with Earth density exist that have even shorter days?

4. Earth and Moon orbit their common center of mass once every 27.3 days. The mass of the Earth is 81 times the mass of the Moon. How far apart are the centers of the Earth and the Moon, and how far is the center of the Moon's orbit from the center of the Earth?

5. To send a spacecraft to one of the outer planets, one should take advantage of the Earth's motion around the Sun. A spacecraft is therefore sent tangentially to the Earth's orbit and brought onto an elliptical orbit around the Sun, which has its perihelion on the Earth's orbit and its aphelion on the orbit of the target planet. It can be shown that the total energy of the probe on its orbit is given by

$$E_{\text{tot}} = -G\frac{M\,m}{2\,a}.$$

11.5 · Potential and Potential Energy

Proof that the speed of the probe depends on its distance r from the Sun as:

$$v = v_F \sqrt{r_E \left(\frac{1}{r} - \frac{1}{2a} \right)}$$

v_F : escape velocity at the Earth's orbit
r_E : radius of the Earth's orbit
a : major semi-axis of the spacecraft's orbit

What is the required launch speed relative to Earth to bring the spacecraft onto an elliptical path to Jupiter ($r_J = 5.3 r_E$)? How big is the advantage compared to the direct path?

The Rigid Body

Contents

Chapter 12   The Rigid Body – 253

Chapter 13   Rotational Movements – 283

Statics of Rigid Bodies

Contents

12.1 Definition – 254

12.2 The Torque – 256

12.3 The Center of Gravity of a Body – 261

12.4 The Theorems of Statics – 264

12.5 Statics of Rigid Bodies – 268

© The Author(s), under exclusive license to Springer-Verlag GmbH, DE, part of Springer Nature 2025
S. Roth and A. Stahl, *Mechanics*,
https://doi.org/10.1007/978-3-662-68079-7_12

12.1 Definition

In our previous considerations, we have simplistically described bodies as point masses. A point mass is
- point-like, that is, it has no extension,
- structureless, that is, it has no internal structure.

Now, we want to abandon this simplification and approach a more realistic description of objects in nature. A rigid body is understood to be an extended body that cannot be deformed, not even by external forces. Microscopically, this means that the positions of the constituents (atoms) that make up the body are fixed relative to each other. They can only be moved through space together.

A rigid body is therefore
- extended, that is, it has a surface, a shape, and a volume,
- structured, that is, it has a density and a composition,
- rigid, that is, shape and structure are unchangeable.

Just like the point mass, the rigid body is also an idealization. There are no truly rigid bodies in nature. The positions of the atoms that make up the body are not immovable. If forces act on the body, the atoms will shift and the body will deform. Here we want to make the approximation that these deformations are negligible. We will revisit them in ▶ Chap. 14.

We have already learned in ▶ Chap. 6 that mass is an important property that describes the inertia of a body. For extended bodies, density and volume determine their mass. If it is homogeneous, which means that the same density prevails everywhere in the body, there is a simple relationship between the quantities mass m, volume V and density ρ, namely:

$$m = \rho V. \tag{12.1}$$

The density is a property of the material from which the body is made. In ◘ Table 12.1 we have compiled some values for you.

If the density varies within the body, we must use an integral to determine the mass. Then it is:

$$m = \int_V \rho(\vec{r})\, d^3\vec{r}. \tag{12.2}$$

12.1 · Definition

Table 12.1 Density of some materials

High vacuum (10^{-5} Pa)	1.3 mg/m^3
Hydrogen (normal pressure)	0.0899 kg/m^3
Air (normal conditions)	1.29 kg/m^3
Aerogel (typical)	20 kg/m^3
Lithium	0.534 kg/dm^3
Ethanol	0.789 kg/dm^3
Heating oil	0.85 kg/dm^3
PE (Polyethylene)	0.95 kg/dm^3
Water	1 kg/dm^3
Aluminum	2.70 kg/dm^3
Steel	7.86 kg/dm^3
Lead	11.34 kg/dm^3
Mercury	13.55 kg/dm^3
Tungsten	19.25 kg/dm^3
Neutron star (in the core)	10^{15} kg/dm^3

We must integrate over the whole volume of the body, indicated by the letter "V" at the bottom right of the integration symbol. Conversely, for the density we have:

$$\rho(\vec{r}) = \frac{d\,m(\vec{r})}{dV}, \qquad (12.3)$$

where $dm(\vec{r})$ represents an infinitesimal volume with mass dm at location $\vec{r}$ and dV denotes its volume.

Example 12.1: Rigid Bodies

The images show examples of bodies which can be treated as rigid bodies.

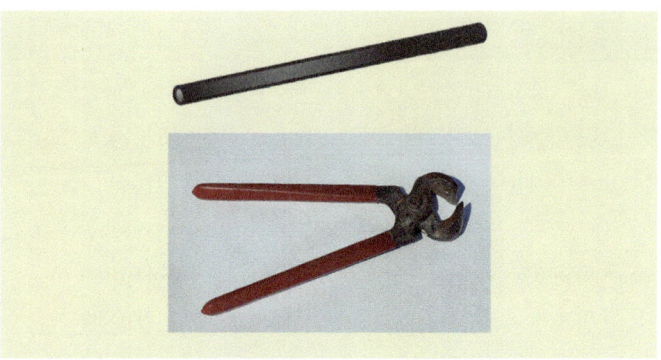

A point mass performs relatively simple movements. It has three independent possibilities of motion, namely motion in the three spatial directions x, y, and z. These motions are called "translations". A rigid body has three more independent motions. A rigid body can turn in motions called "rotations". Again, there are three independent possibilities, namely the rotations around the three coordinate axes. Rotations around other axes can be represented by these and possibly additional translations.

The independent motions of a body are called its "degrees of freedom". A point mass has three degrees of freedom, while a rigid body has six, namely three translations and three rotations.

12.2 The Torque

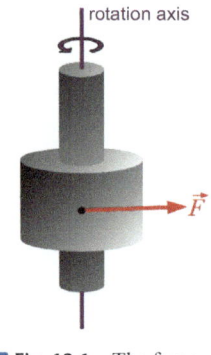

Fig. 12.1 The force causes a torque

We consider a rigid body (■ Fig. 12.1), which is mounted on an axis that allows it to rotate. A force is applied to the body. It tries to rotate the body. This is referred to as a torque.

The connecting line from the pivot point (fulcrum) to the working point of the force is called the lever arm vector. The rotational effect of the force depends on the length of the lever arm vector and the angle at which the force is applied. Relevant is the projection of the lever arm vector onto a line perpendicular to the force and perpendicular to the axis of rotation (■ Fig. 12.2). This projection is called the lever arm. Please recognize, this is the minimal distance between the axis of rotation to the line of action of the force, it is not necessarily the length of the lever arm vector.

12.2 · The Torque

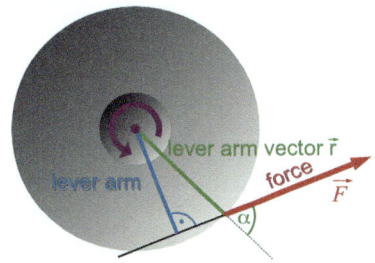

Fig. 12.2 Definition of lever arm and lever arm vector

It is intuitively clear that the rotational effect on the body is proportional to the force $\vec{F}$ and the lever arm vector $\vec{r}$ and also depends on the angle between the lever arm vector and the force as well as the orientation of the axis of rotation. The rotational effect can be expressed by the lever arm l and the force, it is proportional to $lF_\perp$, if $F_\perp$ is the component of the force perpendicular to the axis of rotation. This quantity is called the torque or the moment of force. If you want to avoid the complicated projections, you may express the torque through a cross product. You need to refer to the lever arm vector. The torque $\vec{M}$ is defined as:

$$\vec{M} = \vec{r} \times \vec{F}, \tag{12.4}$$

where $\vec{r}$ is the vector from the axis of rotation to the working point of the force (the lever arm vector) and $\vec{F}$ is the applied force. The torque is an axial vector. The direction of the torque defines the axis of rotation. It generates a rotation in the direction of a right-handed screw (see Appendix A3.16). The magnitude of the torque is

$$|\vec{M}| = |\vec{r}| |\vec{F}| \sin \alpha, \tag{12.5}$$

with the angle α between the lever arm vector and the force.

From the definition, one can read off the unit of torque. It is:

$$[\vec{M}] = 1 \, \text{Nm} = 1 \, \frac{\text{kg m}^2}{\text{s}^2} \tag{12.6}$$

Although it would be mathematically correct to use Joule instead of Nm, this is not common. Please realize, that torque is not a measure of energy.

Example 12.2: Wrench

The wrench generates a torque, which opens or closes the screw.

Example 12.3: Torques in a Curve

When a car drives at high speed through a curve, torques act on the vehicle that can cause it to tip over. The pivot point is the tire on the outside track (orange point in the illustration). If the car tips over, it will rotate around this point. The most important forces are the weight and the centrifugal force, both of which are applied at the center of gravity of the vehicle. The torque $\vec{M}_G$, which is generated by the weight $\vec{F}_G$, is pointing into the plane of the drawing, while the torque $\vec{M}_Z$, which is caused by the centrifugal force $\vec{F}_{ZF}$, points out of the plane. As soon as $\vec{M}_Z$ becomes greater in magnitude than $\vec{M}_G$, the car starts to tip over.

Let's look at an example, a car with a track gauge of $s = 1.5$ m and a mass of $m = 1.6$ t. The lever arm is $l_G = \frac{s}{2} \cos \beta$ and the torque created by the weight is $M_G = mg\frac{s}{2} \cos \beta$. At a tilt angle $\beta = 30°$ this would result in 9558 Nm. The torque M_Z depends on the speed of the car and the radius of the curve. Try to derive the formula yourself! You will see that the vehicle becomes more stable the wider the track gauge and the lower the center of gravity is above the road.

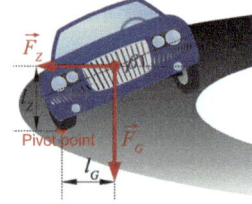

12.2 · The Torque

Example 12.4: The Crankshaft of an Engine

In a piston engine, the connecting rod (length r_1) generates a torque on the crankshaft (lever arm vector $\vec{r}_2$) and thus converts the reciprocating motion of the piston into a rotational motion.

The torque is $\vec{M} = \vec{r}_2 \times \vec{F}$. If F_2 is the vertical component of the force on the crank, then the magnitude of the torque is $M = \sin\varphi \, r_2 F_2$, with the angle φ specifiying the rotation of the crank from its highest position. The forces are derived in the other sketch. If α is the angle of the connecting rod to the vertical, then $F_2 = \cos\alpha \, F_1$ and $F_1 = \cos\alpha \, F_0$. On the other hand, from the two triangles sharing the lever arm ℓ, it follows that $\ell = \sin\varphi \, r_2$ and $\ell = \sin\alpha \, r_1$ and thus $\sin\alpha = r_2/r_1 \sin\varphi$. Inserting all the relations, we get a torque of

$$M = \left(1 - \frac{r_2^2}{r_1^2}\sin^2\varphi\right)\sin\varphi \, r_2 F_0.$$

The graph shows this torque assuming a constant downward force F_0 from the piston.

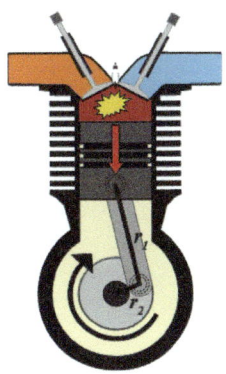

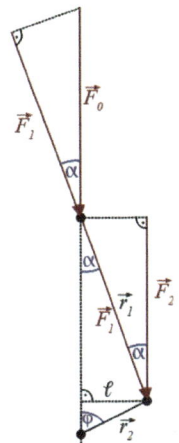

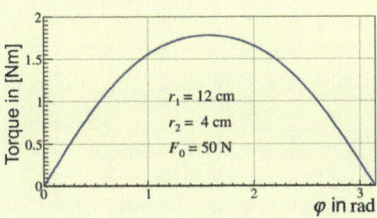

Experiment 12.1: Torque Disk

This experiment demonstrates the balance between torques. A plexiglass disk is mounted on a horizontal axis (photo). Pins are mounted on the disk at equal radial intervals, on which weights can be hung. Each weight causes a torque about the axis. If the torques on the right and left are equal, the disk remains stationary. The photo shows an example. Here, the torque on the left is $M_{\text{left}} = m \cdot 3r + m \cdot r$ and on the right $M_{\text{right}} = 4m \cdot r$.

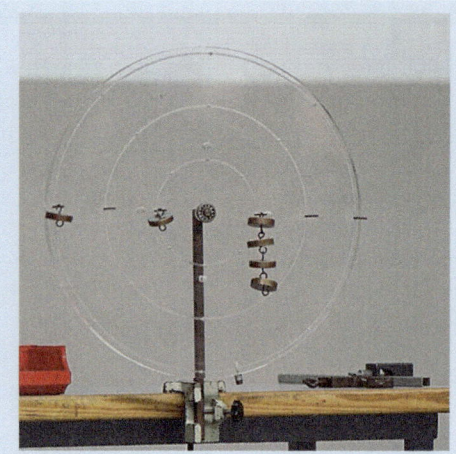

© Photo: Hendrik Brixius

Experiment 12.2: Torque on the Spool of Thread

This is a simple yet interesting experiment. You can try it out yourself. All you need is a spool of thread. In the lecture, we use an oversized 50-cm spool. But it also works with a real one. You unwind a few centimeters of the thread and pull gently on it so that the spool does not slip on the surface. In which direction will it roll?

The answer is not that simple. It depends on how flat you pull. The two sketches show the situation. If you pull very flat, the spool of thread will roll towards you. The lever arm vector points upwards and the torque thus into the plane of the sketch. However, if you pull steeply upwards, the spool rotates in the opposite direction. Now the torque points out of the plane. Note that the pivot point is not given by the axis of the spool, it is rather the point where the spool touches the surface.

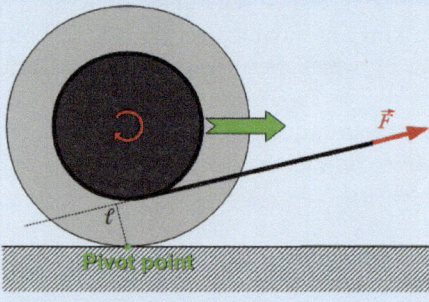

12.3 · The Center of Gravity of a Body

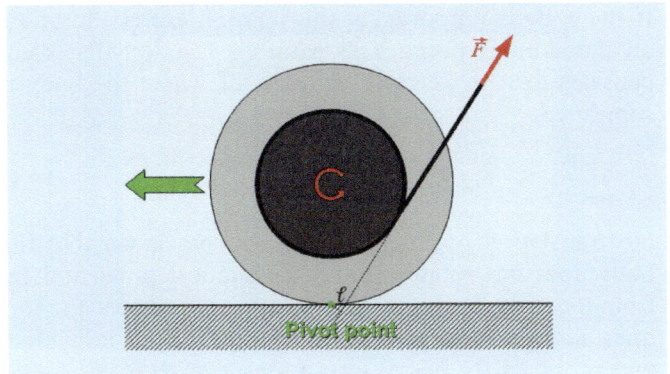

12.3 The Center of Gravity of a Body

In this section, we want to introduce an important concept, that of the center of gravity. We define it as the single point at which one must support a body so that it remains at rest (see ◘ Fig. 12.3) i.e. it neither performs a translation nor a rotation. In Experiment 12.3 we show how to determine the center of gravity of a body experimentally.

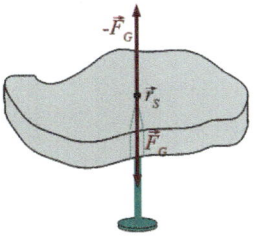

◘ **Fig. 12.3** An irregularly shaped plate supported at the center of gravity. It remains at rest

> **Experiment 12.3: Determination of the Center of Gravity of Plates**
>
> The center of gravity of flat plates can be determined quite easily. We suspend the plate from an arbitrary point and let it find its rest position. At the same point, we attach a plumb line. Now, with a pen, we mark the plumb line on the plate. Then, we repeat the procedure from a few more suspension points. The different lines determined in this way will intersect at a single point. This is the center of gravity of the plate.

We suspect that the center of gravity $\vec{r}_{cg}$ coincides with the center of mass $\vec{r}_{cm}$ which we learned about in ▶ Sect. 8.2. To recognize this, we must show that a body supported at this point neither falls (no translation) nor tips around the support point (no rotation). The center of mass is

© Photo: Hendrik Brixius

$$\vec{r}_{cm} = \frac{\sum_i m_i \vec{r}_i}{\sum_i m_i} = \frac{1}{M} \sum_i m_i \vec{r}_i . \qquad (12.7)$$

If the body is supported at this point, it exerts a force $\vec{F}_G$ on the support point. This causes a reaction that compensates for the weight, regardless of where the body is supported:

$$\sum \vec{F}_{\text{ext}} = \vec{F}_G + \vec{F}_{\text{reactio}} = 0. \tag{12.8}$$

No translation occurs. It remains to be shown that the body does not rotate around the support point and tip from the base. To do so, we show that the sum of the torques about the support point vanishes. Each individual mass point of the body exerts a torque $\vec{M}_i$. It is:

$$\begin{aligned}\vec{M}_i &= (\vec{r}_i - \vec{r}_{cm}) \times m_i \vec{g} \\ &= \vec{r}_i \times m_i \vec{g} - \vec{r}_{cm} \times m_i \vec{g} \\ &= -m_i g (\vec{r}_i \times \hat{e}_z) + m_i g (\vec{r}_{cm} \times \hat{e}_z).\end{aligned} \tag{12.9}$$

The total torque is then:

$$\begin{aligned}\vec{M} &= \sum_i \vec{M}_i = -\sum_i m_i g (\vec{r}_i \times \hat{e}_z) + \sum_i m_i g (\vec{r}_{cm} \times \hat{e}_z) \\ &= -g \sum_i (m_i (\vec{r}_i \times \hat{e}_z)) + g \sum_i (m_i) (\vec{r}_{cm} \times \hat{e}_z) \\ &= -g \left(\sum_i m_i \vec{r}_i\right) \times \hat{e}_z + g M (\vec{r}_{MM} \times \hat{e}_z) \\ &= -g M \vec{r}_{cm} \times \hat{e}_z + g M (\vec{r}_{cm} \times \hat{e}_z) = 0.\end{aligned} \tag{12.10}$$

With this, we have shown that a body supported at the center of mass neither performs a translation nor a rotation. The center of mass therefore coincides with the center of gravity. From now on, we will only use the term "center of gravity" and write $\vec{r}_{cm}$.

For bodies with a continuous mass distribution, the sums must be replaced by integrals:

$$\vec{r}_{cm} = \vec{r}_{cg} = \frac{\int_V \vec{r}\, dm}{\int_V dm} = \frac{1}{M} \int_V \rho(\vec{r}) \vec{r}\, dV. \tag{12.11}$$

12.3 · The Center of Gravity of a Body

Example 12.5: Center of Gravity

Here are some examples of extended bodies. Assuming constant denisity, the center of gravity of the ellipsoid is in its center. The center of gravity of the dumbbell is in the middle of the bar. The octagonal object also has its center of gravity in the middle, which shows that the center of gravity does not necessarily belong to the body.

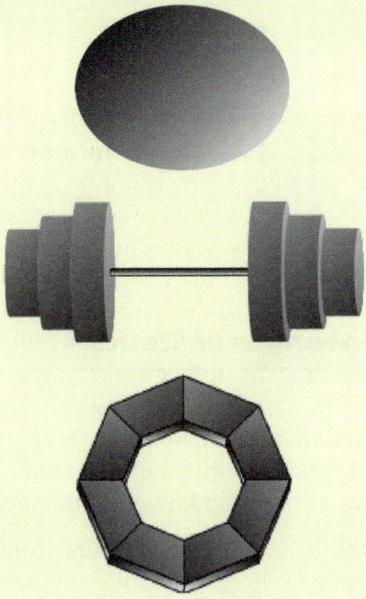

Example 12.6: Center of Gravity of a Hemisphere

We want to determine the center of gravity of a homogeneous hemisphere. For reasons of symmetry, the center of gravity must be on the symmetry axis, which we define as the z-axis. It is therefore sufficient to determine the z-component. The x- and y-components vanish. The z-component is determined from (see mathematical appendix A3.19):

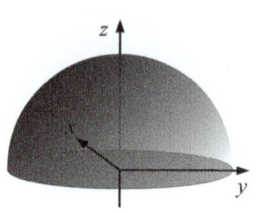

$$(\vec{r}_{cm})_z = \frac{1}{M}\int_V \rho_0 z\, dV = \frac{1}{M}\rho_0 \int_0^R \int_0^{\frac{\pi}{2}} \int_0^{2\pi} r\cos\theta\, r^2 \sin\theta\, d\varphi\, d\theta\, dr$$

$$= \frac{2\pi}{M}\rho_0 \int_0^R \int_0^{\frac{\pi}{2}} r^3 \cos\theta \sin\theta\, d\theta\, dr$$

$$= \frac{2\pi}{M}\rho_0 \int_0^R r^3 \left[-\frac{1}{2}\cos^2\theta\right]_0^{\frac{\pi}{2}} dr = \frac{\pi}{M}\rho_0 \int_0^R r^3\, dr$$

$$= \frac{\pi}{M}\rho_0 \left[\frac{1}{4}r^4\right]_0^R = \frac{\pi}{4M}\rho_0 R^4 .$$

Now, the mass of the hemisphere is $M = \frac{2\pi}{3}R^3 \rho_0$ and thus $(\vec{r}_{cm})_z = \frac{3}{8}R$:

$$\vec{r}_S = \left(0, 0, \frac{3}{8}R\right) .$$

12.4 The Theorems of Statics

Statics deals with the conditions under which bodies are at rest. We have already learned that this is the case when the sum of all external forces and the sum of all external torques is zero. This is also known as the theorem of statics

> **Theorem of Statics (1)**
> A body remains at rest if the sum of all external forces and the sum of all external torques vanish.

For further investigation of the statics of a body, another theorem is often used, which is occasionally also referred to as the theorem of statics. Unfortunately, the convention is ambiguous.

> **Theorem of Statics (2)**
> The effect of all force pairs acting on a body in a plane can be replaced by a single force pair with the center of gravity as their center.

It should not be surprising that several force pairs can be combined into a single one. We have repeatedly added forces. What is surprising is that the center of gravity is exactly in the middle between the contact points of the two forces, as shown in the sketch ◘ Fig. 12.4.

12.4 · The Theorems of Statics

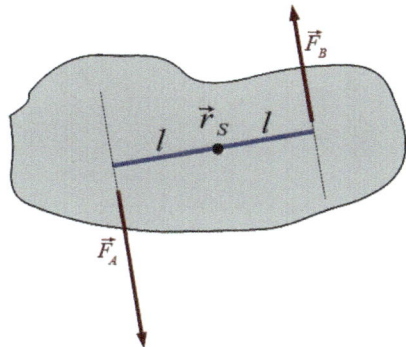

Fig. 12.4 To illustrate the theorem of statics (2)

It is the task of the structural engineers to design a mount for the body that absorbs the two forces and their torque. Please note that these considerations refer to a single plane. Other planes must be considered separately.

We want to take a closer look at the construction of the second theorem. From ■ Fig. 12.4, it is easy to see that the definition of the forces is not unique. For example, one could change the lever arm l as long as the forces are adjusted accordingly. Similarly, one could move the contact points of the forces along their lines of action. This would not change the situation either. The pair of forces described in the sentence is not uniquely defined. The sentence merely states that at least one such pair of forces can be found. This pair is not any arbitrary combination of two forces. The two forces always act along parallel lines of action, which must not coincide.

The second theorem is based on the realization that any motion can be represented by a rotation around the center of gravity and a translation of the body. From the center of mass theorem, we already know that the translational effect of all external forces on a body can be summarized by a single force on the center of gravity. This net force does not exert any torque on the body. In ■ Fig. 12.5, we have chosen the lines of action of the forces to indicate the direction of the net force. The magnitude of the net force is $F_A - F_B$. In this example, it points in the direction of $\vec{F}_A$. For the treatment of rotations, we want to subtract it from the two forces, so that in the following we have $\vec{F}_A = -\vec{F}_B = \vec{F}$.

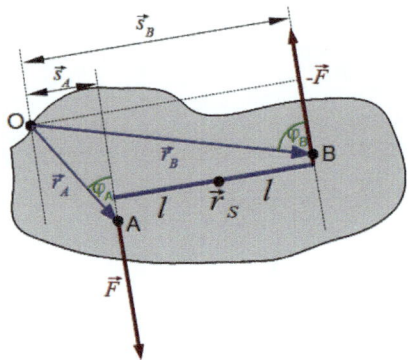

Fig. 12.5 For the parallel displacement of pairs of forces (see text)

Just like the translational effect of the external forces, the rotational effect can be summarized by adding the torques exerted by the individual forces. However, a pivot point must be defined, and it is surprising that it can be shifted to the center of gravity. Let's first assume that the pivot point is an arbitrary point O (Fig. 12.5). The two forces are $\vec{F}$ and $-\vec{F}$. The torque with respect to O is:

$$\left|\vec{M}\right| = \left|\vec{r}_B \times \vec{F} - \vec{r}_A \times \vec{F}\right| = r_B F \sin\varphi_B - r_A F \sin\varphi_A$$
$$= (r_B \sin\varphi_B - r_A \sin\varphi_A) F. \tag{12.12}$$

As can be seen from Fig. 12.5, the bracket can be expressed by the lever arms $\vec{s}_A$ and $\vec{s}_B$:

$$\left|\vec{M}\right| = (s_B - s_A) F = 2lF. \tag{12.13}$$

However, this is precisely the torque that the two forces exert with respect to the center of gravity. We therefore come to the conclusion that the torque generated by a pair of forces with pure rotational effect (equal forces in opposite direction) can be shifted parallel without changing.

> **Example 12.7: Rolling on a Plane**
>
> You pull horizontally on the axis of a wheel. What torque acts on the axis? We first separate the translational motion. The horizontal net force is the difference between the applied force $\vec{F}_0$ and the frictional force $\vec{F}_{fr}$. It causes an acceleration of the wheel. The remaining

12.4 · The Theorems of Statics

part causes a torque around the contact point of the wheel. This is equal to $F_{fr}R$, where R is the radius of the wheel. We can move this parallel to the axis, so it also applies to the axis $M = F_{fr}R$.

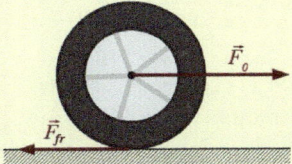

Example 12.8: Rolling of a Wheel, Part 2

We follow the example 12.7 a bit further. In sketch A, the rolling of a wheel on an inclined plane is depicted. Three forces act on the wheel. First, there is the gravitational force $\vec{F}_G$, which acts at the center of gravity of the wheel. It can be broken down into the normal force and the downhill force. The normal force acts perpendicular to the surface and generates the second force, a reaction of the surface on the wheel at the point of contact (red point in the sketch). We have denoted it with $\vec{F}_{reac}$. In addition, there is a frictional force between the wheel and the surface, which prevents the wheel from sliding instead of rolling. It must have the same magnitude as the downhill force, pointing in the opposite direction, so that all forces acting at the point of contact add up to zero and it remains at rest. We have denoted the frictional force in the sketch with $\vec{F}_{fr}$.

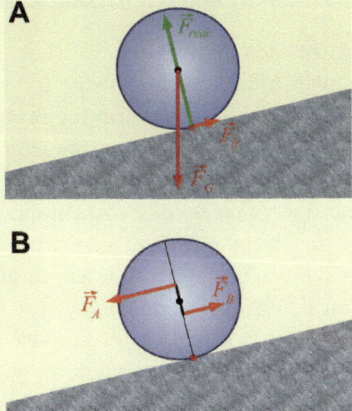

Hence, there are two forces that act at the point of contact ($\vec{F}_{reac}$ and $\vec{F}_{fr}$), and one that acts at the center of gravity ($\vec{F}_G$). According to the second theorem of statics, we can replace these with a pair of forces that act symmetrically at the center of gravity. We have drawn it in sketch B. The two forces are denoted by $\vec{F}_A$ and $\vec{F}_B$. The difference $\vec{F}_A - \vec{F}_B$ corresponds to the downhill-slope force. It accelerates the wheel down the slope (translational motion). In addition, two forces act with a lever arm s at the center of gravity and cause a torque. We have chosen a quarter of the radius r of the wheel as the lever arm and it was rotated so that the two forces are perpendicular to it. They cause a torque of the size $\frac{1}{4} r \left(\vec{F}_A + \vec{F}_B \right)$. This torque accelerates the rotation of the wheel, allowing it to roll down the slope without slipping.

12.5 Statics of Rigid Bodies

As we have seen, a body is at rest when the sum of all external forces and the sum of all external torques vanish.

We first consider a simple example (◘ Fig. 12.6). The cable car is supported by the cable. Torques do not occur here. The weight is compensated by the two forces $\vec{F}_1$ and $\vec{F}_2$ along the cable, which can be determined graphically from the sketch. They keep the cable car at rest.

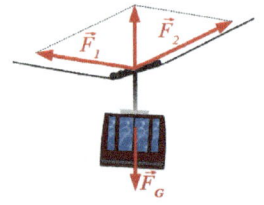

◘ **Fig. 12.6** Forces on a cable car

Example 12.9: Golden Gate Bridge

The roadway of the Golden-Gate Bridge is supported by steel cables, which are attached at the top to two main support cables. Due to the load on the bridge, the main cables sag. By which mathematical function can the sag be described?

We consider the main span between the two pillars. We neglect the weight of the vertical cables, resulting in a constant weight λ per unit length of the bridge. The problem is obviously symmetrical around the central point O, so it is sufficient to consider one half of the bridge. In the cable, the tensile force $\vec{F}_t$ acts, which is compensated in the middle by the tensile force of the other half $\vec{F}_0$. At every point of the bridge, there must

12.5 · Statics of Rigid Bodies

be a balance of forces both in the horizontal and vertical direction. We choose a coordinate system with origin in O. The x-axis points to the right pillar, the y-axis upwards.

If we sum over all elements from the middle of the bridge to an arbitrary point x, it must hold true that

$$\sum F_x = F_t(x) \cos \theta - F_0 = 0$$
$$\sum F_y = F_t(x) \sin \theta - \lambda x = 0.$$

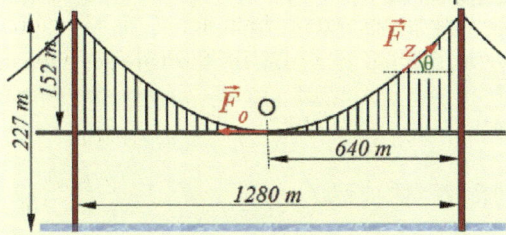

We divide the second equation by the first and obtain

$$\tan \theta = \frac{dy}{dx} = \frac{\lambda x}{F_0},$$

where $\tan \theta$ indicates the slope of the cable at point x. We separate the variables and integrate:

$$\int dy = \int \frac{\lambda x}{F_0} dx \Rightarrow y(x) = \frac{1}{2} \frac{\lambda}{F_0} x^2 + x_0.$$

This results in a parabolic sag of the main cables.

We have now discussed some examples where forces occurred, but no torques. Now we want to deal with torques.

Consider the truck in ◘ Fig. 12.7. We want to calculate the axle loads. The tractor has a weight of 2 t, the trailer including load 12 t. For the tractor, the weight is unevenly distributed. We can approximately assume that the entire weight of the tractor rests on its front axle, while the trailer only loads the rear axle. Then the frontmost axle is loaded with 2 t. The weight of the trailer is distributed on the rear axle of the tractor (A) and on the triple axle of the trailer, which we represent by a single support point at B. For the trailer, it must hold:

$$\vec{F}_G = \vec{F}_A + \vec{F}_B. \tag{12.14}$$

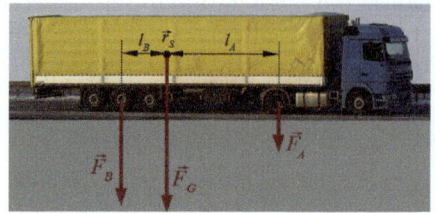

Fig. 12.7 Distribution of the load on the axles of a truck

These forces are compensated by the reaction of the road to the pressure of the tires. But this is not enough to determine the axle loads. We must also consider that the trailer only rests when the sum of the torques vanishes. The weight on the center of gravity causes the two torques $\vec{M}_A$ and $\vec{M}_B$. It must hold:

$$|\vec{M}_A| - |\vec{M}_B| = l_A F_A - l_B F_B = 0 \Rightarrow F_B = \frac{l_A}{l_B} F_A . \quad (12.15)$$

From the two relations, we obtain[1]

$$\begin{aligned} F_A &= \frac{l_B}{l_A + l_B} F_G = 3.4\,\text{t} \\ F_B &= \frac{l_A}{l_A + l_B} F_G = 8.6\,\text{t} . \end{aligned} \quad (12.16)$$

Since the load at B is distributed over three axles, we have per axle
- Front axle of the tractor: approx. 2.0 t,
- Rear axle of the tractor: approx. 3.4 t,
- Trailer per axle: approx. 2.9 t.

Example 12.10: On a Seesaw

Dad is seesawing with his two children, Eve and Max. Where does he need to sit so that the seesaw balances (dimensions and weights see sketch)?
The torques that the two children generate with their weight must be compensated by the father's torque. It must be:

$19\,\text{kg} \cdot 2.8\,\text{m} + 27\,\text{kg} \cdot 2.1\,\text{m} = 69\,\text{kg} \cdot x.$

---

[1] Here we give weight in tons, as is customary in language use. It would be correct to multiply the result by g, which leads to forces of 34 kN and 86 kN.

12.5 · Statics of Rigid Bodies

The result is $x = 1.59$ m. The father must sit at a distance of 1.59 m from the pivot point. By the way, the balance is independent of the inclination of the seesaw. Show it!

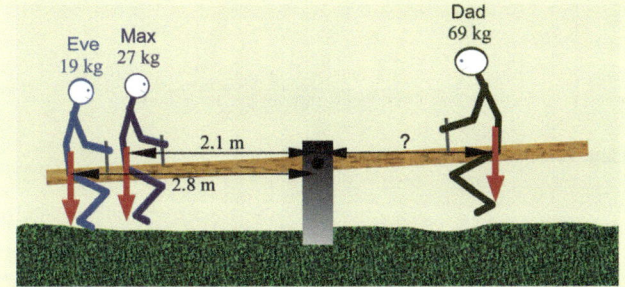

Example 12.11: Jumping off the Diving Board

A diver runs up and jumps off the tip of the board with full force. Let's assume that he achieves an acceleration of $a = 4g$ at takeoff. What forces must be absorbed at the two support points of the diving board?
If we neglect the weight of the diving board, then we have (see sketch):

Balance of forces: $\quad F_1 - F_2 = F_0$
Balance of torques: $\quad l_2 F_2 = l_0 F_0$.

Here, l_0 is the distance of the diver from the right support of the board and l_2 is the distance from the left support to the right. We obtain:

$$F_1 = \left(1 + \frac{l_0}{l_2}\right) F_0 \quad \text{and} \quad F_2 = \frac{l_0}{l_2} F_0.$$

We had assumed a force F_0 of $4mg$, where m is the mass of the diver. The force on the right support can reach values up to 10 kN, which corresponds to a weight of one ton. The force on the left support is indeed lower, but note that it points upwards.

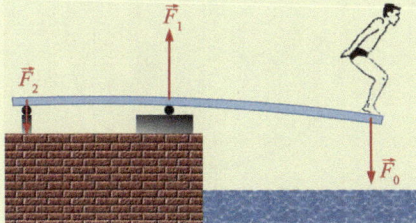

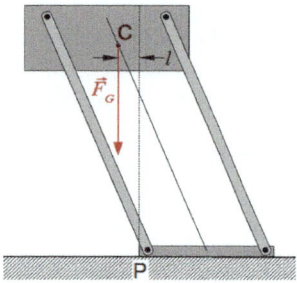

Fig. 12.8 Does the body fall over?

An important application is the statics of bodies resting on a surface. It is about the question of whether a body will remain stable or fall over. As an example, we consider the body in ◘ Fig. 12.8. It is a kind of table with legs attached at a fixed, oblique angle. The point C marks the center of gravity of the table. The weight acts on it. The vertical, dashed line marks the boundary P of the base of the table. If the center of gravity is outside of the base, a torque is created with respect to the point P. The body begins to rotate to the left. It falls over. If the center of gravity were above the base, the torque would point in the opposite direction. It would be compensated by the base. The body would remain stable.

This applies generally to all bodies. As long as the center of gravity is above the footprint, the body remains stable. However, this condition must met not only for one sectional plane through the body, but for all of them, because if the body tips in only one direction, it is already unstable. ◘ Fig. 12.9 shows a three-legged table from above. The three feet are indicated as dark squares. The dashed triangle marks the footprint. On the table is

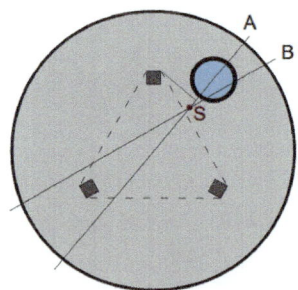

Fig. 12.9 Stability of a three-legged table

12.5 · Statics of Rigid Bodies

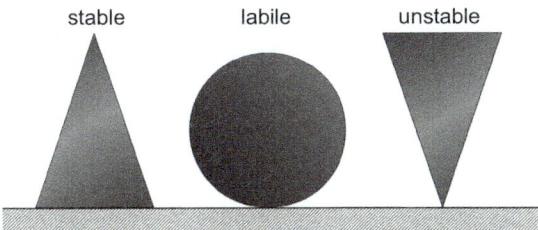

Fig. 12.10 Various equilibrium positions

a heavy object (light blue) that shifts the center of gravity S to the side. In the sectional plane through A, the center of gravity is still within the projected footprint, but not in the sectional plane through B. The table will tip over in the direction of B.

In reality, not every body that has its center of gravity above the footprint is stable. Consider the three examples in Fig. 12.10. In all three cases, the center of gravity is above the base, so the bodies should remain stable. But what happens if the body experiences even a minimal disturbance, e.g., a tiny force to the side? This will have no effect on the body on the left. It is in a stable equilibrium position. But not the body on the right. The disturbance would shift the center of gravity, so that the weight force now exerts a small torque. The body tilts a little, which increases the lever arm of the weight. The torque increases and the disturbance intensifies, causing the body to fall over. This is called an unstable equilibrium. The original equilibrium is unstable against minimal disturbances. The equilibrium of the body in the middle is also not stable against disturbances. However, there is no amplification of the disturbance here. If the body is disturbed, it rolls a little and ends up in a new equilibrium position.

We want to calculate another example (Fig. 12.11). How steep do we have to set up the ladder so that it does not slip? For simplicity, we assume that friction only occurs on the ground, not on the wall.

Here too, we must ensure the balance of forces and torques. The balance of forces is easy to recognize. In the vertical direction, the weight of the ladder $\vec{F}_G$ is compensated by the reaction of the ground on the ladder $\vec{F}_B$. In the horizontal direction, there is the friction on the ground $\vec{F}_{fric}$ and the reaction of the wall $\vec{F}_W$, which compensate each other. To calculate the forces, we con-

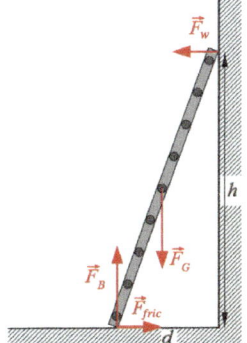

Fig. 12.11 Stability of a ladder

sider the torques with respect to the support point on the ground. The lever arm of the weight is $d/2$, that of the force from the wall is h. The forces $\vec{F}_B$ and $\vec{F}_{fric}$ do not contribute, as their lever arms are zero:

$$M = mg\frac{d}{2} - F_w h = 0. \tag{12.17}$$

We obtain for the magnitude of the frictional force:

$$F_{fric} = F_w = mg\frac{d}{2h} < \mu_H mg \Rightarrow d < 2h\mu_H. \tag{12.18}$$

As long as this condition is met, the ladder remains stable.

> **Experiment 12.4: A Stair of Bricks**
>
> We build a stair of bricks. We stack bricks on a table in such a way that the top brick protrudes as far as possible over the edge of the table. How do we stack the bricks optimally? Is it possible for the top brick to protrude completely over the edge of the table? How large can the overhang become at maximum?
>
> We approach the problem systematically and initially place only one brick on the edge of the table. Optimally placed, the center of gravity of the brick is directly above the edge of the table. The protrusion is $a/2$, where a is the length of a brick.
>
> Now we add a second brick underneath the first. Again, we place the upper brick with its center of gravity above the edge of the lower one, so that it just remains stable. Now we have to determine the common center of gravity of both bricks. It is in the middle, i.e. $\frac{3}{4}a$ away from the right end of the lower brick and just as far from the left end of the upper one. We push this center of gravity to the edge of the table. The lower brick protrudes by $\frac{1}{4}a$, the upper by $\frac{1}{2}a$ over the lower. The total protrusion is $p = \frac{1}{2}a + \frac{1}{4}a$.
>
> Now we lift the two bricks and slide a third one underneath. It can be protruded by $\frac{1}{6}a$ until it becomes instable. We keep adding bricks with the same procedure. For n stones, we get a total protrusion of:
>
> $$p = \sum_{i=1}^{n} \frac{1}{2i} a = \frac{a}{2} \sum_{i=1}^{n} \frac{1}{i}.$$

12.5 · Statics of Rigid Bodies

This results in a harmonic series. The harmonic series diverges, albeit slowly, i.e., it is possible to achieve an arbitrarily large protrusion! Already for four bricks, we get a protrusion of $\frac{25}{12}\frac{a}{2} > 1$. With perfect positioning, the fourth brick is already completely outside the table. In reality we need a few more (see photo).

The photo shows an experiment with wooden blocks. The plumb line indicates that the seventh wooden block is completely outside the bottom one, which here replaces the table top. Try for yourself! Can you stack more precisely?

© RWTH Aachen, Physics Experiment Collection

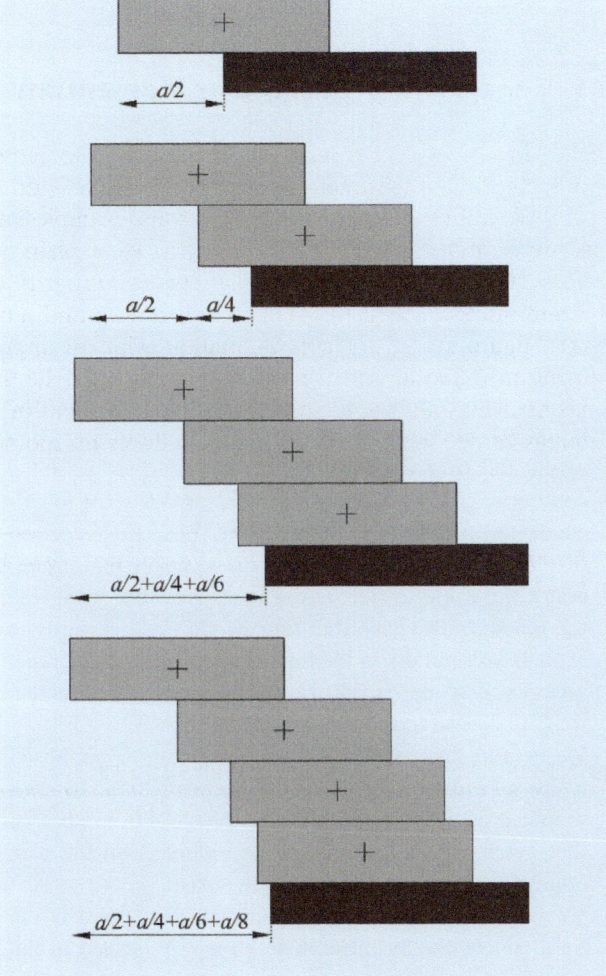

Example 12.12: Force effect of the biceps

Torques also act within our body. The sketch shows the musculature of the upper arm. The biceps generates a torque with its force $\vec{F}_m$ that lifts the forearm (weight $\vec{F}_f \approx 20\,\text{N}$) and a possible weight in the hand $\vec{F}_w$. The pivot point is the elbow. The length of the force arms is approximately $l_m \approx 6\,\text{cm}$ for the biceps, $l_f \approx 18\,\text{cm}$ to the center of gravity of the forearm and approx. $l_w \approx 35\,\text{cm}$ to the weight in the hand. What force must the biceps exert to hold a weight of $10\,\text{kg} \approx 100\,\text{N}$?

We set a balance of torques (φ is the angle between the lower and upper arm):

$$-F_w l_w \sin\varphi - F_f l_f \sin\varphi + F_m l_m \sin\varphi = 0$$

$$\Rightarrow F_m = \frac{F_w l_w + F_f l_f}{l_m} \approx 640\,\text{N}.$$

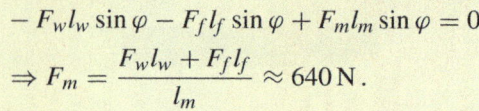

The force that the biceps must exert is significantly greater than the weight of the load. It is independent of the angle at which the weight is held.

Example 12.13: Construction Crane

Construction cranes lift a load over a pulley in the boom. Most cranes have only a limited footprint. They would fall over when lifting the load if there wasn't a counterweight attached at the backside. We want to calculate the necessary counterweight. Your first reaction may be to make the counterweight so heavy that it generates the same torque as the load. But that would be too much. In this case, the crane would tip backwards as soon as you put down the load. To set up a stable crane, we must consider not only the load and the counterweight but also the crane's own weight.

We assume the following data: The crane is supposed to lift a load of $2.2\,\text{t}$ at an outreach of $20\,\text{m}$. Its own weight is $10.5\,\text{t}$, the footprint is $4.5\,\text{m} \times 4\,\text{m}$ as seen in the sketch. The counterweight is mounted on the rear boom at a distance of $8\,\text{m}$ from the mast.

We start with the load on. Then the pivot point is the right end of the footprint. The forces are labeled in the sketch. $\vec{F}_G$ is the own weight of the crane, $\vec{F}_l$ the weight of the load and $\vec{F}_c$ that of the counterweight. At the tipping point, the following must apply:

$$F_l \cdot (20\,\text{m} - 2\,\text{m}) - F_G \cdot 2\,\text{m} - F_{c,\text{min}} \cdot (8\,\text{m} + 2\,\text{m}) = 0$$

$$\Rightarrow m_{c,\text{min}} = \frac{m_l \cdot 18\,\text{m} - m_G \cdot 2\,\text{m}}{10\,\text{m}} = 1860\,\text{kg}.$$

On the other hand, the crane must not tip backwards without a load (pivot point on the left end of the footprint):

$$F_{c,\text{max}} \cdot (8\,\text{m} - 2.5\,\text{m}) - F_G \cdot 2.5\,\text{m} = 0$$

$$\Rightarrow m_{c,\text{max}} = \frac{m_G \cdot 2.5\,\text{m}}{5.5\,\text{m}} = 4773\,\text{kg}.$$

The counterweight must be chosen within this range.

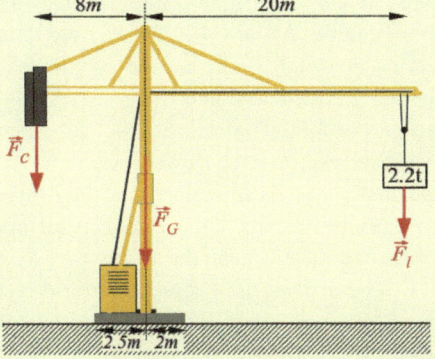

Example 12.14: Statics of a Truss Bridge

You have probably seen a similar bridge to the one in the photo. We want to take a closer look at the statics of such a bridge. It is made up of struts that each form equilateral triangles. This results in identical units (elementary cells), each consisting of four struts. They are drawn in dark in sketch A. The struts are connected to each other at the nodes (red dots). We count two nodes to an elementary cell. These are the two nodes at the left ends of struts A and B. The two nodes at their right ends already belong to the next elementary cell.

We want to neglect the weight of the struts. We only consider the weight of the roadway and any vehicle on it. When the bridge is empty, the center of gravity of each section (between the nodes i and $i+1$) is in the middle between the two ends. If there are vehicles on the road, it becomes more complicated. But we can still determine the weight and center of gravity of each section. From this, we can determine the forces that the section exerts on the nodes i and $i+1$. Please note that at node i a part of the load of the section of the road of the previous cell is added, and likewise at nodes $i+1$ a partial load of the following cell. So, we can calculate the acting weights $\vec{F}_{G,i}$ at each (lower) node.

The crucial step now is to realize that the bridge is only in static equilibrium when the forces at all nodes add up to zero. We have drawn the forces at two nodes in sketch A. The arrows only represent the directions of the forces, not their magnitudes! Individual forces can become negative. Then they point in the opposite direction of the drawn arrow. If we neglect the weights of the struts, the forces that the struts transmit point along their axes[2]. They are equal in opposite directions. Therefore, we label them with the same symbol. For each node on the roadway of the bridge, the following applies:

$$-\vec{F}_{A,i-1} + \vec{F}_{D,i-1} + \vec{F}_{C,i} + \vec{F}_{A,i} + \vec{F}_{G,i} = 0$$

or expressed in components (x-axis horizontal to the right, y-axis upwards):

---

[2] The sums of the forces and the torques that attack each strut must also add up to zero. However, if no other forces act besides the forces that originate from the nodes, this can only be fulfilled if the two forces are equal in opposite directions.

12.5 · Statics of Rigid Bodies

$$-F_{A,i-1} - \cos 60° \, F_{D,i-1} + \cos 60° \, F_{C,i} + F_{A,i} = 0$$
$$\sin 60° \, F_{D,i-1} + \sin 60° \, F_{c,i} - F_{G,i} = 0$$

and for the nodes at the top end of the bridge:
$$F_{B,i} - \cos 60° F_{D,i} + \cos 60° F_{C,i} - F_{B,i-1} = 0$$
$$- \sin 60° F_{C,i} - \sin 60° F_{D,i} = 0.$$

Each cell of the bridge has four struts and thus four forces that we need to determine. If the bridge has N cells, there are $4N$ struts[3]. The bridge rests with its weight on the bridgehead at both ends. A reaction emanates from the bridgehead, carrying the entire bridge. We have denoted it in sketch B with $\vec{F}_{E,1}$. The corresponding force at the other end of the bridge is $\vec{F}_{E,N+1}$. Our task is therefore to determine $4N + 2$ forces. For this, we use the conditions that we have derived at the nodes. There are $2 \cdot (2N + 1)$ conditions. This is exactly the required number to be able to determine all forces. Note that these are linear conditions, which can be solved by introducing a force vector $\vec{G} = (F_{E,1}, F_{A,1}, F_{B,1}, \ldots, F_{D,N}, F_{E,N+1})$, which contains all forces as components. The conditions can be brought into the form $A\vec{G} = 0$, where A is a matrix. The solution can be determined by inverting the matrix.

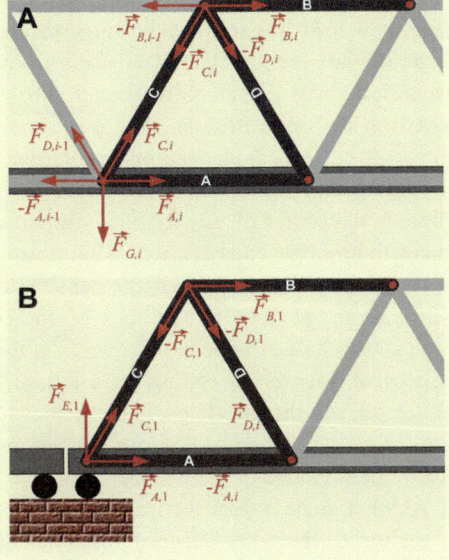

---

3 In the last cell on the right, strut B is actually missing, which must lead to $\vec{F}_{B,N} = 0$.

❓ Problems

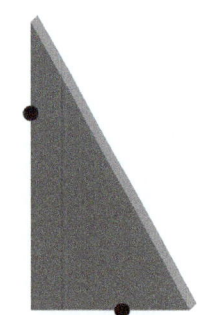

1. In the previous section ( Fig. 12.11), the condition for the stability of an empty ladder was calculated to be $d < 2h\mu_H$. What condition should be met for you to dare to climb the ladder to the top end?

2. Consider a rope that sags under its own weight. The shape in which the rope hangs freely is described by the function $y(x)$. Show that the following applies:

$$y'' = \frac{1}{F_H} \mu \sqrt{1 + (y')^2},$$

where μ represents the rope's own weight per unit length along the rope. Obtain the solution to the differential equation, also known as the catenary. Why is the catenary often used in the construction of arches and vaults?

3. A homogeneous plate has the shape of a right-angled triangle. The plate is to be suspended at two points, which are located on the catheti a and b of the triangle, in such a way that the plate is in unstable (labile) equilibrium, i.e., it does not automatically stand upright. Establish a relationship between the two positions of the suspension points on the catheti axes. Calculate the limiting cases and the symmetric configuration.

4. A homogeneous square plate of mass m and side length a is held horizontally by three threads. Two of the threads are attached at adjacent corners, the suspension point of the third thread can be freely moved on the opposite side. What are the forces in the two fixed threads depending on the position of the suspension point of the third thread?

5. Where is the center of gravity of a straight circular cone with homogeneous density? What position does the center of gravity assume if the density decreases linearly from the base to the apex of the cone, so that it is only half as large at the apex as at the base?

6. A spool of yarn is set rolling on a flat surface by pulling on the thread. The direction in which the spool of yarn rolls away depends on the angle the thread takes relative to the surface.
    (a) At what angle α does the transition from one direction to the other occur, if the outer diameter (roll diameter) of the spool of yarn is $d_a = 3$ cm

and the inner diameter (diameter of the winding) is $d_i = 2.5\,\text{cm}$?

(b) Now the thread is pulled exactly under the transition angle calculated in a) α. With what multiple of the weight of the spool of yarn may then be pulled on the thread at maximum, without the spool of yarn starting to slide? The coefficient of static friction between the rolling surface of the spool of yarn and the surface is $\mu_H = 0.9$.

Rotations

Contents

13.1 The Angular Momentum – 284

13.2 Rotation Around a Fixed Axis – 289

13.3 Conservation of Angular Momentum – 300

13.4 Rolling – 306

13.5 Gyroscopic Motion – 313

13.6 The Inertia Tensor – 327

13.7 Rotation around Free Axes – 338

13.8 Comparison – 340

© The Author(s), under exclusive license to Springer-Verlag GmbH, DE, part of Springer Nature 2025
S. Roth and A. Stahl, *Mechanics*,
https://doi.org/10.1007/978-3-662-68079-7_13

13.1 The Angular Momentum

First, we want to introduce the angular momentum mathematically: We consider a point mass (Fig. 13.1). Its motion is given by:

$$\vec{F} = m\frac{d\vec{v}}{dt} = \frac{d\vec{p}}{dt}. \tag{13.1}$$

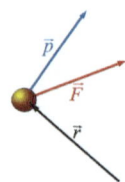

Fig. 13.1 For the definition of angular momentum

We multiply (vector product) from the left with the position vector and obtain the torque:

$$\vec{M} = \vec{r} \times \vec{F} = \vec{r} \times \frac{d\vec{p}}{dt}. \tag{13.2}$$

A short intermediate calculation shows that the right side can be simplified:

$$\frac{d}{dt}(\vec{r} \times \vec{p}) = \frac{d\vec{r}}{dt} \times \vec{p} + \vec{r} \times \frac{d\vec{p}}{dt}. \tag{13.3}$$

The first summand vanishes, since

$$\frac{d\vec{r}}{dt} \times \vec{p} = \vec{v} \times m\vec{v} = m(\vec{v} \times \vec{v}) = 0. \tag{13.4}$$

Therefore, it is

$$\vec{r} \times \frac{d\vec{p}}{dt} = \frac{d}{dt}(\vec{r} \times \vec{p}). \tag{13.5}$$

With this, we obtain:

$$\vec{M} = \frac{d}{dt}(\vec{r} \times \vec{p}). \tag{13.6}$$

For rotations, this expression represents the equivalent to Newton's second axiom for linear movements. It is also called the angular momentum theorem. As you have seen, we derived Eq. 13.1 from Newton's law. So it is not a new axiom. To make the equivalence even clearer, we introduce a new quantity. We call $\vec{L} = \vec{r} \times \vec{p}$ the angular momentum and thus:

$$\vec{M} = \frac{d\vec{L}}{dt}. \tag{13.7}$$

This equation was first established in 1754 by Leonard Euler. It is therefore also called Euler's angular momentum theorem.

13.1 · The Angular Momentum

The angular momentum has a direction given by the cross product (see ◘ Fig. 13.2). It is always perpendicular to the position and momentum vector of the point mass. The right-hand rule indicates the direction.

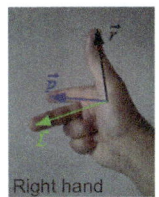

◘ **Fig. 13.2** Right-hand rule for cross products

Example 13.1: Angular Momentum

The two wheels of the bicycle have a large angular momentum around their axes. They stabilize the ride.

Note: Both the angular momentum and the torque refer to a pivot point. The correct specification of an angular momentum and a torque includes the specification of the pivot point to which the quantities refer. They change when you move the pivot point. The pivot point does not necessarily have to be distinguished by an axis. It can be chosen arbitrarily for the description of the movement. The vector $\vec{r}$ is the vector from the pivot point to the point mass. You must choose the same pivot point and thus the same vector $\vec{r}$ for the calculation of torque and angular momentum. The two sketches in ◘ Fig. 13.3 are intended to illustrate how the angular momentum changes by shifting the reference point (pivot point). Both sketches represent the same motion, but both the direction and the magnitude of the angular momentum are different.

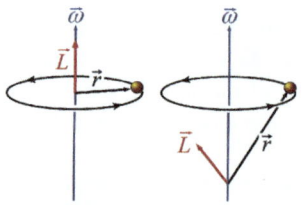

Fig. 13.3 Angular momentum and pivot point

Fig. 13.4 Angular momentum of a linear motion

Even a linear motion can contain an angular momentum. This may sound surprising at first, but consider the example in Fig. 13.4. It is a non-central collision of two balls, with the right ball at rest. We choose the center of gravity of the stationary ball as the pivot point. In this case, the angular momentum points perpendicularly out of the drawing plane. The vector is always drawn at the pivot point.

These considerations again referred to a point mass, but the transition to rigid bodies is not difficult. If we construct a rigid body from point masses, we must extend the definition of angular momentum accordingly. For many point masses, it is:

$$\vec{L} = \sum_i \vec{L}_i = \sum_i \vec{r}_i \times \vec{p}_i, \qquad (13.8)$$

with the angular momenta of all point masses refering to the same pivot point. For a continuous mass distribution, the following applies:

$$\vec{L} = \int_{body} d\vec{L} = \int_{body} (\vec{r} \times \vec{v}(\vec{r})) \, dm$$
$$= \int_{body} (\vec{r} \times \vec{v}(\vec{r})) \rho(\vec{r}) \, dV . \qquad (13.9)$$

> **Example 13.2: Angular Momentum of a Dumbbell**
>
> We consider a dumbbell rotating around its center (pivot point). We approximate it as two point masses at the location of the weights. The distance between the two weights is l, the mass of a weight m. Then the angular momentum is

13.1 · The Angular Momentum

$$|\vec{L}| = \frac{l}{2}mv' + \frac{l}{2}mv' = lm\omega\frac{l}{2} = \frac{1}{2}ml^2\omega.$$

We want to examine the example of the dumbbell (example 13.2) a bit more closely. The dumbbell rotates around its center of gravity, but we choose a different reference point for the calculation of the angular momentum, now. In a first step, we restict the center of gravity; we assume it does not move. The vectors $\vec{r}_1$ and $\vec{r}_2$ point from the reference point to the weights, the corresponding dashed vectors $\vec{r}'_1$ and $\vec{r}'_2$ from the center of gravity to the weights (◘ Fig. 13.5). It is $\vec{r}_1 = \vec{r}_{cm} + \vec{r}'_1$ and $\vec{r}_2 = \vec{r}_{cm} + \vec{r}'_2$. Then it is:

$$\begin{aligned}\vec{L}_{tot} &= \vec{r}_1 \times \vec{p}_1 + \vec{r}_2 \times \vec{p}_2 \\ &= \vec{r}'_1 \times \vec{p}_1 + \vec{r}'_2 \times \vec{p}_2 + \vec{r}_{cm} \times \vec{p}_1 + \vec{r}_{cm} \times \vec{p}_2 \quad (13.10) \\ &= \vec{L}'_{tot} + \vec{r}_{cm} \times (\vec{p}_1 + \vec{p}_2) = \vec{L}'_{tot}.\end{aligned}$$

The first term $\vec{L}'_{tot}$ represents the angular momentum with respect to the center of gravity, the second term vanishes, since with a stationary center of gravity, the sum of the momenta must be zero. Therefore, in the case of a stationary center of gravity, we conclude that the angular momentum does not depend on the reference point.

Now, we consider the general case and allow for movements of the center of mass. We then have:

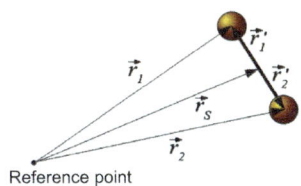

◘ **Fig. 13.5** Angular momentum of a dumbbell

$$\begin{aligned}\vec{L}_{tot} &= \vec{L}'_{tot} + \vec{r}_{cm} \times (\vec{p}_1 + \vec{p}_2) \\ &= \vec{L}'_{tot} + \vec{r}_{cm} \times (m_1 + m_2)\vec{v}_{cm} \\ &= \vec{L}'_{tot} + \vec{r}_{cm} \times M\vec{v}_{cm} \\ &= \vec{L}_{spin} + \vec{L}_{orbit} \,.\end{aligned} \quad (13.11)$$

As you can see, the angular momentum with respect to an arbitrary reference point can be decomposed into two terms. The first is the angular momentum of the rotation around the center of gravity of the body. It is called its "intrinsic angular momentum" or "spin". The second term is the angular momentum of the entire body with respect to the chosen reference point, which is obtained by treating the body as a point mass located at its center of gravity. This term is called the "orbital angular momentum". The orbital angular momentum depends on the reference point, while the intrinsic angular momentum does not.

We have performed this calculation for a rigid body composed of two point masses. You can easily verify that the result applies to any (rigid) body.

You will now learn about another very important conservation law: At the beginning of this chapter, we derived the following relationship from Newton's second axiom (Eq. 13.7)

$$\vec{M} = \frac{d\vec{L}}{dt}\,. \quad (13.12)$$

If no external torque acts on the body, it follows that the time derivative of the angular momentum vanishes, implying that the angular momentum is a conserved quantity. Again, we distinguish between internal and external torques. An internal torque is one that is generated by forces between the point masses of the body. It can have no effect on rigid bodies. An external torque is due to forces attacking from outside. We formulate the law of conservation of angular momentum:

> **Conservation of Angular Momentum**
> In a system where no external torques act, the total angular momentum is conserved.

Like conservation of energy and momentum, the conservation of angular momentum can be traced back to the invariance of the laws of nature under a symmetry trans-

13.2 Rotation Around a Fixed Axis

formation in the sense of Noether's theorem. In this case, it is the rotational symmetry of space, also referred to as isotropy of space, i.e. the observation that no direction in space is distinguished.

13.2 Rotation Around a Fixed Axis

We first deal with rotations where the axis of rotation is fixed by external conditions (mountings).

◻ Fig. 13.6 shows the rotation of a point mass around a fixed axis from two different perspectives: once along the axis of rotation and once perpendicular to it. The vector $\vec{r}_\perp$ falls into the orbital plane of the point mass. We assume that its magnitude remains constant, i.e., the point mass rotates on a circular path.

We calculate the motion:

$$\begin{aligned}\frac{d\vec{L}}{dt} &= \frac{d}{dt}(\vec{r}_\perp \times m\vec{v}) \\ &= \frac{d}{dt}(\vec{r}_\perp \times m(\vec{\omega} \times \vec{r}_\perp)) \\ &= \frac{d}{dt}(m\vec{r}_\perp^2 \vec{\omega}) \\ &= m\vec{r}_\perp^2 \frac{d\vec{\omega}}{dt} = \vec{M}.\end{aligned} \quad (13.13)$$

The time derivative of the angular velocity is called "angular acceleration". It is usually abbreviated with $\vec{\alpha}$. The angular acceleration is a vector. In the case of a fixed axis of rotation, it points along the axis. Thus, our result is $\vec{M} = m\vec{r}_\perp^2 \vec{\alpha}$, which should remind you of Newton's second axiom. To make this analogy even clearer, we introduce the moment of inertia I. We then have:

$$\vec{M} = I\vec{\alpha} \quad \text{with} \quad I = m\vec{r}_\perp^2. \quad (13.14)$$

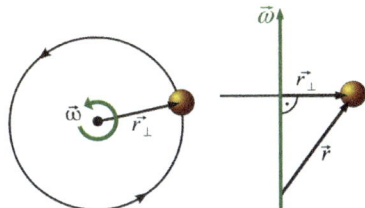

◻ **Fig. 13.6** Rotation of a point mass around a fixed axis

However, there is an important difference between mass and moment of inertia: Mass is a property of the body, the moment of inertia includes in its definition $\vec{r}_\perp$ and thus depends on the position of the axis of rotation.

The definition of the moment of inertia, which we introduced for a single point mass, can in turn be extended to systems of point masses or continuous mass distributions:

$$I = \sum_i m_i \vec{r}_{\perp,i}^2$$

$$I = \int_{body} \vec{r}_\perp^2 \, dm = \int_{body} \rho(\vec{r}) \vec{r}_\perp^2 \, dV. \tag{13.15}$$

We consider the motion of a point mass under the influence of a constant torque. The torque acts along the axis of rotation. Since all vectors point along this axis, we may omit the vector notation and continue using absolute values. It is:

$$M = I\alpha = I\frac{d\omega}{dt} = I\frac{d^2\varphi}{dt^2}$$

$$\Rightarrow \frac{d^2\varphi}{dt^2} = \alpha \quad \text{with} \quad \alpha = \frac{M}{I}, \tag{13.16}$$

where the angle of rotation φ indicates the position of the point mass on the circular path. This differential equation can be easily integrated:

$$\int \frac{d^2\varphi}{dt^2} dt = \int \alpha \, dt$$

$$\frac{d\varphi}{dt} = \omega(t) = \alpha t + \omega_0$$

$$\int \frac{d\varphi}{dt} dt = \int (\alpha t + \omega_0) \, dt \tag{13.17}$$

$$\varphi(t) = \frac{1}{2}\alpha t^2 + \omega_0 t + \varphi_0.$$

As integration constants, we introduced the initial position of the point mass on the circle $\varphi_0 = \varphi(t=0)$ and the initial angular velocity $\omega_0 = \omega(t=0)$. We obtain an accelerated rotation, whose equations remind us of the one-dimensional linear motion at constant acceleration (see Eq. 5.16).

13.2 · Rotation Around a Fixed Axis

> ### Example 13.3: Balance Bar
>
> Have you ever wondered how a balance bar helps in tight walking?
>
> The acrobat stands in a stable equilibrium on the tensioned wire, but the stability range is narrow. If he tilts just a little to the side, he moves towards a point where the position becomes unstable and he can no longer prevent a fall. He must react before he reaches this point. The balance bar increases his moment of inertia, thus giving him time to react to disturbances. Try to estimate the moment of inertia of the balance bar from ◘ Table 13.1 and compare this with the moment of inertia of the acrobat. A 10 m long bar with a weight of 10 kg has a moment of inertia that is greater than that of the acrobat himself! In addition, the balance bar shifts the center of gravity downwards with its downward pointing ends.
>
>
>
> © wikimedia: Sebastiaan ter Burg from Utrecht, The Netherlands[1]

13.2.1 Calculation of Moments of Inertia

To actually calculate rotations, we must know the moments of inertia of the involved bodies. In Experiment 13.1 you will see how to measure a moment of inertia. Here we want to demonstrate the calculation using

---

1 For the licence see: ▶ https://creativecommons.org/licenses/by/2.0/deed.en.

Tab. 13.1 Moments of inertia of some simple bodies with respect to their axes of symmetry

		Moment of inertia	Remarks
	sphere radius R	$\frac{2}{5}MR^2$	
	spherical shell radius R	$\frac{2}{3}MR^2$	infinitesimal thickness of the shell
	cylinder radius R	$\frac{1}{2}MR^2$	
	cylinder radius R, height h	$\frac{1}{4}MR^2 + \frac{1}{12}Mh^2$	
	hollow cylinder radius R	MR^2	
	cone radius R, height h	$\frac{3}{10}MR^2$	
	cuboid edges a and b	$\frac{1}{12}M(a^2+b^2)$	edges perpendicular to the axis
	torus radius R, ring r	$M(\frac{3}{4}r^2+R^2)$	outmost point at $R+r$.

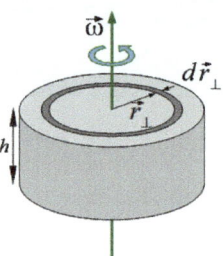

Fig. 13.7 The moment of inertia of a circular disc

the example of a circular disc (Fig. 13.7). We determine the moment of inertia with respect to the axis of symmetry. It is called the polar moment of inertia.

The circular disc has a constant density ρ and a radius R. Then the moment of inertia is:

13.2 · Rotation Around a Fixed Axis

$$I = \rho \int_{body} r_\perp^2 \, dV. \tag{13.18}$$

An infinitesimal ring of the circular disc has the following geometric sizes:

circumference: $2\pi \, r_\perp$
area: $2\pi \, r_\perp \, dr_\perp$ (13.19)
volume: $2\pi \, h \, r_\perp \, dr_\perp$,

$$I = \rho \int_0^R r_\perp^2 \, 2\pi \, h \, r_\perp \, dr_\perp$$

$$= 2\pi \, \rho h \int_0^R r_\perp^3 \, dr_\perp = 2\pi \, \rho h \left(\frac{1}{4} r_\perp^4\right)\Big|_0^R = \frac{1}{2}\pi \, \rho h R^4. \tag{13.20}$$

One can further simplify the result by refering to the mass of the circular disk $M = \pi\rho h R^2$. Then, $I = \frac{1}{2}MR^2$.

The moment of inertia is proportional to the mass of the circular disc. It is—at a fixed mass—independent of the height of the circular disc. Therefore, this result also applies to a cylinder rotating around its axis.

If we add a second circular disc of height h, we can calculate their moments of inertia individually and then add them, or we consider them as a single circular disc with twice the mass M. Since both must lead to the same result, it must be true that $I_{tot} = I_1 + I_2$. This applies generally, as long as all the individual moments of inertia refer to the same axis of rotation.

In ◘ Table 13.1, the moments of inertia of some common bodies with respect to their axes of symmetry are listed. In all cases, the axes of symmetry pass through the center of gravity.

Example 13.4: Moment of Inertia of an Axle with Wheels

An axle with two tires on rims (fig.) is already a quite complex object. The moment of inertia can be composed from individual elements according to ◘ Table 13.1. All moments of inertia refer to the rotation around the axis itself. The moment of inertia of a wheel is composed of the moment of inertia of the rim (radius r) and the tire (inner radius r, outer radius R):

$I_\text{wheel} = I_\text{tire} + I_\text{rim}$. For the rim, we can directly use the entry for a solid cylinder from the table $I_\text{rim} = \frac{1}{2} m_\text{rim} r^2$. For the tire, we use the integral from Eq. 13.20 within the limits r to R: $I_\text{tire} = \frac{1}{2} m_\text{tire}(R^2 + r^2)$. Then, for the entire axle $I_\text{tot} = I_\text{axle} + 2 I_\text{wheel}$, where I_axle should be negligible for most applications.

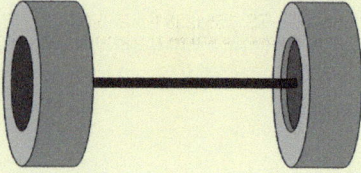

Example 13.5: The moment of inertia of a cone

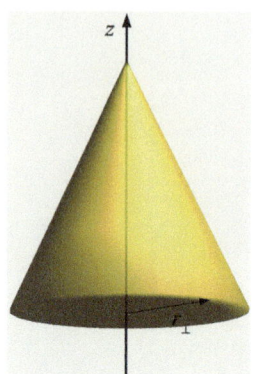

As an example, we want to determine the moment of inertia of a cone with homogeneous density ρ_0 with respect to its axis of symmetry. We choose cylindrical coordinates for the calculation, with the origin in the center of the base of the cone. The radius of the base is R and the height is h. The cylindrical coordinate, which we usually denote with ρ, corresponds here to the perpendicular distance $r_\perp$ from the axis of rotation.

The radius of the cone decreases linearly from the value R at the base of the cone to zero at the tip of the cone. The upper integration limit for $r_\perp$ is therefore $R(1 - z/h)$. The moment of inertia is calculated from:

$$I = \int_V \rho_0 r_\perp^2 \, dV = \rho_0 \int_0^h \int_0^{R(1-z/h)} \int_0^{2\pi} r_\perp^2 \, d\varphi \, r_\perp \, dr_\perp \, dz$$

$$= 2\pi \rho_0 \int_0^h \int_0^{R(1-z/h)} r_\perp^3 \, dr_\perp \, dz = 2\pi \rho_0 \int_0^h \frac{1}{4} r_\perp^4 \Big|_0^{R(1-z/h)} dz$$

$$= \frac{\pi \rho_0 R^4}{2} \int_0^h \left(1 - \frac{z}{h}\right)^4 dz$$

$$= \frac{\pi \rho_0 R^4}{2} \int_0^h \left(1 - 4\frac{z}{h} + 6\frac{z^2}{h^2} - 4\frac{z^3}{h^3} + \frac{z^4}{h^4}\right) dz$$

$$= \frac{\pi \rho_0 R^4}{2} \left(z - 2\frac{z^2}{h} + 2\frac{z^3}{h^2} - \frac{z^4}{h^3} + \frac{1}{5}\frac{z^5}{h^4}\right)\Big|_0^h$$

$$= \frac{\pi \rho_0 R^4 h}{10} = \frac{3}{10} MR^2,$$

where in the last step we substituted the mass $M = \rho_0 V$ with the volume $V = \frac{1}{3}\pi R^2 h$ of the cone. You can find the value in ◼ Table 13.1.

Experiment 13.1: Measurement of Moments of Inertia

A simple method for measuring moments of inertia is a torsional pendulum. A spiral spring is attached to a vertical axis (picture). The object whose moment of inertia is to be measured is attached to the axis. In the picture, this is a weight that rotates around the axis at a distance of about 20 cm. In the picture, the object is attached to the axis through a simple rod. Alternatively, one can attach a table to the axis and place the object on the table, or clamp it onto the table. If you turn the axis out of its equilibrium position and then let go, an oscillations starts. As we will see in ▶ Sect. 17.1, the period of the oscillation is $T = 2\pi\sqrt{I/D}$.

This period is measured. D is the stiffness of the spiral spring, which must be know. First we do a blank measurement T_0 with the apparatus and all holders, but without the object, and then a second measurement T with the object included. Then

$$T^2 - T_0^2 = 2\pi \frac{I - I_0}{D} = 2\pi \frac{I_{\text{object}}}{D}.$$

13.2.2 The Parallel Axis Theorem

Now, you have seen a series of examples in which moments of inertia were determined with respect to axes through the center of gravity of the objects. In principle, the moment of inertia with respect to any axis can be calculated from Eq. 13.15, but the integrals are usually more complicated. Here you will learn an alternative method. If you manage to determine the moment of inertia with respect to an axis through the center of gravity, which is parallel to the desired axis, the moment of inertia to the desired axis can be calculated using the parallel axis theorem, also know as the Huygens-Steiner theorem.

Consider the sketch in ◻ Fig. 13.8. The moment of inertia I_{cm} with respect to an axis through the center of gravity C is known. We want to determine the moment of inertia I_A with respect to an axis parallel to it through point A. The distance y is perpendicular to the line through A and C. We use the Pythagorean theorem:

$$r_{cm}^2 = b^2 + y^2$$
$$r_A^2 = (a+b)^2 + y^2 \qquad (13.21)$$

Then,

$$I_A = \int_V r_A^2 \, dm = \int_V ((a+b)^2 + y^2) \, dm$$
$$= \int_V (a^2 + b^2 + 2ab + y^2) \, dm$$
$$= \int_V a^2 \, dm + \int_V 2ab \, dm + \int_V (b^2 + y^2) \, dm \qquad (13.22)$$
$$= a^2 \int_V dm + 2a \int_V b \, dm + \int_V r_{cm}^2 \, dm$$
$$= a^2 M + 0 + I_{cm}$$

Here, M is the mass of the body. The second integral vanishes, it must be zero due to the definition of the center of gravity. So we get

> **Parallel Axis Theorem**
> If a body of mass M has a moment of inertia I_{cm} with respect to an axis through the center of gravity, then the moment of inertia with respect to an axis shifted parallel by the distance a is
>
> $$I_A = I_{cm} + a^2 M .$$

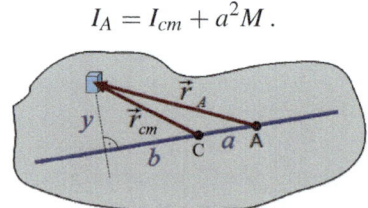

◻ **Fig. 13.8** On the parallel axis theorem

13.2 · Rotation Around a Fixed Axis

> **Experiment 13.2: Parallel Axis Theorem**
>
> With the apparatus from Experiment 13.1, we can verify the parallell axis theorem quantitatively. We place the object on the turntable with its center of gravity directly above the axis of rotation and measures its moment of inertia. Then we shift the body by the distance a, remeasure and compare with the prediction from the theorem.

13.2.3 Rotational Energy

We want to determine the kinetic energy of a rotating body by tracing it back to the known formula for kinetic energy (see ◻ Fig. 13.9). For an infinitesimally small mass element of the body, the kinetic energy is

$$dE_{\text{kin}} = \frac{1}{2} dm\, v^2 = \frac{1}{2} dm\, r_\perp^2 \omega^2, \tag{13.23}$$

which we integrate over the body:

$$E_{\text{rot}} = \frac{1}{2} \omega^2 \int_V r_\perp^2\, dm = \frac{1}{2} I \omega^2. \tag{13.24}$$

This equation for rotation corresponds to the well-known formula $E_{\text{kin}} = \frac{1}{2} mv^2$ for the kinetic energy of a translational motion. It is called the "rotational energy". It is a special form of kinetic energy.

If a torque acts on a body and puts it into rotation, it performs work, even if the center of gravity of the body is not moved and the body "only" rotates (see ◻ Fig. 13.10). If the body is mounted without friction, the work performed is converted into rotational energy. This work can be calculated. For an infinitesimally short distance in the direction of the force, we obtain:

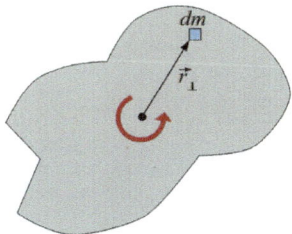

◻ **Fig. 13.9** On the calculation of rotational energy

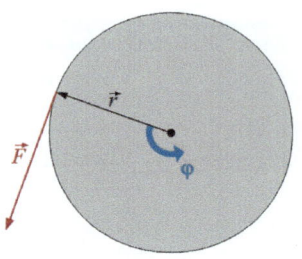

Fig. 13.10 A force performs work

$$W = F\, ds$$
$$\text{with } ds = r\, d\varphi \qquad (13.25)$$
$$W = r F\, d\varphi = M\, d\varphi .$$

In the simplest case, with constant torque, the integration yields $W = M\varphi$. For arbitrary directions of the force and the lever arm, the torque is perpendicular to $\vec{r}$ and $\vec{F}$, but possibly forms an angle with the axis of rotation. Then only the projection of the torque onto the axis of rotation is relevant for the work performed. We obtain the projection through a scalar product:

$$W = \vec{M} \cdot \vec{\varphi} \quad \text{or} \quad W = \int \vec{M} \cdot d\vec{\varphi}, \qquad (13.26)$$

where the second formula is to be used for variable torques. The vector $\vec{\varphi}$ points along the axis of rotation (right-hand rule).

Now, as in the case of a linear motion, we define the power as:

$$P = \frac{dW}{dt} = \vec{M} \cdot \vec{\omega}. \qquad (13.27)$$

Example 13.6: Flywheel Storage

Flywheels can be used to store energy. The sketch shows the structure of a modern flywheel for storage of electrical energy. An electric motor drives the flywheel, converting electrical energy into rotational energy. To recover the energy, one simply has to reverse the electric motor into a generator. Such storage is used to adjust the power of electrical energy sources to the consumer's needs. The rotor is made of carbon fiber-reinforced plastic. Such rotors can reach up to 50,000 revolutions per minute. To reduce friction, the flywheel is built into a vacuum tank. The rotor floats on magnetic bearings. About 90% of the input energy can be retrieved.

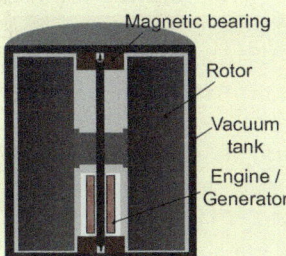

Example 13.7: Rotational Energy in the Acceleration of a Car

When a car accelerates, the engine has to perform work, which does not only flow into the kinetic energy of the car's translational motion. Part of the work is needed to set the wheels and other elements in the engine and transmission into rotation. We want to estimate the ratio of rotational energy to translational energy. We choose the data of the pictured Audi R8 Coupé as an example and consider the acceleration from zero to 100 km/h. The target speed simplifies the calculation, although the ratio of rotational energy to translational energy does not depend on the final speed. Both energies increase quadratically with speed.

(C) wikimedia: Edvvc from London, UK[2]

The kinetic energy is calculated from $\frac{1}{2}mv^2$. With a mass of the car of 1.66 tons, the kinetic energy is $E_{kin} = 640\,\text{kJ}$.

---

[2] License CC-BY, please see: ▶ https://creativecommons.org/licenses/by/2.0/deed.en.

For the calculation of the rotational energy, we take a front wheel of the type 245/35R19. It has a rim diameter of 19 inches (48.26 cm) and a tire height of about 8.6 cm, resulting in a wheel diameter of 65.4 cm and a circumference of 205.5 cm. At 100 km/h the wheel rotates 13.5 times per second, which corresponds to an angular frequency of $\omega = 84.9\,\text{rad/s}$. The rotational energy is calculated according to $E_{\text{rot}} = \frac{1}{2}I\omega^2$. We need to estimate the moment of inertia of the wheel. We estimate the moment of inertia of the rim (mass approx. 12 kg) downwards as a homogeneous circular disk using the formula $I_{rim} = \frac{1}{2}MR^2 \approx 0.35\,\text{kg m}^2$, and that of the tire (mass approx. 11 kg) as a cylindrical shell upwards as $I_{tire} = MR^2 \approx 1.18\,\text{kg m}^2$. In total, this results in approximately $I \approx 1.5\,\text{kg m}^2$. However, the wheel does not rotate about its axis, but about the contact point on the road. According to the parallel axis theorem, we still need to add a contribution of $MR^2 \approx 2.5\,\text{kg m}^2$, which results in a moment of inertia around the contact point of approximately $3.7\,\text{kg m}^2$ and finally in a rotational energy of 13.1 kJ for a front wheel. Considering that the rear wheels are slightly wider and therefore slightly heavier (295 cm at the back, 245 cm at the front), we arrive at a rotational energy of around 60 kJ for all four wheels. This means that between 5 and 10 percent of the work that must be performed to accelerate the car goes into the rotational energy of the wheels. Not an entirely negligible contribution.

13.3 Conservation of Angular Momentum

Already in ▶ Sect. 13.1 we saw that the angular momentum is conserved over time when no external torque acts on the system. We want to take a closer look at some examples here. First, we examine an example from sports. We explain how figure skaters achieve high rotational speeds during pirouettes and jumps (◘ Fig. 13.11). Through run-up and jump-off, the figure skater achieves a rotation with an angular momentum $\vec{L}$ and an initial rotational speed ω_0. If he then brings his arms and legs as close as possible to the axis of rotation, as can be seen in the picture, he reduces his moment

◘ **Fig. 13.11** Andrei Lutai in rotation during a jump. © Wikimedia: Emilia Karbownik[3]

---

3 License CC-BY, please see: ▶ https://creativecommons.org/licenses/by/2.5/deed.de.

13.3 · Conservation of Angular Momentum

of inertia $I = \int r_\perp^2 \, dm$. However, since no external torque acts after the jump-off, the angular momentum $\vec{L} = I\vec{\omega}$ must be conserved. With a reduced moment of inertia I, the rotational speed must therefore increase. The athlete takes advantage of this. The same technique is used by gymnasts or high divers during a somersault.

> **Example 13.8: Flywheel Control**
>
> Astronomical telescopes are moved into the desired direction of observation with motors that move the telescope against the mount. But how is this done with space telescopes? There is no fixed mount against which forces could be exerted on a satellite. Multiple flywheels with differently oriented axes are used. If one of the flywheels is accelerated or decelerated by a motor, the entire satellite rotates around the axis of the wheel, thus keeping the angular momentum constant. The photo shows such a flywheel module.

At the end of its lifetime, when a star has burned (fused) all the matter it is made of, the star implodes. We call this a supernova. The star collapses and the outer shell of the star is blown off. In many cases, a neutron star forms.

During the collapse, the star's moment of inertia is reduced. With the angular momentum preserved, this leads to a strong acceleration of the rotation. A typical mass of such a star is $M = 4 \cdot 10^{30}$ kg (2 solar masses) with a radius prior to the explosion of $R_1 = 7 \cdot 10^5$ km. Such a star typically rotates once every 10 days around its own axis. The angular momentum is:

$$L_1 = I_1 \omega_1 = \frac{2}{5} M R_1^2 \omega_1 . \tag{13.28}$$

During the explosion, the star loses about half of its mass. The core collapses into a neutron star with a radius of about $R_2 = 10$ km. Now the angular momentum is:

$$L_2 = I_2 \omega_2 = \frac{2}{5} \frac{M}{2} R_2^2 \omega_2 . \tag{13.29}$$

If we neglect that the star loses some angular momentum with the ejected shell, the conservation of angular momentum results in

$$\omega_2 = \frac{2 R_1^2}{R_2^2} \omega_1 . \tag{13.30}$$

Fig. 13.12 The Super Nova RCW 103. © NASA/CXC/Penn State/G. Garmire et al.

After the collapse, the neutron star rotates within milliseconds. The photo (Fig. 13.12) shows the Super Nova RCW 103 in X-ray light. The star exploded about 2000 years ago. You can see the ejected shell as gas clouds with the neutron star in the center.

Experiment 13.3: Conservation of Angular Momentum with the Swivel Chair

© Photo: Hendrik Brixius

13.3 · Conservation of Angular Momentum

The pirouette effect and the conservation of angular momentum can be beautifully demonstrated on a swivel chair. In the first experiment, the student holds two weights in his outstretched arms. He is gently turned on by one of us from the outside. If he pulls his arms in and brings the weights to his chest, the rotation speed increases significantly due to the pirouette effect.

In the second experiment, the student is handed a bicycle rim on an axle that has been spun quickly beforehand. The wheel is handed over with the axis of rotation oriented horizontally. The student then tilts the axis vertically, causing the swivel chair to start spinning, in the opposite direction to the wheel. The student can control the speed and direction of the chair's rotation by tilting the axis. When tilting the axis, he feels a significant resistance. He must apply a torque to change the direction of the wheel's angular momentum.

© Photo: Hendrik Brixius

Experiment 13.4: The Gimbal Gyroscope

With the gimbal gyroscope we directly demonstrate the conservation of angular momentum. The axis of the gyroscope is mounted on two nested gimbals in such a way that it can rotate almost without friction. The mount can be seen in the picture. The red arrow marks the axis of rotation of the gyroscope, a heavy metal disk. The holder of the axis can be rotated horizontally and vertically. We spin the gyroscope and then move it through the room. The orientation of the gyroscope's axis remains stable in space, no matter how wildly we turn the mount.

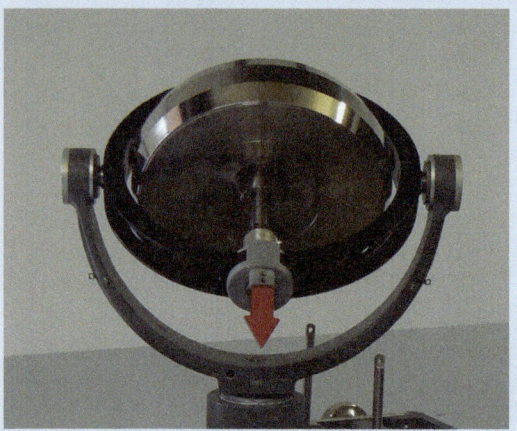

© RWTH Aachen, Physics Experiment Collection

Experiment 13.5: The Stubborn Suitcase

This experiment always causes great amusement among the students in the lecture hall. We organize a small race in which two competitors have to carry a suitcase through a narrow course. Already at the first turn, one of the two suitcases proves to be stubborn. It resists a turn. The runner grips the handle of the suitcase tighter and forces it around the turn, but now the suitcase suddenly flips upwards and almost causes the runner to trip over. Can you guess what's wrong here? One suitcase is completely normal, loaded with a dummy weight. Inside the other suitcase is an electric motor that drives a heavy gyroscope. Initially the axis of the gyroscope is horizontal. If the runner wants to carry the suitcase around a curve, he has to turn the gyroscope's axis and with it the angular momentum. To the surprice of the runner, the gyroscope resists.

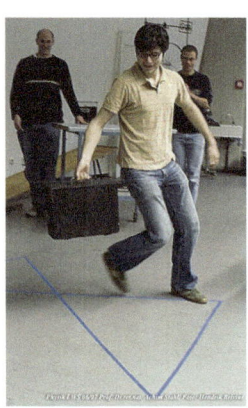

© Photo: Hendrik Brixius

13.3 · Conservation of Angular Momentum

Experiment 13.6: Gyroscope

This is a simple but fascinating experiment. I can still remember how perplexed I was when I first saw it as a young student. A gyroscope—in our case a bicycle rim—is spun vigorously and hung on a rod supported with a thread on each side of the axis. We take a pair of scissors and cut one of the threads. Wouldn't you expect the rim to tip over? But as you can see in the picture, nothing happens. The support is far outside of the center of gravity, but the rim remains horizontal even on one thread. The angular momentum keeps the rotation axis horizontal. Only on closer inspection you notice that the rotation axis slowly changes its orientation. We will come back to this later (Sec. 13.5).

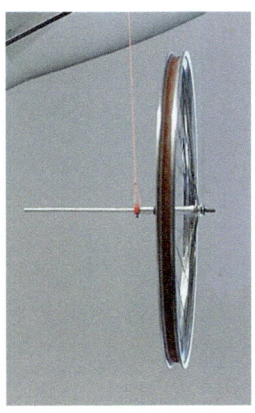

© RWTH Aachen, Physics Experiment Collection

Example 13.9: Discus Throw

A heavy rotating discus maintains its inclination almost over the entire throw due to its large angular momentum. Taking advantage of the lift, it flies significantly further than a non-rotating discus.

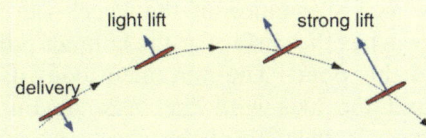

© wikimedia: Bundeswehr-Fotos[4]

Experiment 13.7: Fire Tornado

A fire tornado is a rare natural spectacle (picture). It is created by the rotation of the air in the tornado. If there is a fire in the center of the tornado, it heats the air. The air rises and an underpressure is created below. Rotating air flows from the outside to the center. Due to the conservation of angular momentum, the rotation increases significantly. The air is strongly swirled, which fans the flame and creates the fire column.
We recreate the fire tornado in an experiment. A flame (a few cm high) is placed on a turntable. A wire mesh surrounds the turntable. We spin the turntable and the

---

[4] License CC-BY, please see: ▶ https://creativecommons.org/licenses/by/2.0/deed.de.

mesh sets the air in rotation, causing the fire tornado to start.

© RWTH Aachen, Physics Experiment Collection

© 2012 Chris Tangey

13.4 Rolling

We now come to a rotational motion that is familiar to us from everyday from vehicles of all kinds: rolling.

▪ Fig. 13.13 shows the rolling of a wheel on a solid surface. The axis of rotation of the wheel is in its center. The wheel touches the surface at the contact point A and rolls around this point. The movement consists of a rotation around the axis with the instananeous velocity $\vec{v} = \vec{\omega} \times \vec{r}$ and a translation of the wheel with the velocity of the center of gravity $\vec{v}_{cm}$.

At the point of contact, a frictional force must act to prevent the wheel from sliding on the surface. For the contact point A, the instantaneous velocity is $\vec{v}_A = \vec{\omega} \times \vec{r}_A$. If the wheel rolls without slipping, the contact point must

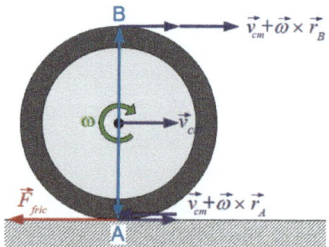

▪ **Fig. 13.13** A wheel rolls on a surface

13.4 · Rolling

remain stationary. From this, we obtain the condition of rolling, $v_{cm} = R\omega$, which sets the rotational speed ω of the wheel in relation to its translational speed v_{cm}. Here, R is the radius of the wheel. The highest point of the wheel B moves at double the speed.

As an example, we want to examine the rolling of a cylinder on an inclined plane (◘ Fig. 13.14). We try to calculate the acceleration of the cylinder. We decompose the weight into the downhill-slope force and the normal force, as indicated in the sketch. The downhill force creates a torque around the contact point A of the cylinder. It has the size (M: torque; m: mass of the cylinder):

$$M = F_{slope} R = m g R \sin\alpha. \tag{13.31}$$

From this we determine the angular acceleration

$$M = I \frac{d\omega}{dt} \Rightarrow \frac{d\omega}{dt} = \frac{M}{I}. \tag{13.32}$$

From ◘ Table 13.1 we read the moment of inertia of a solid cylinder about its axis. It is $I_{cm} = \frac{1}{2}mR^2$. This value refers to a rotation about an axis through the center of gravity. But we are interested in the moment of inertia with respect to an axis through the contact point. We use the parallel axis theorem, to determine it:

$$I_A = I_{cm} + mR^2 = \frac{3}{2}mR^2. \tag{13.33}$$

Thus, we obtain for the angular acceleration:

$$\frac{d\omega}{dt} = \frac{m g R \sin\alpha}{\frac{3}{2}mR^2} = \frac{2}{3}\frac{g}{R}\sin\alpha. \tag{13.34}$$

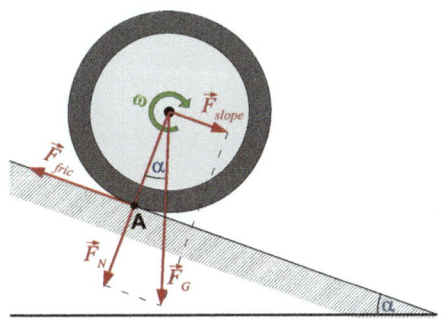

◘ **Fig. 13.14** A wheel rolls on an inclined plane

Now, we can calculate the linear acceleration from the condition of rolling:

$$v_{cm} = R\omega \Rightarrow a_{cm} = R\frac{d\omega}{dt} = \frac{2}{3} g \sin\alpha . \qquad (13.35)$$

The acceleration
- does not depend on the mass (as in free fall),
- does not depend on the radius of the cylinder,
- depends on the distribution of mass in the cylinder (through I_{cm}).

Experiment 13.8: Cylinder rolls on an inclined plane

We want to investigate the conditions that determine the linear acceleration of a rolling cylinder on an inclined plane For this purpose we conduct a small race between a red and a white cylinder (picture). The cylinders have the same radius and the same mass, verified with a scale. Nevertheless, the white cylinder always wins. The cylinders consist of an aluminum shell and the two lids, also made of aluminum, on which they roll. The reason why the white cylinder is faster only becomes apparent once we unscrew the lids. Inside the white cylinder is a solid wooden cylinder, while in the red cylinder we find a lead sheet, bent into a tube that just fits the aluminum shell. The interior of the red cylinder is empty. The moment of inertia of the white cylinder is approximately $\frac{1}{2}MR^2$ and thus smaller than the approximate moment of inertia of the red hollow cylinder (MR^2). Therefore, the white cylinder is faster.

© RWTH Aachen, Physics Experiment Collection

Now, we investigate the condition under which a solid cylinder rolls down the slope without slippage. The normal force is compensated by the support of the plane. In the

13.4 · Rolling

direction of motion, the downhill force and the frictional force act on accelerating the cylinder. We must have:

$$\vec{F}_{tot} = \vec{F}_{slope} + \vec{F}_{fric} = m\vec{a}_{cm}$$

$$mg\sin\alpha - \mu_s F_N = \frac{2}{3} mg\sin\alpha$$

$$mg\sin\alpha - \mu_s mg\cos\alpha = \frac{2}{3} mg\sin\alpha$$

$$\sin\alpha - \mu_s \cos\alpha = \frac{2}{3} \sin\alpha \qquad (13.36)$$

$$\mu_s \cos\alpha = \frac{1}{3} \sin\alpha$$

$$\mu_s \geq \frac{1}{3} \tan\alpha \,,$$

where we have used the formula for static friction known from ▶ Sect. 9.2. We have measured the static friction coefficient for aluminium on wood to be 0.18. If we insert this value into the equation, we see that the cylinder should roll without slipping up to an angle of inclination of 28°. At larger angles, it will slide down the slope. This can be demonstrated with experiment 13.8. If we tilt the plane more, we can see that the cylinder starts to slide on the plane at about 28°.

We want to discuss the rolling of a wheel a bit more generally. We assume that the axle of the wheel is accelerated by a force F_0, which acts parallel to the surface. This could be a force that pulls externally on the axle, or a force $F_0 = M_0/R$, generated by a motor that turns the axle with a torque M_0. The condition of rolling determines the velocity of the axle $v_{cm} = R\omega$, from which the acceleration

$$a_{cm} = R\frac{d\omega}{dt} \qquad (13.37)$$

follows. The torque M, which sets the wheel in rotation, is:

$$M = F_0 R = I_A \frac{d\omega}{dt}, \qquad (13.38)$$

from which an acceleration of

$$a_S = \frac{R^2}{I_A} F_0 \qquad (13.39)$$

results. Here, I_A is the moment of inertia of the wheel with respect to the contact point A. It is obtained with the help of the parallel axis theorem from the moment of inertia I_{cm} about its center of gravity: $I_A = m R^2 + I_{cm}$. Mathematically speaking, the moment of inertia I_{cm} must take a value that falls somewhere between zero, in the case that the entire mass of the wheel is concentrated on the axle, and $m R^2$, in the case that all mass is located on the outermost radius R. For realistic wheels, the value will probably lie somewhere between $\frac{1}{2}mR^2$, which corresponds to a wheel approximated as a homogeneous circular disc, and mR^2. We introduce a constant k and write the moment of inertia as $I_{cm} = k\,mR^2$. The factor k describes the distribution of mass on the wheel. For realistic wheels, it will be between 0.5 and 1. Then $I_A = (1 + k)\,m R^2$. Finally we get, for the linear acceleration of the wheel:

$$a_{cm} = \frac{1}{1+k}\frac{F_0}{m}. \tag{13.40}$$

Due to the rotation, the acceleration of the wheel is reduced by a factor $\frac{1}{1+k}$ compared to a point mass of the same mass m.

The condition for rolling without slippage, including the static friction F_{fric} now reads:

$$F_0 - F_{fric} = m\,a_{cm} \tag{13.41}$$

from which we derive a condition for the coefficient of static friction μ_H:

$$\mu_s \geq \frac{k}{1+k}\frac{F_0}{F_N} \tag{13.42}$$

Here, F_N is the normal force on the support.

Example 13.10: A Braking Car

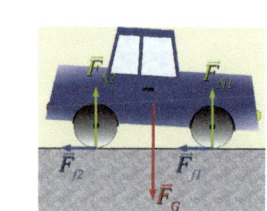

When braking, a car tends to tilt forward. We want to try to explain this. Please, consider the sketch. It shows the forces that occur when braking.

The weight is distributed between the two contact points of the wheels. For simplicity, we assume that the center of gravity is in the middle between the two axles, each at a distance l from them. The reaction to the weight carries the car, the frictional forces cause the braking ($a < 0$):

$$m g = F_{N1} + F_{N2},$$
$$m a = F_{f1} + F_{f2}.$$

13.4 · Rolling

Assuming that the wheels do not lock, the frictional forces are

$$F_{f1} = \mu_s F_{N1}$$
$$F_{f2} = \mu_s F_{N2}$$
$$F_{f1} + F_{f2} = \mu_s(F_{N1} + F_{N2}).$$

This results in

$$F_{N1} + F_{N2} = mg = m\frac{a}{\mu_s}$$

$$\Rightarrow a = \mu_s g.$$

We now consider the individual forces on the axles and calculate the torques with respect to the center of gravity. With uniform braking, a balance is established, in which the left and right turning torques compensate each other:

$$M_{N2} + M_{R1} + M_{R2} = M_{N1}$$
$$lF_{N2} + hF_{R1} + hF_{R2} = lF_{N1}$$
$$lF_{N2} + h\mu_s F_{N1} + h\mu_s F_{N2} = lF_{N1}$$
$$lF_{N2} + h\mu_s F_{N2} = lF_{N1} - h\mu_s F_{N1}$$
$$(l + h\mu_s)F_{N2} = (l - h\mu_s)F_{N1}$$
$$\Rightarrow F_{N1} = \frac{(l + h\mu_s)}{(l - h\mu_s)}F_{N2}.$$

Here, l is the horizontal distance of the axles from the center of gravity and h is the height of the center of gravity above the ground. We insert typical values for a car:

$l = 1.5\,\text{m}$
$h = 0.75\,\text{m}$
$\mu_s = 0.6$
$m = 800\,\text{kg}$

and obtain

$F_{N1} = 520\,\text{kg}\,g \quad F_{R1} = 3060\,\text{N}$
$F_{N2} = 280\,\text{kg}\,g \quad F_{R2} = 1650\,\text{N}.$

As you can see, the normal force on the front axle is significantly larger, causing the car to tilt forward. In addition, the braking force on the front axle is also larger, so the brakes on the front wheel must be correspondingly stronger than the rear brakes. You may be familiar with the different braking effects of the front and rear wheels from cycling.

Experiment 13.9: Maxwell's Wheel

We all know Maxwell's Wheel as a toy, usually called a Yo-Yo. It consists of a flywheel, here a spoked wheel. The wheel is hung on two strings that are wound around its axis. If you let it go, the strings unwind and set the wheel in rotation. We have mounted the wheel on a beam balance, which we zeroed to the weight of the wheel. We let go and the scale shows a reduced weight of the wheel. To allow a steady display, the scale is damped in an oil bath. Once the strings are completely unwound, you can see a jerk on the scale. The wheel continues to turn, winds the strings back up, and rises. The scale shows that even when rising, the weight of the wheel is reduced.

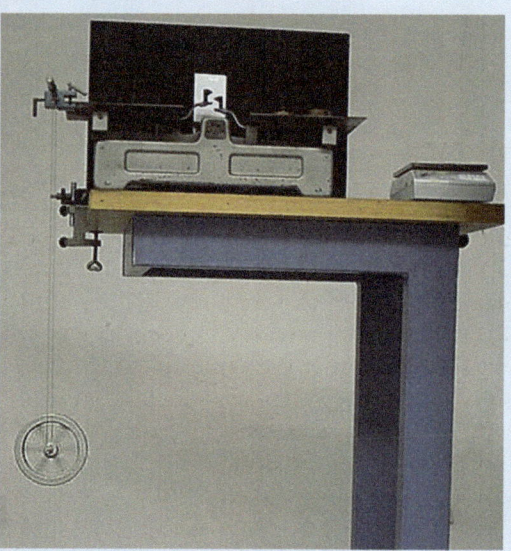

© RWTH Aachen, Physics Experiment Collection

> **Example 13.11: Maxwell's Wheel**
>
> We want to calculate the motion of Maxwell's Wheel from Experiment 13.9. The weight acts on the wheel. We decompose it into the component that causes the translational acceleration F_a, and the component that drives the rotation F_φ:
>
> $$F_G = F_a + F_\varphi.$$
>
> We express the angular acceleration α through the torque $M = rF_\varphi$, translated into $M = I\alpha$ with the condition of rolling $a = r\alpha$. Here r is the radius of the axle on which the strings are wound.
>
> $$F_G = ma + \frac{M}{r} = ma + \frac{I\alpha}{r} = ma + \frac{Ia}{r^2}.$$
>
> The moment of inertia is dominated by the wheel, which we want to approximate as a disc. It is $I = \frac{1}{2}mR^2$, with the mass m and the radius R of the wheel. Thus,
>
> $$F_G = mg = ma + \frac{1}{2}\frac{mR^2 a}{r^2} = \left(1 + \frac{1}{2}\frac{R^2}{r^2}\right)ma$$
>
> $$\Rightarrow a = \frac{1}{1 + \frac{1}{2}\frac{R^2}{r^2}} g.$$
>
> The force in the strings, which is indicated by the scale, is
>
> $$F_{\text{strings}} = m(g - a) < mg.$$
>
> Note that the wheel gets an upward impulse when reversing, which gives the wheel an initial velocity for ascending. This velocity is reduced during the ascent by F_a. It points downwards when winding, which also reduces the force in the string during ascent compared to a stationary wheel.

13.5 Gyroscopic Motion

We have discussed rotational movements around a fixed axis (▶ Sect. 13.2) and around parallel movable axes (▶ Sect. 13.4). The next step deals with axes that are only supported at a single point, before we want to deal with rotation around completely free axes in the last section.

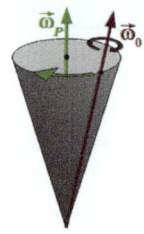

Fig. 13.15 Precession of a gyroscope

13.5.1 Precession of the Gyroscope

We start this section by taking a closer look at the movement of the gyroscope. You have seen a version in Experiment 13.6. A similar apparatus is shown in Experiment 13.10, which is even better suited for our investigations. In this experiment, a flywheel rotates at high speed $\vec{\omega}_0$ around an axis. If a torque acts on the axis, the experiment shows that the axis is no longer stationary, but rotates around the vertical (Fig. 13.15). This can be described as follows: The axis of the wheel moves around on the surface of a cone, the axis of which is vertical. This circling of the axis is called the "precession" $\vec{\omega}_p$, in contrast to the rapid rotation of the flywheel around its own axis $\vec{\omega}_0$, which is sometimes called the "rotation" of the wheel. In general, we speak here of "gyroscopic motion" or "gyrating". The Greek word "gyroscope" translates to "spinning top instrument".

Experiment 13.10: Vertical Gyroscope—Part 1

This gyroscope consists of a flywheel (gyro), whose axis is supported at a single point (see picture). The gyro axis is vertical, unlike the gyroscope in Experiment 13.6, which had a horizontal axis. The holder of the gyro can be seen in the zoom. The gyro rotates in a ball bearing around the axis of a stationary holder. It has a tip at the lower end, with which it stands on a rod that is attached to the table. The center of gravity of the gyro is just below the tip, so it stands stable. We spin it and observe the motion.
If the gyro is well adjusted, it rotates around the vertical axis, which does not move. We attribute this, as in Experiment 13.6, to the conservation of angular momen-

13.5 • Gyroscopic Motion

tum. What happens, if we apply a torque to the axis? For this purpose, a lateral hook is attached above the holder, to which we hang a weight. Since it hangs a few centimeters away from the axis, it causes a torque. We can adjust the size of the torque via the weight. In Experiment 13.6, the torque is generated by the weight of the gyro and is not adjustable. The torque causes the gyro axis to tilt to the side. But it does not flip to the side, as would be the case without the rotation of the gyro. The gyro axis only tilts a little and then slowly rotates around the vertical. The faster the gyro spins, the less the axis tilts and the slower it rotates around the vertical.

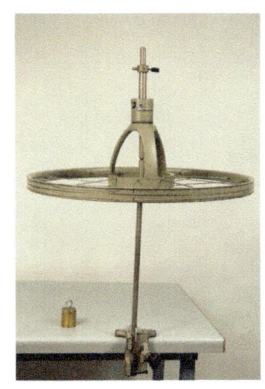

We examine the precession more closely: The attached weight causes a torque $\vec{M}$ around the mounting of the gyroscope. The force $\vec{F}$ and torque are shown in ◘ Fig. 13.16. Please check the orientations! The torque points horizontally and causes a change in angular momentum (Eq. 13.7)

$$\frac{d\vec{L}}{dt} = \vec{M} \qquad (13.43)$$

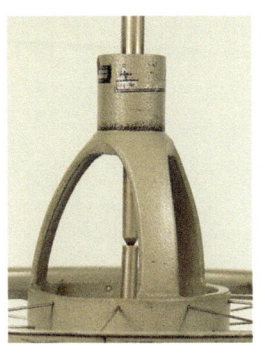

in the direction of the torque. In Experiment 13.10 we described that the center of gravity of the gyroscope is just below the support point. If the gyroscope tilts, its weight causes another torque. We want to neglect this. We assume that the gyroscope is adjusted so that its center of gravity is exactly at the support point, so this torque vanishes.

© RWTH Aachen, Physics Experiment Collection

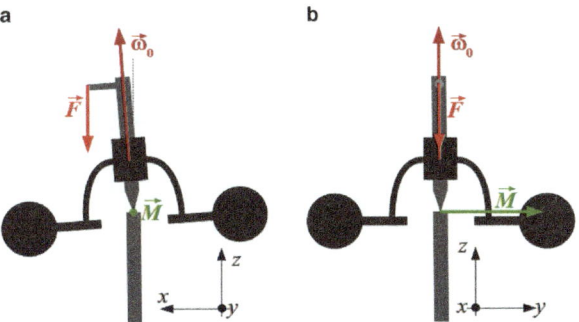

◘ **Fig. 13.16** Side view of the gyroscope from Experiment 13.10. **a** perpendicular to the lever arm; **b** in the direction of the lever arm

In ◯ Fig. 13.17 the angular momenta are shown in different projections. In the top view (b), you can see the change in angular momentum in the direction of $\vec{M}$. A short time Δt results in a change of ΔL. If you add this vectorially to the projection of the angular momentum $\vec{L}$, which the gyroscope had immediately after tilting, you can see how the axis of rotation precesses. From the sketch, we read off

$$\Delta \varphi = \frac{\Delta L}{\left(\vec{L}(0)\right)_{xy}}, \tag{13.44}$$

where $(\vec{L}(0))_{xy}$ is the projection of $\vec{L}(0)$ into the horizontal plane. We can read it again from sketch a in ◯ Fig. 13.17. It is $(\vec{L}(0))_{xy} = L_0 \sin \alpha$. Here, L_0 is the magnitude of the angular momentum, which does not change when tilting from the vertical by the angle α:

$$\Delta \varphi = \frac{\Delta L}{L_0 \sin \alpha}. \tag{13.45}$$

The tilting of the axis of rotation by the angle α reduces the component of the angular momentum in z-direction from originally L_0 to $L_0 \cos \alpha$. However, since there is no torque acting in this direction, this component of the angular momentum cannot change. In fact, it does not change. The apparent change is compensated by the angular momentum $\vec{L}_p$, which is associated with the precession. It also points in the z-direction and corresponds to the angular momentum of the tilted, but non-rotating wheel on the cone around the vertical. Consequently, it must apply:

$$L_0 = L_0 \cos \alpha + L_p \Rightarrow L_p = (1 - \cos \alpha) L_0. \tag{13.46}$$

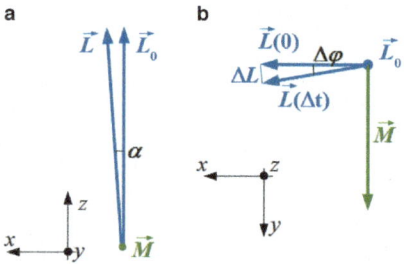

◯ **Fig. 13.17** Angular momentum on the gyroscope; **a** view perpendicular to the lever arm; **b** view from above

13.5 · Gyroscopic Motion

So we can write:

$$\Delta L = \sin \alpha \, \Delta\varphi \, L_0$$
$$\frac{\Delta L}{\Delta t} = \sin \alpha \, \frac{\Delta\varphi}{\Delta t} \, L_0. \tag{13.47}$$

If we now move to infinitesimally small time steps, we obtain:

$$\frac{dL}{dt} = \sin \alpha \, \omega_p \, L_0, \tag{13.48}$$

where the right side can also be written considering the direction of the vectors as

$$\frac{d\vec{L}}{dt} = \vec{\omega}_P \times \vec{L}. \tag{13.49}$$

This is called the "gyroscopic equation".

For a special case of a gyroscope, the angular velocity of the precession $\vec{\omega}_P$ can be easily determined. In this case, there is no additional weight on the axis of the gyroscope. Instead, the torque that drives the precession is generated by the weight of the gyroscope itself, by adjusting the support point so that the center of gravity of the gyroscope is above the support point. If the gyroscope tilts to the side, a torque is created.

$$\vec{M} = \vec{r} \times m\vec{g} = \frac{d\vec{L}}{dt} = \vec{\omega}_P \times \vec{L}, \tag{13.50}$$

where $\vec{r}$ is the vector from the support point of the gyroscope to its center of gravity. It forms an angle α with the vertical. The mass of the gyroscope is m. Then, in terms of magnitude, it follows that

$$r\,m\,g \sin \alpha = \sin \alpha \, \omega_p \, L_0$$
$$\Rightarrow \omega_p = \frac{r\,m\,g}{L_0} = \frac{r\,m\,g}{I\,\omega_0}. \tag{13.51}$$

It can be seen that the precession is faster the greater the torque that drives the gyroscope. It slows down with a larger moment of inertia and faster spinning of the gyroscope.

Experiment 13.11: Humming Top

The humming top is an old tin toy, very similar to a gyroscope. One can be seen in the picture. You pump on the blue button at the top and the body starts spinning. If it spins fast enough (the humming top then hums), you let it go. At the beginning, no precession can be seen. The tilt angle α is so small and the precession so slow that the top seems to stand still. Due to the inevitable friction ω gradually decreases, the precession frequency increases according to Eq. 13.51 and at the same time the tilt angle increases. Now you can clearly see the precession. The more ω decreases, the larger ω_p and α become, until the precession finally turns into a tumble and the top bumps somewhere. These observations can be made with all toy tops.

13.5 · Gyroscopic Motion

Example 13.12: Sakai Top

You can bent a Sakai top from a piece of wire (see photo and sketch). However, it only runs stably at a certain value of the angle β. The angle must be chosen so that the center of gravity is above the pivot point. To determine the angle, calculate the x-component of the center of gravity. It is composed of the centers of gravity of the circular segment (x_B, m_B) and the two legs (x_A, m_A). With the radius r and the mass distribution ρ we find

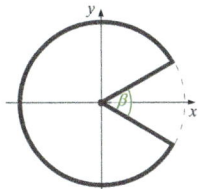

$$x_A = \frac{r}{2} \cos \frac{\beta}{2}$$

$$m_A = 2r\rho$$

$$x_B = \frac{1}{r(2\pi - \beta)} \int_{\beta/2}^{2\pi - \beta/2} x r \, d\varphi = \frac{1}{r(2\pi - \beta)} \int_{\beta/2}^{2\pi - \beta/2} r^2 \cos\varphi \, d\varphi$$

$$= \frac{-2r}{2\pi - \beta} \sin \frac{\beta}{2}$$

$$m_B = r(2\pi - \beta)\rho.$$

From the condition $x_B m_B + x_A m_A = 0$ we finally get $\tan \frac{\beta}{2} = \frac{1}{2}$ or $\beta = 53°$. Try bending one yourself, e.g., from a paperclip.

Example 13.13: Saros Series

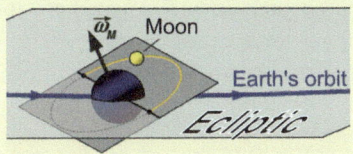

The orbit of the Moon around the Earth is tilted by about $\alpha = 5°$ compared to the orbit of the earth around the Sun, the so-called ecliptic, (see sketch). The Earth-Moon system forms a gyroscope that is influenced by the forces of the Sun. These create a torque that tries to turn the Earth-Moon gyroscope into the ecliptic. However, this torque changes with the position of the Moon in its orbit. We will therefore average it.

First, we want to determine the torque. Note that strictly speaking, the combined center of gravity of the Earth and Moon orbits around the Sun. The Earth and Moon revolve around this point, the much heavier Earth with a correspondingly smaller radius. If the Earth and

Moon were moving in a homogeneous gravitational potential, the torque on the Earth would exactly compensate that on the Moon. But the gravitational potential is not homogeneous. At new moon, the Moon is closer to the Sun and therefore its torque predominates. The variation of the distance of the Earth to the Sun is negligible in comparison. The additional force on the Moon is

$$\Delta F = G \frac{m_M m_S}{(r_E - r_M)^2} - G \frac{m_M m_S}{r_E^2} \approx 2 \frac{r_M}{r_E} G \frac{m_M m_S}{r_E^2}$$

with the masses of the Sun and Moon m_M, m_S and the radii of the Earth's and Moon's orbits r_E, r_M. The associated lever arm is $r_M \sin \alpha$. In the full moon position, the force on the Moon is weaker by a similar amount, resulting in a corresponding torque in the same direction. In the positions in between, both ΔF, as well as the lever arm, are smaller by a factor of $\cos \varphi$ when φ is the angle to the new moon position. The averaging results in $\overline{\cos^2 \varphi} = 1/2$. Over the course of a year, the Earth-Moon system also rotates around the Sun once, which leads to a further reduction of the torque by a factor of 2. Thus, the average torque is:

$$\overline{M} = \frac{1}{2} \frac{r_M}{r_E} G \frac{m_M m_S}{r_E^2} r_M \sin \alpha = \frac{1}{2} r_M^2 m_M \omega_E^2 \sin \alpha ,$$

where in the last step we have equated gravitational force and centrifugal force on the Earth's orbit. The angular momentum of the Earth-Moon gyroscope is

$$L_M = r_M m_M v_M = r_M^2 m_M \omega_M .$$

From Eq. 13.50 it follows that

$$\omega_p = \frac{\overline{M}}{L_M \sin \alpha} = \frac{1}{2} \frac{\omega_E^2}{\omega_M} .$$

From the equation, a precession period of 26 years is derived. In fact, the period only lasts 18.5 years. The difference lies in the approximations we have made.

The precession of the Earth-Moon gyroscope determines the occurrence of solar eclipses. At the two so-called nodes, the Moon's orbit pierces the ecliptic. Due

13.5 · Gyroscopic Motion

to the precession, the nodes move around the Earth. Only when the Moon is in its orbit around the Earth at one of the nodes and this node is exactly between Earth and Sun, a solar eclipse occurs. The name Saros series goes back to the calculation of solar eclipses.

13.5.2 Nutation of the Gyroscope

We discussed the rotation and precession of a gyroscope. However, these do not yet describe the full complexity of a gyroscope's motion. Another motion, called nutation, superimposes the presession of the gyroscope's axis. Let's go back to the gyroscope in Experiment 13.12:

Experiment 13.12: Vertical Gyroscope—Part 2

To make the movement of the gyroscope's axis visible, we attach a red LED to the tip of the axis, darken the room, and capture the movement in a long exposure with a camera. Two images are shown. The first was taken with a high rotational speed of the gyroscope. The tip of the rotation axis moves in a reasonably smooth circle. Not so in the second image. In this shot, the rotational speed was significantly lower. The rotation axis goes through so-called epicycles. The slow rotation due to precession is overlaid by a faster rotation on a circle with a significantly smaller radius. This is the nutation.

Upon closer inspection, we realize that the amplitude of the nutation (the radius of the circle) decreases over time due to friction. Such a damping of the nutation is observed in most gyroscopes. The nutation is initiated at the beginning of the movement (e.g., by the attachment of the weight) and then fades away.

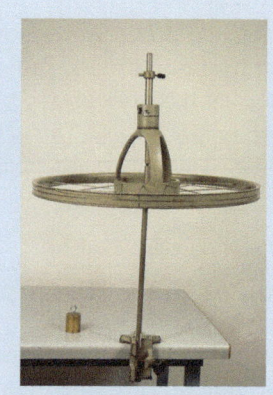

© RWTH Aachen, Physics Experiment Collection

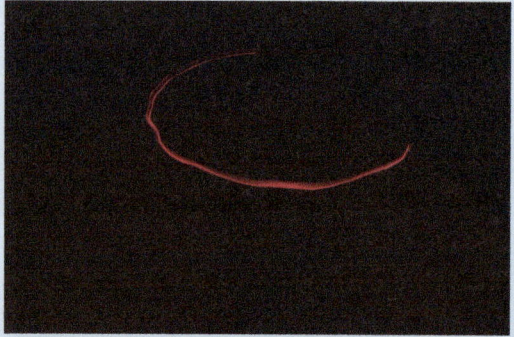

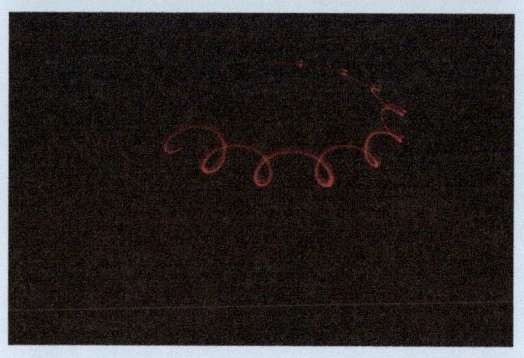

13.5 · Gyroscopic Motion

As we have already seen, after release, the axis of the gyroscope will tilt downwards by the small angle α until the vertical components of L_0 and L_p correspond to the original angular momentum. However, the axis will not remain in this equilibrium position. Due to the initial downward movement, a periodic up and down of the axis around the angle α of the equilibrium position starts. In the experiment, we observe a periodic rocking of the axis. This movement, which is superimposed on the rotation of the wheel around its axis and the precession, is called "nutation". The nutation is usually strongly damped by frictional forces, so it can only be observed at the beginning of the movement.

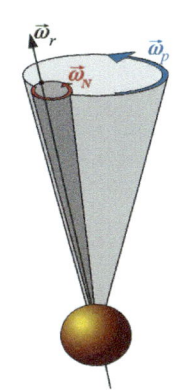

Fig. 13.18 shows the complete motion of a rotating gyroscope under the influence of a torque. $\vec{\omega}_r$ denotes the rotation of the body about its axis, $\vec{\omega}_p$ the precession of the rotation axis about the vertical, and $\vec{\omega}_N$ the nutation. The amplitude and speed of the three movements depend on the boundary conditions (nature of the rotating body, position of the rotation axis, ω_r). The calculation of the nutation is reserved for the courses in theoretical physics.

Fig. 13.18 Rotation, precession, and nutation of the gyroscope

Example 13.14: Movement of the Earth's Axis

The rotation of the Earth is an excellent example of gyroscopic motion. The Earth is not perfectly round. It has an equatorial bulge. At the equator, its diameter is 12,756 km, from pole to pole only 12,713 km. Now, the part of the equatorial bulge close to the Sun is attracted more strongly by the Sun than the sun-far part creating a torque on the Earth's axis. The Earth's axis is tilted by 23.44° relative to the normal of the orbital plane. This leads to a precession of the Earth around the vertical. The Earth rotates once around its own axis (sidereal day) in 23 h 56 min 4.1 s. A complete precession takes about 26,000 years.
The precession caused by the Sun's torque is overlaid with further disturbances by the Moon and the other planets. These are erroneously referred to as nutation in astronomy. A small nutation in our sense also occurs. It has a period of about 304 days.

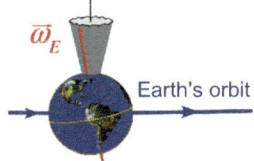

Experiment 13.13: A Ball Gyroscope

This gyroscope consists of a solid steel ball (radius approx. 25 cm) from a ball bearing of a lignite excavator. Due to the large mass, all processes proceed quite slowly and can therefore be observed precisely. The gyroscope is supported on an air cushion, so it rotates almost without friction. Again, a rod is attached at the top with a small weight. If the rod is tilted from the vertical, it exerts a torque on the gyroscope, which drives a precession. With light bumps against the rod, we can stimulate a nutation, which only slowly decays in this gyroscope.

© RWTH Aachen, Physics Experiment Collection

Experiment 13.14: Levitron

This is an exciting little top that floats on a magnetic field. The magnetic field is generated by permanent magnets, which are stacked in a box with alternating polarity. The top itself is made of ferromagnetic material. It is repelled by the magnets.

First, we align the base horizontally. Otherwise, the top will later run off the base. Then we place a plexiglass disc on the base and spin the top on it. Once the top is running stably, we carefully lift the plexiglass disc and the top with it. This requires a bit of finesse. Once the top is floating, we can pull the plexiglass disc out from under it. In principle, you can also demonstrate precession and nutation here, but it is also nice to simply watch the floating top.

© RWTH Aachen, Physics Experiment Collection

13.5.3 The Gyrocompass

The gyrocompass indicates the north-south direction. You spin it and after some time it aligns its axis of rotation accordingly. It does not use magnetic forces and is therefore not affected by magnetic disturbances.

Experiment 13.15: Magnus Gyrocompass

We use the gimbal gyroscope that we already know from Experiment 13.4 to built a gyrocompass. The axis that makes the gyroskope vertically rotatable in its suspension is blocked, the horizontal direction is provided with damping. The Earth's rotation is too slow to be detected with this compass. It has no motor and stops spinning after about a minute due to friction. However, aligning with the Earth's axis takes a few minutes. Therefore, we place the top on a swivel chair that simulates the Earth's rotation. We tilt the top, as seen in the picture, to indicate a position in the northern hemisphere. The rotation axis of the top is again marked with a red arrow.
Now we spin the gyroscope in an arbitrary orientation and then slowly and as evenly as possible turn the swivel chair. The axis of rotation of the gyroscope immedi-

© RWTH Aachen, Physics Experiment Collection

ately turns towards the fictitious North Pole. It oscillates around this direction until the damping has calmed the axis down. Now it is aligned in the north-south direction. Depending on the initial condition, it can also happen that the arrow points in the opposite direction towards the fictitious South Pole. The gyrocompass only indicates the north-south direction, it cannot distinguish between the North and South Pole.

● Fig. 13.19 shows a commercial gyrocompass. The housing is cut open so that you can look inside. You can see the flywheel driven by an electric motor. The axis of the gyroscope lies horizontally. It is mounted in such a way that the axis can rotate freely in the horizontal plane. A damping of this rotation reduces the settling time of the compass after starting the motor.

To understand the function of the gyrocompass, let's imagine for simplicity's sake that it is set up at the equator. The gyro axis points east-west at the start, so initially in the wrong direction (● Fig. 13.20). We look "from above" onto the Earth. The black double arrow indicates the initial gyro axis. Due to the rotation of the earth, the gyro axis must also rotate around the Earth's axis. A torque acts on the gyro, pointing north (indicated in the figure by the green dot). Due to $\frac{d\vec{L}}{dt} = \vec{M}$ the component of the gyro's angular momentum pointing north increases, which eventually aligns it in a north-south direction.

● Fig. 13.19 A gyrocompass from an airplane. © Wikimedia: Hannes Grobe/AWI[5]

---

5   License CC-A 3.0, see ▶ https://creativecommons.org/licenses/by/3.0/deed.en.

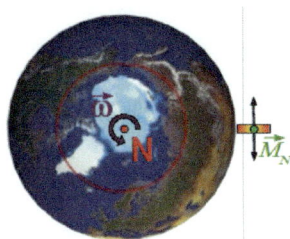

Fig. 13.20 Schematic representation of a gyrocompass at the equator. View from the north onto the Earth

> **Example 13.15: Cycling**
>
> When cycling hands-free, you steer the front wheel by leaning into the curve. This can be explained as follows: By shifting your weight, you create a torque on the front wheel (sketch). This causes a precession of the wheel's axis in the horizontal plane, which turns it in the desired direction. However, this simple explanation is controversially discussed.[6]

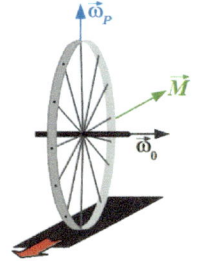

13.6 The Inertia Tensor

To conclude this chapter on the rotational motion of rigid bodies, we want to deal with rotations where the axis of rotation is not subject to any constraints. It can freely align itself in space. To handle such situations, we first want to expand the concept of the moment of inertia. We have introduced the moment of inertia of a body as a scalar quantity depending on the position of the axis of rotation. We had to recalculate it for each axis. It should be clear that the moments of inertia of closely spaced axes are related to each other. If we limit ourselves to axes through the center of gravity[7], we can specify the moment of inertia for any direction of the axis in closed form, but not by a scalar quantity, but by a second-order tensor, a 3×3-matrix. We indicate this by put-

---

6 A detailed presentation can be found, for example, in: Rainer Müller, Classical Physics, Chap. 13.8
7 Once we have captured the moments of inertia through the center of gravity, we can calculate moments of inertia with respect to any axis using the parallel axis theorem.

ting a tilde above the symbol: $\tilde{I}$. It is called the inertia tensor (for tensors see Appendix A3.20).

We once again consider the relation $\vec{L} = I\vec{\omega}$. We derived it from the orbital angular momentum of individual mass elements:

$$\vec{L}_i = \vec{r}_i \times m_i \vec{v}_i = \vec{r}_i \times m_i(\vec{\omega} \times \vec{r}_i) \tag{13.52}$$

Here, we can choose the vector $\vec{\omega}$ arbitrarily. When introducing the moment of inertia in ▶ Sect. 13.2, we pulled $\vec{\omega}$ out of the expression and combined the rest into the moment of inertia. Now we proceed differently. We calculate the expression component by component ($\vec{\omega} = (\omega_x\ \omega_y\ \omega_z)$, $\vec{r}_i = (x_i\ y_i\ z_i)$):

$$\begin{aligned}
\vec{L}_i &= \vec{r}_i \times m_i \begin{pmatrix} \omega_y z_i - \omega_z y_i \\ \omega_z x_i - \omega_x z_i \\ \omega_x y_i - \omega_y x_i \end{pmatrix} \\
&= m_i \begin{pmatrix} y_i^2 \omega_x - x_i y_i \omega_y - x_i z_i \omega_z + z_i^2 \omega_x \\ z_i^2 \omega_y - y_i z_i \omega_z - x_i y_i \omega_x + x_i^2 \omega_y \\ x_i^2 \omega_z - x_i z_i \omega_x - y_i z_i \omega_y + y_i^2 \omega_z \end{pmatrix} \\
&= m_i \begin{pmatrix} (y_i^2 + z_i^2)\omega_x - x_i y_i \omega_y - x_i z_i \omega_z \\ -x_i y_i \omega_x + (x_i^2 + z_i^2)\omega_y - y_i z_i \omega_z \\ -x_i z_i \omega_x - y_i z_i \omega_y + (x_i^2 + y_i^2)\omega_z \end{pmatrix} \\
&= m_i \begin{pmatrix} y_i^2 + z_i^2 & -x_i y_i & -x_i z_i \\ -x_i y_i & x_i^2 + z_i^2 & -y_i z_i \\ -x_i z_i & -y_i z_i & x_i^2 + y_i^2 \end{pmatrix} \cdot \begin{pmatrix} \omega_x \\ \omega_y \\ \omega_z \end{pmatrix}.
\end{aligned} \tag{13.53}$$

In the third line, we rearranged the terms and rewrote the result in the last line in matrix notation. Now we can read off the inertia tensor. We just need to do the transition from individual mass points to the integral over the body. It is then $\vec{L} = \tilde{I}\vec{\omega}$ with:

$$\tilde{I} = \begin{pmatrix} I_{xx} & I_{xy} & I_{xz} \\ I_{yx} & I_{yy} & I_{yz} \\ I_{zx} & I_{zy} & I_{zz} \end{pmatrix} \tag{13.54}$$

13.6 · The Inertia Tensor

and

$$I_{xx} = \int_V (y^2 + z^2)\,dm \quad I_{xy} = -\int_V xy\,dm \quad I_{xz} = -\int_V xz\,dm$$

$$I_{yx} = -\int_V xy\,dm \quad I_{yy} = \int_V (x^2 + z^2)\,dm \quad I_{yz} = -\int_V yz\,dm$$

$$I_{zx} = -\int_V xz\,dm \quad I_{zy} = -\int_V yz\,dm \quad I_{zz} = \int_V (x^2 + y^2)\,dm.$$

(13.55)

Example 13.16: Inertia Tensor of a Cuboid

As an example, we calculate the inertia tensor of a cuboid with homogeneous density ρ, the square base area $a \times a$ and the height b. The sketch shows the coordinate system, whose origin coincides with the center of gravity of the cuboid. Now we have to calculate the nine integrals from Eq. 13.55.
The first two are

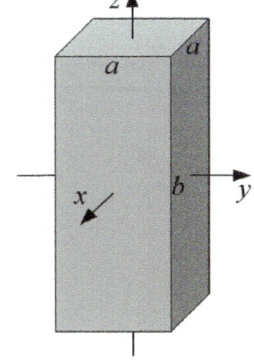

$$I_{xx} = \int_V \left(y^2 + z^2\right)dm = \rho \int_{-b/2}^{+b/2}\int_{-a/2}^{+a/2}\int_{-a/2}^{+a/2} \left(x^2 + z^2\right)dx\,dy\,dz$$

$$= \int_{-b/2}^{+b/2}\int_{-a/2}^{+a/2} \left[\frac{1}{3}x^3 + z^2 x\right]_{-a/2}^{+a/2} dy\,dz$$

$$= \int_{-b/2}^{+b/2}\int_{-a/2}^{+a/2} \left(\frac{1}{12}a^3 + z^2 a\right) dy\,dz = \int_{-b/2}^{+b/2} \left(\frac{1}{12}a^4 + z^2 a^2\right) dz$$

$$= \frac{1}{12}a^4 b + \frac{1}{12}b^3 a^2,$$

$$I_{xy} = -\int_V xy\,dm = -\rho \int_{-\frac{b}{2}}^{+\frac{b}{2}}\int_{-\frac{a}{2}}^{+\frac{a}{2}}\int_{-\frac{a}{2}}^{+\frac{a}{2}} xy\,dx\,dy\,dz$$

$$= \int_{-\frac{b}{2}}^{+\frac{b}{2}}\int_{-\frac{a}{2}}^{+\frac{a}{2}} \left[\frac{1}{2}x^2 y\right]_{-\frac{a}{2}}^{+\frac{a}{2}} dy\,dz = 0.$$

Overall, it results in

$$\tilde{I}_{\text{cuboid}} = \frac{1}{12}a^2 b \begin{pmatrix} a^2 + b^2 & 0 & 0 \\ 0 & a^2 + b^2 & 0 \\ 0 & 0 & 2a^2 \end{pmatrix}.$$

This result can be compared with the entry in ◨ Table 13.1. I_{xx} corresponds to the moment of inertia given there. The off-diagonal elements vanish here, because we had initially oriented the coordinate axes along the symmetry axes of the cuboid.

One can now replace the moment of inertia with the inertia tensor in the different relations that we have introduced before. For example in:

$$E_{\text{rot}} = \frac{1}{2} \vec{\omega}^T \tilde{I} \vec{\omega}. \tag{13.56}$$

The inertia tensor can be graphically illustrated by the inertia ellipsoid. An example is shown in ◨ Fig. 13.21. The center of the ellipsoid is located at the center of gravity C of the body. The axes of the ellipsoid are called the "principal axes of inertia". They are determined by the symmetry axes of the rotating body and do not necessarily coincide with the coordinate axes, as these depend on the orientation of the body in the coordinate system. Therefore, we have labeled the principal axes of inertia in the figure with x', y' and z'.[8] Now, any arbitrary axis of rotation (through the center of gravity) can be drawn. Its point of penetration through the surface of the ellipsoid is labeled P in ◨ Fig. 13.21. The length of the line CP indicates the moment of inertia with respect to this axis:

$$|SP| = \frac{1}{\sqrt{I}}. \tag{13.57}$$

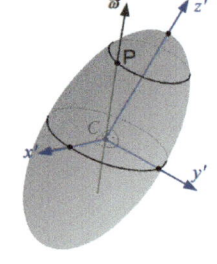

◨ **Fig. 13.21** The inertia ellipsoid

Note: Contrary to intuition, the moment of inertia is particularly small for the directions for which the semi-axis of the inertia ellipsoid is particularly large. Moments of inertia with respect to axes that do not pass through the center of gravity can be calculated from the parallel axis theorem.

---

5 In general, the principal axes of inertia are given by the three eigenvectors of the inertia tensor. If these are used as coordinate axes, the inertia tensor becomes a diagonal matrix.

13.6 · The Inertia Tensor

> **Example 13.17: Inertia tensor**
>
> We introduced the inertia tensor via the angular momentum (Eq. 13.53). We want to briefly show that this is in line with our previous definition of the moment of inertia (Eq. 13.15). We had:
>
> $$I = \int_{body} r_\perp^2 \, dm,$$
>
> with the perpendicular distance to the axis of rotation, which we defined from $r_\perp^2 + r_\parallel^2 = r^2$. The component parallel to the axis of rotation is determined by a scalar product: $r_\parallel = \vec{r} \cdot \hat{\omega} = \vec{r} \cdot \vec{\omega}/\omega$. With this we have:
>
> $$I = \int_{body} \left(r^2 - \left(\frac{\vec{r} \cdot \vec{\omega}}{\omega}\right)^2 \right) dm$$
>
> $$= \frac{1}{\omega^2} \int_{body} \left(\vec{r}^2 \vec{\omega}^2 - (\vec{r} \cdot \vec{\omega})^2 \right) dm.$$
>
> If we now write out the components, we get:
>
> $$I = \frac{1}{\omega^2} \sum_{i,j=1}^{3} \left[\int_{body} (\vec{r}^2 \delta_{ij} - r_i r_j) \, dm \right] \omega_i \omega_j$$
>
> $$= \frac{1}{\omega^2} \sum_{i,j=1}^{3} I_{ij} \, \omega_i \omega_j \quad \text{with} \quad I_{ij} = \int_{body} (\vec{r}^2 \delta_{ij} - r_i r_j) \, dm.$$
>
> Finally we replace $r_1 = x, r_2 = y, r_3 = z$ and we see the result that we already know from Eq. 13.53.

The moments of inertia with respect to the three main axes are called the principal moments of inertia I_a, I_b, I_c. They are usually sorted by size: $I_a \leq I_b \leq I_c$. If all three principal moments of inertia are different from each other, we speak of an "asymmetric spheroid". If at least two are the same, the top is called "symmetric". In ◘ Fig. 13.22, two molecules are shown. In rotation H_2O is an asymmetric spheroid, while CO_2 is symmetric (the moments of inertia about the vertical and the axis perpendicular to the plane of the drawing are the same). Among the symmetric spheroids, we further distinguish prolate ($I_a < I_b = I_c$) and oblate spheroids ($I_a = I_b < I_c$).

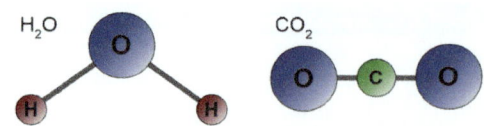

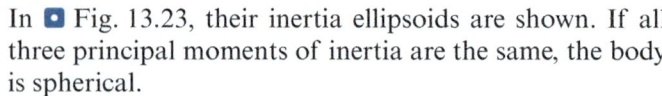

Fig. 13.22 Examples: Molecular tops

In ■ Fig. 13.23, their inertia ellipsoids are shown. If all three principal moments of inertia are the same, the body is spherical.

If we choose the principal axes of inertia as the coordinate axes, the inertia tensor is a diagonal matrix:

$$\tilde{I} = \begin{pmatrix} I_a & 0 & 0 \\ 0 & I_b & 0 \\ 0 & 0 & I_c \end{pmatrix}. \tag{13.58}$$

Then the angular momentum during rotation around the axis $\vec{\omega}$ is:

$$\vec{L} = \tilde{I}\,\vec{\omega} = (I_a \omega_{x'}, I_b \omega_{y'}, I_c \omega_{z'}). \tag{13.59}$$

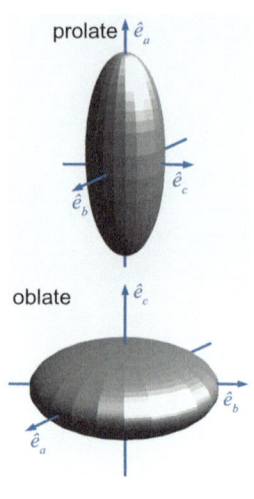

Fig. 13.23 Inertia ellipsoids of a prolate and oblate spheroids

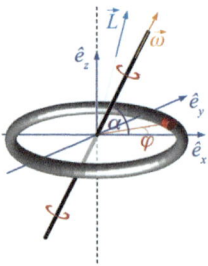

> **Example 13.18: Skewed Rotation**
>
> Here we discuss an example of a rotation, where the angular momentum does not fall on the axis of rotation, a situation that may surprise you. We consider the rotation of a tire. The tire is shown in the illustration. We have aligned it so that its axis of symmetry (dashed line) is vertically oriented. The tire is supposed to rotate around an axis that does not coincide with the axis of symmetry. The axis of rotation is tilted against the axis of symmetry by the angle $90° - \alpha$. In the coordinate system indicated in the figure, it lies in the x-z-plane and forms an angle α with the x-axis.
>
> We calculate the angular momentum by dividing the tire into infinitesimally small pieces. In the position depicted in the sketch, the tire lies in the x-y-plane. An infinitesimal element has the position $\vec{r} = R(\cos\varphi, \sin\varphi, 0)$, where R is the radius of the ring and φ determines the position of the element along the ring. We determine the tangential velocity $\vec{v}$, with which the element rotates around the axis, from $\vec{v} = \vec{\omega} \times \vec{r}$, with $\vec{\omega} = \omega_0 (\cos\alpha, 0, \sin\alpha)$:
>
> $\vec{v} = R\,\omega_0 (\sin\alpha \sin\varphi, -\sin\alpha \cos\varphi, -\cos\alpha \sin\varphi)$.

13.6 · The Inertia Tensor

In addition to the tangential velocity, velocity components also occur in the radial direction and along the axis of rotation, but these are irrelevant here as they do not contribute to the angular momentum. From the velocity, we now determine the angular momentum $\vec{L} = \vec{r} \times m\vec{v}$:

$$d\vec{L} = dm\, R^2\, \omega_0 \left(\cos\alpha \sin^2\varphi, -\cos\alpha \sin\varphi \cos\varphi, \sin\alpha\right),$$

with the mass $dm = m\, d\varphi/2\pi$. Finally we integrate the entire tire ($\vec{L} = \int_0^{2\pi} d\vec{L}$) and obtain:

$$\vec{L} = mR^2\, \omega_0 \left(\frac{1}{2}\cos\alpha, 0, \sin\alpha\right),$$

which is not parallel to the vector $\vec{\omega}$. Now, if the tire rotates around the axis of rotation, the axis of symmetry rotates on a cone around the axis of rotation. The angular momentum rotates with it. It will always stand between the axis of rotation and the axis of symmetry in the plane spanned by these two vectors.

If you fix the axis with bearings and force the tire to rotate around this oblique axis, the angular momentum must rotate on a cone around the axis of rotation, too. To achieve this, torques are necessary, which are transferred through the bearings to the axis. The direction of the torque rotates with the rotation of the tire. You can feel it as a shaking at the bearings, provided the tire is tilted as in our example.

Our small calculation can also be viewed differently: If the axis of rotation is free, no torques can be exerted on it. The angular momentum is fixed in space. But since the axis of rotation includes an angle with the angular momentum, the axis of rotation now rotates on a cone around the stationary angular momentum. The body tumbles. This is always the case when the axis of rotation does not coincide with one of the axes of symmetry of the body.

Let's finally have a look at the formulas in this example. At the beginning of this section, we introduced the moment of inertia via the relation $\vec{L} = I\vec{\omega}$ which we cannot apply here, because it assumes a parallelism between $\vec{L}$ and $\vec{\omega}$. Instead, we must use the relation $\vec{L} = \tilde{I}\vec{\omega}$ with the inertia tensor $\tilde{I}$. If we neglect the volume of the tire,

> as we have done in the calculation above, we need the inertia tensor of a circular ring. This is
>
> $$\tilde{I} = \begin{pmatrix} \frac{1}{2}mR^2 & 0 & 0 \\ 0 & \frac{1}{2}mR^2 & 0 \\ 0 & 0 & mR^2 \end{pmatrix},$$
>
> which, with $\vec{\omega} = \omega_0 (\cos\alpha, 0, \sin\alpha)$ gives the correct angular momentum vector.

Obviously, the principal axes of inertia offer a certain simplification of the description of rotational movements. In a coordinate system S', which is defined by the principal axes of inertia $\hat{e}_{x'}, \hat{e}_{y'}$ and $\hat{e}_{z'}$, the inertia tensor has a diagonal form with time-constant elements. However, it should be noted that in some situations the body-fixed system of the main inertia axes S' rotates relative to a space-fixed system S and thus does not represent an inertial system. If we transform the basic equation $\vec{M} = d\vec{L}/dt$ into the system S', we must consider that the time derivative receives a contribution that is due to the rotation of S' relative to S. We do not want to perform the derivation here. The result is:

$$\begin{aligned} \vec{M} &= \frac{d\vec{L}}{dt} \quad &\text{in } S \\ \vec{M} &= \frac{d\vec{L}}{dt} + \left(\vec{\omega} \times \vec{L}\right) \quad &\text{in } S'. \end{aligned} \quad (13.60)$$

The relationship between $\vec{L}$ and $\vec{\omega}$ in S' is simple. It applies:

$$\vec{L} = I_a \omega_{x'} \hat{e}_{x'} + I_b \omega_{y'} \hat{e}_{y'} + I_c \omega_{z'} \hat{e}_{z'} \quad (13.61)$$

with the principal moments of inertia I_a, I_b and I_c with respect to the principal axes of inertia $\hat{e}_{x'}$, $\hat{e}_{y'}$ and $\hat{e}_{z'}$. The ▶ equation 13.60 in the system S' is called the Euler's equations of rotation or simply Euler's equations. These are three coupled differential equations. Written out, they are:

13.6 · The Inertia Tensor

$$M_{x'} = I_a \frac{d\omega_{x'}}{dt} + (I_c - I_b)\omega_{y'}\omega_{z'}$$
$$M_{y'} = I_b \frac{d\omega_{y'}}{dt} + (I_a - I_c)\omega_{z'}\omega_{x'} \qquad (13.62)$$
$$M_{z'} = I_c \frac{d\omega_{z'}}{dt} + (I_b - I_a)\omega_{x'}\omega_{y'}.$$

However, note that to solve these differential equations, the torque $\vec{M}$ must first be transformed into the system S'.

Example 13.19: Slip-free billiard shot

You hit a billiard ball (mass m, radius R, moment of inertia I) parallel to the table. At what height h do you have to hit the ball so that it starts to roll without slipping, even without friction to the table?
A homogeneous ball is a spherical gyroscope with three identical principal moments of inertia:

$$\tilde{I} = \begin{pmatrix} \frac{2}{5}mR^2 & 0 & 0 \\ 0 & \frac{2}{5}mR^2 & 0 \\ 0 & 0 & \frac{2}{5}mR^2 \end{pmatrix}.$$

The impact generates a torque of magnitude $\vec{M} = -F(h-R)\hat{e}_z$. Thus, the Euler equations are:

$$0 = I_a \frac{d^2\varphi_x}{dt^2}$$
$$0 = I_b \frac{d^2\varphi_y}{dt^2}$$
$$-F(h-R) = I_c \frac{d^2\varphi_z}{dt^2}$$

with the rotation angles φ_i around the i-th axis. The ball is set into motion by the impact in x-direction. So, $F = m a_x$ must apply. The acceleration a_x is connected to the angular acceleration through the rolling condition i.e., $a_x = -R d^2\varphi_x/dt^2$ must apply. This results in:

$$-F(h-R) = I_c \frac{d^2\varphi_z}{dt^2} = -I_c \frac{a_x}{R} = -I_c \frac{F}{mR}$$

from which

$$h = R + \frac{I_c}{mR} = \frac{7}{5}R$$

follows. If the ball is hit at this height, it will roll without slipping.

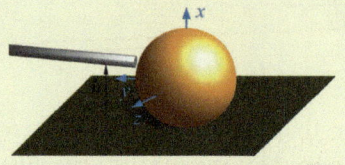

Example 13.20: Force-free symmetric gyroscope

As another application of Euler's equations, we want to discuss the motion of a force-free symmetric gyroscope. We assume that it is an oblate gyroscope, i.e. $I_a = I_b < I_c$. The Euler equations are:

$$0 = I_a \frac{d\omega_{x'}}{dt} + (I_c - I_a)\,\omega_{y'}\,\omega_{z'}$$

$$0 = I_a \frac{d\omega_{y'}}{dt} + (I_a - I_c)\,\omega_{x'}\,\omega_{z'}$$

$$0 = I_c \frac{d\omega_{z'}}{dt},$$

where we have already replaced I_b by I_a. The third equation states that $\omega_{z'}$ is constant. To solve the first two differential equations, we introduce a new quantity Ω:

$$\Omega = \frac{I_c - I_a}{I_a}\,\omega_{z'}.$$

This is a constant frequency, the meaning of which we will clarify below. The differential equations then take the following form:

$$\frac{d\omega_{x'}}{dt} + \Omega\,\omega_{y'} = 0$$

$$\frac{d\omega_{y'}}{dt} - \Omega\,\omega_{x'} = 0.$$

The general solution is:

$$\omega_{x'}(t) = \cos(\Omega t)\,\omega_{x'}(0) - \sin(\Omega t)\,\omega_{y'}(0)$$
$$\omega_{y'}(t) = \sin(\Omega t)\,\omega_{x'}(0) + \cos(\Omega t)\,\omega_{y'}(0).$$

In the case of an oblate gyroscope, the position of the z'-axis is given as the axis of symmetry of the gyroscope, also called the figure axis. The other two principal axes are perpendicular to it, but due to $I_a = I_b$ they are not uniquely determined. We can rotate them so that $\omega_{y'}(0)$ becomes zero. Thus, the axis of rotation $\vec{\omega}$ initially lies in the x'-z'-plane. The angle of $\vec{\omega}$ to the z'-axis is θ. Note that $\omega_\perp = \sqrt{\omega_{x'}^2 + \omega_{y'}^2} = \omega_{x'}(0)$ is as constant just as $\omega_{z'}$. This implies that the angle θ must be constant. The gyroscope's axis of rotation moves with the angular velocity Ω on a cone around the figure axis, unless in addition to $\omega_{y'}(0)$ also $\omega_{x'}(0)$ is zero. The movement for $\omega_{x'}(0) \neq 0$ we have called the tumbling of the gyro axis in example 13.18. With the Euler equations, we were able to determine the movement explicitly.

We have calculated the motion in the body-fixed system of the principle inertia axes S', but usually we are interested in the space-fixed system S. We still need to derive the coordinate transformation from S' to S. Fortunately, there is a simple approach here via the angular momentum. This must be preserved in the space-fixed system S. It is calculated according to

$$\vec{L} = \tilde{I}\vec{\omega} = I_a \omega_{x'} \hat{e}_{x'} + I_a \omega_{y'} \hat{e}_{y'} + I_c \omega_{z'} \hat{e}_{z'}$$
$$= I_a (\vec{\omega} - \omega_{z'} \hat{e}_{z'}) + I_c \omega_{z'} \hat{e}_{z'} = I_a \vec{\omega} + (I_c - I_a) \omega_{z'} \hat{e}_{z'},$$

which shows that the angular momentum always lies in the plane spanned by $\vec{\omega}$ and the figure axis $\hat{e}_{z'}$. Furthermore, we see that in addition to $\omega_{z'}$ also ω_z is constant when we consider the rotational energy. This is calculated as

$$E_{\text{rot}} = \frac{1}{2} \vec{\omega}^T \tilde{I} \vec{\omega} = \frac{1}{2} \vec{\omega}^T L \hat{e}_z = \frac{1}{2} \omega_z L.$$

Since both L and also E_{rot} being constant, ω_z must also be constant. However, this means that $\vec{\omega}$ revolves around the angular momentum on a cone, and since the angular momentum, the vector $\vec{\omega}$ and the figure axis lie in one plane, the figure axis must revolve around the angular momentum at the same speed. Although there are many more details of this motion to discuss, we want to end the discussion at this point.

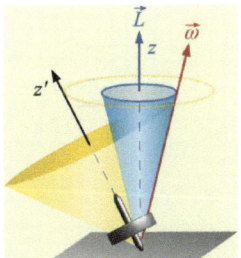

13.7 Rotation around Free Axes

We now turn to free rotations, i.e., rotational motions where the axis of rotation can move freely in space. Think, for example, of a juggler who throws balls or cones and sets them in rotation. These are also rigid bodies, which we refer to here as gyroscopes. The trajectory of the center of gravity is given by the throw, but how does the axis of rotation change on the path?

In the simplest case, no external torque acts on the gyroscope. The angular momentum vector is constant in this case, but as can be seen from Eq. 13.59, the axis of rotation of the body does not generally coincide with the direction of the angular momentum. The axis of rotation is then no longer fixed in space, it rotates around the direction of the angular momentum. The gyroscope tumbles. A rotation with a fixed axis of rotation is only obtained under certain conditions, which we will not derive here, but at least want to indicate. A fixed axis of rotation is obtained

a. in the case of a spherical gyroscope, i.e. if $I_a = I_b = I_c$.
b. in the case of a symmetric gyroscope when rotating about an axis perpendicular to the axis of symmetry, i.e. if $I_a = I_b$ and $\omega_{z'} = 0$ or $I_b = I_c$ and $\omega_{x'} = 0$.
c. for any gyroscope, when the axis of rotation coincides with one of the principal axes. Then only one of the three components of $\vec{\omega}$ is different from zero.

You can find case c. in example 13.20 if you set $\omega_{x'} = \omega_{y'} = 0$ there. Then the angle θ becomes zero and the axis of rotation and the figure axis coincide.

If one of the three conditions is met, the axis of rotation coincides with the angular momentum. It is constant and the gyroscope rotates as if around a fixed axis. However, for practical application, we must study whether the rotation around the corresponding axis is stable, i.e., we must investigate what happens when the axis of rotation is slightly deflected from the direction of angular momentum due to a small disturbance (imbalance in the gyroscope, air currents, disturbances when starting, etc.).

It turns out that

a) in the case of a spherical gyroscope, all axes are stable,
b) in the case of a symmetric gyroscope, the two configurations in which $\vec{L}$ and $\vec{\omega}$ coincide, are stable,

13.7 · Rotation Around Free Axes

c) in the general case, only the rotation around the principal axes, which belong to the smallest and largest moment of inertia, are stable. If you rotate the gyroscope around the middle axis, a small disturbance causes the gyroscope to tumble and eventually settle on one of the two stable axes.

Experiment 13.16: Rotation around Free Axes

The conditions under which a body stably rotates around a free axis may initially seem complex. However, the results are quite intuitive, as this experiment shows.
Different objects can be hung on a motor and set into rotation around the vertical axis. We first consider a rod. A rod is a symmetric, prolate gyroscope with $I_a < I_b = I_c$. Where I_a is the moment of inertia concerning the rotation around the longitudinal axis, and I_b and I_c around the axes perpendicular to it. The rod is flexibly connected to the motor via another thin and light rod. When the motor is switched off, the rod hangs in position A due to gravity (see figure). This is still the case when the motor is slowly rotating. We now rotate around $\omega_{x'}$, which is not stable according to the discussion in this chapter, but is possible with slow rotation due to gravity. If the motor starts to spin, the rod comes into positions like B shown in the figure due to disturbances. It moves irregularly. If the rotation speed is finally high enough, the rod jumps into position C, where it rotates stably. Now the rotation around the longitudinal axis of the rod ($\omega_{x'}$) vanishes, as should be the case according to our considerations.
You can attach other objects and observe what happens. The photo shows a chain that hangs down when the motor is switched off. At high rotation speed, it will form a circle, with a vertically aligned axis around which the chain rotates stably.

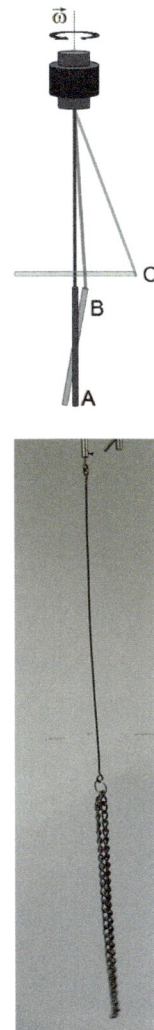

© RWTH Aachen, Physics Experiment Collection

Experiment 13.17: Juggling with Cigar Boxes

This experiment is a game with a surprising outcome: Two students are invited onto the stage to toss each other the cigar boxes depicted in the photo. Opposite sides of the cigar boxes are colored identically to make it easier to identify these sides. The two throwers are positioned in such a way that the audience can see them from the side, and they are asked to put the cigar boxes into as fast a rotation as possible when throwing. The cigar boxes should initially be thrown so that the audience sees the yellow side. The cigar boxes then rotate around an axis perpendicular to the yellow surface. This works very well.

© RWTH Aachen, Physics Experiment Collection

In the next round, the cigar boxes are turned so that the small, blue surface faces the audience and the cigar boxes rotate around an axis perpendicular to this. In this orientation, they are a little more difficult to throw, but this usually works quite well too.

Finally, the students are asked to throw the cigar boxes with the red side facing the audience. Surprisingly, this is not possible. No matter how often the students try, the cigar boxes tumble briefly, veer to the side, and at the end of the flight path rotate around the yellow or sometimes the blue side. The cigar box is an asymmetric top. The red end face corresponds to the intermediate moment of inertia I_b, and as we have learned, only the rotation around the axes with the greatest and smallest moment of inertia are stable, not the rotation around the intermediate one.

13.8 Comparison

Over the course of this chapter, in addition to the already known translational motions, you have learned about rotations. We have tried to build the description of rota-

13.8 · Comparison

tions similar to that of translations. At the end of this chapter, we would like to clarify this analogy once more.

In ◘ Table 13.2, the physical quantities that we use to describe motions are compared. Translational motions are described through distance, speed, and acceleration. In the case of rotations, these are replaced by the angle of rotation, the angular velocity, and the angular acceleration. Once an axis of rotation has been determined, the moment of inertia in rotations corresponds to the mass in translations. If you do not want to decide on an axis of rotation, you must work with the inertia tensor and the parallel axis theorem. Angular momentum and torque take on the roles that momentum and force have in translations. However, you should note that the analogy that the table shows is limited. The quantities of rotation refer to a specific axis, whereas the quantities of translation do not require such a reference. This difference is most obvious in the case of mass and moment of inertia. The mass is a property of the body that can be specified independently of a spatial reference, but its moment of inertia cannot. It changes depending on the orientation or position of the axis of rotation. The same applies to all other quantities in the table.

In ◘ Table 13.3, the most important relations for calculating translational and rotational motions are compared. Here too, the limited analogy that we have already seen in ◘ Table 13.2 is recognizable. Finally, in ◘ Table 13.4, the most important conversions from translational to rotational quantities are compiled.

◘ **Table 13.2** Physical quantities for describing translational and rotational motions

Translation	Rotation
Path $\vec{s}$	Rotation angle $\vec{\varphi}$
Speed $\vec{v}$	Angular speed $\vec{\omega}$
Acceleration $\vec{a}$	Angular acceleration $\vec{\alpha}$
Mass m	Moment of inertia I
Momentum $\vec{p}$	Angular momentum $\vec{L}$
Force $\vec{F}$	Torque $\vec{M}$

Table 13.3 Physical relations for translational and rotational motions

Translation	Rotation
$\vec{p} = m\vec{v}$	$\vec{L} = I\vec{\omega}$
$\vec{F} = m\vec{a} = \frac{d\vec{p}}{dt}$	$\vec{M} = I\vec{\alpha} = \frac{d\vec{L}}{dt}$
$W = \int \vec{F}\, d\vec{s}$	$W = \int \vec{M}\, d\vec{\varphi}$
$E_{kin} = \frac{1}{2}mv^2$	$E_{rot} = \frac{1}{2}I\omega^2$

Table 13.4 Conversion of translational quantities into rotation

$\vec{v} = \vec{\omega} \times \vec{r}$

$\vec{L} = \vec{r} \times \vec{p}$

$\vec{M} = \vec{r} \times \vec{F}$

$I = \int r_\perp^2\, dm$

❓ Problems

1. A cyclist accelerates from rest to a final velocity of 30 km/h within 5 s. Each wheel has a mass of 1 kg and a diameter of 0.62 m. Calculate the angular acceleration of a wheel and the force the cyclist must exert to accelerate a wheel. Assume that the entire mass of the wheel is concentrated on its tread. Calculate once with the wheel hub and once with the contact point of the tire as the rotation point. Compare both results!

2. A wire with a mass per unit length ρ is bent into a Sakai gyroscope, where the circular arc receives a curvature radius R. Calculate the moment of inertia with respect to the gyroscope axis, i.e., the principal axis of inertia. Compare it with the moment of inertia of a circular ring of the same radius.

3. A homogeneous disc with radius R, rotating with angular velocity ω and with the axis of rotation perpendicular to the disc surface, is placed on a flat surface. The coefficient of friction between the surface and the disc is μ. What is the angular deceleration with which the disc is slowed down? Apply the result to a hockey puck (diameter 76.2 mm), which starts at an initial speed of 6 m/s and covers a distance of 22 m on

13.8 · Comparison

the ice. If the puck is set spinning at a rotation frequency of $40\,\frac{1}{s}$, how long does it continue to spin?

4. At what speed does the tip of a homogeneous rod of length l, which topples from a vertical resting position, reach the ground? Assume that the foot of the rod does not slip, but is mounted to rotate without friction.

5. A homogeneous disc with mass M and radius R is mounted on a pair of rails via an axis assumed to be massless with radius r. A thread is wound around the axis, pulling a freely hanging weight of mass m. Calculate the time it takes for the center of the disc, which is initially at rest, to cover the distance l!

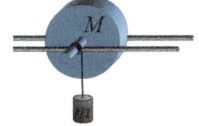

6. Calculate the acceleration of a solid cylinder on the one hand and a hollow cylinder on the other rolling down a slope with a gradient of 10%! Compare with the acceleration of a sliding cylinder. How large must the coefficient of static friction be between the outer surface of the rolling cylinders and the surface of the inclined plane so that they do not start to slide?

7. A boy throws a bouncy ball with radius R giving it some spin and the initial speed v_0 at an angle θ with respect to the horizontal on the ground. How large must the angular velocity ω_0 of the bouncy ball be at the time of release so that it flies directly back towards the hand? Assume an elastic bounce without slippage. Apply the result to the situation of a soccer ball (hollow sphere!) that bounces down from the goal's crossbar, just hits behind the goal line (ball fully behind the line) and then jumps back up to the crossbar. How many revolutions per second must the ball have for this (elastic bounce without slippage on the synthetic turf, goal line width 10 cm, ball diameter 22 cm, height of the goal's crossbar 2.44 m, ball speed 120 km/h)?

8. A roller mill is a type of mill, that was primarily used for oil mills in the past. Here, the millstone, whose axis of rotation is rotatably mounted on a vertical rod, is rolled over the ground material. What is the total force with which the millstone presses on the ground material (own weight plus the force due to precession), if its radius is $r = 0.5$ m and its mass $m = 500$ kg are and it is moved around the vertical driving rod at a distance of $R = 1$ m with a rotation

© Wikimedia: Dorothea Witter-Rieder

frequency of $f = 1\,\frac{1}{s}$? First, transfer the sketch and enter the required vectors.

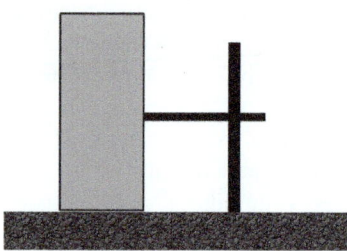

9. Through tidal forces, the Moon slows down the rotation of the Earth around its own axis, just as the Moon's own rotation has already been slowed down so much by tidal forces that it now coincides with the orbital rotation and therefore the Moon always faces the Earth with the same side. How large is the distance between the Earth and the Moon and how long is an Earth day when the tidal deceleration of Earth's rotation is also completed? Take advantage of the fact that although rotational energy is lost due to the deceleration, the total angular momentum of the system remains constant. Neglect the inclination of the Earth's axis to the Moon's orbit and the angular momentum that remains in the Earth's and Moon's own rotation after the completion of tidal deceleration. Consider the Earth as a homogeneous sphere with a moment of inertia $\frac{2}{5}mR^2$.

Elastic Bodies

In the previous chapter, we dealt with rigid bodies. Rigid bodies are an idealization. We have discussed this in detail. Microscopically, bodies consist of atoms or molecules. Atoms have a typical size of 0.1 nm. Molecules can be larger. From a microscopic perspective, the idealization of the rigid body implies that the atoms or molecules are located in a fixed position in relation to their neighbors. They do not change the distance or position to their neighbors. Such a body retains its external shape. It is a solid. We now want to abandon the idealization of the rigid body. We will allow the atoms and molecules to shift against each other under the influence of external forces. This will make it possible to deal with liquids and gases in addition to solids. The most important properties of these three states of matter are briefly summarized below.

Solids

Solids usually have a crystalline structure. Elastic forces hold the atoms (or ions or molecules) at their lattice positions. Under the influence of external forces, their distances and the directions between them change only slightly. The distance of the atoms or molecules in the lattice varies between 0.2 and 50 nm depending on the solid. We distinguish:
- Single crystals: The entire body is made up of a regular lattice that extends over the entire body.
- Polycrystalline bodies: The body is made up of many small single crystals. The orientation of the single crystals changes from one to the next.
- Amorphous solids: These do not have a regular lattice, but the position of the atoms or molecules relative to each other is nevertheless (almost) fixed, so that the body does not change its shape.

Liquids

The atoms or molecules that make up a liquid are not bound to fixed positions. They can slide against each other arbitrarily. Their distance is similar to that in solids. A liquid adapts its shape to the shape of the vessel in which it is kept. The liquid completely fills the vessel up to the filling level.

Gases

In gases, the atoms or molecules are almost freely movable. A gas completely occupies the space available to it. The distance of the atoms in the gas is strongly dependent on the pressure; no typical value can be given.

Example IV.1: Silicon

The illustrations show silicon as a solid in its different forms. The first image shows a silicon single crystal, as it is used in the semiconductor industry (a so-called ingot). The external shape originates from the manufacturing process. The surface is structureless, shiny. The next image shows polycrystalline silicon, with an enlargement of the surface in the third image. The individual crystals are clearly visible. Furthermore, there is amorphous silicon. Its surface is dull and rough.

© Wikimedia: Stahlkocher

© Wikimedia: Warut Roonguthai

© Wikimedia: Armin Kübelbeck

Contents

Chapter 14 Elastomechanics – 349

Chapter 15 Hydro- and Aerostatics – 369

Chapter 16 Hydro- and Aerodynamics – 407

Elastomechanics

Contents

14.1 Strain – 350

14.2 Bending – 355

14.3 Compression – 360

14.4 Shearing – 362

14.5 Finite Element Method – 364

© The Author(s), under exclusive license to Springer-Verlag GmbH, DE, part of Springer Nature 2023
S. Roth and A. Stahl, *Mechanics*,
https://doi.org/10.1007/978-3-662-68079-7_14

We first want to investigate how solids deform under the influence of external forces. We will only consider liquids and gases in the next chapter (▶ Chap. 15).

14.1 Strain

The deformation of a solid under external forces can be demonstrated with a wire. It lengthens when you pull on it.

Experiment 14.1: Straining a Copper Wire

A copper wire about 5 meters long with a diameter of 0.5 mm is attached at one end to the wall. The other end is guided over a pulley with a bucket attached to it, into which weights can be filled. A pointer is mounted on the pulley, which indicates a strain of the wire.

If we fill weights into the bucket, the wire stretches. We observe that the strain is proportional to the weight in the bucket. If we remove the weights from the bucket, the wire shortens to its original length.

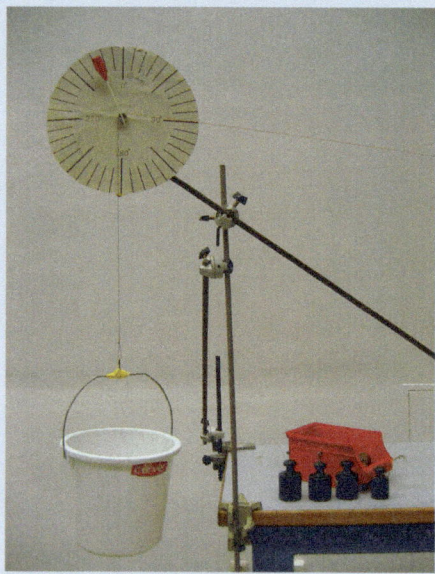

© RWTH Aachen, Physics Experiment Collection

14.1 · Strain

In ◘ Fig. 14.1, the strain of a wire is schematically represented. In the elastic range, the strain is proportional to the applied force $\vec{F}$. A thicker wire would stretch less under the same force. The strain must be inversely proportional to the cross-sectional area A of the wire. The inverse proportionality can be understood through a thought experiment: Imagine you want to stretch not just one wire with the cross-sectional area A with the force $\vec{F}$, but two of these wires in parallel. The force then distributes evenly between the wires and each is only stretched by half. But now you can also consider the two parallel wires as one wire with double the cross-sectional area, which is only stretched by half. The strain is therefore inversely proportional to the cross-sectional area.

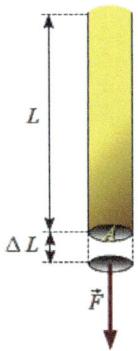

◘ **Fig. 14.1** Strain of a wire

We define the (tensile) stress σ as an auxiliary quantity for describing the strain:

$$\sigma = \frac{F}{A}, \qquad (14.1)$$

where the force acts in the direction of the wire and the area is measured perpendicular to the force (if the wire should be cut at an angle). We measure the strain ε itself as the relative change in length of the wire $\varepsilon = \Delta L/L$.

The relationships observed in experiments can be expressed by the simple relation $\varepsilon \propto \sigma$. We introduce a proportionality constant E and obtain:

$$\sigma = E\varepsilon. \qquad (14.2)$$

This is called Hooke's law.

As we will see later, the strict proportionality of Hooke's law is an approximation that only applies to very small strains. The proportionality constant E is called the Young's modulus. It is a material property of the wire. In ◘ Table 14.1 you will find the values of some materials as examples. The unit of Young's modulus is N/m^2, it is also abbreviated as Pascal (P_a).

Equation 14.2 represents only a special case of the Hooke's Law. In general, the law, which was first published as an anagram by the English physicist Robert Hooke in 1676, states that the deformation of elastic bodies is proportional to the applied stresses.

❯ Hooke's Law

The deformation of an elastic body is proportional to the applied stress.

Table 14.1 Elastic constants of some materials

Material	Young's modulus GPa	Shear modulus GPa	Poisson's ratio
Tungsten	400	150	
Steel	200	80	0.3
Copper	110	40	0.36
Aluminium	70	25	0.35
Carbon fiber	150	–	0.1
Glass (fiber)	50 … 90	20	0.2 … 0.3
Wood		1	
(along the fibers)	10 … 20		
(across the fibers)	0.2 … 2		
Plastics (PE, PP, PA)	1 … 5	0.1	0.3 … 0.5
Rubber	0.01 … 0.1	0.0002	0.5
Graphene	~ 1000		

We have introduced Young's modulus as a material constant, which must be determined experimentally for each wire. It is obtained from the quantitative evaluation of an experiment of the type of Experiment 14.1. For simple materials, the Young's modulus can also be calculated from the properties of the solid (from the properties of the bonds between the atoms). You will learn about it in solid state physics. The calculation is complicated and usually of limited accuracy, so it is of little importance for the practice of elastomechanics.

Example 14.1: Stiffness of a Wire

In ▶ Sect. 6.5 we introduced the stiffness. It indicates how much force must be applied to a spring to stretch it by a certain length: $F = k\,s$. The stiffness of a wire or an elastic thread can be defined accordingly. This stiffness is determined by Young's modulus:

$$k = \frac{F}{s} = \frac{F}{\Delta L} = \frac{\sigma A}{\varepsilon L} = E\frac{A}{L}.$$

14.1 · Strain

A wire that is stretched changes its shape. It will become thinner. We demonstrate it with a model in Experiment 14.2. This phenomennon is referred to as a "lateral strain of the wire". The more the wire is stretched, the thinner it will become. The lateral strain is proportional to the lengthening of the wire

$$\frac{\Delta d}{d} = -\mu \frac{\Delta L}{L}. \tag{14.3}$$

Here, d is the diameter of the wire. The proportionality constant μ is called "Poisson's ratio".

Experiment 14.2: Lateral Strain of a Rubber Band

Lateral strain does not only occur in wires. Here we demonstrate it using a rubber band. A white circle is drawn on a short and wide rubber band. When the rubber band is manually stretched, the circle deforms into an ellipse. It lengthens in the direction of the tensile stress and narrows perpendicular to it. As a reference, a transparent foil hangs in front of the rubber band, on which a circle (red dashed) in the original shape is drawn. The foil is only attached on one side and is therefore not stretched.

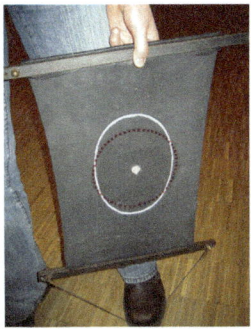

© RWTH Aachen, Physics Experiment Collection

Experiment 14.3: Elongation beyond the Elastic Limit

For this experiment, we use the same setup as in Experiment 14.1. We put a weight in the bucket, observe the lengthening of the wire, and remove it again. Then we gradually increase the weight. Initially, the elongation increases proportionally to the weight and disappears completely after the weight has been removed but the copper wire has only a limited tensile strength. The linear range of Hooke's Law ends for this wire at about 1.5 kg. With higher weights, the lengthening of the wire does not completely return to zero after the weight has been removed. At about 4 kg the wire begins to flow. After the weight has been put in the bucket, the wire gradually continues to lengthen. It gets longer and longer until it finally snaps and the bucket with the weights crashes on the floor. Sometimes you are lucky and the flow of the wire comes to a halt but if you then add even a small additional weight, it begins to flow again and snaps.

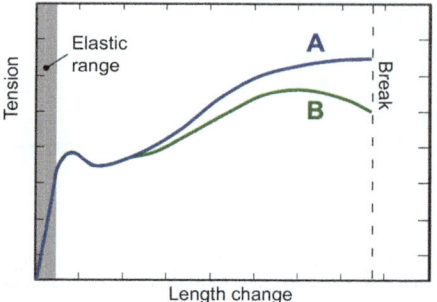

Fig. 14.2 Elongation of a wire under tensile stress. Curve B: Tensile stress corrected for cross-sectional change due to lateral contraction

We consider once again the elongation of a wire. We continue to increase the tensile stress until the wire finally snaps (Experiment 14.3). In Fig. 14.2, we show the change in length of the wire. The figure is called a stress-strain diagram. The elastic region, in which Hooke's law applies, is highlighted in gray. The end of the elastic region is called the "yield point". A change in length in the elastic region is only of a temporary nature. If the stress is removed from the wire, it returns to its original length. This is no longer the case if the yield point is exceeded. If the stress is removed in this case, the wire does shorten, but not back to its original length. A permanent deformation (elongation) remains. Beyond the yield point, an increase in tensile stress causes a significantly larger change in length. If the stress is increased further, the wire finally snaps. The maximal stress the wire sustains without snapping is called its "tensile strength". For a steel wire, the yield point is about $2 \cdot 10^8 \, \text{N/m}^2$ and the tensile strength is about $5 \cdot 10^8 \, \text{N/m}^2$.

The stress values read from curve A of the stress-strain diagram (tensile strength, yield point) do not correspond to the true stress in the material. The stress is calculated from the tensile force and the initial cross-section of the wire. However, lateral contraction reduces the cross-section of the wire during the tensile test. If the tensile stress is corrected for the actual cross-section of the wire at a given stress, curve B in Fig. 14.2 results.

14.2 · Bending

> **Example 14.2: Tensile Strength of a Steel Wire**
>
> A steel wire with a cross-sectional area of $1\,\text{mm}^2$ (0.56 mm radius) can carry a weight of up to 50 kg. The tensile strength of steel wires is $5 \cdot 10^8\,\text{N/m}^2$. From Eq. 14.1 a maximum tensile force of $F = 5 \cdot 10^8\,\frac{\text{N}}{\text{m}^2} \cdot 1\,\text{mm}^2 = 500\,\text{N}$, which corresponds to a weight of approximately 50 kg. However, even at 20 kg load, a significant extension of the wire can be seen.

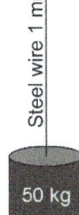

14.2 Bending

In ▶ Sect. 14.1 we discussed the simplest elastic property of a solid, the strain. Now, we want to deal with another elastic process, the bending or flexure. It occurs, for example, when beams are loaded. It causes stresses in the body, some areas of the body are stretched, while others are compressed. In experiment 14.4, these stresses are visualized.

> **Experiment 14.4: Visualization of Stresses**
>
>
>
> © RWTH Aachen, Physics Experiment Collection

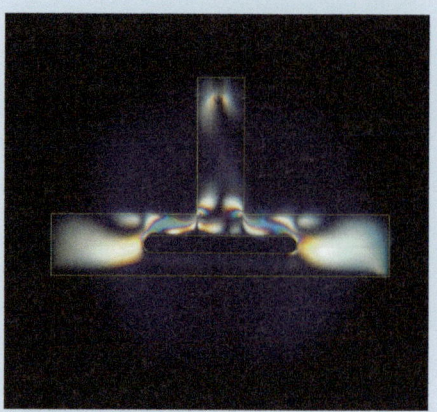

© RWTH Aachen, Physics Experiment Collection

We can make stress in a material visible through the effect of stress-induced birefringence. The two pictures show a part of the optical setup (from left: projection lamp, first polarizer, test object) and the result. The transparent part (plastic) is illuminated with polarized light. Mechanical stresses create an anisotropy in the material, which depends on the wavelength. The phase velocity of light with polarization in the direction of the stress will differ from that of light with perpendicular polarization. This rotates the polarization vector. Since the effect is wavelength-dependent, a colored image of the stresses is created.

We want to quantitatively capture a simple example of a bend, a loaded beam fixed on one end. The situation is sketched in ◘ Fig. 14.3. The beam is firmly attached to a wall on the left. At the right end, it is loaded with a force $\vec{F}_0$. The length of the beam is L_0. It has a rectangular cross-section with height a and width b. The beam deforms under the load. It is stretched on its upper side and compressed on its lower side. In the middle of the beam, there is an area that retains its original length. It is called the "neutral axis".

We consider a section of length L from the bent beam (see ◘ Fig. 14.3, second sketch). The bending radius ρ and the bending angle φ are linked together via $L = \rho \cdot \varphi$. The upper end of the beam is stretched by ΔL:

$$L + \Delta L = \left(\rho + \frac{a}{2}\right)\varphi = \rho\varphi + \frac{1}{2}a\varphi. \tag{14.4}$$

14.2 · Bending

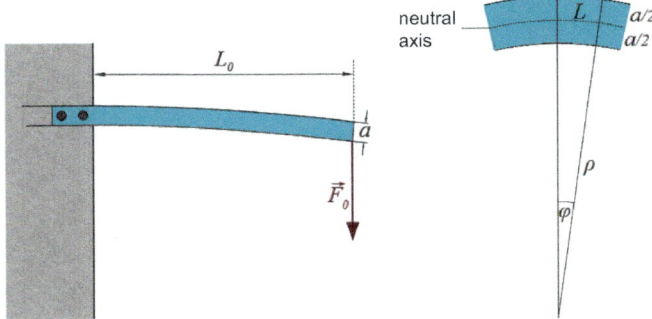

Fig. 14.3 A beam under a load $\vec{F}_0$

For the relative change in length, we get:

$$\frac{\Delta L}{L} = \frac{\frac{1}{2} a \varphi}{\rho \varphi} = \frac{a}{2\rho}. \tag{14.5}$$

From Hooke's law we obtain the occurring stresses. In terms of magnitude, we get

$$\begin{aligned} \text{above:} \quad & \sigma_a = E \frac{a}{2\rho} \\ \text{middle:} \quad & \sigma_m = 0 \\ \text{below:} \quad & \sigma_b = E \frac{a}{2\rho}. \end{aligned} \tag{14.6}$$

At any arbitrary radial distance x to the neutral axis, there is a corresponding stress $\sigma = E \frac{x}{\rho}$.

Forces result from these stresses, which in turn generate a torque in the beam (Fig. 14.4). The torque is opposed to the torque generated by the loading of the beam. The beam will deform until the torques compensate each other.

From the stress in an infinitesimally small piece dx of the beam, a force

$$dF = \sigma \, dA = E \frac{x}{\rho} b \, dx \tag{14.7}$$

results, from which an infinitesimal torque arises:

$$dM = x \, dF = \frac{Eb}{\rho} x^2 dx. \tag{14.8}$$

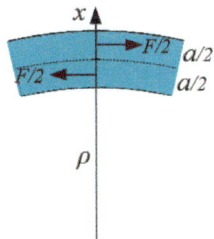

Fig. 14.4 Forces in the curved beam

Now, we integrate over the cross-section of the beam from the neutral axis to the beam's edge ($x = a/2$). A factor of 2 takes into account that from the upper and lower half of the beam a torque of the same size and direction results. Then it is:

$$M = 2 \int_0^{a/2} \frac{Eb}{\rho} x^2 dx = 2 \frac{Eb}{\rho} \left[\frac{1}{3} x^3 \right]_0^{a/2} = \frac{a^3 b}{12} \frac{E}{\rho}. \quad (14.9)$$

This torque must compensate the external torque $M_0 = F_0 L_0$. So, it must be:

$$M = \frac{a^3 b}{12} \frac{E}{\rho} = F_0 L_0 = M_0 \Rightarrow \rho = \frac{a^3 b}{12} \frac{E}{F_0 L_0}, \quad (14.10)$$

which determines the bending of the beam.

We want to examine this result more closely. A large bending radius ρ means that the beam bends only slightly. This is achieved by a larger modulus of elasticity E or by a short beam L_0. The longer the beam, the more it bends. Obviously, the bending also increases with increasing load. The dependence on the cross-section of the beam is interesting. Here, it's not simply the cross-sectional area that comes into play, as you might naively expect. Given a certain cross-sectional area, a beam with a large height (a^3), but a small width (b) bends significantly less. This is utilized in statics by using beams that have a rectangular cross-section, with a greater height than width.

Example 14.3: Cross-Sections of Steel Beams

We want to compare the stability of three steel beams. The three beams have the same cross-sectional area of $a^2/4$. The cross-sections are shown in the figure with their dimensions. We want to determine the bending radius under vertical and horizontal load. The calculation for A and B results directly from Eq. 14.10.

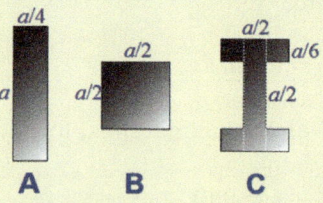

To calculate the double-T beam, we note that our calculation is linear in thickness perpendicular to the load-bearing force. In our minds, we can cut the beam into layers parallel to the neutral axis and then simply add the contributions of the individual layers to the bending radius. This can be easily verified using the example of the square beam B. If, for example, it is divided into three layers each with the thickness $\frac{1}{6}a$, the bending radius for each individual disk is 1/3 of the total value. If you add the bending radii of the three layers, you get back to the correct value of the entire beam. Physically, this is because no forces occur between the layers, so it doesn't matter whether the individual layers are combined into a beam or not. This linearity does not apply in the vertical direction!

In this way, we can calculate the double-T beam. The division into three layers for the calculation of the vertical load is indicated in the sketch by white dashed lines. You get

$$\rho = \frac{1}{12}\frac{E}{F_0 L_0}\left(\left(\frac{a}{2}+2\frac{a}{6}\right)^3\frac{a}{6}+4\left(\frac{a}{6}\right)^3\frac{a}{6}\right).$$

The results are summarized in the following table.

Beam	Vertical	Horizontal
A	$\dfrac{1}{48}\dfrac{E}{F_0 L_0}a^4$	$\dfrac{1}{768}\dfrac{E}{F_0 L_0}a^4$
B	$\dfrac{1}{192}\dfrac{E}{F_0 L_0}a^4$	$\dfrac{1}{192}\dfrac{E}{F_0 L_0}a^4$
C	$\dfrac{43}{5184}\dfrac{E}{F_0 L_0}a^4$	$\dfrac{19}{5184}\dfrac{E}{F_0 L_0}a^4$

The beam A has the highest stiffness under vertical load, but also the lowest under horizontal load. This should not be surprising. The comparison between B and C is interesting: Under vertical load, the double-T beam bends only about half as much as the square one, while under lateral load they are almost equally stiff. The stability of the double-T beam can be further increased by making the plates from which it is constructed even thinner and taller (while maintaining a constant cross-sectional area).

Finally, we want to examine the load limit of the beam. The beam breaks as soon as the tensile strength of the material is exceeded at any point in the beam. The stress is maximum at the top (or due to compression at the bottom) surface. At the top we have (Eq. 14.5), $\Delta L/L = a/2\rho$. Thus, the maximum stress in the material is

$$\sigma_{\max} = E \frac{a}{2\rho} = E \frac{a}{2} \frac{12}{a^3 b} \frac{F_0 L_0}{E} = \frac{6 F_0 L_0}{a^2 b}. \quad (14.11)$$

If the tensile stress induced by the load exceeds the tensile strength of the material, the beam breaks.

14.3 Compression

Under "compression", we understand the reduction of the volume of a body under the influence of external (pressure) forces. Compression is another elastic property of solid bodies that we want to study in this chapter. A reduction in volume already occurs simply by reversing the direction of the force straining a body. We want to start with that.

We consider a cuboid as a simple example (◘ Fig. 14.5). The volume of the cuboid is Ld^2. The change in volume can be traced back to the change in length during compression, which we have already discussed (▶ Sect. 14.1). However, it should be noted that the change in length goes hand in hand with a lateral contraction. In our case, with a negative change in length, the lateral contraction is positive, i.e. the cuboid widens transversely to the direction of compression. We determined the change of length (Eq. 14.2) and the lateral contraction (Eq. 14.3) to be

$$\frac{\Delta L}{L} = \frac{\sigma}{E},$$
$$\frac{\Delta d}{d} = -\mu \frac{\Delta L}{L}. \quad (14.12)$$

In the case of compression (◘ Fig. 14.6), we do not refer to σ as stress, but as pressure or pressure change, if

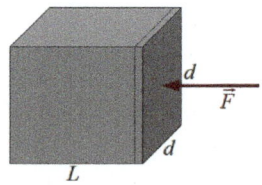

◘ **Fig. 14.5** Compression of a cuboid

14.3 · Compression

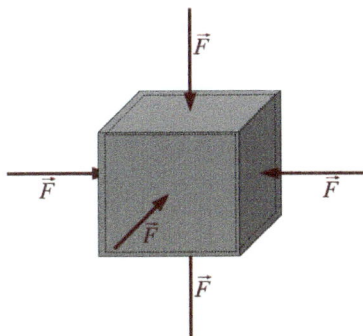

Fig. 14.6 Compression of a body by pressure from all sides

one wants to take into account that the body already felt the ambient pressure before an addional external pressure was applied.

We calculate the change in volume:

$$\begin{aligned}
\Delta V &= (L + \Delta L)(d + \Delta d)^2 - Ld^2 \\
&= (L + \Delta L)(d^2 + \Delta d^2 + 2\Delta dd) - Ld^2 \\
&= (Ld^2 + L\Delta d^2 + 2L\Delta dd + \Delta Ld^2 + \Delta L\Delta d^2 \\
&\quad + 2\Delta L\Delta dd) - Ld^2 \\
&= L\Delta d^2 + 2L\Delta dd + \Delta Ld^2 + \Delta L\Delta d^2 + 2\Delta L\Delta dd \\
&\approx 2L\Delta dd + \Delta Ld^2 \\
&= \left(2\frac{\Delta d}{d} + \frac{\Delta L}{L}\right)Ld^2 = \left(2\frac{\Delta d}{d} + \frac{\Delta L}{L}\right)V.
\end{aligned}$$
(14.13)

Please note that the formulas for the change in length and the transverse contraction are only valid for small changes of length. They are linear approximations. Therefore, we have neglected in our calculation all terms that go beyond a linear approximation, i.e., contain more than one factor Δ. We did this only at a relatively late stage in the calculation, where it was foreseeable that the leading terms would be of first order in Δ.

We now express the change of volume through Young's modulus:

$$\left(\frac{\Delta V}{V}\right)_{\text{strain}} = \left(-2\mu\frac{\sigma}{E} + \frac{\sigma}{E}\right) = (1 - 2\mu)\frac{\sigma}{E}. \quad (14.14)$$

In our first example, the force only acted from one side. We now want to assume that a change in ambient pressure causes a force to act from all sides. This would be the case, for example, if the body is submerged in water and then feels the water pressure.

The body now undergoes a compression in all three spatial directions. This is called compression or volume strain. The volume is thereby reduced threefold:

$$\left(\frac{\Delta V}{V}\right)_{\text{compression}} = 3\left(\frac{\Delta V}{V}\right)_{\text{strain}} = 3(1-2\mu)\frac{\Delta p}{E}. \tag{14.15}$$

14.4 Shearing

As another elastic property of solid bodies, we want to examine shearing. It occurs when forces act parallel to the surfaces of the body. In Fig. 14.7, a simple example can be seen. The body is fixed at its bottom surface, a force acts on the top parallel to the surface. Note that the force on the top not only applies a force, but also a torque to the body. Both the force and the torque must be compensated by the fixture.

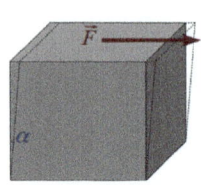

Fig. 14.7 Shearing of a cuboid

We define the shear stress τ as:

$$\tau = \frac{|\vec{F}_\parallel|}{A}. \tag{14.16}$$

The definition of shear stress is similar to the definition of tensile stress in Eq. 14.1. Both are defined by a ratio of force and area and therefore carry the unit Pascal, but the tensile stress referes to the force component perpendicular to the area, while the shear stress is formed from the parallel components.

The procedure is now analogous to strain or compression. For small stress, the shear angle α (see Fig. 14.7 for the definition) turns out to be proportional to the shear stress:

$$\tau = G\alpha. \tag{14.17}$$

Again, we have introduced a proportionality constant. It is called the modulus of shear or modulus of rigidity. Like Young's modulus, it is a material constant. Some values were already given in Table 14.1.

14.4 · Shearing

The various elastic constants are not independent of each other. For example, it can be shown (elasticity theory) that:

$$\frac{E}{2G} = 1 + \mu. \tag{14.18}$$

Example 14.4: Torsion

The twisting or torsion represents a special form of shear. It occurs in wires when one end is fixed and the other is twisted. A twist on a short piece of wire is shown in the sketch. In such a situation, a special form of Hooke's law comes into play. The applied force generates a torque, and this causes a proportional twist of the wire:

$$M = D\alpha.$$

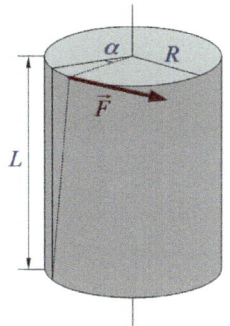

The twisting of thin wires is a very sensitive means of measuring small torques. For an example look back at the gravitational balance in Experiment 11.1 The elastic constant D describing the angular stiffness can be traced back to the shear modulus G considering the dimensions of the wire. We want to derive the relationship. We divide the cylinder shown in the sketch into infinitesimally thin, coaxial hollow cylinders with radius r and wall thickness dr. If the force $d\vec{F}$ is applied to one of the hollow cylinders, a shear stress $d\tau = dF/dA$ arises, where $dA = 2\pi r\, dr$ corresponds to the end face of the hollow cylinder. The shear angle α can be expressed by the angle φ, which describes the twist of the cylinder. From the sketch we read $r\varphi = L\alpha$. Thus, such a hollow cylinder generates a torque of magnitude

$$dM = r\, dF = r\tau\, 2\pi r\, dr = G\alpha\, 2\pi r^2\, dr = \frac{2\pi G}{L}\varphi r^3\, dr.$$

By integrating, we obtain the total torque:

$$M = \frac{2\pi G}{L}\varphi \int_0^R r^3\, dr = \frac{\pi G R^4}{2L}\varphi.$$

With this, we related the elastic constant D to the modulus of shear G. We got:

$$D = \frac{\pi R^4}{2L}G.$$

With this, we have discussed the most important elastic properties of solid bodies. More can be found in books of statics.

14.5 Finite Element Method

You have now learned about some elastic properties of solid bodies. They can be used to determine the deformation of bodies under the influence of external forces and torques. A simple example was discussed in ▶ Sect. 14.2: the bending of a beam under a load at its free end. We were able to determine the stresses in the beam and to derive the load limit at which the beam breaks. Statics is based on such calculations. They are indispensable for determining, for example, the load-bearing capacity of a bridge or a roof truss, or the deformation of a machine element in operation. Unfortunately, the analytical methods presented can handle only simple applications. In practice, they are not sufficient. We must resort to numerical approximation methods. We briefly outline the most widely used method, the finite element method. A real introduction would, however, go beyond the scope of this text.

The Finite Element Method allows solving partial differential equations under boundary conditions. This can be used, among other things, to determine the deformation of bodies under external loads. The calculation is performed on a computer and can be roughly divided into three steps.

1. In the preparation phase, the body to be examined is broken down into simple subparts. In the case of a three-dimensional calculation, these are volume elements. They are called "finite elements" because of their finite size, distinguishing them from infinitesimally small elements that would occur in a mathematically exact solution. Their corners, where adjacent elements touch, are called "nodes". The approximate determination of the displacement of these nodes is the goal of the calculation. The division of the body into finite elements is automatically performed by a computer program. This is usually the most difficult and time-consuming part of the calculation.
2. The forces acting on the nodes are linked to the displacements via a differential equation. The elastic

14.5 · Finite Element Method

properties of the body are incorporated into this differential equation. The differential equations are approximately solved for each volume element individually (*splines*). At each node, the solution must meet boundary conditions. At nodes inside the body, the forces must add up to zero, at nodes on the outer edge, they must match the external forces. The boundary conditions lead to a (very large) linear system of equations, which is solved numerically. The solution sought is the one that minimizes the energy stored in the system. At the end of the calculation, the position of the nodes has been approximately determined and the stresses can be determined from them.

3. Finally, the result is graphically processed so that the deformation of the body becomes visible and one can see, for example, how large the stresses in the body become at different points. Again, if the stresses exceed the load limit of the body at even one point only, it will break or tear. The deformation of the body can be read from the displacement of the nodes compared to the unloaded initial state.

Example 14.5: Car Accident

The illustration shows the result of a calculation using finite elements of the deformation of a car body upon impact on a wall edge. Shown are the deformation of the body after the impact and the stresses in the material of the body through a color code (green: high stresses, blue: low stresses).

Example 14.6: Bend of a Beam

We want to revisit the example of a beam that is fixed at one end (◘ Fig. 14.3). The figure shows the result of a calculation of the stresses using the finite element method. The beam is fixed on the left in the wall. In the calculation, it was assumed that this fixation is perfectly rigid. At the right end, it is loaded with a force $\vec{F}_0$ vertically downwards. The stresses are color-coded. The scale ranges from 0 to 90 MPa. The neutral fiber discussed in ▶ Sect. 14.2 is clearly visible in the middle of the beam where no stresses occur. The stresses increase non-linearly towards the top and bottom edge of the beam. The stresses are greatest directly at the fixation. They are still significantly below the load limit of a steel beam.

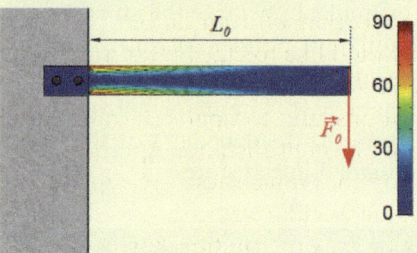

❓ Problems

1. A railway overhead wire made of copper has a length 1500 m and is tensioned by the force 10 kN. As a result, it extends by 1.1 m. How large is the cross section of the wire originally and by how much is it reduced by the tensioning? (Young's modulus of copper: $E = 1.1 \cdot 10^{11}$ N/m², shear modulus: $G = 4.0 \cdot 10^{10}$ N/m²)
2. An aluminum rod with a cross-sectional area of 1 cm² is stretched by a force of 10 kN. What change in material density of the rod is observed? (Young's modulus of aluminum: $E = 0.7 \cdot 10^{11}$ N/m², Poisson's ratio: $\mu = 0.35$)
3. A steel wire is hung from one end. How long can the wire be at most so that it does not tear under its own weight? (Density of steel: $\rho = 7.8 \cdot 10^3$ kg/m³, tensile strength: $5.0 \cdot 10^8$ N/m²)

14.5 · Finite Element Method

4. A rope of length L made of a material with Young's modulus E has a circular cross-section, whose diameter decreases linearly along the rope from d_1 to d_2. By how much does the rope extend when one pulls at the thinner end with the force F?

5. In the following we investigate the expansion of a spoked wheel:
   a) A straight thin rod of length L made of a material with density ρ and Young's modulus E rotates at the angular velocity ω around an axis through the end of the rod and perpendicular to the rod axis. How large is the resulting elongation of the rod?
   b) A thin ring with radius R made of a material with density ρ and Young's modulus E rotates at the angular velocity ω around an axis through the center of the ring perpendicular to the ring surface. How large is the resulting elongation of the ring's circumference?
   c) Compare the results from a) and b) in the context of the rotation of a spoked wheel!

6. A rigid beam of mass m is horizontally suspended by three wires (two at the ends, one in the middle), which are exactly the same length when not loaded. The two outer wires are identical, while the middle wire differs from the two outer wires in Young's modulus E and the wire diameter d. With what force are the wires tensioned? How should the diameters of the wires d_m d_a be set for a given ratio E_m/E_a, so that each wire carries the beam with the same force?

7. A horizontal rod of mass m is held at both ends by two wires of equal length made of identical material, but with different cross-sections A_1 and A_2. Where along the rod can a mass $M \gg m$ be hung so that the rod remains horizontal?

8. Consider a hollow wire with the length L, the radius R and a very thin wall of thickness d. The material has the shear modulus G. What torque must be applied to achieve a twist of the wire by the angle φ? Determine from this the angular stiffness D of the hollow wire. How large is this for a solid wire with the same outer radius?

9. The bending radius of a rod with any cross-section is given by

$$\rho = \frac{E}{F_0 L_0} \int x^2 \, dA.$$

The integral is also called the "second moment of area". Derive a formula for the curvature of a tube with inner radius R_i and outer radius R_a.

Hydro- and Aerostatics

Contents

15.1 Pressure – 370

15.2 Compressibility – 377

15.3 Hadrostatic Pressure – 381

15.4 Buoyancy – 388

15.5 Interfaces – 393

© The Author(s), under exclusive license to Springer-Verlag GmbH, DE, part of Springer Nature 2025
S. Roth and A. Stahl, *Mechanics*,
https://doi.org/10.1007/978-3-662-68079-7_15

15.1 Pressure

We have concluded the discussion of solids and now want to turn to gases and liquids. Gases and liquids resemble each other in many aspects. For this reason, we want to discuss them together. To simplify the description, we introduce a new umbrella term: the fluid. It is intended to denote both gases and liquids.

One of the most important quantities for dealing with fluids is the pressure. A fluid experiences pressure when forces act on it. Pressure is defined by the force that acts on an area inside the fluid or at its boundary. The pressure p is defined as the ratio of the component of the force perpendicular to the area, divided by the size of the area itself:

$$p = \frac{|\vec{F}_\perp|}{A}. \qquad (15.1)$$

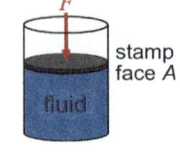

Fig. 15.1 On the definition of pressure

Pressure is a scalar quantity, even though it is often marked as an arrow in sketches (see Fig. 15.1). The arrow then indicates the direction of the force.

The unit of pressure in the SI system is the Pascal:

$$[p] = 1\,\frac{\text{N}}{\text{m}^2} = 1\,\text{Pa}. \qquad (15.2)$$

A number of non-Si units are still in use, which you can find in Table 15.1.

Table 15.1 Common pressure units

Unit	10^5 Pa correspond to
bar	1 bar
mm of mercury	750 torr
physical atmosphere	0.98692 atm
technical atmosphere	1.0197 at
pounds per square inch	14.50 psi

15.1 · Pressure

Example 15.1: The Weight of Air

An air pressure of 100,000 Pa implies that the weight of about ten metric tons of air rests on each square meter ($p = mg/A$). We do not feel this weight because of an identical pressure inside our bodies. If instead, we consider the vacuum container sketched in the figure (diameter 20 cm), a weight of about 300 kg of air rests on the lid. In a ventilated state, this is not a problem, as the air inside presses from below against the lid with the same force, thus carrying the weight of the air column above. But, this changes when the container is evacuated. Now the lid actually has to carry the air column. It must be designed to resist this load.

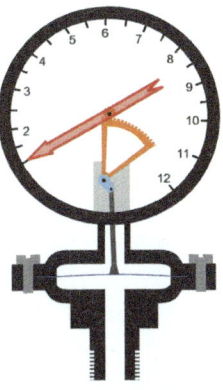

Example 15.2: Diaphragm Gauge

The diaphragm or membrane gauge is a simple device for measuring pressure. It is used, for example, on gas cylinders. It measures the pressure difference between the flange and the inside of the manometer, which contains air at ambient pressure. The two areas are separated by a membrane (blue in the sketch), which moves when the pressure at the flange increases. The movement is transferred to the pointer by a mechanism.

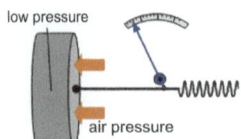

> **Example 15.3: Aneroid Barometer**
>
> We distinguish gauges that measure a pressure against another pressure (usually ambient pressure), and barometers, which determine the pressure absolutely by measuring against vacuum (or at least greatly reduced pressure). Aneroid barometers can be found in many devices such as weather stations or altimeters. The lid of the can consists of an elastic membrane. During manufacture, a vacuum is created inside, against which the ambient pressure is measured. Depending on the ambient pressure, the membrane is pressed in to varying degrees, which is transferred to a pointer via a mechanism and displayed as pressure.

> **Example 15.4: Torricelli's Barometer**
>
> Evangelista Torricelli served as the successor to the physicist Galileo Galilei at the court of the Medici. He invented the barometer, the principle of which we still use today. The illustration shows a barometer based on Torricelli's principle. The longer leg of the U-tube is closed at the top. In the shorter leg, which is open at the top, there is a mercury reservoir. The U-tube is completely filled with mercury (without air bubbles) and only then is it erected. This creates a vacuum at the upper end of the longer leg. The difference between the mercury level in the two legs is about 760 mm under normal air pressure. This is the basis for the unit Torr. If the air pressure rises or falls, the mercury level in the longer leg rises or falls accordingly. For accurate measurements, temperature fluctuations must be corrected. Therefore, a small thermometer is mounted on the barometer.

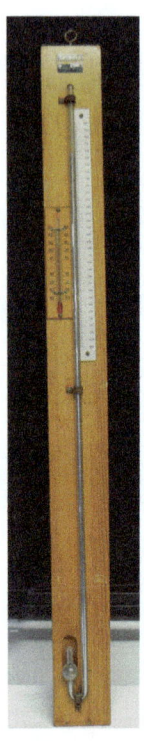

© RWTH Aachen, Physics Experiment Collection

We have defined pressure as a scalar quantity, i.e., without reference to a direction. This is a sensible definition, as the pressure in a fluid actually does not depend on the direction. It is isotropic. We want to explain this assertion using a thought experiment. Consider ◼ Fig. 15.2. A prism floats in a fluid. The side lengths are l_1 and l_2, the corresponding surfaces A_1 and A_2. Forces act on the three

15.1 · Pressure

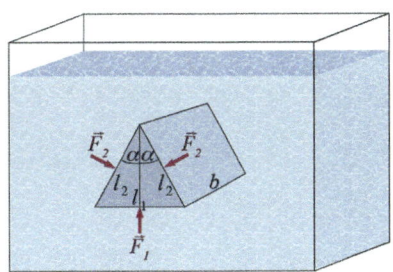

Fig. 15.2 On the isotropy of pressure

sides. Since the body floats, the forces balance each other. We want to quantify these.

The tip of the triangle encloses twice the angle α. We calculate it as:

$$\sin \alpha = \frac{l_1/2}{l_2}. \tag{15.3}$$

The force $\vec{F}_1$ acts from below. A force $\vec{F}_2$ acts on each of the inclined surfaces. It has a vertical component that points downwards and compensates the force $\vec{F}_1$. This component is

$$F_{2,v} = \sin \alpha \, F_2. \tag{15.4}$$

From the condition that the body is floating, it follows that

$$F_1 = 2 \sin \alpha \, F_2 = 2 \frac{l_1/2}{l_2} F_2 = \frac{l_1}{l_2} F_2$$
$$\frac{F_1}{l_1} = \frac{F_2}{l_2}$$
$$\frac{F_1}{l_1 b} = \frac{F_2}{l_2 b} \tag{15.5}$$
$$\frac{F_1}{A_1} = \frac{F_2}{A_2}$$
$$p_1 = p_2.$$

Here, p_1 is the pressure on the base surface from below and p_2 is the pressure on the slanted side surfaces. The calculation has shown that the pressure on both surfaces is the same, regardless of how the surface is oriented, i.e., regardless of the angle α.

These considerations go back to Blaise Pascal, after whom we name the unit of pressure.

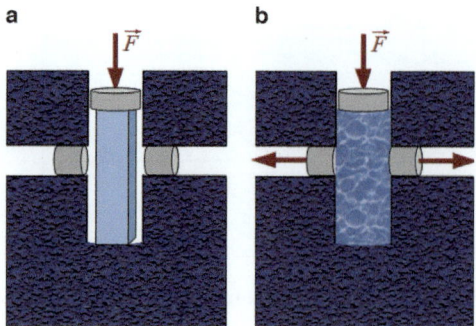

Fig. 15.3 Pressure distribution in a solid (**a**) and a fluid (**b**)

Fig. 15.3 illustrates how isotropic pressure distribution occurs. In this thought experiment, a force is exerted on the medium from above. In sketch a, the medium is a solid, in sketch b, a fluid. The stamp presses from above on the medium. Two horizontal stamps show the spread of pressure. If you press on the solid (a), it retains its shape and only passes the force onto the bottom. In a fluid, however, (b) the molecules can move against each other. The fluid also passes the pressure to the side and thus changes its shape.

> **Experiment 15.1: Isotropy of Pressure**
>
> The isotropy of pressure can be demonstrated with a simple experiment. We measure the pressure in a water vessel with an aneroid manometer. A can is sealed on one of its sides with a membrane. The can contains air, initially at normal pressure. When the manometer is put under water, the water pressure slightly indents the membrane, the air pressure in the can rises to the value of the surrounding water pressure. The air pressure is transferred to an electronic gauge through a hose, where it is measured.
>
> The can is mounted turnable on a rod so that the membrane can be oriented in different directions (in the picture it points upwards). We demonstrate that we get the same pressure at any orientation of the membrane (although it does depend on the depth in the vessel, as we will see in
> ▶ Sect. 15.3).

15.1 · Pressure

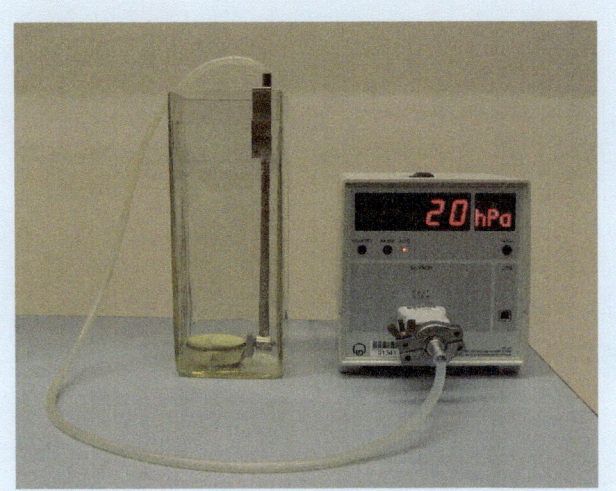

© RWTH Aachen, Physics Experiment Collection

When pressure is applied to a plunger, a force acts on it. If the plunger moves under the pressure, work is performed. It can be easily calculated as:

$$dW = F\,dx = pA\,dx = p\,dV. \tag{15.6}$$

We will use this relation repeatedly. The work performed is given by the pressure and the volume change of the fluid.

The technology of hydraulics is based on the isotropic propagation of pressure. This principle is illustrated in the sketch in ◘ Fig. 15.4.

The hydraulic pump generates a pressure $p = F_1/A_1$ in the right piston through a force $\vec{F}_1$. The pump is connected to the hydraulic cylinder on the left via hoses. Since the pressure spreads equally in all directions, the pressure p also prevails in the left hydraulic cylinder. This generates a force $F_2 = pA_2 = F_1A_2/A_1$, which can be much larger than the generating force in the pump.

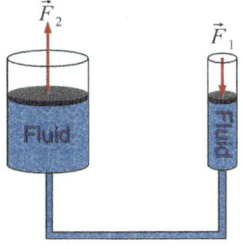

◘ Fig. 15.4 A simple hydraulic system with two cylinders

Example 15.5: Hydraulic Brake

A hydraulic braking mechanism transfers the force from the pedal through the pressure in the hydraulic lines to the pistons in the brake, which then press the brake pads against the brake disc. A hydraulic brake does not require a complicated system of linkages.

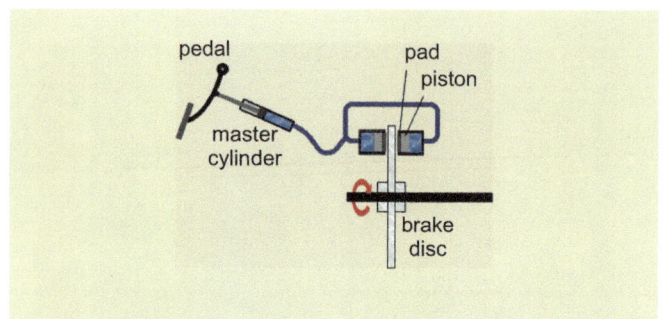

Example 15.6: Hydraulics

Many machines are powered by hydraulics, especially when large forces are involved. Oil is used as a fluid pressurised by an engine-driven pump. The pressure is transmitted to the cylinders that move the machine through hoses (hydraulic lines). The movement can be controlled with valves. The excavator in the figure is just one example of many.

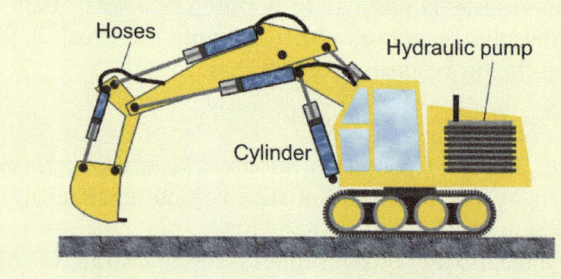

Experiment 15.2: Hydraulic Press

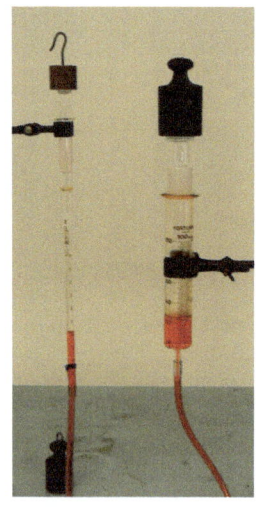

© RWTH Aachen, Physics Experiment Collection

This experiment demonstrates the operation of a hydraulic press. The press consists of two medical syringes of different sizes (10 ml and 100 ml). They are connected by a hose and filled with colored water. Now, weights are carefully placed on the syringe plunger. Depending on the weight, the plungers move up or down. The force on the large syringe is greater in proportion to the plunger areas than the force on the small syringe. In equilibrium, it carries a correspondingly larger weight.

15.2 · Compressibility

Finally, in this introduction to pressure, we want to deal with the question of how the pressure impacts the fluid itself. We consider a small volume element in the fluid and ask ourselves what force the pressure exerts on this volume. The situation is outlined in ◘ Fig. 15.5 . First, we see that under a spatially constant pressure, the forces on the opposing surfaces just compensate each other. In this case, the volume element experiences no force. A net force only arises when we consider local changes in the pressure in the fluid. Then the force on the volume element is

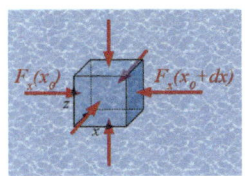

◘ **Fig. 15.5** Forces on a small element in a fluid

$$\vec{F} = \begin{pmatrix} F_x(x_0) - F_x(x_0 + dx) \\ F_y(y_0) - F_y(y_0 + dy) \\ F_z(z_0) - F_z(z_0 + dz) \end{pmatrix}$$
$$= \begin{pmatrix} (p_x(x_0) - p_x(x_0 + dx))\, dy\, dz \\ (p_y(y_0) - p_y(y_0 + dy))\, dx\, dz \\ (p_z(z_0) - p_z(z_0 + dz))\, dx\, dy \end{pmatrix} . \tag{15.7}$$

If we make the volume element small enough, we can expand the pressure around the point (x_0, y_0, z_0) according to Taylor's theorem:

$$p(x_0 + dx) = p(x_0) + \left.\frac{\partial p}{\partial x}\right|_{x=x_0} dx + \ldots \tag{15.8}$$

and correspondingly for the other directions. We will neglect terms of the order $(dx)^2$ or $dx\, dy$ in the following. Thus, we obtain:

$$\vec{F} = \begin{pmatrix} -\left.\frac{\partial p}{\partial x}\right|_{x=x_0} dx\, dy\, dz \\ -\left.\frac{\partial p}{\partial y}\right|_{y=y_0} dx\, dy\, dz \\ -\left.\frac{\partial p}{\partial z}\right|_{z=z_0} dx\, dy\, dz \end{pmatrix} = -\operatorname{grad} p\, dV . \tag{15.9}$$

This is called the pressure force. It leads to a movement of the volume element within the fluid. In the static case, which we refer to in this chapter, it must vanish. We will use it later to describe flows.

15.2 Compressibility

We now want to deal with the elastic deformations of fluids. With solids, we have learned about a variety of different deformations, e.g., elongation, compression, bending, shear. Since the molecules in fluids are freely mova-

Table 15.2 Compressibility of some liquids at room temperature and, for comparison, the compressibility of iron

Material	Compressibility in 10^{-11}/Pa
Acetone	120
Ethanol	110
Petroleum	70
Water	46
Mercury	4
Iron	0.6

ble against each other, these deformations do not lead to an elastic reaction. If you try to shear a liquid, e.g., by shearing the vessel (at constant volume), the molecules in the liquid will rearrange. There is no restoring force that counteracts the shear, and therefore no elastic deformation. The same applies to the other examples.

The only elastic deformation that can be observed in fluids is associated with a change of volume, the compression. As with solids, it turns out to be proportional to pressure (▶ Sect. 14.3).

$$\frac{\Delta V}{V} = \frac{1}{\bar{\kappa}} \Delta p \quad \text{or} \quad \frac{\Delta V}{V} = \kappa \, \Delta p. \tag{15.10}$$

The constant $\bar{\kappa}$ in the first relation is called "bulk modulus", the constant κ in the second "compressibility". ◘ Table 15.2 shows the values of some liquids and for comparison the value of iron. For liquids and solids, the bulk moduli are very small, i.e., even with high pressure, the volume changes only slightly. For gases, they are much larger. They can be calculated for gases from Boyle-Mariotte's law. More on this will be covered in the volume on thermodynamics.

We demonstrate the low compressibility and the omnidirectional pressure propagation in a fluid with an experiment (Experiment 15.3).

Example 15.7: Volume Change of Different Materials

When external pressure is applied to a body, it compresses decreasing its volume. However, the change in volume depends strongly on the properties of the body.

15.2 · Compressibility

Imagine we take various bodies from the water surface to a depth of 10 m where the pressure on the body is increased by about 100,000 Pa. The volume of one liter of a steel cube will not even decrease by 0.1 mm^3 (compression modulus 160 GPa). If we take a liter of water enclosed in a thin plastic bottle to the same depth, its volume decreases by about 5 mm^3 (compression modulus 2 GPa), still difficult to observe. However, if we take a bottle filled with air to this depth, we see a significant effect. Its volume reduces by about 10% (compression modulus 0.14 MPa).

Experiment 15.3: Compressibility of Water

We demonstrate the low compressibility of water in the following experiment. With an air rifle (caliber 4.5 mm), we shoot at a Styrofoam coffee cup. We shoot three times at different cups: one cup is empty (filled with air), one is filled with sand, and the third with water.

The first sequence of images (Fig. 15.6) shows shots taken with a high-speed camera (10,000 frames per second) at the empty cup. The displayed images are separated by 50 ms. The projectile comes from the right. It can be seen on the right edge of the first image. It penetrates the walls of the cup. A hole is created in the cup at the entry and exit points. At the exit hole, some plastic pieces are torn out, making the exit hole slightly larger. The pieces can be seen in images 4 and 5, and the exiting projectile can be seen in images 3 and 4.

In the second round, we shoot at the sand-filled cup. (We have omitted the sequence of images.) The projectile enters the cup and gets stuck in the sand. We tip the cup over and fish out the projectile.

In the third round, we shoot at the water-filled cup (Fig. 15.6 second sequence of images). In the first image, you can see the projectile, which again hits from the right. The entry of the projectile into the cup not only punches a hole in the cup from which water leaks out, but the cup bursts. In images 3 to 5, you can see how the cup explosively tears apart and water sprays out. An inspection of the cup after the shot shows several cracks across the cup.

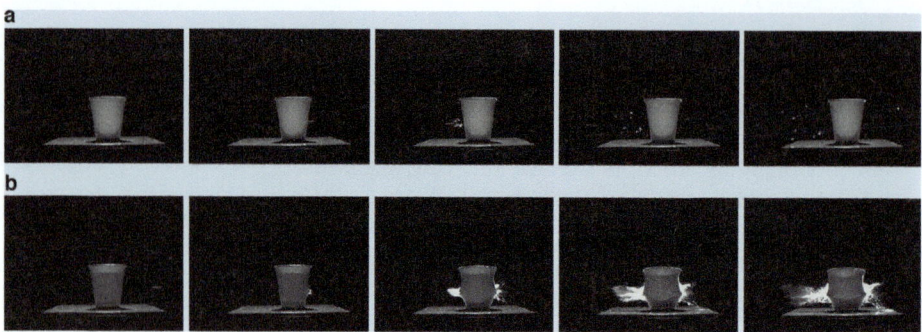

Fig. 15.6 © RWTH Aachen, Physics Experiment Collection

How can the different results be explained? Let's start with the sand-filled cup. When the projectile enters, it penetrates the wall of the cup and exerts pressure on the contents of the cup. In the case of sand, the pressure only spreads locally in front of the projectile tip. There is high friction, which quickly slows down the projectile. The situation is different in the case of a cup filled with air or water. In the fluid, the pressure spreads isotropically. With the entry of the projectile, pressure is created everywhere in the cup. In the case of the air-filled cup, this leads to a compression of the air in the cup, which eventually escapes upwards. The pressure on the walls remains low throughout the entire time, so that no damage to the cup occurs in addition to the entry and exit holes. The situation is different with the water-filled cup. Water can be considered approximately incompressible. A pressure wave propagates from the tip of the projectile, which runs almost undiminished in all directions and hits the walls. This pressure wave causes the walls to burst and sprays the water out.

You could also seal the cup at the top, e.g. by pouring liquid wax onto the water. But this makes no significant difference. The pressure wave in the water spreads at the corresponding speed of sound. It hits the walls and tears them apart long before water can escape upwards.

We roughly estimate the forces on the walls. The volume of the water in the cup is about $0.2\,l = 200\,\text{cm}^3$, the projectile has a volume of approx $1\,\text{cm}^3$. Immediately after its penetration into the cup, the volume available to the water is reduced by $\Delta V = 1\,\text{cm}^3$,

$$\frac{\Delta V}{V} = \frac{1\,\text{cm}^3}{200\,\text{cm}^3} = 0.005\,,$$

15.3 · Hadrostatic Pressure

which leads to a pressure increase due to the low compressibility:

$$\Delta p = \overline{\kappa} \frac{\Delta V}{V} = \frac{1}{46 \cdot 10^{-11} \, \text{Pa}^{-1}} 0.005 \approx 10^7 \, \text{Pa} \approx 100 \, \text{bar}.$$

The shell of the cup has an area of approximately $100 \, \text{cm}^2$, from which a force on the mantle of

$$F = pA = 10^7 \, \text{Pa} \cdot 10^{-2} \, \text{m}^2 = 100 \, \text{kN}$$

results.

Even though this is only a rough estimate and the actual pressure is reduced by friction and other effects, it is clear that the walls of the cup cannot withstand such pressure. The pressure causes the cup to burst.

The last sequence of images shows a similar experiment. Instead of the cup, we shoot at a raw egg. Despite the air bubble in the egg, the projectile causes the egg to burst.

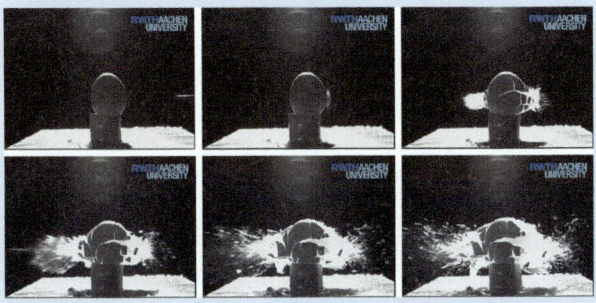

© RWTH Aachen, Physics Experiment Collection

15.3 Hadrostatic Pressure

In the previous considerations, we have neglected external forces on the fluids. We now want to turn to an important influence of such external forces, the influence of gravity. Gravity creates a pressure on lower layers of a fluid due to the weight of the layers above The pressure increases from top to bottom in the fluid. This pressure is called the "hydrostatic pressure". We want to determine it first for a liquid, which we assume to be almost incompressible.

The sketch in ◘ Fig. 15.7 shows a volume element in a liquid at the depth z_0. The fluid is supposed to be in static equilibrium, which implies that the element is

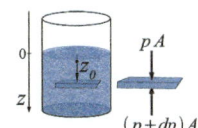

◘ **Fig. 15.7** On the determination of the hydrostatic pressure

at rest, the forces on the element add up to zero. In addition to the forces indicated in the sketch by the ambient pressure, the weight acts as an additional force. It is given by:

$$F_G = mg = \rho V g = \rho A g\, dz, \tag{15.11}$$

if dz is the height of the element and A its area. This force must be compensated by an increase in pressure with depth:

$$\begin{aligned} pA + \rho A g\, dz &= (p + dp)A \\ p + \rho g\, dz &= p + dp \\ \rho g\, dz &= dp \\ \tfrac{dp}{dz} &= \rho g \\ \Rightarrow p(z) &= \rho g z + p_0\,. \end{aligned} \tag{15.12}$$

The quantity p_0, which appeared as an integration constant, represents the ambient pressure at the surface. The term $\rho g z$ is the hydrostatic pressure. It increases linearly with depth. For water, the coefficient ρg is approximately 1 bar/10 m, i.e., the water pressure increases by 1 bar for every 10 meters of depth. This pressure has to be added to the pressure at the surface.

Experiment 15.4: Hydrostatic Pressure in Water

With the setup from Experiment 15.1, the hydrostatic pressure in water can be quantitatively determined. We compare the increase in pressure with the water depth in which the can is located and find a linear increase in pressure with depth. As can be seen in the picture from Experiment 15.1, at the bottom of the vessel at a depth of about 20 cm we measure the expected increase in pressure of about 2000 Pa compared to the surface.

Experiment 15.5: Communicating Vessels

In Eq. 15.12 we have learned that the hydrostatic pressure in a vessel filled with liquid depends solely on the density of the fluid and the liquid level, but not on the dimensions or orientation of the vessel. We demonstrate this finding with an arrangement called communicating vessels (see figure). The illustration shows three such communicating vessels. Due to the connection at the bot-

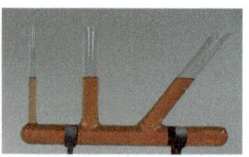

© RWTH Aachen, Physics Experiment Collection

tom, a uniform pressure p is established (hence the name "communicating"), i.e., all three tubes reach the same hydrostatic pressure at the bottom. You can see in the photo that this creates equal filling heights in all three tubes.

Experiment 15.6: Hydrostatic Paradox

As we have seen, the hydrostatic pressure in some depth in a liquid is generated by the weight of the liquid column above it. The hydrostatic paradox refers to this finding. The two images show two different vessels filled with colored water. They are inserted into a holder that seals the open-bottomed vessels with a membrane. The pressure at the bottom of the vessels can be measured via this membrane and the red pointer.
In the conical vessel, there is more liquid. One might naively expect that the higher weight would create greater hydrostatic pressure at depth. But this is not the case. The pointer shows the same pressure for both vessels. The pressure depends solely on the height of the water column, not on its shape (see Eq. 15.12).

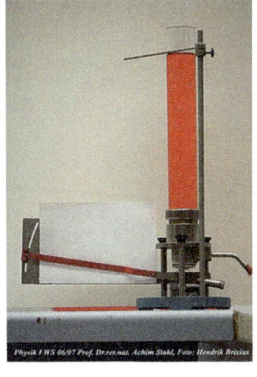

© Photos: Hendrik Brixius

Experiment 15.7: Density Balance

With the simple apparatus shown in the figure, we determine the density of unknown liquids. The U-tube contains yellow-colored water, with a density of $\rho_w = 1\,\text{g/cm}^3$. We pour a small amount of the liquid with unknown density ρ_x (here petroleum) on top of the water. Now we measure the filling levels h_x and h_w, as indicated in the figure. Since at the height of the lower dashed line the same pressure must prevail in both legs of the U-tube, the weight of the columns of liquid resting above this line must be equal. Consequently, it must hold (A is the cross-section of the tube):

$$\rho_w h_w A g = \rho_x h_x A g \Rightarrow \rho_x = \rho_w \frac{h_w}{h_x}.$$

A measurement yielded $h_w = 3.2\,\text{cm}$ and $h_x = 5.1\,\text{cm}$, from which we calculate a density of $\rho_x = 0.62\,\text{g/cm}^3$ (literature value of the density of petroleum: $0.66\,\text{g/cm}^3$).

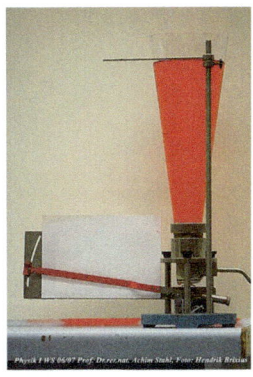

© Photos: Hendrik Brixius

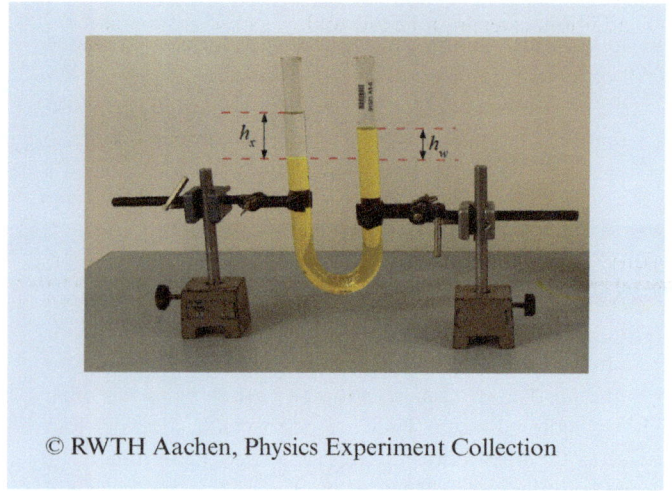

© RWTH Aachen, Physics Experiment Collection

Considering gases instead of liquids, we can no longer assume a constant density. Gases are easily compressible. The weight of the upper layers compresses the layers underneath, so their density increases downwards. This is the case, for example, in the Earth's atmosphere. At the Earth's surface, the density of the air is higher than on a mountain. The actual variation of pressure with height is shown in ◘ Fig. 15.8.

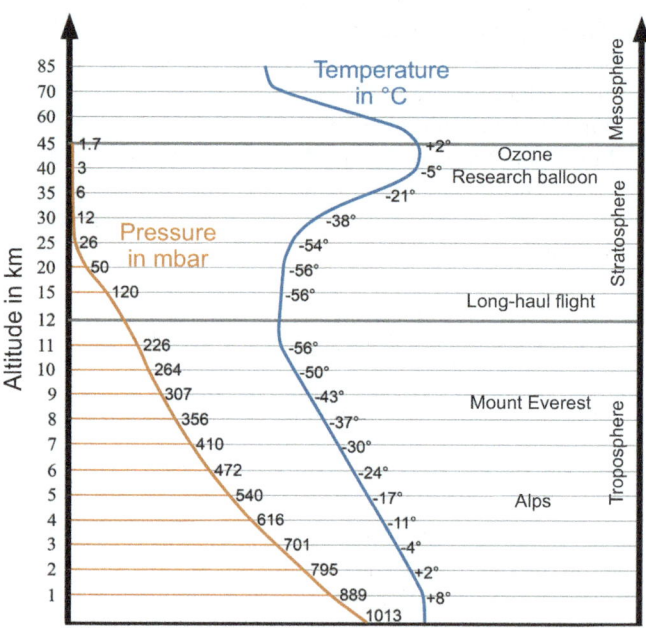

◘ **Fig. 15.8** Temperature and pressure in the atmosphere (heights not to scale)

15.3 · Hadrostatic Pressure

We calculate the change in air pressure with height h. According to Boyle-Mariotte's law, the following applies to ideal gases:

$$pV = p\frac{m}{\rho} = \text{konst.} \tag{15.13}$$

Even though air is not an ideal gas, we will use this relation as an approximation. We denote the quantities at sea level with an index 0 and the quantity at the height h without an index. Then:

$$p_0 \frac{m}{\rho_0} = p\frac{m}{\rho}$$
$$\frac{p_0}{\rho_0} = \frac{p}{\rho} \tag{15.14}$$
$$\rho(p) = \rho_0 \frac{p}{p_0}.$$

As with incompressible fluids, the relation $\frac{dp}{dz} = \rho(z)g$ (Eq. 15.12) also applies here, only now the density varies with height. We replace the depth z (directed downwards) with the height h (directed upwards):

$$-\frac{dp}{dh} = \rho(p)g$$

$$\frac{1}{\rho(p)}dp = -g\,dh$$

$$\int_{p_0}^{p} \frac{1}{\rho(p')}dp' = \int_{0}^{h}(-g)\,dh'$$

$$\int_{p_0}^{p} \frac{1}{\rho_0 \frac{p'}{p_0}}dp' = -g\int_{0}^{h}dh' \tag{15.15}$$

$$\frac{p_0}{\rho_0}\ln p'\Big|_{p_0}^{p} = -g\,h'\Big|_{0}^{h}$$

$$\frac{p_0}{\rho_0}(\ln p - \ln p_0) = -g\,h$$

$$\ln \frac{p}{p_0} = -\frac{\rho_0}{p_0}g\,h$$

$$\frac{p}{p_0} = \exp\left(-\frac{\rho_0}{p_0}g\,h\right)$$

$$p(h) = p_0 \exp\left(-\frac{\rho_0}{p_0}g\,h\right).$$

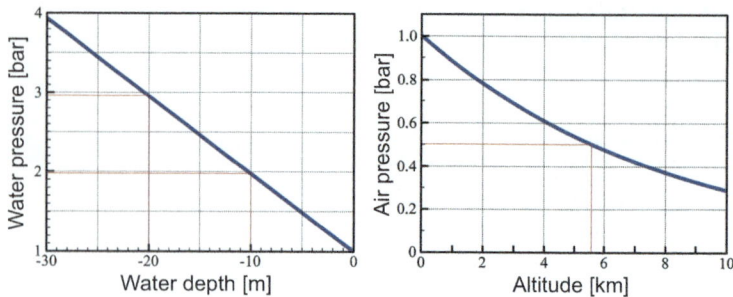

Fig. 15.9 Variation of pressure in air and water with height or depth

This is called the "barometric formula". If you insert the numerical values for our atmosphere, you get

$$p(h) = 1.013 \cdot 10^5 \, \text{Pa} \, e^{-h/8\,\text{km}}. \tag{15.16}$$

The air pressure decreases exponentially with increasing altitude, at a rate of about 12% every 1000 m.

The two graphs in Fig. 15.9 show air and water pressure in comparison. With air, we see an exponential progression, while we have derived a linear progression for water. This difference is due to our assumption regarding the compressibility of the fluid. We can consider liquids as approximately incompressible, which leads to the linear progression, while this assumption is not justifiable for a gas.

© Wikimedia: Cpl. Christopher O'Quin

Example 15.8: Altimeter

An altimeter is based on the barometric formula. It measures the air pressure, which is then displayed as height on a suitable scale. In the picture, an altimeter can be seen on the hand of a sky-diver. Similar altimeters are used in mountaineering. However, for measurements over longer times, corrections must be applied for variations in air pressure due to weather changes.

Example 15.9: Gravitational pressure in the Sun

The Sun is a gigantic gas ball held together by gravity. The weight of the outer layers creates a gravitational pressure that increases towards the center of the Sun.

15.3 · Hadrostatic Pressure

In a very simple model, we can assume that gravity and gravitational pressure keep the gas in balance. We try to estimate this pressure. In fact, the radiation pressure from nuclear fusion inside the Sun and the centrifugal forces due to the rotation of the Sun should be added, but we neglect these here for simplicity.

We start by determining the mass dm of a spherical shell with radius r. This is:

$$dm(r) = 4\pi r^2 \rho(r)\, dr.$$

The mass $m(r)$ enclosed within the radius r is given by:

$$m(r) = 4\pi \int_0^r \rho(r')\, r'^2\, dr'.$$

At the radius r the gravitational pressure $p(r)$ prevails. This pressure decreases by $dp = dF_G/A$ with increasing radius, where dF_G is the gravitational force on the spherical shell with surface A. So:

$$dp(r) = \frac{-G\frac{m(r)\,dm(r)}{r^2}}{4\pi r^2} = -G\frac{m(r)\rho(r)}{r^2}\, dr.$$

To determine $m(r)$ and $p(r)$, the density profile $\rho(r)$ must be known. This is complicated because the temperature inside the sun increases significantly, the composition of the gas changes (helium is added), and the gas is far outside the range of ideal gases. To still achieve a rough estimate, we assume a constant density ρ_0. Then

$$m(r) = \frac{4\pi}{3}\rho_0 r^3$$

and

$$p(r) = -G\int \frac{\frac{4\pi}{3}\rho_0 r'^3\, \rho_0}{r'^2}\, dr'$$

$$= -\frac{4\pi}{3} G\rho_0^2 \int r'\, dr' = -\frac{2\pi}{3} G\rho_0^2 r^2 + C.$$

The integration constant C is determined from the condition that the gravitational pressure at the outer radius of the star ($r = R$) must vanish. We obtain:

$$p(r) = \frac{2\pi}{3} G\rho_0^2 \left(R^2 - r^2\right) = \frac{3}{8\pi} G \frac{M^2}{R^4}\left(1 - \frac{r^2}{R^2}\right).$$

In the last step, we have inserted the mass of the Sun $M = \frac{4\pi}{3}\rho_0 R^3$ with a value of $M = 1.989 \cdot 10^{30}$ kg and a radius of $R = 696 \cdot 10^6$ m. We estimate in this greatly simplified model at the center of the Sun a density of $\rho_c = \rho_0 = 1.4\,\text{g/cm}^3$ and a pressure of $p_c = 1.3\,10^9$ bar. The proper values are about $\rho_c = 160\,\text{g/cm}^3$ and $p_c = 2.5\,10^{11}$ bar significantly higher, which is primarily due to the compression of the gas inside the Sun.

15.4 Buoyancy

A body that is fully or partially immersed in a fluid experiences a force opposing gravity due to the pressure in the fluid. This force is responsible for some bodies floating, other bodies hovering at a fixed depth in the fluid or sinking slowly to the bottom. This force is called "(static) buoyancy".

Example 15.10: Dead Man

Due to the high salt content in the Dead Sea, the buoyancy there is greater than in other bodies of water. It is so large that it can comfortably keep a person afloat.

(C) Wikimedia: Ingo Carl

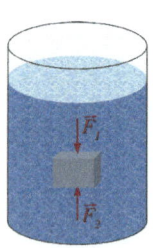

Fig. 15.10 Buoyancy on a cuboid

15.4 · Buoyancy

We want to try to quantify the buoyancy F_{buo}. For this, we consider a cuboid that is fully immersed in a fluid (◘ Fig. 15.10). The buoyancy results from the difference in forces on the lower (F_2) and upper surface (F_1). The base area of the cuboid is A. The upper surface is at a depth of h_1, the lower surface is at a depth of h_2. Then it is:

$$\begin{aligned}
F_{\text{buo}} &= F_2 - F_1 \\
&= A p_2 - A p_1 \\
&= A\,(\rho_{fl}\, g\, h_2 + p_0) - A\,(\rho_{fl}\, g\, h_1 + p_0) \quad (15.17)\\
&= \rho_{fl}\, g\, A\, (h_2 - h_1) \\
&= \rho_{fl}\, V_{\text{body}}\, g \, .
\end{aligned}$$

with the density of the fluid ρ_{fl} and the volume of the body V_{body}. The product $\rho_{fl} V_{\text{body}}$ yields the mass of the fluid displaced by the body and $\rho_{fl} V_{\text{body}}\, g$ its weight. It can therefore be said that the buoyancy corresponds to the weight of the displaced fluid, also known as the "Archimedes' principle". According to legend, it was discovered by Archimedes as early as the third century BC when he was asked to verify whether a crown ordered by his king was actually made of pure gold.

© RWTH Aachen, Physics Experiment Collection

Experiment 15.8: Buoyancy

Buoyancy can be measured by tracking the weight of a body as it is submerged in water. The weight then reduces by the buoyancy.
An aluminum cylinder is suspended from a spring scale. It shows 6.2 N hanging in air. Then, using the lifting table, we raise the water cup until the cylinder is fully submerged. Now the spring scale only shows 3.7 N. The buoyancy is 2.5 N.

Whether a body floats, hovers, or sinks in a liquid depends on its weight and buoyancy (◘ Fig. 15.11). If the

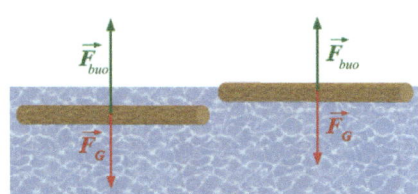

◘ **Fig. 15.11** Forces during floating

© RWTH Aachen, Physics Experiment Collection

buoyancy of the body is less than its weight, it sinks to the bottom. This is the case when the average density of the body is greater than the density of the liquid. The weight is given by $F_G = V_{body}\, \rho_{body}\, g$, while the buoyancy is given by $F_{buo} = V_{body}\, \rho_{fluid}\, g$. If buoyancy and weight balance each other, the body hovers in the liquid. If it is lighter, it rises to the surface until it eventually protrudes from the fluid. If it continues to rise, the displaced volume and thus the buoyancy decrease until a balance between F_{buo} and F_G is achieved. Now the body floats on the surface.

Please note that buoyancy and weight always act in opposite directions, but they do not necessarily apply at the same point. The weight acts on the center of gravity of the body, while the buoyancy acts on the center of gravity of the displaced fluid. If the mass in the body is unevenly distributed, its center of gravity might deviate from that of the displaced fluid. This creates a torque that can lead to instabilities in the position of the body.

Example 15.11: Scuba Diving

© Wikimedia: Stephan Borchert

Controlling buoyancy is an important aspect of scuba diving. A sport diver wears a neoprene suit (typically two layers of 7 mm) to protect himself from the cold water. This suit increases his volume and thus his buoyancy, which makes him float. To submerge in the water, he carries lead weights on a belt or in his jacket. Underwater, he regulates the buoyancy with a buoyancy control device (BCD), a jacket with chambers he can fill with air from the compressed air bottle. The chambers inflate, increase the volume of the diver, and therefore the buoyancy. He regulates the pressure in the chambers with a valve such that he floats in the water.

Once the diver has balanced himself, he is in an unstable equilibrium. If the buoyancy is even slightly too low, the diver slowly sinks. The hydrostatic pressure increases and compresses the air in the chambers of the BCD. This reduces the buoyancy and the diver sinks even faster. He must adapt the buoyancy. Experienced divers fine-regulate the buoyancy through their breathing technique, by holding more or less air in their lungs.

Example 15.12: The Keel of a Sailing Yacht

At the bottom of a sailboat keel is a counterweight. This shifts the yacht's center of gravity downwards, below the center of gravity of the displaced water. If the boat heels to the side, a torque $\vec{r} \times \vec{F}_G$ around the center of gravity of the displaced water (pivot point) arises from the counterweight, which stabilizes the boat. Even a capsized boat can right itself in this way.

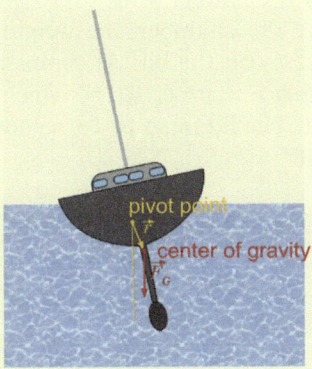

Experiment 15.9: Cartesian Diver

The Cartesian diver is a children's toy based on buoyancy. Inside a water-filled bottle is a figure with just enough buoyancy to float. The bottle is sealed with a rubber cap. Inside the figure is an air bubble. When you press on the rubber cap, you compress the air bubble in

© Wikimedia: Gustav Broennimann

the figure and thereby reduce its buoyancy. It sinks. By varying the pressure, you can make the figure hover. If you reduce the pressure, liquid leaks out of the figure and the air bubble grows. Often the figure is shaped so that the liquid exits through a small opening on the side. This sets the figure in rotation appearing to dance in the bottle.

Example 15.13: Balloon Flight

Buoyancy occurs in all fluids, not just in liquids. The picture shows a hot air balloon. The density of the hot air in the envelope is less than the density of the surrounding air. Therefore, the weight of the displaced air is greater than the weight of the balloon. It flies.

There is also an ambient pressure in the atmosphere that constantly acts on us from all sides. We do not feel it because the same pressure exists inside our bodies. We recognize it in an experiment when we eliminate the counter pressure inside a vessel by evacuating it. Imagine a cylindrical vacuum chamber with a cross-section of $1\,\text{m}^2$. If you evacuate the chamber, the weight of the air column above it rests on the lid. At a pressure of 1 bar, the force on the lid is $F = pA = 10^5\,\text{N}$, which corresponds to a weight of 10 t. The lid must be correspondingly thick to bear this weight.

Experiment 15.10: Magdeburg Hemispheres

The physicist and mayor of Magdeburg Otto von Guericke demonstrated atmospheric pressure on an evacuated sphere in 1657 (diameter 42 cm). It consisted of two hemispheres, which were pressed together by the atmospheric pressure after he evacuated the sphere. Even 16 horses were unable to separate them, as can be seen in the historical engraving.

We replicate the experiment with two smaller spheres (radius 6 cm). After evacuation, the force on one of the hemispheres is given by $F = pA = 10^5\,\text{Pa} \times \pi r^2 \approx 1.1\,\text{kN}$. The sphere should

© RWTH Aachen, Physics Experiment Collection

be able to hold 110 kg. The demonstration can be seen in the second image.

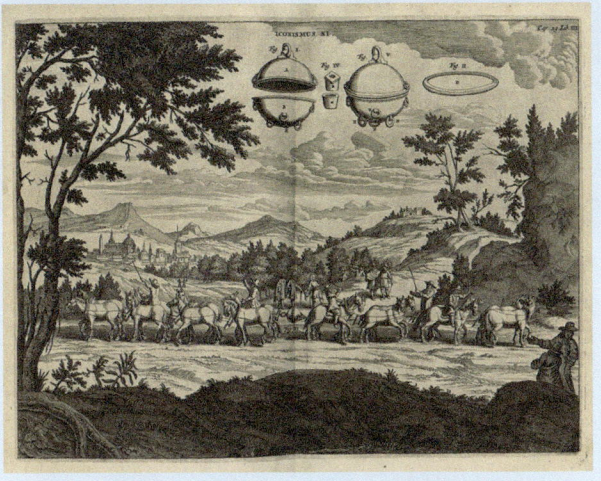

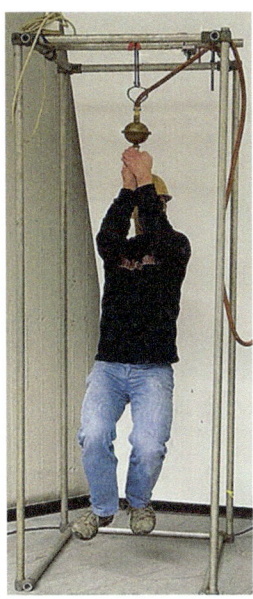

© Photo: Hendrik Brixius

15.5 Interfaces

15.5.1 Surface Tension

Surface tension is a property of the surface of a liquid, which can be considered as the interface between a liquid and a gas such as air. The surface of a liquid behaves similarly to a stretched, elastic film. Surface tension is the cause of this elastic behavior. It is known from everyday life. Liquids like water take on the energetically favorable shape of a spherical droplet to form as small a surface as possible. The higher the surface tension, the more spherical the droplet becomes. This is counteracted by the weight, which presses the liquid downwards into a flat "puddle". Despite the high density, the droplet shape is particularly pronounced in mercury. Mercury has a very high surface tension (see ◘ Fig. 15.12).

Surface tension arises from cohesion, that is, the attractive force between molecules in a fluid. If we consider a molecule inside the fluid, the forces it experiences from its neighbors add up vectorially to zero (◘ Fig. 15.13). However, this is not the case for a molecule at the surface, as neighbors are missing on one side. Such a molecule ex-

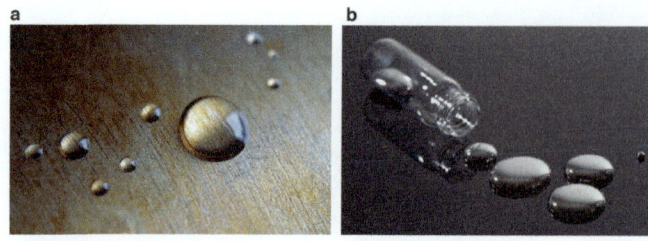

Fig. 15.12 Formation of droplets. **a** Water droplet on a metal surface. © Wikimedia: Brocken Inaglory; **b** Mercury droplet © Fotolia.com: marcel

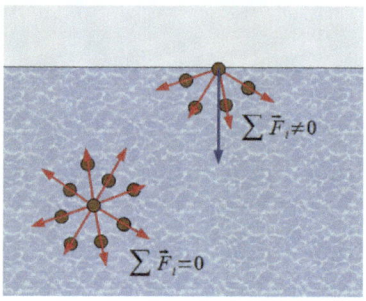

Fig. 15.13 Forces on a molecule inside a liquid and at the interface

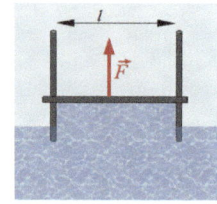

Fig. 15.14 Measurement of the surface tension

periences a net force that points into the interior of the fluid. To increase the surface of a fluid, additional molecules must be brought to the surface and work must be performed against the cohesive forces.

Consider the simple situation in **Fig. 15.14**. A liquid film has formed between the surface of a fluid and a wire. The film is laterally bounded by two additional wires. We enlarge the film by pulling the horizontal wire upwards. The increase in surface area is $\Delta A = 2l\,ds$, if we move the wire upwards by the distance ds. The factor 2 takes into account the front and back of the film. The work ΔW, which we must perform, is proportional to the increase in surface area. It is

$$\Delta W = \sigma \, \Delta A. \tag{15.18}$$

15.5 · Interfaces

The quantity σ is called "specific surface energy" or "surface tension". It can be determined from ◯ Fig. 15.14. It is

$$\sigma = \frac{\Delta W}{\Delta A} = \frac{F\,ds}{2l\,ds} = \frac{F}{2l} = \frac{F}{L}, \qquad (15.19)$$

where we have introduced L as the length over which the surface is increased (front plus back). This result explains the designation of σ as tension.

Experiment 15.11: Measurement of Surface Tension

To measure the surface tension we refer to Eq. 15.19. However, we choose a slightly different arrangement. Instead of a wire, we use an aluminum ring with a radius of 3 cm. It is supported with some strings from a spring scale. The spring scale shows the force $F = 5.0$ cN. The water-filled dish below is sitting on a lifting table. We lift it up until the ring dips into the water, and then carefully lower it again. Now the displayed force increases due to the surface tension to $F = 5.5$ cN. A liquid film forms on the inside and outside over a length of $L = 2 \cdot 2\pi r$. The surface tension is $\sigma = 0.5 \text{ cN}/4\pi r$, approximately 10 cN/m (literature value 7.2 cN/m at 20 °C).

© RWTH Aachen, Physics Experiment Collection

Example 15.14: Water Striders

Water striders are a family of insects that can walk on the surface of water with the help of the surface tension. The animals are 8 to 20 mm long (without legs) and walk on six legs, which at the end – as can be seen in the picture – lie flat on the water. How much weight can the legs carry?

We want to give an estimate: The legs touch the water at least over a length of 2 mm. At the limit of the supported weight, the surface is pressed in so deep that the force on the legs points vertically upwards. It is $F = \sigma_{\text{water}} L$, where L is the length of the support around the leg, 2 mm on each side. We neglect the short distances, which correspond to the thickness of the legs. So $L = 4$ mm per leg and thus $F = 6 \cdot 0.072 \frac{\text{N}}{\text{m}} \cdot 4 \text{ mm} = 1.7$ mN. They can carry up to $m = F/g = 0.17$ g. The water strider actually weighs significantly less.

© Wikimedia: Tim Vickers

Experiment 15.12: Floating Paperclip

As children, you learned that iron does not float. But here we make an exception: We make a paperclip float. If you place it carefully on the surface of water, the surface tension is large enough to keep the clip floating (see figure).

Try it yourself! The experiment is easier with a small paperclip. It's best to lay it flat on the tip of a knife and carefully place it on the water surface with the knife. Once it floats, you push the knife into the water and remove it. The paperclip floats. Rinse the glass thoroughly with fresh water beforehand and rinse off the paperclip. Dish soap contains substances that reduce the surface tension of the water and thus improve the wetting of your dishes. Only in this way can the cleaning water penetrate into small crevices. For our experiment, this is counterproductive. If you take a tiny drop of dish soap on the tip of the knife and touch the surface of the water with it, the paperclip will immediately sink.

Alternatively, you can also make a razor blade or a piece of aluminum foil float.

Experiment 15.13: A Minimal Surface through Surface Tension

Surface tension leads to the formation of droplets, because in the shape of the droplet, the surface is minimal in relation to the volume. Here we demonstrate the corresponding effect in two dimensions. We use a loop of a plastic thread with an irregular shape. We attach it with three more threads to a plastic ring. Now we wet the entire ring with soap bubble solution from the toy store. A flat surface forms, filling the entire interior of the plastic ring. The inner thread still has an irregular shape. With a little practice, we manage to break the soap bubble skin in the inner ring while maitaining it on the outside. Now the soap bubble has a hole along the inner thread. The surface tension minimizes the length of the border. The formerly irregular thread is forced into an almost perfect circle, which can be faintly seen in the photo.

© RWTH Aachen, Physics Experiment Collection

15.5 · Interfaces

Experiment 15.14: Pressure in a Soap Bubble

Inside a soap bubble, there is a slight overpressure that gives the soap bubble its shape. We want to investigate it. We dip a funnel into soap solution and wet the opening of the funnel. We gently blow into the funnel, to form a soap bubble. We stop blowing just in time before the soap bubble detaches from the funnel. We use some tubing and a three-way valve as shown in the figure. We first inflate the left soap bubble to a radius r_0. Then we switch the valve to the right soap bubble and inflate it to about 80% of r_0. Now we want to compare the pressure in the two differently sized soap bubbles. To do this, we set the valve so that it connects the two soap bubbles and equalizes the pressure between the two soap bubbles. We observed that the small soap bubble shrinks and inflates the larger bubble further.

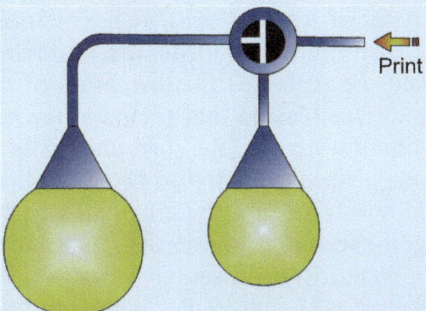

How can this be? Naively, one might expect that the pressure in the soap bubble is greater the larger its radius. But the opposite is the case. You have probably noticed the same effect inflating a balloon with your mouth. Initially, inflating is very difficult. For the first few centimeters, you have to blow hard against the pressure in the balloon. As the size increases, the pressure in the balloon decreases and inflating becomes easier. (To avoid misunderstandings: The size of the balloon does not only depend on the pressure in the balloon, but also significantly on the amount of air you blow into it.)

Let us quantify the energetic conditions: The surface of the soap bubble with radius r is A. Work is performed (energy is released) by the expansion of the air in the soap bubble. The force is $F = pA$ and the work performed during a small expansion of the soap

> bubble is $dW = p A\, dr = p\, 4\pi\, r^2 dr$. On the other hand, work must be performed to increase the surface of the soap bubble against the surface tension (energy is consumed). According to Eq. 15.18 the work is $dW = \sigma\, dA = 2\sigma\, 8\pi\, r\, dr$, where the factor 2 takes into account that both the inner and outer surface of the soap bubble are enlarged. In the equilibrium established, both energy contributions balance each other. Consequently, $p\, 4\pi\, r^2\, dr = \sigma\, 8\pi\, r\, dr$, from which follows that $p = 4\frac{\sigma}{r}$. The pressure in equilibrium is inversely proportional to the radius. As suspected, it decreases with increasing radius of the soap bubble.

15.5.2 Adhesive Tension

Different types of tensions arise at the interface between two media even if the two media are in different phases. For example, we explained the formation of water droplets by the surface tension that occurs at the interface between the liquid water and the surrounding air.

However, surface tension also occurs at the boundary to the solid substrate, and there can also be surface tension between two liquid phases, e.g., with an oil droplet floating on water. In all cases, surface tension is created by forces between the molecules within a medium (here in the water). These are called "cohesive forces". We have neglected forces that act between the molecules of different media across the interface. These are called "adhesive forces". In the case of a raindrop, these would be, for example, forces between water molecules and the molecules of the surrounding air. However, there are cases where these forces are relevant. The meniscus of a pipette (example 15.15) is such a case, created by the forces that act between the water molecules and the glass wall. The tension that occurs in this process is called the "adhesive tension". We will examine it a little more closely in the following.

15.5 · Interfaces

Example 15.15: Meniscus

A water surface in a glass is never perfectly flat. It curves upwards towards the wall due to the adhesive tension. This creates a characteristic shape of the surface, which is particularly noticeable in narrow glass tubes. It is called the meniscus (from the Greek word *meniskos*, which means "crescent"). The illustration shows the meniscus in a laboratory pipette.

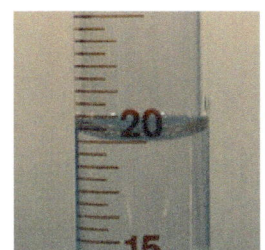

To understand the effects of adhesive tension, we first return to the microscopic image. If we consider a molecule at an interface far away from its edge, the components of the forces acting on the molecule parallel to the interface will cancel each other for reasons of symmetry. For a molecule at the edge of the interface, this is no longer the case. A net force parallel to the interface arises here.

As an example, let's consider the surface of water in a glass (◘ Fig. 15.15). Three media adjoin each other: The glass wall (index 1), the water (index 2), and the air above the water (index 3). Between each pair of media, a force arises at the interface, which has a component parallel to the respective interface at its edge. These are denoted in ◘ Fig. 15.15 with $\vec{F}_{ij}$. The forces are proportional to the length l of the edge over which they act. We can therefore write

$$F_{ij} = \sigma_{ij} l. \tag{15.20}$$

The quantities σ_{ij} are called the "adhesion tensions".

As indicated in ◘ Fig. 15.15, the adhesion tensions alter the surface of the water. We want to determine the critical angle, i. e. the angle α, which the surface encloses with the wall of the glass. A static state will be establish in the glass of water, in which the vertical components of the forces compensate each other. (The horizontal components compensate each other, too, but are of less relevance here.) It must apply (z-direction vertical):

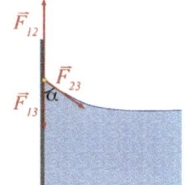

Fig. 15.15 Forces at the interfaces in a glass of water

$$\sum_{i,j}(\vec{F}_{ij})_z = 0. \tag{15.21}$$

These components can now be determined individually:

$$\begin{aligned} F_{12} - F_{13} - F_{23} \cos\alpha &= 0 \\ \sigma_{12} l - \sigma_{13} l - \sigma_{23} l \cos\alpha &= 0 \\ \cos\alpha &= \frac{\sigma_{12} - \sigma_{13}}{\sigma_{23}}. \end{aligned} \tag{15.22}$$

Depending on the relative size and directions of these tensions, the surface will either curve upwards or downwards.

> **Example 15.16: Adhering Dewdrops**
>
>
>
> On an early, sunny morning, one can observe dewdrops clinging to a spider's web and sparkling in the sun. Adhesive forces and resulting adhesive tensions keep the water droplets in the web.

> **Experiment 15.15: Critical Angle in a Wedge Glass**
>
> In a wedge glass, the formation of a meniscus can be easily observed. At the pointed end of the glass, you look across the surface. As described in the text, the curvature of the surface depends on the adhesive tension and thus on the materials. The image shows colored water on the right and mercury on the left, each in a wedge glass, the pointed ends facing each other. The surface of the colored water is curved upwards, that of the mercury downwards.
>
>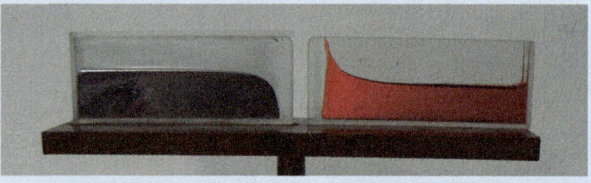
>
> © RWTH Aachen, Physics Experiment Collection

15.5.3 Capillarity

Let's look again at the calculation of the critical angle of a liquid surface, which led to Eq. 15.22. A balance of forces cannot be established in all cases. There are situations where F_{12} is greater than the sum of the other two forces, even in the limit of $\alpha = 0°$. In this case, the liquid will rise along the vessel wall, wetting it. However, additional forces come into play in this case, the weight of the rising liquid and the surface tension. If the liquid rises

15.5 · Interfaces

at the edge, the surface increases and work must be performed against the forces resulting from the surface tension. If the surface tension is large enough, the entire liquid level will be raised despite its weight. This occurs especially in narrow tubes (capillaries). It is called "capillarity". The lifting force at the edge of the liquid, along which adhesive tension occurs to the vessel wall, extends linearly with the radius of a capillary, whereas the surface of the liquid and thus the mass that must be raised increases quadratically with the radius. Therefore, the effect is stronger in narrow capillaries than in wide ones.

The more fluid rises in a capillary, the greater the weight that must be overcome by the adhesive tension. Beyond a certain height, this can no longer work, so the column of fluid remains at this height h_{max}. Since the forces from the adhesive tension, increase linearly with the radius of the capillary, but the weight quadratically, the maximum height must be inversely proportional to the radius of the capillaries. This is the statement of the capillary law. It states

$$h_{max} = \frac{2\sigma}{\rho g} \frac{\cos \alpha}{r}, \qquad (15.23)$$

where ρ is the density of the fluid, r is the radius of the capillary, and α is the boundary angle to the edge of the capillary.

Intuitive Derivation of the Formulas

Have you noticed: Instead of mathematically deriving Eq. 15.23 as you were used to, we simply wrote down some arguments about the dependence of the maximum height on the radius of the capillary. Of course, the mathematical derivation of the formula is the correct way to get it, but it is still useful to think about the main relationships of a formula. The mathematical derivation does provide the desired result, but not always the necessary understanding. If you have grasped the arguments from above, you have gained an understanding of why the maximum height increases with decreasing radius (because the weight to be lifted decreases with the radius faster than the adhesive tension that lifts it). Try to justify the main relationships of other laws in your own words. If you succeed in this, the step to a mathematical derivation is usually not far off.

In our case, we have a balance between adhesive tension and weight at the maximum height: $F_G = F_{\text{adhesive}} = \sigma d$, where F_{adhesive} is the force from the adhesive tension and d is the circumference of the capillary. Now, $F_G \propto h_{\max} r^2$ and $d \propto r$, from which the r-dependence follows. If you carry out the proportions, you get the full capillary law.

Experiment 15.16: Capillarity

The capillary law can be demonstrated with a series of glass capillaries of different diameters. We dip them into colored water and observe that the water rises further in the thinner capillaries. The diameter of the capillaries is given in the picture in mm.

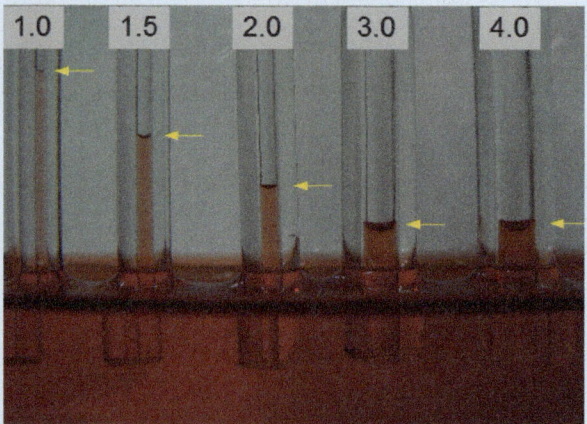

© RWTH Aachen, Physics Experiment Collection

Example 15.17: Water Transport in Trees

Trees absorb water with their roots and transport it to the leaves, where it is used for photosynthesis or evaporates. The transport of water against gravity is a complicated process in which several effects come into play. Capillarity is one of these effects, but also an osmotic effect in the roots (root pressure) contributes and a suction created by the evaporation of water in the leaves (transpiration pull).

The taller the tree, the more difficult the transport of water. The maximum height a tree can reach is probably limited by water transport. The illustration shows the tallest known tree named Hyperion, a Sequoia (*coastal redwood*) with a height of 115.6 m. Trees probably cannot grow much taller.

Example 15.18: Paper Chromatography

Paper absorbs liquids through capillarity. This effect is used in paper chromatography for the identification of substances. A drop of the substance to be examined and further drops of comparison substances are applied to a filter paper and dried. Then the paper is placed in a container with some solvent so that only the lower end dips into the solvent. The solvent rises due to capillarity and takes the substances from the drops with it, with these rising at different speeds. If the experiment is stopped after a certain time, characteristic height patterns are created.

The illustration shows a paper chromatogram of felt-tip pens of different colors.

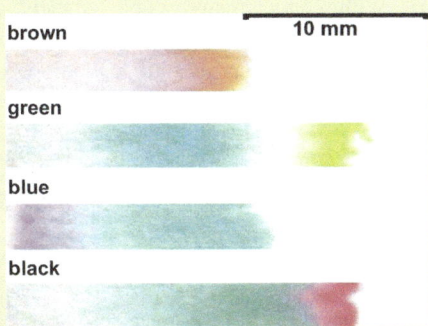

© Ulfbastel from the German-speaking Wikipedia

Example 15.19: Fountain Pen

Capillarity is important for the function of a fountain pen. A drop of ink travels from the tank into a small hole under the nib. From there, a gap leads to the tip of the nib. It is clearly visible in the illustration. The capillarity transports the ink through the gap to the tip of the nib, where it is used for writing.

❓ Problems

1. A table tennis ball (mass 2.7 g, diameter 40 mm) is pushed to a depth of $d = 10$ cm under water. To what height h does it jump out of the water after being released? Neglect friction, surface tension, and the buoyancy in the air.

2. A piece of ice floats on the interface between water and oil. What volume fraction of the ice piece is below the water surface ($\rho_\text{Water} = 1.0\,\text{g/cm}^3$, $\rho_\text{Oil} = 0.75\,\text{g/cm}^3$, $\rho_\text{Ice} = 0.92\,\text{g/cm}^3$)? Calculate for the case of an ice cube with 2 cm edge length in a cylindrical glass of 4 cm diameter, how the liquid levels of water and oil have changed after the ice cube has melted!

3. A stratospheric balloon is supposed to carry a payload of 1200 kg to an altitude of 33 km.

   a) With how many cubic meters of helium must the balloon be filled at the start? Assume a start under standard conditions (pressure 1013 hPa, temperature 0°C, $\rho_\text{He} = 0.18\,\text{kg/m}^3$, $\rho_\text{Air} = 1.3\,\text{kg/m}^3$) and neglect the decrease in temperature with increasing altitude as well as the balloon's own weight! What is the diameter of the balloon at the target altitude?

   b) The material of the stratospheric balloon is made of 20 µm thick polyethylene ($\rho_\text{PE} = 0.9 \cdot 10^3\,\text{kg/m}^3$). The temperature at 33 km altitude is -40 °C. The density of the gases can be assumed to be inversely proportional to the absolute temperature. Calculate corrections to the result of subtask a) from this!

   c) Should the balloon be filled exactly with the calculated amount of helium?

4. Consider a vertical dam with arc length L and curvature radius R, behind which water is filled up to a height H. Calculate the average amount of force

acting on the dam over the surface of the dam, and from this, the force transferred on each of the two mountain walls!

5. Two soap bubbles with diameters 4 cm and 6 cm collide and merge into a large soap bubble. Calculate the internal pressure in the two soap bubbles before the merging and in the large soap bubble afterwards (surface tension of the soap solution: $\sigma = 0.030 \, \text{N/m}$)! Assume the air to be incompressible. How good is this approximation?

6. How high does water rise in a capillary tube with a diameter of 1 mm ($\sigma = 0.073 \, \text{N/m}$)? Is it possible that solely due to capillary forces, the sap in the 0.01 mm thick channels of a tree is transported to its crown? In both cases, neglect the correction due to the boundary angle α.

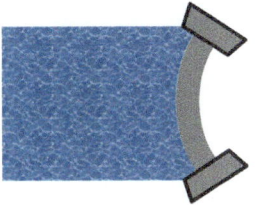

Hydro- and Aerodynamics

Contents

16.1　Description of Flows – 408

16.2　The Continuity Equation – 413

16.3　The Flow of Ideal Fluids – 417

16.4　Internal Friction – 432

16.5　Laminar Flows – 437

16.6　Turbulent Flows – 446

16.7　Drag – 447

16.8　Dynamic Lift – 455

© The Author(s), under exclusive license to Springer-Verlag GmbH, DE, part of Springer Nature 2025
S. Roth and A. Stahl, *Mechanics*,
https://doi.org/10.1007/978-3-662-68079-7_16

16.1 Description of Flows

After dealing with stationary fluids, we now want to get to know the phenomena that are connected with moving (flowing) fluids. We start with a description and classification of flows before we try to explain them in later chapters. But first, we need a way to make flows visible. Such a possibility is presented in Experiment 16.1. We want to consider it further as an example.

Experiment 16.1: Water Tunnel

The first illustration shows the apparatus. Water is pumped at high speed through a channel (arrows). In the central area, we can look through the channel through glass panes. From above, we can introduce various objects into the channel. In the pump (left), the water is swirled. Air bubbles are introduced, which can be recognized as white dots with suitable lighting. These make the flow visible. The flow is recorded with a high-speed camera at 500 frames per second. In the pictures, you can see short white lines that result from motion blur from the images of the air bubbles.

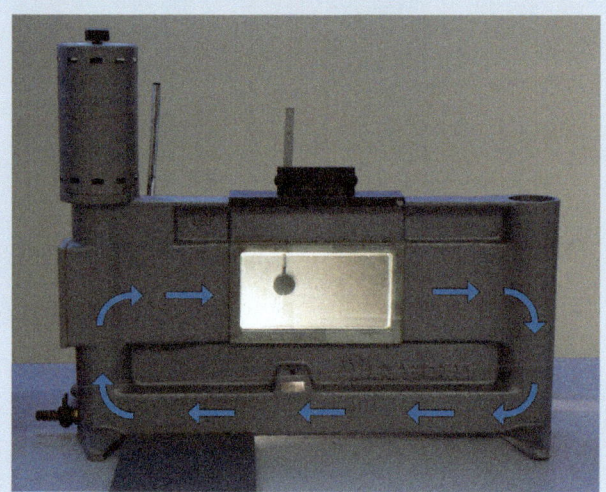

© RWTH Aachen, Physics Experiment Collection

The following illustrations each show a snapshot from the films of the flow around a sphere, an open half-cylinder, and an airfoil. In all images, the flow comes from the left.

© RWTH Aachen, Physics Experiment Collection

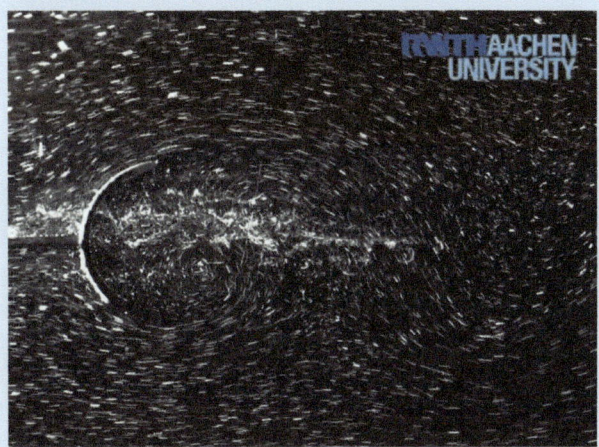

© RWTH Aachen, Physics Experiment Collection

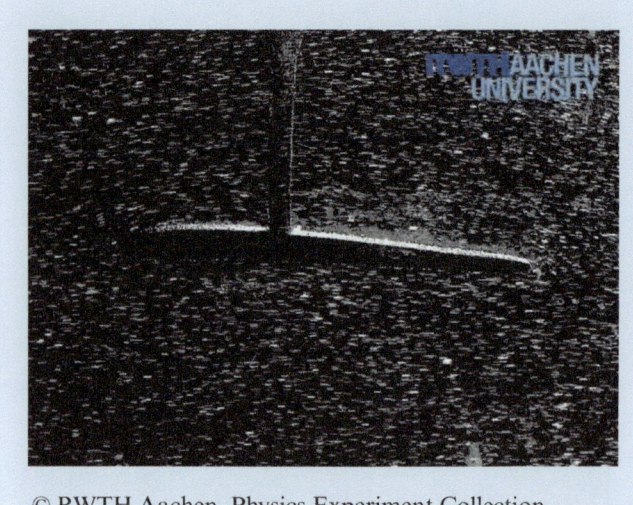

© RWTH Aachen, Physics Experiment Collection

To describe flows, we first need to quantify them. Consider ◘ Fig. 16.1, which shows a section from the first recording of Experiment 16.1. The motion blur in the image creates small lines from the air bubbles. From the direction and length of the lines, we can determine the velocity of the air bubble at the moment of recording. We now proceed as follows. We divide the flow into volume elements (bluish rectangles). For each volume element, we determine the velocity of the fluid $\vec{v}(\vec{r}_0, t_0)$. If necessary, we may have to interpolate between air bubbles. This gives us for the entire flow $\vec{v}(\vec{r}, t_0)$ at the time of recording and, if we finally evaluate the entire film, $\vec{v}(\vec{r}, t)$. This is called the "velocity field of the flow". It describes the flow mathematically and contains the complete information about the flow. Generally, the velocity of the fluid depends not only on the location, but it can also change over time. If this is not the case, we speak of a "steady flow".

Unfortunately, in many situations the velocity field is not accessible, so we have to make do with simpler representations of the flow. We want to introduce two important ones here. If we use a simple camera with a longer exposure time instead of the high-speed camera from experiment 16.1, we see the air bubbles as lines. We have visualized the trajectories of the bubbles. In ◘ Fig. 16.2, the example with the hollow cylinder from the experiment can be seen. Each line represents the path of a bubble through the flow channel. This is called the

16.1 · Description of Flows

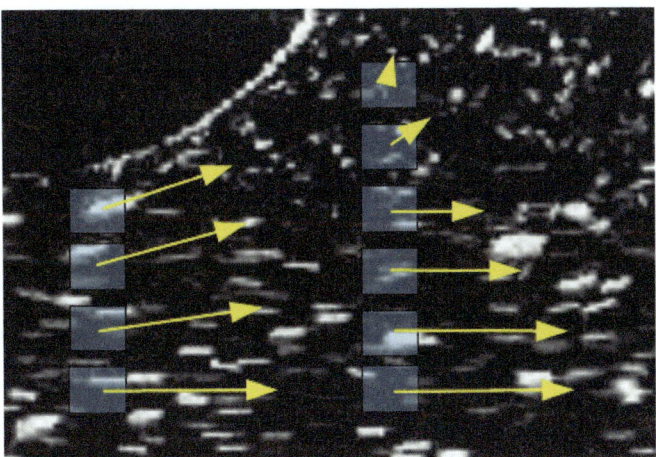

Fig. 16.1 Capturing a flow as a velocity field

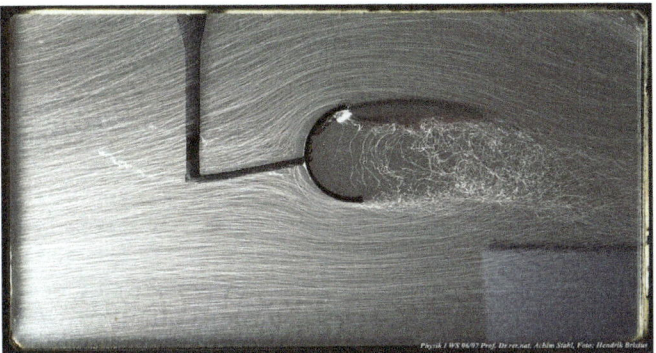

Fig. 16.2 Long-term exposure of a flow. © Photo: Hendrik Brixius

"trajectory of the flow". The velocity field can be recovered from the trajectories, provided the flow is stationary. The direction of the velocity vector must be tangential to the trajectory at every point. The magnitude of the velocity can only be inferred indirectly: in areas where the trajectories are closer together, the velocity is higher and vice versa.

Example 16.1: Wind Tunnel

Flows in air can be visualized by smoke in a wind tunnel. The illustration shows the flow on a car with roof load at a speed of 60 km/h. Behind the roof load, the flow is turbulent, so no lines are recognizable anymore.

© michael.grossmann@autobild.de

Another way to visualize a velocity field is through streamlines. They represent snapshots of the flow. The streamlines are chosen so that at every point the tangent of the streamline coincides with the direction of the velocity vector of the flow. The streamlines therefore indicate the direction of movement of the volume elements of the flow at a certain point in time. In contrast, the trajectories show the movement of individual volume elements over time. In the case of a stationary flow, the trajectories and streamlines coincide.

There are two different types of flows: laminar flows and turbulent flows. A laminar flow is always uniform, i. e. adjacent layers of the fluid glide smoothly along each other. In ◘ Fig. 16.2, you can see areas of laminar flow to the left of the hollow cylinder, as well as above and below it. The trajectories of individual volume elements do not intersect. The velocity field is time-independent, provided the experimental conditions are not changed. Laminar flow is most likely to be found at low flow velocities. In contrast, turbulent flows are characterized by vortices, as we find, for example, in ◘ Fig. 16.2 to the right of the hollow cylinder. The flow often changes very rapidly over time. The trajectories of adjacent volume elements can differ macroscopically even at infinitesimal separations, showing chaotic characteristics. Turbulent flows occur at high velocities, especially at corners and edges of obstacles.

In a flow, adjacent layers will generally move at different velocity, creating friction between the layers and leading to energy loss. We refer to this phenomena as "internal friction" or "viscosity".

If we disregard friction, neglect gravity, and assume the fluid to be incompressible, we reach a simplified situation referred to as an "ideal fluid". Ideal fluids are relatively easy to describe, which is their significance, although, there are no ideal fluids in nature. We use ideal fluids in some situations as an approximation of real fluids. For example, flows in gases can be approximated quite well by an ideal fluid, provided the flow velocity is small compared to the speed of sound everywhere in the gas. We will learn more about friction in fluids in ▶ Sect. 16.4.

16.2 The Continuity Equation

First, we want to define the volumetric flow rate. It is a measure of the amount of fluid that is transported through the flow within a certain time. In ◘ Fig. 16.3, the definition is illustrated using a flow in a pipe. The volumetric flow rate I indicates how much fluid flows through the area A per unit of time. It is:

$$I = \frac{\Delta V}{\Delta t}. \tag{16.1}$$

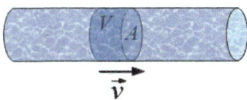

◘ Fig. 16.3 Definition of the volumentric flow rate

The volumetric flow rate can be determined from the sketch. All molecules located in the marked volume V will cross the area A in the time Δt if the volume is chosen such that the base area of the cylinder is exactly A and its height is $v \Delta t$. So, it is:

$$I = \frac{\Delta V}{\Delta t} = \frac{A v \Delta t}{\Delta t} = A v. \tag{16.2}$$

If the flow is none-stationary, we must switch to infinitesimal quantities. Then the height of the cylinder is dx and it follows that

$$I = \frac{dV}{dt} = \frac{A\, dx}{dt} = A \frac{dx}{dt} = A v. \tag{16.3}$$

Instead of the volumetric flow rate, which describes the flowing volume, one can also specify the mass flow rate, which refers to the moving mass. It indicates the mass of the fluid that crosses the area A in the time Δt. Because of $\rho = m/V$:

$$J = \frac{\Delta m}{\Delta t} = \frac{\rho \, \Delta V}{\Delta t} = \rho I = \rho A v. \tag{16.4}$$

If the cross-section of a pipe through which a fluid flows changes, the mass or volumetric flow rates must nevertheless remain constant, because what enters the pipe from one end must eventually exit on the other end. An example is sketched in ◘ Fig. 16.4, which also introduces the designations. The following must apply for the volumetric flow rate:

$$\begin{aligned} I_1 &= I_2 \\ A_1 v_1 &= A_2 v_2 \\ \frac{v_1}{v_2} &= \frac{A_2}{A_1}, \end{aligned} \tag{16.5}$$

i.e., the velocities behave inversely to the available cross-sectional areas of the pipe. Obviously, this does not only apply to cylindrical pipes, but to any flow, whether it is an air flow in a wind tunnel or the flow of a river in its bed. What enters at one end must exit at the other end, as long as we can assume the fluid to be incompressible. After all, the fluid cannot disappear in the pipe. However, there can be delays in time-varying flows. If the inflow is increased, it will take some time before the outflow follows. During this time, the mass in the flow increases, in gaseous flows by increasing the pressure and thus the density of the fluid, in liquid flows by an increase in the fill height in the channel. Example 16.2 shows an illustration.

Example 16.2: Flood wave below a reservoir

Imagine a river that is regulated by a reservoir above. At the outflow of the reservoir, the volumetric flow I_1 is measured, further down the river we get I_2. In the steady state, the outflow I_1 is constant and I_2 will display the same volumetric flow as I_1. Between I_1 and I_2 there is always the same amount of water in the river bed. If the

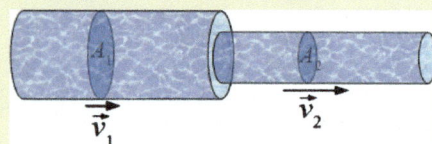

◘ **Fig. 16.4** Volume flow at the narrowing of a pipe

16.2 · The Continuity Equation

outflow of the reservoir is increased, I_1 increases immediately. A flood wave runs down the river and the amount of water in the river increases, but only when the flood wave has reached measuring station 2 will I_2 increase, too.

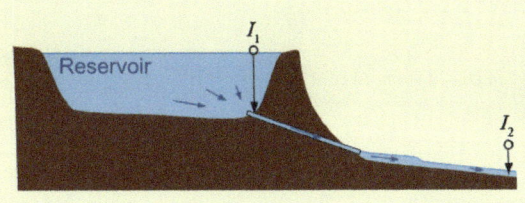

We now want to capture this insight mathematically. We consider the mass flow through an infinitesimal cube (● Fig. 16.5). First, we reduce the problem to one dimension, the flow points exclusively in the x-direction. The mass flow J_1 enters the cube on the left and exits on the right as J_2. Since the cube should only be infinitesimally small, J_2 can only differ slightly from J_1. Therefore, we may expand the mass flow in a Taylor series:

$$J_1 = J(x_0)$$
$$J_2 = J(x_0) + \frac{\partial J(x,t)}{\partial x} dx \, . \quad (16.6)$$

If J_1 differs from J_2, the mass of the fluid, which is in our cube, changes. Since the volume of the cube is supposed to be constant, the density of the fluid must change:

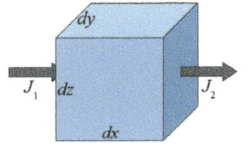

● **Fig. 16.5** For the derivation of the continuity equation

$$J_2 - J_1 = \frac{\partial J(x,t)}{\partial x} dx = -\frac{\partial \rho(x,t)}{\partial t} dV$$
$$= -\frac{\partial \rho(x,t)}{\partial t} dx \, dy \, dz$$
$$\frac{\partial}{\partial x}(\rho(x,t) \, v(x,t) \, dA)dx = \frac{\partial}{\partial x}(\rho(x,t) \, v(x,t) \, dx \, dy \, dz)$$
$$= -\frac{\partial \rho(x,t)}{\partial t} dx \, dy \, dz$$
$$\Rightarrow -\frac{\partial}{\partial x}(\rho(x,t) \, v(x,t)) = \frac{\partial \rho(x,t)}{\partial t} \, .$$
$$(16.7)$$

The quantity $j = \rho v$ is called the "mass flux", indicating how much mass a flow moves per unit of time and per unit of area. Now, we need to take into account that the flow can spread in all three spatial directions, i.e., in addition to the mass flows J_1 and J_2 through the left and right surface, there can still be mass flows through the front/back and top/bottom surfaces, and of course also across them. We have to add these:

$$-\frac{\partial}{\partial x}(\rho(\vec{r},t)\vec{v}(\vec{r},t)) - \frac{\partial}{\partial y}(\rho(\vec{r},t)\vec{v}(\vec{r},t))$$
$$-\frac{\partial}{\partial z}(\rho(\vec{r},t)\vec{v}(\vec{r},t)) = \frac{\partial \rho(\vec{r},t)}{\partial t} \quad (16.8)$$
$$-\frac{\partial}{\partial x}\vec{j}(\vec{r},t) - \frac{\partial}{\partial y}\vec{j}(\vec{r},t) - \frac{\partial}{\partial z}\vec{j}(\vec{r},t) = \frac{\partial \rho(\vec{r},t)}{\partial t}.$$

The result is called the "continuity equation". It can be expressed a bit more elegantly in mathematical terms by using the divergence operator. Then it reads:

> **Continuity Equation**

$$-\operatorname{div}\vec{j}(\vec{r},t) = \frac{\partial \rho(\vec{r},t)}{\partial t} \quad \text{with} \quad \vec{j}(\vec{r},t) = \rho(\vec{r},t)\,\vec{v}(\vec{r},t).$$

For incompressible fluids ($\rho = $ const.) it reduces to:

$$\operatorname{div}\vec{v}(\vec{r},t) = 0. \quad (16.9)$$

Continuity equations always arise in physics when a certain property cannot be lost. Here, it is about the molecules of the fluid. You will encounter another example in electricity, where there is a continuity equation for the electric charge. Furthermore, continuity equations can be established for the energy. In general, the divergence of the current density indicates how much of this property flows out of or into the infinitesimal volume, the difference manifesting itself in a change in density of the property in the volume.

Example 16.3: Rapids

If the cross-section of a river narrows, the velocity of the flow must increase. You can see this impressively in a whitewater, when rocks partially obstruct the river course. As a result, the velocity of the water between the rocks increases significantly, putting the canoeist in distress.

16.3 · Flow of Ideal Fluids

Example 16.4: Junction Rule

If the volumetric flow to a volume element is limited to certain paths, a junction rule can be derived from the continuity equation. This is the case when the flow is restricted to pipes. An example is shown in the sketch. Five pipes of different diameter lead to a common point, which is called a junction. The volumetric flows I_1 and I_5 through the lower two pipes point into the junction, while the fluid flows out again through the other three pipes. If we assume the fluid to be incompressible, then $I_1 + I_5 = I_2 + I_3 + I_4$ must apply. If we define inflowing volumetric flows as positive and outflowing as negative, the following must apply:

$$\sum_i I_i = 0.$$

This is called a junction rule. In electricity, you will encounter a corresponding junction rule for electrical currents, which plays an important role in electronics.

16.3 The Flow of Ideal Fluids

16.3.1 Bernoulli's Principle

We had defined ideal fluids as incompressible fluids in which no friction occurs and for which the influence of gravity is negligible. We want to deal with their behavior, first.

We consider the flow through a pipe, whose cross-section narrows from A_1 to A_2 (◼ Fig. 16.6). From the continuity equation, we know that at the narrowing, the velocity of the flow must increase, as:

$$A_1 v_1 = A_2 v_2. \tag{16.10}$$

Since the velocity in the smaller pipe is higher, the fluid must be accelerated at the boundary layer. With no external forces present, only a pressure difference at the narrowing can be responsible for the acceleration. On the boundary layer, which is marked in ◼ Fig. 16.6 with the thickness dx, a force acts:

$$F = p_1 A_2 - p_2 A_2 = (p_1 - p_2) A_2$$
$$= m \frac{dv}{dt} = m \frac{v_2 - v_1}{dt}. \tag{16.11}$$

With $m = \rho A_2 \, dx$ we get:

$$(p_1 - p_2) A_2 = \rho A_2 \frac{dx}{dt} (v_2 - v_1)$$
$$p_1 - p_2 = \rho \frac{dx}{dt} (v_2 - v_1), \tag{16.12}$$

where $\frac{dx}{dt}$ is the average velocity in the boundary layer. It is:

$$\frac{dx}{dt} = \frac{v_2 + v_1}{2}$$
$$\Rightarrow p_1 - p_2 = \rho \frac{v_2 + v_1}{2} (v_2 - v_1) = \frac{1}{2} \rho \left(v_2^2 - v_1^2 \right)$$
$$p_1 + \frac{1}{2} \rho v_1^2 = p_2 + \frac{1}{2} \rho v_2^2 \tag{16.13}$$

or

$$p + \frac{1}{2} \rho v^2 = \text{const.} \tag{16.14}$$

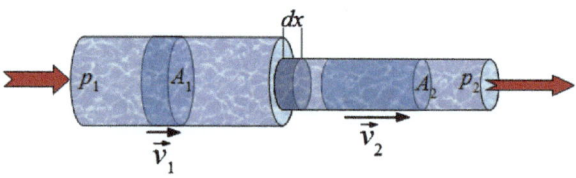

◼ **Fig. 16.6** On the flow of an ideal fluid

16.3 · Flow of Ideal Fluids

This is called "Bernoulli's equation" after the physicist Daniel Bernoulli. It establishes a relationship between the velocity of flow and the pressure in the pipe.

We derived Bernoulli's equation dynamically, i.e., from Newton's axioms. It can also be obtained from the law of conservation of energy. At the interface, volume work is converted into kinetic energy. Therefore, the following applies:

$$pV + \frac{1}{2}mv^2 = E_{tot} = \text{const.}$$
$$pV + \frac{1}{2}\rho V v^2 = E_{tot} = \text{const.} \quad (16.15)$$
$$p + \frac{1}{2}\rho v^2 = \text{const.}$$

Now, we have converted an energy balance into a pressure balance. The quantities on the left side carry the unit of pressure, so the constant on the right side must carry the same unit. It is called the "stagnation pressure" p_{stag}. Let's have a closer look at the left side: there we find the quantity p, which occurs independently of the velocity, and the term $\frac{1}{2}\rho v^2$, which increases strongly with the velocity of the flow. The first term is called the "static pressure", the second the "dynamic pressure". Thus, Bernoulli's equation reads

▶ Bernoulli Equation

$$p_{stat} + \frac{1}{2}\rho v^2 = p_{stag} = \text{const.}$$

◘ Figure 16.7 shows a simple arrangement for determining the static component of the pressure. We want to measure the static pressure in a pipe, in which the fluid

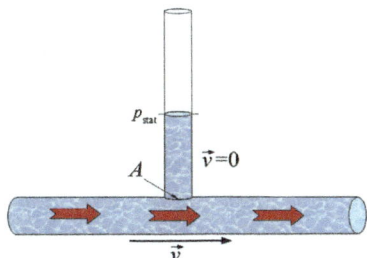

◘ **Fig. 16.7** The water column in the standpipe indicates the static pressure

flows at a velocity *v*. A vertical standpipe (open at the top) is attached to the pipe and connected to the flow via the opening *A*. The height of the fluid column in the standpipe indicates the static pressure in the fluid. At the interface *A*, a balance is established between the pressure in the flowing fluid and the gravitational pressure in the fluid column of the standpipe. Since the fluid flows parallel to the interface, the dynamic pressure is ineffective here, because the velocity *v* in the term of the dynamic pressure denotes a velocity perpendicular to the area *A*. The height of the vertical column is determined solely by the static pressure. The fluid in the column rises until the gravitational pressure at the top has become zero and the surface is in equilibrium with the ambient pressure.

The influence of the dynamic pressure on the height of water columns leads to the hydrodynamic paradox, which we demonstrate in Experiment 16.2.

© RWTH Aachen, Physics Experiment Collection

Experiment 16.2: Hydrodynamic Paradox

As we have seen, the pressure in vertical standpipes depends on the flow rate in the pipe below. This is the basis of the hydrodynamic paradox. Our pipe has a bulbous widening in the middle. We pump water through the pipe, which flows slower in the wider section in the middle than in the rest of the pipe. Because of

$$p_{\text{stat}} + \tfrac{1}{2}\rho v^2 = p_{\text{stag}} \quad \text{or} \quad p_{\text{stat}} = p_{\text{stag}} - \tfrac{1}{2}\rho v^2$$

the static pressure is higher there, which leads to a higher pressure in the central standpipe. Our standpipes here are so short that the water column rises above the end of the pipes and water fountains emerge from the pipes. As can be clearly seen in the photograph, the fountain at the bulbous widening is higher than at the adjacent pipe, contrary to the naive expectation that a higher flow rate would be caused a higher pressure and would generate higher fountains ($\Rightarrow$ Paradox).

The following two photos show another demonstration of the hydrodynamic paradox. Water flows from a container through a horizontal pipe. The pipe in the first photo has a constant diameter. The four standpipes show the pressure loss due to friction in the liquid, which we have so far neglected. In the second photo, we see a pipe with three attached standpipes and a narrowing of the horizontal pipe directly under the central standpipe. It can be seen that there the height of the water column is significantly reduced compared to the pipe with a constant diameter.

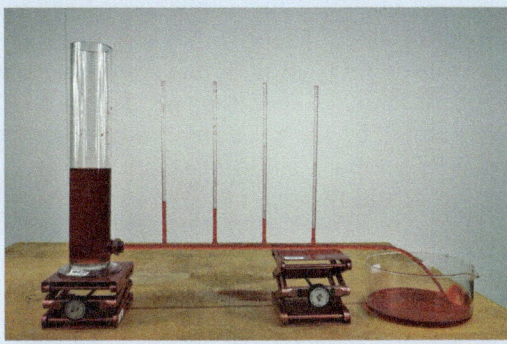

© RWTH Aachen, Physics Experiment Collection

Initially, we neglected the influence of gravity on the flu`energy conservation in Eq. 16.15, where we did not take the potential energy in the gravitational field of the Earth into account. Now, we want to extend our consideration to include the influence of gravity. We expand the energy balance to include potential energy:

$$\begin{aligned} pV + \frac{1}{2}\rho V v^2 + \rho V g h &= E_{\text{tot}} \\ p + \frac{1}{2}\rho v^2 + \rho g h &= p_{\text{stag}} \\ p_{\text{stat}} + p_{\text{dyn}} + p_{\text{grav}} &= p_{\text{stag}} \end{aligned} \quad (16.16)$$

and obtain a version of Bernoulli's equation that includes the gravitational pressure. The height h is a relative height within the fluid, as the zero point of potential energy is freely selectable.

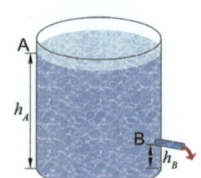

Fig. 16.8 Pressure conditions in a container with gravitational pressure

We consider an example (Fig. 16.8). A liquid container has an spout at a height h_B above the ground. We want to determine the velocity v_B at which the liquid flows out. The open-top container is filled up to the height h_A with water. The gravitational pressure on the surface of the liquid is given by the atmospheric pressure p_atm. From Bernoulli's equation, we determine the various components of the pressure and from this the exit velocity. We demand a constant stagnation pressure between points A and B:

$$p_\text{atm} + \frac{1}{2}\rho v_a^2 + \rho g h_a = p_\text{atm} + \frac{1}{2}\rho v_b^2 + \rho g h_b$$
$$\rho g h_a = \frac{1}{2}\rho v_b^2 + \rho g h_b,$$
(16.17)

where we have already taken into account in the second step that the flow velocity at the surface of the water vanishes. This leads to:

$$\frac{1}{2}\rho v_b^2 = \rho g(h_a - h_b)$$
$$\frac{1}{2} v_b^2 = g \Delta h$$
$$v_b = \sqrt{2 g \Delta h}.$$
(16.18)

Experiment 16.3: A Bucket with a Hole

© RWTH Aachen, Physics Experiment Collection

We have recreated the situation from Fig. 16.8. The vessel has three spouts at different heights, so we can study the change of the exit velocity with depth. The exit velocity can be determined from the distance from the vessel at which the water jet hits the ground. If we place the vessel on a flat surface (not at the table edge, as in the picture), the free fall of the water results in a distance of $s = v_b\sqrt{2 h_b/g}$. If we apply our result for v_b from Eq. 16.18, we obtain

$$s = \sqrt{2 g(h_a - h_b)}\sqrt{\frac{2 h_b}{g}} = 2\sqrt{(h_a - h_b)h_b}.$$

Here we find a nice optimization task: At what height h_b should one attach the spout to achieve a maximum range of the water jet? We determine the maximum of s:

$$\frac{\partial s}{\partial h_b} = 2\frac{1}{2}\frac{1}{\sqrt{(h_a - h_b)h_b}}(h_a - 2h_b) = 0 \Rightarrow h_b = \frac{1}{2}h_a.$$

One might guess this result from the photo.

16.3.2 Pressure Probes

In ▶ Sect. 15.1 we got to know a series of manometers and barometers, used to measure pressure in the static case. In the case of flowing fluids the pressure depends on the orientation of the sensitive surface to the flow, whether we measure the static, the dynamic or the stagnation pressure. In ◘ Fig. 16.9 the two essential orientations are shown. In the upper case, the flow is parallel to the surface. In this case, only the static pressure acts on the surface A. The dynamic pressure is ineffective. Not so in the lower case (pitot tube), where the surface is oriented perpendicular to the flow indicating the stagnation pressure. This can be understood as follows: Immediately in front of the sensitive surface, the flow velocity is zero, so there is $p_{stag} = p_{stat}$. Since there is an equilibrium with the surrounding flow, the stagnation pressure in front of the surface must correspond to the stagnation pressure in the flow. However, the manometer must not be too large, as it would otherwise disturb and change the flow (in ◘ Fig. 16.9 this would probably be the case).

The behavior of the manometers can also be understood on a microscopic basis (◘ Fig. 16.10). The force acting on the surface A is caused by molecules that hit the surface and bounce off it. Each molecule exerts an impulse on the surface with the pressure (force per area) resulting from the time average over the force impulses of all molecules.

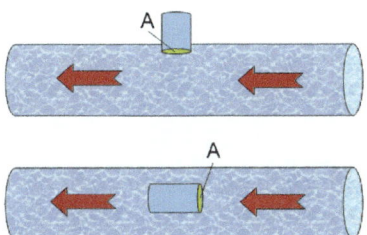

◘ **Fig. 16.9** The orientation of the sensitive surface determines which pressure is measured

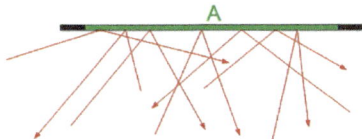

◘ **Fig. 16.10** Microscopic explanation of pressure

If the surface is oriented parallel to the flow, only the velocities perpendicular to the flow direction are relevant. These are statistically distributed and represent the static pressure. However, if the fluid moves towards the surface (flow velocity perpendicular to the surface), the momenta of the molecules relative to the surface are larger, the impulses are larger when bouncing off, and consequently the pressure is higher. Now the surface feels the stagnation pressure, including the dynamic pressure generated by the movement of the fluid.

With the arrangements from ◘ Fig. 16.9, the static pressure and the stagnation pressure can be measured but the dynamic pressure cannot be determined directly. We derive it from a differential measurement: $p_{dyn} = p_{stag} - p_{stat}$. This difference can be evaluated directly with a pitot-static tube also called a Prandtl tube. The basic structure is outlined in ◘ Fig. 16.11. The pitot tube has two openings. One is oriented in the middle against the flow (gray) and measures the stagnation pressure p_{stag}. Additional openings are located on the side, which are connected to the outer space (light blue) of the pitot tube. These openings are oriented parallel to the flow, so that they only feel the static pressure p_{stat}. The central opening is connected to the airtight housing of the display instrument via a tube. In our example, the pressure difference is determined by a can manometer. The expansion of the pressure can, which leads to the deflection of the pointer, is caused by the difference between stagnation pressure and static pressure.

In ◘ Fig. 16.12, the different probes are compared against each other.

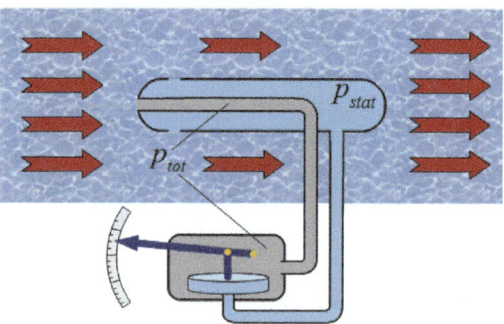

◘ **Fig. 16.11** Principle of the Prandtl's pitot tube

16.3 · Flow of Ideal Fluids

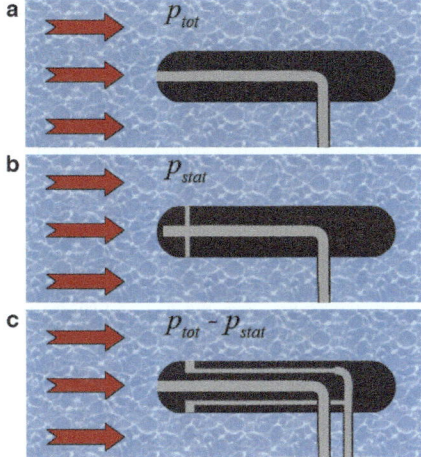

Fig. 16.12 Different pressure probes, **a** simple pitot tube, **b** static probe, **c** Prandtl tube

Experiment 16.4: Prandtl Tube

In the experiment, a Prandtl tube can be seen in a wind tunnel. The differential measurement of p_{stag} and p_{stat} is done here by a U-tube manometer. Different objects can be introduced into the wind tunnel in front of the Prandtl tube and then the velocity profile can be measured behind the obstacles.

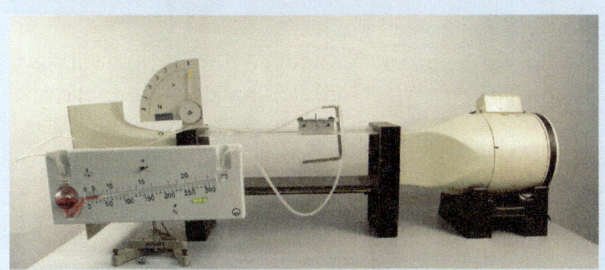

© RWTH Aachen, Physics Experiment Collection

Example 16.5: Speed Measurement on Airplanes

The speed of airplanes is measured with a Prandtl tubes. The differential pressure depends on the velocity of the

© Wikimedia: Kārlis Dambrāns[3]

---

1 License CC_BY, see https://creative commons.org/licenses/by/2.0/deed.de

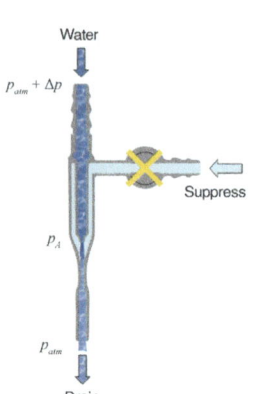

© Wikimedia: Muraer[4]

incoming air and thus gives a measure of the speed of the aircraft. The first image shows two Prandtl tubes on the fuselage of an Antonov An-225. The second image shows a Prandtl tube of an F15. Clearly visible are the opening at the front for the transmission of the stagnation pressure and five holes as side openings for the static pressure. With the Prandtl tube, the speed of the aircraft is measured relative to the surrounding air flow. This is supplemented today by a measurement of the absolute speed with GPS.

Experiment 16.5: Water Jet Pump

Water is pushed through a nozzle with an overpressure Δp, after which it relaxes to atmospheric pressure. In the nozzle, the water reaches high speeds (v_A). In the surrounding (stationary) air, the same pressure (p_A) is established. This is

$$p_A = p_\text{atm} + \Delta p - \frac{1}{2}\rho v_A^2.$$

If the water pressure Δp is high enough, the speed reaches values that are so large that p_A becomes less than atmospheric pressure. A vacuum (underpressure) is created. The pump sucks in air.

© Photo: Hendrik Brixius

Example 16.6: Carburetor

In a classical Otto engine without direct injection, the gasoline must be atomized and mixed with air in the car-

---

2 License CC_BY, see https://creative commons.org/licenses/by/2.0/deed.de

burator before combustion. During the intake stroke, the cylinder sucks air through the carburetor. Similar to the water jet pump, a static vacuum is exploited, which arises when the air flows through a constriction in the carburetor (Venturi tube). The intake pipe ends in the constriction. The vacuum sucks in gasoline through the fuel nozzle and atomizes it. The throttle valve controls the fuel supply. It is connected to the accelerator pedal.

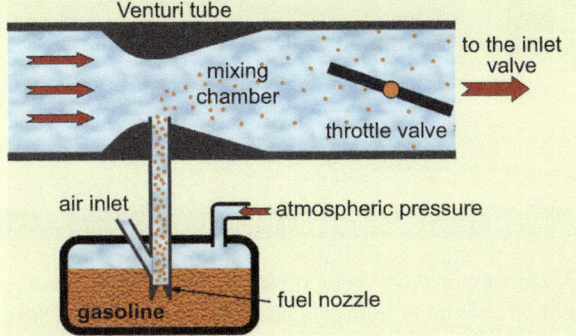

Simple atomizers, like the one on a perfume bottle use the same principle. By pressing the spray button, an air stream is pushed through a Venturi tube, sucks in perfume, atomizes it, and expels it with the air stream. The photo in Exp. 16.5 shows a Venturi tube that sucks in colored water by a water jet.

Example 16.7: Bunsen Burner

When the gas flows out of the nozzle, it creates a dynamic vacuum. This causes the Bunsen burner to suck in air and mix it with the gas. The strength of the flame is controlled by regulating the air supply.

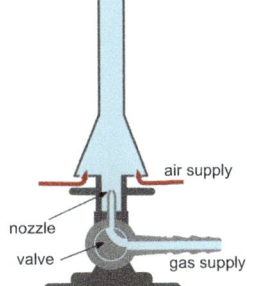

Example 16.8: Convertible Top

When a convertible (with the top closed) is driven at high speed, an underpressure is created over the top due to the wind. This causes the top to bulge outwards, contrary to intuition. The force on the top is not insignificant. It can tear off a poorly secured top.
At a speed of $v = 120\,\text{km/h}$ the pressure difference between the interior (p_i) and the exterior (p_e) is:

$$p_e + \frac{1}{2}\rho v^2 = p_i \Rightarrow p_i - p_e = \frac{1}{2}\rho v^2 \approx 1400\,\frac{\text{N}}{\text{m}^2},$$

where we use the numerical value for the density of air of approximately $\rho = 1.3\,\text{kg/m}^3$. For a cover with an area of about $2\,\text{m}^2$ this corresponds to a force of 2800 N or a load of 280 kg, which the cover must withstand.

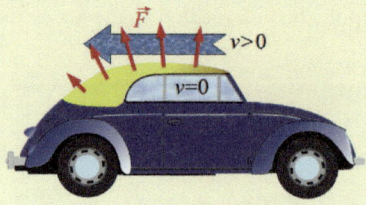

Example 16.9: Storm Striping Roofs

A storm with high wind speeds can strip roofs. It does not, as one might naively expect, push the roofs in. No, it untiles them. The tiles are torn away by the storm from the outside, which can be easily explained with the Bernoulli's equation. Inside the house, there is no wind and thus atmospheric pressure p_0. Outside, there are high wind speeds, which are further increased by the house representing an obstacle to the wind. On the outside, the static pressure $p_{\text{stat}} = p_0 - 1/2\,\rho v^2$ prevails leading to a pressure difference and a force on the tiles that points outward and can lift the tiles off the roof.

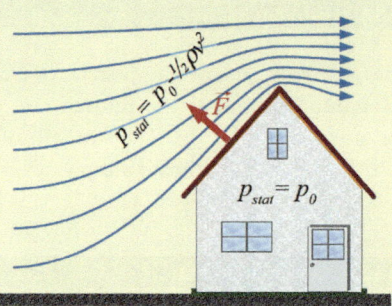

Example 16.10: Wind Turbine

Today, electrical power from wind plays an increasing role. We first want to determine how much wind power is actually available. The kinetic energy of the air in a volume dV is:

16.3 · Flow of Ideal Fluids

$$E_{\text{wind}} = \frac{1}{2} \rho_{\text{air}} v_{\text{wind}}^2 \, dV.$$

The volume $V = v_{\text{wind}} A \, \Delta t$ reachs the surface A covered by the three rotors within the unit of time Δt, where A refers to the circular disk that the rotors sweep over. Then the available power is:

$$P_{\text{wind}} = \frac{1}{2} \rho_{\text{air}} v_{\text{wind}}^3 A.$$

It changes with the cube of the wind speed creating a major logistical problem. On many days, the wind speed is so low that operating the turbine is not worthwhile and it is shut down (typically at wind speeds below $4 \, \text{m/s}$). On days with medium to high wind speeds, the plant generates electrical energy that then needs to be stored. At wind speeds above about $25 \, \text{m/s}$ the turbine must be turned out of the wind, to avoid damage of the rotor.

Now, we look at the power that can be transferred to the rotor, which will be converted into electrical power with high efficiency using a generator. It can be calculated as:

$$P_{\text{rotor}} = F v,$$

where the force F on the rotor is given by Bernoulli's equation

$$F = \frac{1}{2} \rho_{\text{air}} \left(v_1^2 - v_2^2 \right).$$

Here, v_1 is the wind speed of the approaching air and v_2 the wind speed behind the rotor. To determine P_{rotor}, we need the actual amount of air that flows around the rotor. It is not given by v_{wind}, as a static pressure is created in front of the rotor, which reduces the speed of the air. We estimate v_{wind} by the average of the velocity of the air in front of and behind the rotor[3]. Then

---

[3] The slowing down of the air seems to violate the continuity equation. But this is not the case. The rotor displaces a portion of the air flow sideways. The cross-sectional area through which the air flows, increases, so that vA remains constant.

© Marco Verch via ccnull.de

$$P_{rotor} = \frac{1}{2}\rho_{air}\left(v_1^2 - v_2^2\right)\frac{(v_1+v_2)}{2}$$
$$= \frac{(v_1^2 - v_2^2)(v_1+v_2)}{2v_1^3}P_{wind} = fP_{wind},$$

with $f = \dfrac{(v_1^2 - v_2^2)(v_1+v_2)}{2v_1^3}.$

Like the power of the wind, the power transferred to the rotor is also proportional to the cube of the wind speed. The factor f indicates the efficiency of the conversion. A closer inspection shows a maximum for $v_2 = \frac{1}{3}v_1$, for which the rotors are optimized for. Contrary to intuition, a complete deceleration of the air to $v_2 = 0$ does not produce the highest rotor power. The maximum efficiency is 0.59. Real wind turbines have typical efficiencies of 0.4.

Experiment 16.6: Floating Balls in an Air Stream

© Photo: Hendrik Brixius

Most readers have probably seen this experiment somewhere before. It is shown in many technology museums or science shows: A ball (or several) dances in the air stream of a blower. The fact that the ball floats is quite easy to understand. The air blowing from below pushes the ball upwards, thereby compensating for its weight. But how can its position be stable? Several balls dance wildly in the air stream. Why aren't they pushed to the side when they accidentally get a little off-center? And why does the ball still float when the blower is tilted, as in the second picture?

It is stabalized in a position in the middle of the stream by the dynamic pressure, as we know it from Bernoulli's equation. The blower creates an air stream with a velocity profile, which is highest in the middle and decreases towards the edge. If the ball gets a little off-center due to a small disturbance, the speed of the air on the inside is higher than on the outside. This creates a pressure difference that pushes the ball back to the center. This even works when the blower is tilted. Due to its gravity, the ball slips a little out of the center downwards, but the pressure difference, which now results from to the dynamic pressure, keeps it in place, as long as the disturbances are not too large.

© Photo: Hendrik Brixius

16.3 · Flow of Ideal Fluids

> **Experiment 16.7: Aerodynamic Paradox**
>
> The aerodynamic paradox is a phenomenon that occurs with strong air currents that are trapped between two surfaces. The dynamic pressure reduces the static pressure between the surfaces, causing them to move together. The paradox exists in many variants. The illustration shows a variant that you can easily perform yourself with only two sheets of paper. Hold them a finger's width apart in front of your mouth and blow hard between the sheets. Naively, one might expect the sheets to be pushed apart, but they move together.

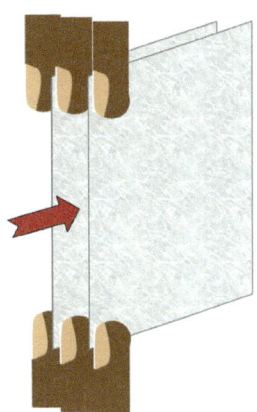

16.3.3 The Equation of Motion

At the end of this section, we want to pick up the line of thought from Eq. 15.9 again. We had determined the force on an infinitesimally small volume element in the fluid due to the pressure to be $\vec{F}_p = -\operatorname{grad} p \, dV$. With this, we now want to set up the equation of motion for such a volume element. We start from Newton's equation of motion, but, here we are less interested in the motion of a single volume element than in the change of the flow field $\vec{v}(\vec{r}, t)$ under the influence of the forces. The following equation must apply for a volume element of mass dm:

$$\vec{F} = dm \, \frac{d\vec{v}}{dt}. \tag{16.19}$$

We consider the force derived from the pressure in Eq. 15.9 $\vec{F}_p$, as well as possibly other external forces on the fluid $\vec{F}_{\text{ext}}$. Such an external force could be, for example, the gravitational force $\vec{F}_G = dm \, \vec{g}$. Equation 16.19 requires a total derivative, which is given by:

$$\begin{aligned}\frac{d}{dt}\vec{v}(x(t), y(t), z(t), t) &= \frac{\partial \vec{v}}{\partial t} + \frac{\partial \vec{v}}{\partial x} \cdot \frac{\partial x}{\partial t} + \frac{\partial \vec{v}}{\partial y} \cdot \frac{\partial y}{\partial t} + \frac{\partial \vec{v}}{\partial z} \cdot \frac{\partial z}{\partial t} \\ &= \frac{\partial \vec{v}}{\partial t} + \vec{v} \cdot \operatorname{grad} \vec{v} \\ &= \frac{\partial \vec{v}}{\partial t} + \operatorname{grad} \frac{\vec{v}^2}{2} - \vec{v} \times \operatorname{rot} \vec{v}.\end{aligned} \tag{16.20}$$

You will learn about the relation used in the last transformation step in vector analysis. You can convince yourself of the correctness by writing out the components. The term $\frac{\partial \vec{v}}{\partial t}$ describes accelerations that arise because the speed at a fixed location may change over time. The remaining term (or both terms) takes into account that even in a static velocity field, acceleration occur when a volume element of the fluid moves from one position to another position with a different intrinsic velocity. This change in velocity also represents an acceleration. Now, if we replace the mass in Eq. 16.20 with the density ($dm = \rho\, dV$), we get

$$\rho\, dV \left(\frac{\partial \vec{v}}{\partial t} + (\vec{v} \cdot \mathrm{grad})\vec{v} \right) = -\,\mathrm{grad}\, p\, dV + \rho \vec{g}\, dV \quad (16.21)$$

or

$$\rho \left(\frac{\partial \vec{v}}{\partial t} + (\vec{v} \cdot \mathrm{grad})\vec{v} \right) = -\,\mathrm{grad}\, p + \rho \vec{g}. \quad (16.22)$$

When written out in components, these are three partial differential equations that are coupled via the term $(\vec{v} \cdot \mathrm{grad})\vec{v}$, which allow to calculate the velocity profile of a flow. They are called the "Euler equations". We will revisit them at the end of the following chapter to extend them to a fluid with friction.

16.4 Internal Friction

Figure 16.13 shows the reaction of a cuboid to a shear force. The first image shows a solid body. The shear force acts on a plate that is attached to the surface and pulls it to the right. This results in an elastic reaction, which we have learned about in ▶ Sect. 14.4. The shear stress generates a restoring force $F_\parallel = \tau A$. In the second image, the same setup is shown on a fluid (here a liquid). The plate floats on the fluid. But, when the force $F_\parallel$ is applied, there is no elastic reaction $F_\parallel$. In the fluid, the molecules move until a force-free state is restored. The plate keeps moving as long as the force acts.

However, this only applies to a static case. If you keep pulling a plate with a constant force $F_\parallel$ over the surface of a liquid (▶ Fig. 16.14), a reactive force does occur, in the form of friction between the liquid and the plate,

16.4 · Internal Friction

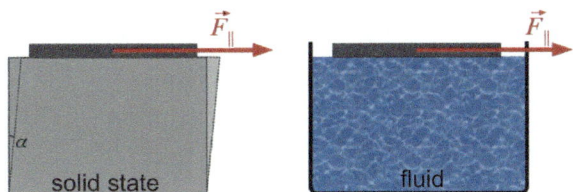

Fig. 16.13 Shear of a solid body and attempted shear of a fluid

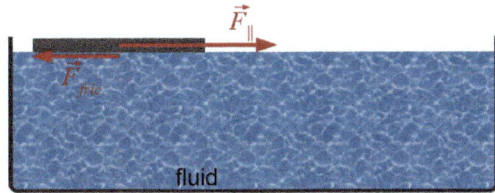

Fig. 16.14 A fluid showing friction

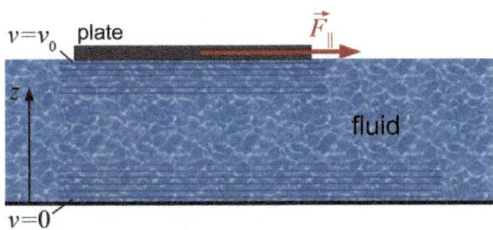

Fig. 16.15 Layered structure of friction

proportional to the velocity at which the plate moves. The plate is accelerated until the applied force and the frictional force balance each other and a constant velocity is established.

To further investigate friction, we divide the fluid into thin horizontal layers (Fig. 16.15). The z-axis in the figure indicates the height in the fluid, with the origin at the bottom of the vessel. A plate, floating on the fluid, is again pulled by a horizontal force. The vessel is large enough for a constant velocity to be established. It can be assumed that adhesive forces link the top and bottom layers of the fluid to the vessel or the plate, resepctively. The top layer thus moves at the velocity v_0 of the plate, the bottom layer is at rest ($v = 0$). A frictional force arises between any two adjacent layers, which is proportional to the area of contact. Furthermore, the frictional force depends linearly on the velocity difference between the two layers, the greater it is, the greater the frictional force. Therefore, we can write:

$$\frac{F_{\text{fric}}}{A} = \eta \frac{dv}{dz}. \tag{16.23}$$

The constant of proportionality η, which we have inserted into the equation, is called (dynamic) viscosity. The unit of viscosity is:

$$[\eta] = 1\,\frac{\text{kg}}{\text{m s}} = 1\,\frac{\text{N s}}{\text{m}^2} = 1\,\text{Pa s}. \tag{16.24}$$

Sometimes, the unit poise is still in use: 1 Poise $= 0.1\,\text{Pa s} = 1\,\text{g cm}^{-1}\text{s}^{-1}$. The quantity η/ρ is called "kinematic viscosity". The reciprocal $1/\eta$ is called the "fluidity". In ◘ Table 16.1, the viscosities of some fluids are listed as examples.

You might wonder why we listed glass in this table. It is an amorphous solid with a molecular structure is similar to that of a liquid. The molecules are not arranged in a regular lattice. In fact, an amorphous solid can flow, but, the viscosity of glass is so high that the flow cannot be observed on human time scales. Here and there, one reads that one can recognize the flow in medieval church windows, as they are often thicker at the bottom than at the top, although this is probably more a result of the manufacturing process.

◘ **Table 16.1** Viscosity of some substances

	Viscosity η in mPa s
Glass	$\approx 10^{19}$
Pitch	$\approx 10^{11}$
Honey	3000
Oil	100 … 1000
Alcohol	1.2
Water	1.0
Ethyl ether	0.2
Air	0.018
Methane	0.011
Hydrogen	0.009

16.4 · Internal Friction

The viscosity is a material property that must be determined experimentally, for example with a rotational viscometer (Fig. 16.16). It is used to determine the viscosity of liquids. The test liquid is filled into a narrow gap between two coaxial cylinders. One of the two cylinders is driven by a motor (Searle system with rotating inner cylinder, Couette system with rotating outer cylinder), the other is stationary. In the narrow gap between the walls of the cylinders, friction occurs that drags the rotation. This resistive force F_{fric} is recorded e.g., via the electrical power of the motor. Furthermore, we must measure the rotational speed ω of the motor. Then we get:

$$\eta = \frac{F_{fric}}{2\pi \bar{r} h} \left(\frac{\omega \bar{r}}{r_2 - r_1} \right)^{-1} \quad \text{with} \quad \bar{r} = \frac{r_1 + r_2}{2}. \tag{16.25}$$

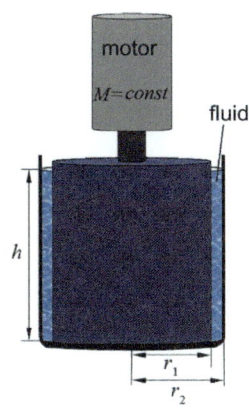

Fig. 16.16 Viscometer (Searle system)

The viscosity is highly temperature dependent. Therefore, it is important to ensure stable temperature conditions during the measurement.

We introduced viscosity through the friction between layers in the fluid. For the general treatment of flows, we will need to translate the frictional forces from layers to infinitesimally small volume elements in the fluid. We address this here already.

Consider the sketch in Fig. 16.17. It shows a volume element of the size dx, dy, dz, starting from the lower left corner at (x_0, y_0, z_0). If the fluid's velocity is constant everywhere, our volume element will be carried along by the flow. Friction only occurs when layers move differently, i.e., when there is a non-constant velocity profile in the fluid. Initially, we assume that the fluid flows in the z-direction and that the velocity only changes with the x-direction. Then we have to consider the friction on the two boundary surfaces shown in the sketch, which move at the velocities $\vec{v}(x_0)$ and $\vec{v}(x_0 + dx)$. Let's assume that the velocity along the x-direction increases. Then the left boundary layer will move slower and the right one faster than our volume element. The friction with the left boundary layer slows down the movement of our volume element, the one with the right side accelerates it. The net force is:

$$(\Delta F_{fric})_z = dF_{fric}(x_0 + dx) - dF_{fric}(x_0). \tag{16.26}$$

We use Eq. 16.23 to express the frictional forces $F_{fric} = \eta A \frac{dv}{dr}$. Since the volume element is infinitesimally

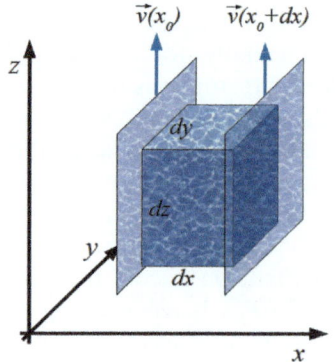

Fig. 16.17 Friction on an infinitesimal volume element

small, we may expand the velocity profile according to Taylor. So:

$$\vec{v}(x_0 + dx) = v_z(x_0 + dx)\hat{e}_z = \left(v_z(x_0) + \frac{\partial v_z}{\partial x} dx\right)\hat{e}_z.$$
(16.27)

Then:

$$\begin{aligned}(\Delta F_{\text{fric}})_z &= \eta\, dy\, dz\, \frac{\partial v_z(x_0 + dx)}{\partial x} - \eta\, dy\, dz\, \frac{\partial v_z(x_0)}{\partial x} \\ &= \eta\, dy\, dz \left(\frac{\partial v_z(x_0)}{\partial x} + \frac{\partial^2 v_z(x_0)}{\partial x^2} dx\right) - \eta\, dy\, dz\, \frac{\partial v_z(x_0)}{\partial x} \\ &= \eta\, \frac{\partial^2 v_z(x_0)}{\partial x^2}\, dx\, dy\, dz = \eta\, \frac{\partial^2 v_z(x_0)}{\partial x^2}\, dV.\end{aligned}$$
(16.28)

If we no longer limit the velocity profile to the x-direction, we obtain accordingly:

$$(\Delta F_{\text{fric}})_z = \eta\left(\frac{\partial^2 v_z}{\partial x^2} + \frac{\partial^2 v_z}{\partial y^2} + \frac{\partial^2 v_z}{\partial z^2}\right) dV = \eta\, \nabla^2 v_z\, dV.$$
(16.29)

In the last step, we inserted the Laplace operator $\nabla^2 = \left(\frac{\partial^2}{\partial x^2} + \frac{\partial^2}{\partial y^2} + \frac{\partial^2}{\partial z^2}\right)$[4]. Finally, if we allow for an arbitrary direction of flow, we obtain:

$$d\vec{F}_{\text{fric}} = \eta\left(\nabla^2 v_x, \nabla^2 v_y, \nabla^2 v_z\right) dV = \eta\, \nabla^2 \vec{v}\, dV.$$
(16.30)

---

4 The Laplace operator is often denoted by Δ. We use the notation ∇^2 to distinguish it from the symbol Δ, which we often use to express differences (as here ΔF_r).

Example 16.11: Pitch-Drop Experiment

This experiment probably belongs more to the class of curiosities than scientific experiments. It was started in 1930 by Professor Thomas Parnell at the University of Queensland in Brisbane, Australia. There is a funnel containing pitch which flows through the funnel, albeit very slowly. At the bottom of the funnel, drops form, which eventually detach and fall into the beaker below the funnel. Professor Parnell had to wait eight years for the first drop to fall. More followed in 1947, 1954, 1962, 1970, 1979, 1988, 2000 and 2014.

You may wonder how long it took Mr. Parnell to fill the pitch into the funnel. He used a trick. He heated the pitch, making it more fluid. The viscosity drops significantly and he was able to fill in the pitch. Now he just had to wait until the pitch had solidified again, for which he estimated three years.

Viscosity can be understood and partially quantified through an atomic image. However, this would go beyond the scope of this book. Even in a fluid, forces act between the molecules, that are not strong enough to bind a molecule to its neighbors, but they generate a force that opposes the mutual displacement of the molecules. This is the microscopic cause of the friction.

16.5 Laminar Flows

16.5.1 Hagen-Poiseuille Law

At the beginning of this chapter, we described laminar flows as uniform, vortex-free flows that primarily occur at low flow velocities. Friction can play a role in laminar flows, with external forces performing work against the frictional forces. The external forces are typically caused by pressure differences along the flow. Apart from overcoming friction, the work is converted into the kinetic energy of the fluid, although this is still negligibly small in laminar flows.

We introduced friction in the fluid in the last chapter, where we assumed laminar flows. A division of the flow into stable layers is meaningful only in laminar flows. Turbulences cannot be divided into layers! We want to continue these considerations and apply them to a particularly important example: the laminar flow through a cylindrical pipe.

We divide the flow in the pipe into individual, concentric layers (◘ Fig. 16.18). On the left side, the fluid enters under a pressure p_1, on the right side it exits with p_2. We assume a laminar stationary flow. Work to overcome the friction is performed on the fluid due to the pressure difference between the two ends of the pipe. At the left end of the pipe, the pressure performs work, at the right end, a part of the energy is released again. The balance for a fixed time interval for a cylinder of the fluid with radius r and cross-sectional area A_{cross} results in:

$$\Delta W = F_1 \Delta s - F_2 \Delta s = p_1 A_{\text{cross}} \Delta s - p_2 A_{\text{cross}} \Delta s = \Delta p \, \Delta V, \tag{16.31}$$

where Δs is the distance the fluid moves through the pipe during the fixed time interval and $\Delta V = A \Delta s$ is the volume that has moved. In a steady state, for each cylinder, the force $F_{\Delta p}$, exerted by the pressure difference, must balance the frictional force F_{fric} acting at the boundary layer. We denote with R the inner radius of the tube and the length of the tube by L. Then the forces on a cylindrical layer are:

$$F_{\Delta p} = \Delta p \, A_{\text{cross}} = \Delta p \, \pi r^2$$
$$F_{\text{fric}} = -A_{\text{lat}} \eta \frac{dv}{dr} = -2\pi r L \eta \frac{dv}{dr}. \tag{16.32}$$

We have taken the formula for the frictional force from Eq. 16.23. Please note that we had defined the coordinate z (replaced here by r) in the direction of increasing velocity z. Here, the velocity is maximal at the center of the pipe and zero at the outer diameter. It decreases with the coordinate r. Therefore, an additional minus sign appears here, turning the frictional force positive in the end. We insert the two equations into the balance of forces and integrate:

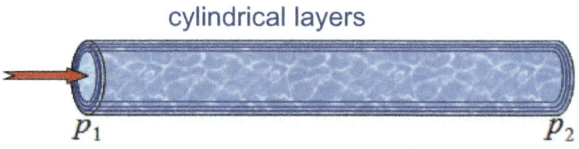

◘ **Fig. 16.18** Layers of equal flow in a tube

16.5 · Laminar Flows

$$F_{\Delta p} = F_{\text{fric}}$$

$$\Delta p \, \pi r^2 = -2\pi r L \eta \frac{dv}{dr}$$

$$\Delta p \, r = -2 L \eta \frac{dv}{dr}$$

$$\frac{\Delta p}{2 L \eta} r = -\frac{dv}{dr} \qquad (16.33)$$

$$\frac{\Delta p}{2 L \eta} \int_r^R r' dr' = -\int_r^R \frac{dv}{dr'} dr'$$

$$\frac{\Delta p}{2 L \eta} \frac{1}{2}\left(R^2 - r^2\right) = -(v(R) - v(r))$$

$$v(r) = \frac{1}{4\eta} \frac{\Delta p}{L}\left(R^2 - r^2\right).$$

In the next-to-last step, we took into account that the velocity must vanish at the outer diameter of the pipe ($v(R) = 0$), where the flow touches the wall. We obtain a velocity profile that increases quadratically towards the center of the pipe.

Now, we calculate how much fluid flows through the tube in a fixed time. We look at the volumetric flow I, which we introduced in ▶ Sect. 16.2. We had:

$$I = \frac{dV}{dt} = A_{\text{cross}} \, v. \qquad (16.34)$$

For an infinitesimally thin cylindrical layer in the tube, this results in a volume flow dI of the size

$$dI = v(r) \, dA_{\text{cross}} = v(r) \, 2\pi \, r \, dr = \frac{1}{4\eta}\frac{\Delta p}{L}\left(R^2 - r^2\right) 2\pi \, r \, dr. \qquad (16.35)$$

Integrating the total flow results in:

$$I = \int_0^R \frac{1}{4\eta}\frac{\Delta p}{L}\left(R^2 - r^2\right) 2\pi r \, dr$$

$$= 2\pi \frac{1}{4\eta}\frac{\Delta p}{L}\int_0^R \left(R^2 - r^2\right) r \, dr$$

$$= 2\pi \frac{1}{4\eta}\frac{\Delta p}{L}\left[\frac{1}{2}r^2 R^2 - \frac{1}{4}r^4\right]_0^R \qquad (16.36)$$

$$= 2\pi \frac{1}{4\eta}\frac{\Delta p}{L}\left(\frac{1}{2}R^4 - \frac{1}{4}R^4\right) = \frac{\pi}{8\eta}\frac{\Delta p}{L} R^4.$$

We finally obtained the Hagen-Poiseuille law.

> **Hagen-Poiseuille Law**

$$I = \frac{\pi}{8\eta} \frac{\Delta p}{L} R^4.$$

The volumetric flow through a pipe is very sensitive to the radius of the pipe. It increases with its fourth power! In contrast, the pressure difference and the length of the pipe enter linearly and in direct proportion to each other. If you extend a pipe, you must increase the pressure difference over the pipe accordingly to maintain a constant flow. You can increase the volumetric flow linearly by increasing the pressure on the pipe. If you want to double the volumetric flow, you have to double the pressure. Much more efficient is an increase in the diameter of the pipe. An increase in the radius by about 20% has the same impact.

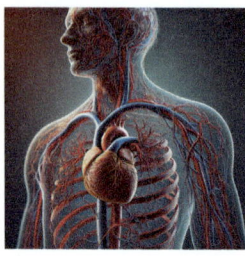

Example 16.12: Blood Circulation

The Hagen-Poiseuille law is of crucial importance for the regulation of human blood circulation. The arteries, through which the blood flows from the heart into the body, are surrounded by muscles that can change their cross-section. As we have seen, according to the Hagen-Poiseuille law, already a slight change in the radius of the arteries will result in a noticeable change in blood flow.

In old age, the diameter of the arteries can decrease due to calcification. For example, a decrease by 10%, will reduce the blood flow by $1^4 - 0.9^4 = 34\%$. To maintain blood flow, the body must increase blood pressure by 34%, leading to hypertension.

Experiment 16.8: Hagen-Poiseuille

The experiment quantitatively shows the dependence of the volumetric flow through a pipe on its diameter, illustrating the Hagen-Poiseuille law. Two pipes with different diameters are attached to the opening of a bottle via a T-piece and the bottle is turned upside down. The two pipe sections have the same length and the setup ensures that the same pressure difference prevails over both

16.5 · Laminar Flows

pipes. If we open the outlet valve, we can read and compare the amount of liquid that has drained off into the beakers below. We choose two pipes that differ in diameter by a factor of two. Indeed, we find that the amount of liquid that drains from the thicker pipe is 16 times the amount of the smaller pipe.

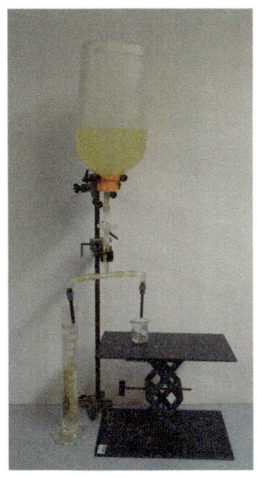

© RWTH Aachen, Physics Experiment Collection

16.5.2 Piping

We have now learned about the flow conditions in single pipes. Now we calculate more complicated networks of pipes, from the Hagen-Poiseuille law. The procedure is equivalent to the one you will learn later on in electricity when calculating complicated networks of resistors. One can virtually represent electrical circuits through water flows in pipes. The volumetric flow of the fluid corresponds to the electric current and the pressure corresponds to the voltage. The conditions at an electrical resistor and Ohm's law, which you probably still remember from highschool, are shown in ◻ Fig. 16.19.

Now, we derive a corresponding law for the volumetric flow through a pipe. From Eq. 16.36 it follows:

$$\Delta p = \frac{8 \eta L}{\pi} \frac{I}{R^4} . \tag{16.37}$$

We rewrite this as:

$$\Delta p = p_1 - p_2 = W I , \tag{16.38}$$

with the pipe resistance

$$W = \frac{8 \eta L}{\pi} \frac{1}{R^4} . \tag{16.39}$$

The situation is illustrated in ◻ Fig. 16.20.

But what happens if the radius of the pipe changes? We separate the pipe into sections of constant radius and combine the sections in a network to the original pipe.

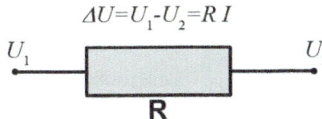

◻ **Fig. 16.19** Ohm's Law

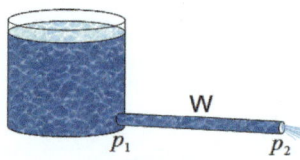

Fig. 16.20 The pipe resistance W

We may involve either series or parallel circuits. Let's start with a series circuit of two pipes with different pipe resistance as depicted in ◘ Fig. 16.21. Due to the continuity equation, it must hold:

$$I_1 = I_2 = I, \qquad (16.40)$$

where I_1 is the volumetric flow through pipe 1 and I_2 is the corresponding volumetric flow through pipe 2. The pressure differences must add up over the two pipes:

$$\Delta p_{\text{tot}} = \Delta p_1 + \Delta p_2. \qquad (16.41)$$

With this, the total flow can be calculated:

$$\begin{aligned}\Delta p_{\text{tot}} &= \Delta p_1 + \Delta p_2 = W_1 I + W_2 I = (W_1 + W_2)I \\ &= W_{\text{tot}} I \quad \text{with} \quad W_{\text{tot}} = W_1 + W_2. \end{aligned} \qquad (16.42)$$

We can use this procedure successively for several pipes, so we come to the conclusion: If the fluid flows through several pipes one after the other, the pipe resistances must be added.

The conditions are different when the fluid flows through two pipes in parallel (◘ Fig. 16.22). Now the pressure difference for both pipes is the same:

$$p_1 = p_2 = p_{\text{tot}}, \qquad (16.43)$$

whereas the volumetric flow through both pipes is additive:

$$I_{\text{tot}} = I_1 + I_2. \qquad (16.44)$$

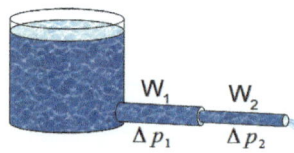

Fig. 16.21 Series connection of two pipes

16.5 · Laminar Flows

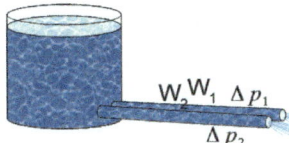

Fig. 16.22 Parallel connection of two pipes

We calculate the total flow:

$$I_{tot} = I_1 + I_2 = \frac{\Delta p_{tot}}{W_1} + \frac{\Delta p_{tot}}{W_2} = \left(\frac{1}{W_1} + \frac{1}{W_2}\right)\Delta p_{tot} \quad (16.45)$$

$$= \frac{1}{W_{tot}}\Delta p_{tot} \quad \text{with} \quad \frac{1}{W_{tot}} = \frac{1}{W_1} + \frac{1}{W_2}.$$

If the fluid flows through several pipes in parallel, the reciprocals of the individual pipe resistances must be added to the reciprocal of the total resistance.

In determining the pipe resistance, we started from the Hagen-Poiseuille law, which applies to circular pipe cross-sections. For other shapes, corrections arise, which can be derived through a calculation similar to that for round pipes. If the width and height of the cross-section are not too different, the radius must be replaced by $2A/U$ where A is the cross-sectional area and U is the wetted perimeter of the pipe (in the case of open channels, only the part of the perimeter that is wetted by the fluid).

16.5.3 Stokes' Drag

Finally, we want to consider the movement of a body through a fluid taking friction into account. For simplicity, we limit our discussion to spheres (◘ Fig. 16.23). Under the influence of gravity $\vec{F}_G$ such a sphere falls through the fluid. It accelerates until the friction $\vec{F}_{fric} = \vec{F}_S$ balances the weight reduced by the buoyancy $\vec{F}_{buo}$ and a constant sinking speed is established.

It has been shown experimentally that the frictional force is proportional to the surface area A of the sphere and to the viscosity η of the fluid. Furthermore, it depends on the velocity gradient in the surrounding flow. The boundary layer will stick to the sphere and sink with it, while hardly any movement in the fluid will be observed at a greater distance from the sphere:

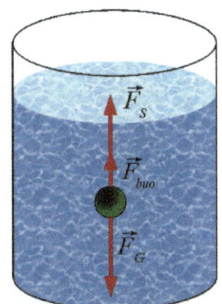

Fig. 16.23 Motion with friction in a fluid

$$F_S = -\eta A \frac{dv}{dr} = -4\pi R^2 \eta \frac{dv}{dr}. \qquad (16.46)$$

Now, we should actually calculate the velocity profile to determine the term $\frac{dv}{dr}$. But we do not want to do this exactly. Instead, we estimate the gradient by assuming that the velocity reduces from the velocity of the sphere v to zero over the distance of a readius of the sphere, so:

$$\frac{dv}{dr} \approx -\frac{v}{R}. \qquad (16.47)$$

We insert this estimate and obtain:

$$F_S \approx 4\pi R^2 \eta \frac{v}{R} = 4\pi \eta R v. \qquad (16.48)$$

A more precise calculation of the velocity profile would have lead us to Stokes's law:

> **Stokes's Law**
>
> $$F_S = 6\pi \eta R v.$$

It describes the friction of a sphere in a laminar flow. Please remember that the drag increases proportionally with the relative velocity between the body and the fluid and is also proportional to the viscosity. These dependencies are not only found with the friction of a sphere, but always with laminar flows. The factor $6\pi R$ relates to the geometry of the sphere and needs to be adapted for bodies of different shapes.

> **Experiment 16.9: Stokes's Law**
>
> In this experiment, we observe how spheres sink through a fluid. We use castor oil because of its high viscosity which makes the spheres sink slowly. After an initial acceleration, a constant sinking speed is reached, given by the equilibrium of weight, buoyancy, and friction:
>
> $$F_G - F_{\text{buo}} = F_S$$
> $$\rho_{\text{sphere}} V_{\text{sphere}} g - \rho_{\text{oil}} V_{\text{sphere}} g = 6\pi \eta_{\text{oil}} R v$$
> $$\frac{4\pi}{3} R^3 g (\rho_{\text{sphere}} - \rho_{\text{oil}}) = 6\pi \eta_{\text{oil}} R v$$
> $$\Rightarrow v = \frac{2}{9} g \frac{\rho_{\text{sphere}} - \rho_{\text{oil}}}{\eta_{\text{oil}}} R^2.$$

16.5 · Laminar Flows

The sinking speed should increase quadratically with the radius of the spheres. This can be quantitatively verified by a measurement of the time for a certain sinking. It is important to ensure that the spheres are completely wetted with oil (possibly wet beforehand) and that no air bubbles form. The diagram shows the measurements in comparison to the values calculated from the above formula. For the large measuring cylinder (diameter 5 cm), the measurements agree very well with the expectations. On the small cylinder (diameter 3 cm), there are deviations for large spheres (up to 1 cm in diameter). These are probably caused by a cutoff of the velocity profile at the walls of the cylinder. The liquid dragged along by the sphere reaches the walls, where additional friction occurs.

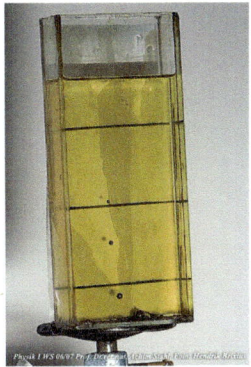

© Photo: Hendrik Brixius

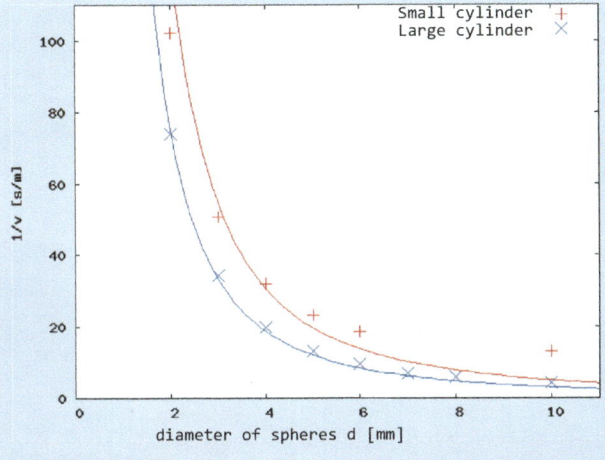

Example 16.13: Falling-Sphere Viscometer

This type of viscometer, used to determine the viscosity of liquids, based on Stokes' law. A cylinder is filled with the liquid under investigation. A small metal sphere (radius R) falls through the liquid and the sinking velocity v is determined. According to Stokes' law, the viscosity can be calculated as:

$$\eta = \frac{2gR^2}{9v}\left(\rho_{\text{sphere}} - \rho_{\text{fluid}}\right).$$

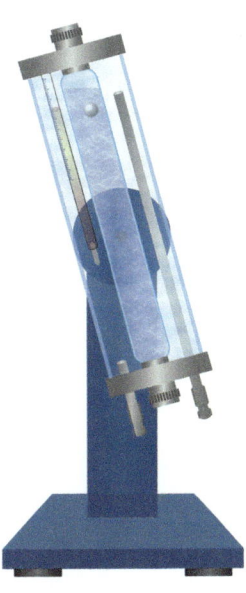

16.6 Turbulent Flows

As we have seen, flows at low velocities are mostly laminar. If the velocity is increased, a point is reached at which vortex formation sets in. The images in ◘ Fig. 16.24 show this transition for the example of the flow around a cylinder at five different flow velocities. At high velocities we speak about turbulent flow.

The flow velocity increases from (a) to (e). In (a), we see the uniform streamlines of a laminar flow. Already in (c), we can see the first vortices emerging behind the sphere, which are clearly formed in (d). At even higher flow velocities, as in (e), individual vortices detach from the obstacle and can still be observed at a great distance from it.

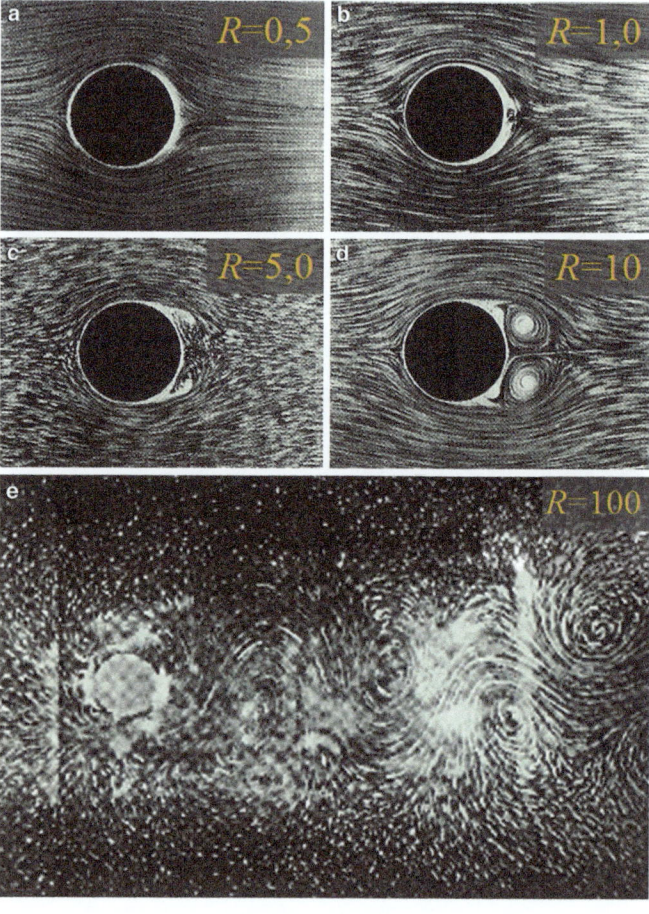

◘ **Fig. 16.24** Flow around a sphere with increasing velocity

16.7 · Drag

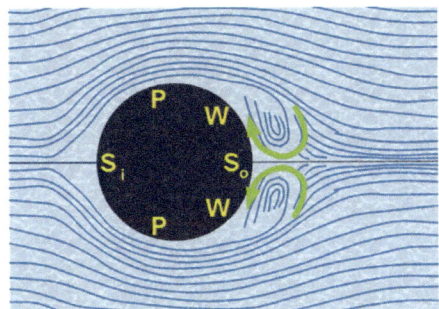

Fig. 16.25 The formation of vortices behind a sphere

But how are vortices formed? To this end, we take a closer look at the flow pattern in Fig. 16.25. It shows the flow around a sphere at a velocity at which the first two vortices form behind the sphere. The fluid is streaming in from the left. We recognize two stagnation points, S_i and S_o, in the incoming and outgoing flow, where the fluid is at rest. We seem to identify a flow line that leads from the left to S_i and ends there as well as a line that starts from S_o. At P, the flow velocity is at its maximum, which can be recognized by the density of the streamlines. The fluid is accelerated from S_i to P, reaches its maximum speed at P and is decelerated again towards S_o. However, due to friction, it already comes to a halt before it reaches S_o at the points W. At the stagnation points we always have $p_{\text{stat}} = p_{\text{tot}}$, whereas the static pressure at P is lower. A pressure difference arises, pushing the fluid from S_o back to P via W. This pressure difference causes a movement that runs against the main flow. On the surface of the sphere, the velocity differences and thus the pressure differences are largest. Further away from the surface, they decrease rapidly. This creates a torque that in turn generates the two vortices. At edges, steps or other transitions, the flow velocity often changes sharply. There, the acceleration and also the pressure differences are particularly large. Vortex formation first begins at these points.

16.7 Drag

As we have seen, when a body is circulated around by a fluid, a resistive force is created, called fluid resistance or simply drag. We want to redisucss it in a more general context.

We already dealt with the case of laminar flows in ▶ Sect. 16.5. Stokes' law describes the drag:

$$F_S = 6\pi \eta R v. \tag{16.50}$$

In the case of turbulent flows, another effect comes into play, which even dominates over Stokes' friction. In front of the object, the static pressure p_1 prevails, which is maximal here, as the fluid is at rest at this point (stagnation point). Behind the object, the static pressure is reduced by the flow in the vortices. Therefore,

$$p_{\text{stag}} = p_1 = p_2 + \frac{\rho}{2} v_v^2, \tag{16.51}$$

applies, where v_v is the velocity of the fluid in the vortex. There is a static pressure difference between the front and back of the object, which leads to a force against the direction of motion. This results in an additional resistance. It is

$$\Delta p = p_2 - p_1 = \frac{\rho}{2} v_v^2 \tag{16.52}$$

and the additional resistive force is:

$$F_{\text{fric}} = F_D = \Delta p A = \frac{\rho}{2} v_v^2 A. \tag{16.53}$$

Please remember, in this formula, v_v is not the relative velocity between the object and the fluid, but the local velocity of the fluid in the vortices behind the body. It varies greatly from position to position in the vortex field. Nevertheless, it is correlated with the velocity v of the body. With increasing flow velocity v the local velocity in the vortices also increases. We replace in the formula v_v by v and introduce an additional factor C_D that corrects for the difference between v_v and v. We end up with the following relation:

$$F_D = C_D \frac{\rho}{2} v^2 A. \tag{16.54}$$

The constant C_D, called the drag coefficient, depends on the shape of the body. Drag coefficients for some shapes are given in ◘ Table 16.2.

At low flow velocities, we are in the laminar range, where the drag (frictional force) initially increases linearly with the velocity. As the velocity increases, the first vortices form, and the dynamics of the vortices generate an additional drag, which now increases quadratically with

16.7 · Drag

Table 16.2 Drag coefficients for some body shapes

Body		c_W-Value
Drop shape		0.04
Wing		0.1
Wing (flat bottom)		0.2
Sphere		0.4
Standing human		0.7 …0.8
Semi-sphere (massive)		0.8
Disk or square plate		1.2
Hollow half sphere		1.4
Rectangular plate (long)		2.0

the flow velocity. Since it grows faster with velocity than Stokes' friction, it will dominate at high velocities. Therefore, we find a quadratic dependence on the velocity in a turbulent flow.

We want to point out, that the drag must be larger in the turbulent flow than in the laminar range. The fluid is strongly accelerated and then decelerated again in the vortices, which extracts energy from the flow and increases the resistance.

Experiment 16.10: Drag

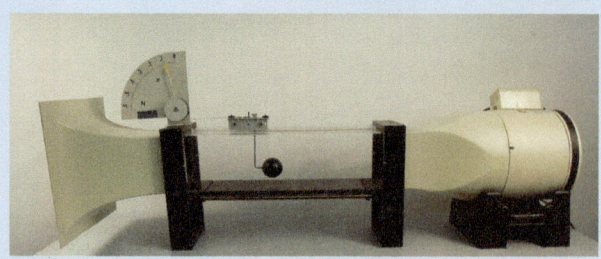

© RWTH Aachen, Physics Experiment Collection

The drag of a body in a turbulent flow is hardly calculable. It must be determined experimentally. The illustration shows a wind tunnel set up for measuring drag in air. The fan (on the right) sucks air through the tunnel. In the middle area, the object under study (here a sphere) can be inserted. It is suspended from a movable sled, which is connected by a string to a force meter (above the left end of the channel). The force meter directly indicates the drag at a given flow velocity.

Experiment 16.11: Smoke Rings

This is an atmospheric experiment for blowing off candles. It demonstrates the high wind speed in vortices. We cut a 15 cm large round hole into the front side of a cardboard box with an approximate size of 70 cm. We set up a row of burning candles a few meters away. Now we hit the sides of the cardboard box hard, so that it expels air through the hole. Upon exit, a ring of vortices is created at the edge of the hole, which then floats stably towards the candles. When we demonstrate the experiment, we fill the box with smoke or fog, e.g., from a fog machine or by burning damp tobacco in the box, to turn the vortices visible. We might enhance the effect closing the box until it is filled with smoke and only open it immediately before the experiment.

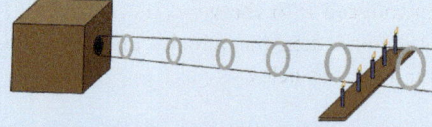

After a strong hit on the cardboard walls, you will see a smoke ring floating towards the candles. If a smoke ring

16.7 · Drag

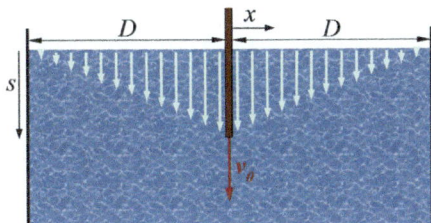

Fig. 16.26 Thought experiment for deriving the Reynolds number

> hits a candle dead on, the candle will flicker, but it will not go out. Only when the vortices of the ring directly hit the flame is the wind speed so high that the candle is blown out. Aiming is not easy, it takes some practice.

We try to approach the complicated conditions in a general flow in a more quantitative way through a thought experiment (Fig. 16.26). A flat plate sinks into a long basin of width $2D$. Edge effects at the end of the plate are negligible. We think of the fluid divided into parallel layers that are dragged along with the plate (in the direction s). The two layers touching the plate move at the same speed v_0 as the plate itself. Layers further away move slower. The two layers on the left and right end of the basin adhere to the walls and do not move at all. You can easily verify, that a linear velocity profile is established:

$$v(x) = \frac{|x|}{D} v_0. \tag{16.55}$$

First, we determine the frictional force F_{fric} that must be overcome to immerse the plate in the fluid and the work W_{fric} performed against it:

$$\begin{aligned} F_{\text{fric}} &= 2\eta A \frac{dv}{dx} = 2\eta A \frac{v_0}{D} \\ W_{\text{fric}} &= F_{\text{fric}} s = 2\eta A s \frac{v_0}{D}. \end{aligned} \tag{16.56}$$

The factor 2 arises because friction occurs on either side of the plate. The deeper the plate is immersed, the larger the friction. With the immersion, more and more fluid is dragged along, accelerated to the velocity corresponding to the distance $|x|$ from the plate. The work required is:

$$W_{\text{acc}} = \int_V \frac{1}{2} v^2(|x|)\, dm = \int_V \frac{1}{2} \rho v^2(|x|)\, dV \quad \text{with } dm = \rho\, dV$$

$$= 2A \int_0^D \frac{1}{2}\rho\left(\frac{x}{D}v_0\right)^2 dx$$

$$= 2A \frac{\rho}{2}\left(\frac{v_0}{D}\right)^2 \left[\frac{1}{3}x^3\right]_0^D = \frac{1}{3}\rho v_0^2 A D.$$

(16.57)

Now, in your mind, let the width D of the basin approach infinity. Then, the frictional force tends to zero. This is to be expected, as the velocity gradient $\frac{dv}{dx}$ continues to decrease, and thus the frictional forces become smaller and smaller. In contrary, the work required for acceleration, diverges because a larger and larger mass of fluid must be accelerated with increasing width. This is an unphysical result! It would imply that the plate could no longer submerge in the fluid in a large basin because the energy to accelerate the fluid is lacking. We must conclude that our approach already contained incorrect assumptions and this is indeed the case. We assumed that the velocity profile spreads from the plate to the edge of the basin, regardless of how wide the basin is. This cannot be correct. In reality, the dragging of the fluid layers is limited to a finite area around the plate. The size of this area adjusts so that the work on acceleration does not exceed the work on friction. Only this limited area is moving along with the plate. Therefore, it must hold:

$$\frac{1}{3}\rho v_0^2 A D \leq 2\eta A s \frac{v_0}{D}$$

$$\frac{\rho v_0}{\eta}\frac{D^2}{s} \leq 6 \qquad (16.58)$$

$$\frac{\rho v_0 D}{\eta}\frac{D}{s} \approx 1,$$

where the last line is only a rough estimate. The first factor in this estimate contains the quantities ρ, η, v_0, D, which describe the fluid and the flow. It is called the Reynolds number Re:

$$Re = \frac{\rho v_0 D}{\eta}. \qquad (16.59)$$

16.7 · Drag

The second term is a purely geometric factor that describes the dimensions of our thought experiment.

We can distinguish three special cases:

$Re \ll 1$ In this case, D/s must be much larger than 1, i.e., the extent of the layers dragged by the plate is much larger than the plate itself. This in turn implies that the velocity of the fluid changes only slowly over a large distance. The gradient $\frac{dv}{ds} = \frac{v_0}{D}$ of the velocity profile is low. A laminar flow is created.

$Re \approx 1$ In this case, $D \approx s$. The boundary layer is about as wide as the plate is high. This is the transition area to the next special case.

$Re \gg 1$ In this case, the boundary layer must be much narrower than the dimensions of the plate. The speed changes over a very short distance from v_0 to zero. The gradient is large. Vortices are created. The flow is turbulent.

The Reynolds number thus distinguishes between laminar and turbulent flows. It is common to speak of "turbulent flows" for Reynolds numbers above 1200 although the limit $Re \approx 1200$ is only an approximate guideline. The exact limit at which vortex formation begins depends on the shape of the body. For streamlined bodies, vortices only start much later than for bodies with corners and edges.

Example 16.14: Reynolds Numbers

From the Reynolds numbers, limit velocities can be estimated, at which the formation of vortices is to be expected. From $Re \approx 1200$ at the boundary to turbulent flow, it follows that $v_{\lim} \approx 1200\, \eta/\rho$. In air, this speed is low, approximately 0.01 m/s. Even a raindrop falling through the air generates vortices, let alone airplanes and other flying objects.

To conclude this section, we want to revisit the general calculation of flows. Starting from Newton's equation of motion, we arrived at the Euler equations (Eq. 16.22).

$$\rho\left(\frac{\partial}{\partial t}\vec{v} + (\vec{v} \cdot \mathrm{grad})\vec{v}\right) = -\,\mathrm{grad}\,p + \rho\vec{g}. \qquad (16.60)$$

On the right side we find the external forces, to which we now need to add the frictional force from Eq. 16.30. We get

$$\rho\left(\frac{\partial}{\partial t}\vec{v} + (\vec{v} \cdot \mathrm{grad})\vec{v}\right) = -\mathrm{grad}\, p + \rho \vec{g} + \eta\,\Delta \vec{v}. \quad (16.61)$$

They are called the Navier-Stokes equations. They allow to calculate even compressible fluids including friction. Unfortunately, the equations are so complex that they usually cannot be solved analytically. They are the basis for the numerical simulation of flows.

Example 16.15: Flow Models

Flows (especially turbulent ones) are very difficult to calculate. When planning flow channels, models (so-called hydraulic models) of the river course, the dam or the wind tunnel on a reduced scale are built to optimize the flow before the actual construction begins. However, one cannot use the final fluid in the model. The fluid must be chosen so that its density and viscosity result in the same Reynolds number as expected in the later channel. This is the only way to ensure that similar flow conditions arise in the model.

Example 16.16: The Drag Coefficient of a Car

© Mercedes-Benz Group AG

The drag of a car at high speed is to a large amount determined by the size of the car (cross-sectional area A)

and its shape. The influence of the shape is measured by the drag coefficient C_D. The resisitve force is:

$$F_{\text{fric}} = C_D \frac{\rho_{\text{air}}}{2} v^2 A \, .$$

The drag increases quadratically with the speed. Starting from a speed of around 50 km/h, the drag dominates over other frictional losses such as the rolling resistance of the wheels or friction in the engine. The drag is essential for the fuel consumption of a car and is therefore reduced as much as possible by the designers.

In the early days of automobiles, the speed was low and fuel consumption was insignificant. Automobiles from these times have a drag coefficients of around 1. Optimization only began in recent decades. Today, cars can achieve C_D values that are just above 0.2. The illustration shows a special development by Mercedes-Benz, the Vision EQXX, an electrical car with a drag coefficient of 0.13, the lowest value of a street-legal car, at the time of writing of this text.

© Wikimedia: US government

© Wikimedia: EveryPicture

16.8 Dynamic Lift

In ▶ Sect. 15.4 we discussed Archimedes' principle, the static lift (buoncy) in a stationary fluid. In flowing fluids, another buoyant force appears, which is called the "dynamic lift". It always occurs when bodies are asymmetrically flowed around. We want to discuss it here using the example of an airplane wing. The dynamic lift is responsible for airplanes flying, but also occurs in many other examples. A small selection can be seen in Example 16.17.

Example 16.17: Dynamic Lift

All the objects depicted utilize dynamic lift: sails, ship propellers, wind turbines, spoilers.

© Wikimedia: Ahunt

© Wikimedia: Strubbl

The dynamic lift on the wings and the fuselage of an airplane keeps it in the air. Let us start the discussion with a warning: A conclusive explanation of flying is unfortunately not that simple. If we are honest, we have to admit that although we can develop and build even larger and more complex airplanes, we do not have a real explanation as to why they fly. We will now introduce you to three different models of dynamic lift. In the end, we will return to the basic question of why an airplane flies. But first, take a look at Experiment 16.12. It shows the dynamic lift in a wind tunnel and demonstrates that it is the cause that keeps airplanes in the air.

> **Experiment 16.12: Dynamic Lift on a Wing**
>
> We will once again use the wind tunnel that you already know from Experiment 16.4 and Experiment 16.10. Now, we introduced a section of a model wing with its characteristic profile into the wind tunnel. The wing is attached to two rods so that the angle of attack can be adjusted from the outside. We use again the measuring device for air drag from Experiment 16.10, and included a spring scale in the fixture of the wing, which allows to measure the lift directly. With the fan switched off, the pointer of the spring scale is balanced to zero, allowing us to observe the change in weight due to the flow. To measure the flow velocity, a Prandtl pitot tube can be introduced additionally. The illustration shows the wing, the support with the integrated measuring device for the lift and the Prandtl tube.

16.8 · Dynamic Lift

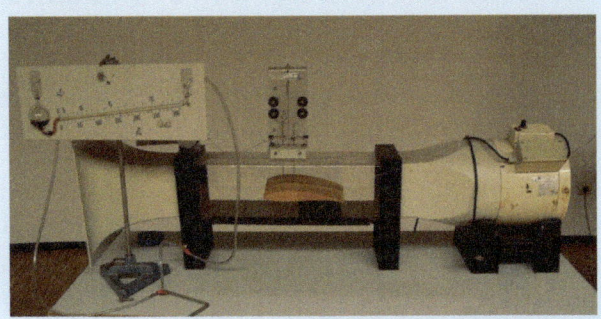

© RWTH Aachen, Physics Experiment Collection

Even at angles of attack just below 0°, a positive lift is observed, which originates from the end of the profile that is still clearly directed downwards even at an angle of attack of 0°. It increases steadily with angles of attack up to around 20°. If the angle of attack is increased further, however, the lift quickly decreases. The flow tears off. At negative angles of attack, a negative lift (downward force) is observed, i.e., an increase in the weight of the wing due to the flow. If the flow velocity is varied, a quadratic dependence can be approximately recognized.

The measuring device for drag, shows a signigicant increase of drag with increasing angle of attack, both for positive and negative angles.

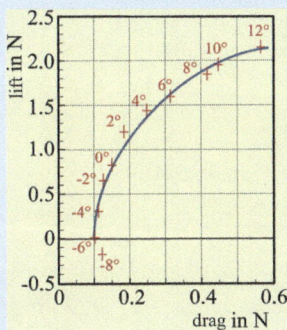

16.8.1 1. Model: Conservation of Momentum, Impulse

The aerodynamic shape of a wing significantly reduces the air drag of an airplane, thus facilitating flight. However, for the explanation of lift, we want to abstract from the complicated shape here and assume a simple board as the wing profile, see ◘ Fig. 16.27. In the sketch, you can see the angle of attack α, the total force on the wing $\vec{F}_{tot}$ and its decomposition into the frictional force $\vec{F}_{drag}$ and the dynamic lift $\vec{F}_{lift}$. We can neglect the static lift here.

If the wing is exposed to a flow, the air molecules collide with the underneath of the wing. We want to assume that they bounce off the wing elastically. Then:

$$|\vec{v}_i| = |\vec{v}_f| = v. \tag{16.62}$$

From the sketch, we read:

$$|\Delta \vec{v}| = 2v \sin \alpha. \tag{16.63}$$

Now we calculate the force from Newton's second axiom:

$$F_{tot} = \frac{\sum \Delta p}{\Delta t} = \frac{\sum m \, \Delta v}{\Delta t} = \frac{\Delta M}{\Delta t} \Delta v = \frac{\Delta M}{\Delta t} 2v \sin \alpha, \tag{16.64}$$

where $\frac{\Delta M}{\Delta t}$ is the mass of air that hits the wing in the time span Δt. We still need to relate it to the speed of the plane (◘ Fig. 16.28). With

$$\frac{\Delta M}{\Delta t} = \rho \frac{\Delta V}{\Delta t} \tag{16.65}$$

and the relation

$$\Delta V = h_0 A = v \, \Delta t \sin \alpha \, A, \tag{16.66}$$

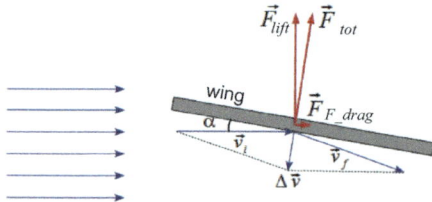

◘ Fig. 16.27 Flying: Model 1

16.8 · Dynamic Lift

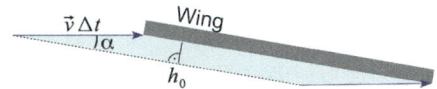

Fig. 16.28 To calculate $\Delta M/\Delta t$

which can be read from the sketch, we obtain:

$$F_{\text{tot}} = \rho \frac{\Delta V}{\Delta t} 2v \sin\alpha = 2\rho A v^2 \sin^2\alpha. \quad (16.67)$$

Finally, we decompose the force into the lift and the drag force. The decomposition can be read from ◘ Fig. 16.27. It results in:

$$\begin{aligned} F_{\text{lift}} &= F_{\text{tot}} \cos\alpha = 2\rho A v^2 \sin^2\alpha \cos\alpha, \\ F_{\text{drag}} &= F_{\text{tot}} \sin\alpha = 2\rho A v^2 \sin^3\alpha. \end{aligned} \quad (16.68)$$

At this point, we can initially state that this model correctly reflects the basic dependencies that we saw in Experiment 16.12. The dynamic lift is proportional to the area of the wing, it depends quadratically on the speed of the plane and the dependence on the angle of attack is reasonably described.

We dare to make a quantitative comparison. For a Boeing 747, the following approximate values apply:

$$\begin{aligned} A &= 500\,\text{m}^2, \\ v &= 900\,\text{km/h}, \\ \rho_{\text{air}} &= 1.3\,\text{kg/m}^3, \\ \alpha &\approx 5°. \end{aligned}$$

Substituting these values gives:

$$\begin{aligned} F_{\text{lift}} &= 6 \cdot 10^5\,\text{N}, \\ F_{\text{drag}} &= 5 \cdot 10^4\,\text{N}. \end{aligned} \quad (16.69)$$

The actual values are:

$$\begin{aligned} F_{\text{lift}} &= 3 \cdot 10^6\,\text{N}, \\ F_{\text{drag}} &= 1.5 \cdot 10^5\,\text{N}. \end{aligned} \quad (16.70)$$

The approach of modelling the wings as rectangular plates was very rough, and perhaps you should not expect too much from this model, but at least it correctly reflects the order of magnitude of lift and drag. The lift

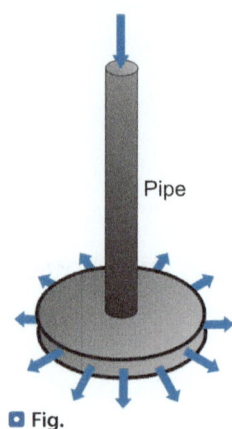

Fig. 16.29 Aerodynamic paradox

comes out a factor of 5 too small. The result is sensitive to the angle of attack. If we use 6° instead of the given 5°, the deviation is already reduced to a factor of 4. Furthermore, we ignored the contribution of the fuselage of about 20% to the lift. Overall, we should be satisfied with the results from this simple model.

However, the model itself is inadequate, crude, and many of the details of a real airplane are not taken into account. But even worse, there is a fundamental problem with this model. Consider again the aerodynamic paradox (Experiment 16.7) and try to explain it with this model. The failure of the model is even more obvious in another version of the aerodynamic paradox, shown in ◘ Fig. 16.29. Air is blown through a tube into the gap between the two discs, where the lower disc is only loosely attached. According to our model, the pressure of the incoming molecules should push the lower disc downwards, but if you perform the experiment, you will find that the lower disc is lifted upwards. Our model fails, and based on everything we have learned in ▶ Sect. 4.2, we should therefore reject the model.

16.8.2 2. Model: Bernoulli's Equation

With our second model, we try to explain the dynamic lift on the wing through the dynamic pressure from Bernoulli's equation. For this model, we need to consider the flow around the wing's profile.

We assume that a laminar flow forms around the wing. It is shown in ◘ Fig. 16.30. The air streams in from the left, where it is divided by the wing into a stream above and below the wing. Behind the wing, the two streams reunite. Due to the curvature of the wing, the path of the molecules on the top is longer than on the bottom so that the molecules flow faster on top ($v_o > v_u$). Then, from Bernoulli's equation, we get:

$$p_o + \frac{\rho}{2} v_o^2 = \text{const} = p_u + \frac{\rho}{2} v_u^2 \Rightarrow p_o < p_u. \quad (16.71)$$

This results in a dynamic lift:

$$F_{\text{lift}} = (p_u - p_o)A = \frac{\rho}{2}\left(v_o^2 - v_u^2\right)A = \frac{1}{2}\rho v^2 A\, C_{\text{lift}}, \quad (16.72)$$

16.8 · Dynamic Lift

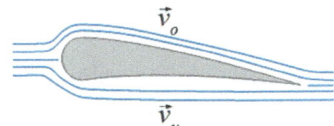

Fig. 16.30 Flow around a wing profile

where C_{lift} is a geometric factor that indicates how much faster the flow is above than below. Again, we find a proportionality of the dynamic lift to $\rho v^2 A$, which we have already seen in the first model. This is a success. The numerical agreement with the data of a real airplane depends significantly on the value of C_{lift}, which we could not determine here.

However, there is also criticism of this model:

We assumed that the volume of air divided in front of the wing reunites identically behind the wing. This was the main assumption for the difference in flow velocity above than below the wing. But this assumption is not fullfilled, the air takes longer to flow around the wing above. As a result, the flow behind the wing is directed downwards.

This model of flying relies heavily on the shape of the wing. An airplane with flat wings should not be able to fly, but this is not correct. Many of the early biplanes (see ◘ Fig. 16.31) had flat wings and could still fly. The same line of thoughts can be applied to airplanes flying upside down, which is possible with many airplanes, although with a larger angle of attack. According to our

Fig. 16.31 Historical biplane

model, the lift should then point downwards and the airplane should crash.

And finally, we should note that the model started from Bernoulli's equation, which describes laminar flow but, the flow around the wing of an airplane is strongly turbulent. We should not apply Bernoulli's equation at all.

16.8.3 3. Model: Vortex formation

As we have already mentioned, vortices form on the wing of an airplane. These are important for the flight characteristics and the explanation of flying. With the third model, we want to consider them explicitly. The most important vortex is the so-called starting vortex. It is shown schematically in ◘ Fig. 16.32 (in the figure on the right at the end of the wing).

The starting vortex is already created when the airplane accelerates before takeoff. Additional vortices are created at the lateral ends of the wings, at the tail, and at the engines. Under special weather conditions, the vortices can trigger condensation in the air and thus become visible. In ◘ Fig. 16.33, the vortices that detach from the wingtips are clearly visible.

The rotating air in the vortices contains angular momentum which has to be compensated by a counter-rotating vortex when they are created. For the starting vortex, the counter-rotating vortex forms around the wing. It is schematically represented in ◘ Fig. 16.34 (red). On the upper side of the wing, the flow velocities of the regular flow and the counter vortex add up, underneath they are opposite. This increases the flow velocity on top and reduces it underneath the wing, which - according to this model - leads to a dynamic lift due to Bernoulli's equation.

But this third model must also be criticized:

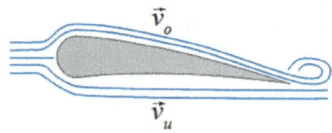

◘ **Fig. 16.32** Flow around a wing with starting vortex

16.8 · Dynamic Lift

Fig. 16.33 Vortices on a Boeing 727

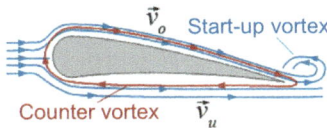

Fig. 16.34 Flow around a wing with starting vortex and counter vortex

First of all, it should be noted that the model is not concrete enough to derive analytical predictions from it. It has its significance as an explanation of simulations or measurements. Again, the model refers to Bernoulli's equation, although it only applies to laminar flows, while the model explicitly refers to vortex formation. This is a fundamental inconsistency of this model.

But there are also empirical problems. If the starting vortex were to detach from the airplane, according to this model the counter vortex would disappear and the airplane would no longer fly. The starting vortex can be dragged hundreds of meters behind the airplane. If a second airplane intersects between the wing and the starting vortex, it should cut it off and cause the first airplane to crash. This was actually tested. Luckily nothing bad happened, contrary to the prediction of our third model.

If you are now completely confused about flying, you should read the methodological insert.

> **Models**
>
> What have we learned about flying? Perhaps first of all, that flying is not that easy to explain. We have discussed three models of varying complexity. All three have their pros, the first one is simple (collisions of molecules on the wing) and explains some essential aspects correctly. The second model (Bernoulli) arrives at a very similar result in a more elegant and convincing way. The third model is particularly suitable for numerical simulations. But all three have their cons, too. We have shown that all three models fail at some important points. In ▶ Sect. 4.2 we discussed the concept of falsification. We have seen that a single incorrect prediction of a model is enough to refute it. In this sense, we would have to reject all three models presented here. What should we adopt as a model instead? Well, we would have to calculate the motion of all air molecules and that of the airplane using Newton's equations of motion. We are convinced that this would lead to a correct description of flying, but unfortunately, this calculation is far too complicated to actually carry it out. Flying is such a complex process that we cannot reduce it to the elementary laws of mechanics. We have no choice but to use models of the kind we have presented. However, we must be aware that these types of models (they are not theories in the sense of ▶ Sect. 4.2) have a limited scope. It is so narrow that these models only correctly represent some aspects of flying and not others. This may now also make it understandable why it might make sense to use more than one model. Different models describe different aspects of flying correctly and fail at other points. Not one of the models is correct and the others are wrong, but with several together, we might approach an understanding of the phenomenon of flying.

16.8.4 Magnus Effect

There is another form of dynamic lift that we want to discuss briefly at the end of this section. It is called the "Magnus effect".

If a symmetric body, such as a cylinder, is flowed around, a force is created on the body due to friction.

16.8 · Dynamic Lift

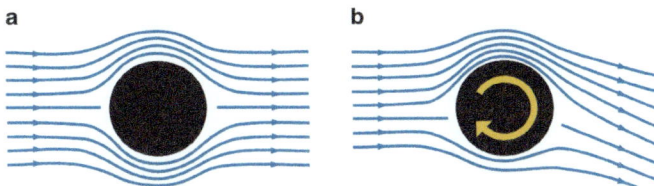

Fig. 16.35 Flow around a stationary (**a**) and rotating (**b**) cylinder

For reasons of symmetry, this must point in the direction of the flow (Fig. 16.35a). The situation changes when we put the cylinder into rotation. The layer immediately around the cylinder adheres to it and is dragged along in the rotation. The motion of the fluid is transferred to adjacent layers through friction. As a result, a large part of the flow is deflected past the top of the cylinder (Fig. 16.35b). The velocity of the flow increases at the top and decreases at the bottom. According to Bernoulli's equation, a pressure difference arises, which leads to an upward force. The rotating cylinder is deflected sideways in the flow. This is called the "Magnus effect".

The force $\vec{F}_M$ from the Magnus effect can be quantified:

$$\vec{F}_M = \frac{1}{2} C_M \, \rho_{Fl} \, A \, \vec{\omega} \times \vec{v} \, . \tag{16.73}$$

A constant C_M is introduced which describes the influence of the shape of the rotating body and the details of the flow. It is difficult to predict and usually determined experimentally. The force is proportional to the density of the fluid ρ_{Fl}, to the cross-sectional area A of the rotating body, as well as to the flow velocity $\vec{v}$ and to the rotational speed $\vec{\omega}$. It is perpendicular to both.

Experiment 16.13: Magnus Effect

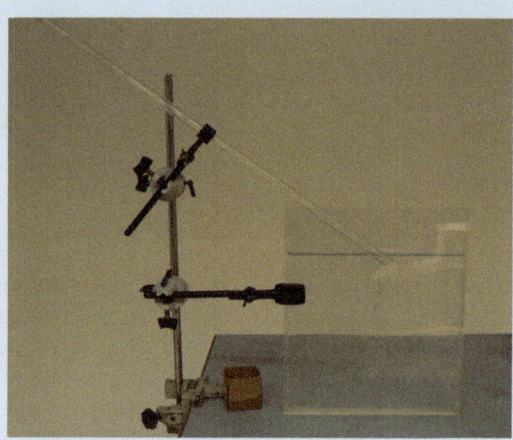

© RWTH Aachen, Physics Experiment Collection

The experiment demonstrates the Magnus effect in water. Due to the higher viscosity, the Magnus effect in water is larger and easier to visualize than in air. Different objects slide through a glass tube into a water basin. At the end of the tube, the bodies fall through the water, a situation equivalant to water flowing (upwards) past the objects. The objects, made of solid metal, with a diameter of only about 5 mm, fall quite quickly though the water. We record their motion with a high-speed camera. First, we let a cylinder slide through the tube with the front face ahead. Its path follows a projectile motion (with friction) in the water ending at point A. In the second run, we use a sphere, not sliding through the tube, but rolling. This puts it into a rapid rotation. At the end of the tube, the sphere falls into the water and it is noticeably deflected to the left by the Magnus effect. On the left side of the sphere, the flow caused by the fall and the local flow caused by the rotation point in the same direction. They add up and reduce the static pressure on the left and therefore the ball is deflected to the left.

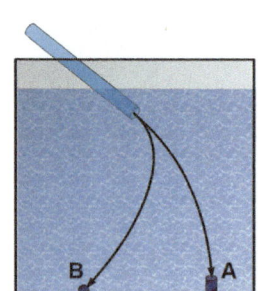

Example 16.18: Magnus Effect in Sports

The Magnus effect plays an important role in some ball sports, e.g., in top-spin in tennis or table tennis.

16.8 · Dynamic Lift

© Wikimedia: Dr. Stephan Roscher

In top-spin, the table tennis player moves the racket from bottom to top during the stroke (as seen in the picture of Timo Bolt) and thus imparts a strong forward rotation to the ball. The Magnus effect now results in a downward force that pushes the ball (see sketch) down onto the table. This allows for harder strokes, where the ball would miss the end of the table without the Magnus effect.

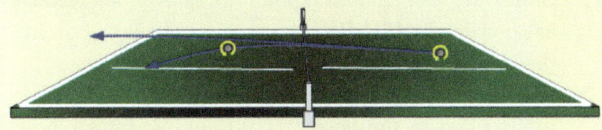

Example 16.19: Ship Propulsion by Magnus Effect

The Magnus effect can be used to propel a special type of sailing ships developed by Anton Flettner. Large rotors are set up on the ship, which are called Flettner rotors. The illustration shows the sailing schooner Buckau, converted by Flettner in 1924. After a series of tests, it successfully crossed the Atlantic in 1926. Even today, ships with Flettner rotors are still being built, primarily to use wind energy and thereby reduce fuel consumption.

The propulsion only works with wind, blowing from the side. Then the Flettner rotors, which all have to rotate in the same direction, generate a forward force that propels the ship.

Problems

1. A pumped storage plant has an upper reservoir that is 240 m above the turbine house. Calculate, neglecting friction losses, at what speed the water arrives at the turbines. What should be the total cross-sectional area of the fall pipes to deliver an electrical power of 600 MW? Assume an efficiency of the plant of 75%.
2. How long does it take to empty a cylindrical tank initially filled to the height H through a circular hole with the diameter D located in the bottom area of the tank?
3. A container is filled with oil up to a height of 0.5 m ($\rho = 900\,\text{kg/m}^3$, $\eta = 0.1\,\text{Pa s}$). The fill level is kept constant by refilling. At the bottom of the container, oil is drained through a thin horizontal hose with a length of 10 m and a diameter of 10 mm. What is the volumetric flow rate? Assume laminar flow. Is this assumption justified?
4. In a glass of champagne, one observes carbon dioxide bubbles ascending at a speed of 10 cm/s. What is the diameter of the bubbles? (Viscosity of the champagne: $2.0 \cdot 10^{-3}$ Pa s, density: $990\,\text{kg/m}^3$.)

16.8 · Dynamic Lift

5. The lift of an airplane F_{lift} can be described by the following formula:

$$F_{\text{lift}} = \frac{1}{2} \rho v^2 A \, c_A,$$

where ρ the density of the air, A the wing area, and c_A the lift coefficient. The Airbus A 380 has a mass of 560 t, possesses an effective wing area of 850 m² with a c_A value of 1.2 and is powered by four engines each with a thrust of 310 kN. What is the minimum runway length required for an A 380 to take off in Frankfurt on the Main? Ignore friction and air resistance. What about a takeoff in Mexico City? The maximum speed of the A 380 is 1100 km/h. Up to what height could the A 380 climb at this speed (ignore the temperature gradient)?

6. The power of a wind turbine can be estimated by the following formula:

$$\frac{P}{W} = \frac{1}{4} \cdot \frac{A}{m^2} \cdot \left(\frac{v}{m/s}\right)^3$$

Here, A is the area swept by the blades, and v is the wind speed. What drag coefficient C_D of the wind turbine is assumed in the above formula? How large must a wind power plant be dimensioned according to this formula in order to deliver approximately 2 MW power at wind strength 4 Bft ($v = 8$ m/s)?

Oscillations and Waves

Oscillations and waves are a central theme of all physics. Important examples of oscillations can be found not only in mechanics, but also in other areas of physics. Light is explained as an oscillation, electrons perform oscillations in atoms, heat in solids is due to oscillations, etc. Even here in mechanics, we have already relied on oscillations, e.g., in the determination of moments of inertia (Experiment 13.1). We now want to systematically deal with oscillations and the waves based on them.

Contents

Chapter 17   Oscillations – 473

Chapter 18   Waves – 533

Chapter 19   Acoustics – 561

Oscillations

Contents

17.1 Harmonic Oscillations – 474

17.2 Damped Oscillations – 487

17.3 Forced Oscillations – 495

17.4 Coupled Oscillations – 507

17.5 Standing Waves – 517

© The Author(s), under exclusive license to Springer-Verlag GmbH, DE, part of Springer Nature 2025
S. Roth and A. Stahl, *Mechanics*,
https://doi.org/10.1007/978-3-662-68079-7_17

17.1 Harmonic Oscillations

17.1.1 Simple Oscillations

Oscillations are periodic movements of a body, repeating itself at regular intervals (period). For every oscillation, there is an equilibrium position, in which the total force on the body, called the bob, vanishes. If you deflect the bob from the equilibrium, a motion arises from the interplay of a restoring force, which wants to return the bob to the equilibrium position, and the inertia of the bob.

Let us start the discussion with an example of oscillations in Experiment 17.1.

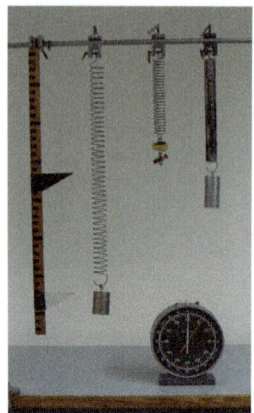

© RWTH Aachen, Physics Experiment Collection

> **Experiment 17.1: Spring Pendulum**
>
> The setup shows some spring pendulums with different masses and different springs. We measure the period with a stopwatch and the amplitude with the ruler on the left and see that they differ between the pendulums. The amplitude is the maximum distance that the bobs move out of the equilibrium position. It depends on how strongly we initiate the oscillations at the beginning. Within the accuracy of these measurements, we observe that the period does not change for different amplitudes. Furthermore, we notice that the amplitude gradually decreases due to friction.

Now, we look at the example of the spring pendulum from Experiment 17.1 in more detail (see ◘ Fig. 17.1):

The spring pendulum settles into a equilibrium position x_{eq} that is determined by the balance between the weight F_G and spring force F_S:

$$F_S = k\, x_{eq} = m g = F_G. \tag{17.1}$$

If the bob is displaced from its equilibrium position, a restoring force arises from the now altered balance of gravity and spring force. If we denote the displacement from the equilibrium position with x, then the restoring force is:

$$F_r = k\,(x_{eq} - x) - m g = -k\, x. \tag{17.2}$$

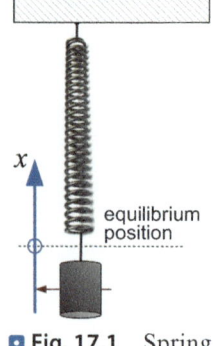

◘ Fig. 17.1 Spring pendulum

The minus sign indicates that the restoring force is directed against the displacement. If the bob is lifted upwards, the restoring force points downwards and vice versa. In general, we can state that the restoring force al-

17.1 · Harmonic Oscillations

ways points towards the equilibrium position and increases with the displacement. Here we have approximated the spring force by Hooke's law and obtained a proportional increase in the restoring force with the displacement. Later, we will later call it the "harmonic approximation".

Now, we calculate the motion of the pendulum mass from Newton's second law:

$$m \frac{d^2x(t)}{dt^2} = -k\,x(t)$$
$$\frac{d^2x(t)}{dt^2} + \frac{k}{m} x(t) = 0. \tag{17.3}$$

We define a constant $\omega_0 = \sqrt{k/m}$. It remains to be shown that this quantity is indeed an angular velocity as indicated by the choice of the symbol. Now, we have:

$$\frac{d^2x(t)}{dt^2} + \omega_0^2 x(t) = 0. \tag{17.4}$$

This is the differential equation of harmonic oscillation. The solution to this differential equation is:

$$x(t) = x_0 \sin(\omega_0 t + \varphi_0) \quad \text{with} \quad \omega_0 = \sqrt{\frac{k}{m}}. \tag{17.5}$$
$$ = x_0 \cos(\omega_0 t + \varphi_0')$$

Now we see that the constant ω_0 indeed represents the angular frequency with which the bob oscillates. It is related to the frequency f_0 and the period T_0 of the oscillation through:

$$\omega_0 = 2\pi f_0 \quad \text{and} \quad f_0 = \frac{1}{T_0}. \tag{17.6}$$

The unit of frequency is Hertz $[f] = 1\,\text{Hz} = \frac{1}{\text{s}}$.

We recognize the correlations that we have already seen in Experiment 17.1: The period of oscillation depends on the spring constant and the mass of the bob, but not on the amplitude x_0 of the oscillation.

Perhaps a remark on the solution of the differential equation is appropriate here. We did not determine the solution of the differential equation through a proper mathematical procedure, we essentially guessed it. Now, we have to proof that this is indeed a proper solution of the differential equation. To do this, we calculate the derivatives:

$$x(t) = x_0 \sin(\omega_0 t + \varphi_0)$$
$$v(t) = \frac{dx(t)}{dt} = x_0 \omega_0 \cos(\omega_0 t + \varphi_0)$$
$$a(t) = \frac{d^2x(t)}{dt^2} = -x_0 \omega_0^2 \sin(\omega_0 t + \varphi_0).$$
(17.7)

We insert these into the differential equation:

$$\frac{d^2x(t)}{dt^2} + \omega_0^2 x(t) = 0$$
$$-x_0 \omega_0^2 \sin(\omega_0 t + \varphi_0) + \omega_0^2 x_0 \sin(\omega_0 t + \varphi_0) = 0. \checkmark$$
(17.8)

The solution contains two integration constants, which we may choose freely. The first constant, x_0, indicates the amplitude of the oscillation, the other, φ_0, the phase of the oscillation at the time $t = 0$. Depending on the value of φ_0, the bob is at $t = 0$ either at rest, at a point of maximum displacement, or somewhere in between. It should also be mentioned that the points of maximum displacement are called the "turning points" of the oscillation, as the direction of movement of the bob reverses at these points.

17.1.2 Energy Considerations

How much energy is stored in an oscillation and in what form? There is potential energy stored in the spring (spring energy E_{spring}) and the kinetic energy of the motion E_{kin}:

$$E_{\text{spring}} = \frac{1}{2} k x^2$$
$$E_{\text{kin}} = \frac{1}{2} m v^2.$$
(17.9)

The total energy is then:

$$E_{\text{tot}} = \frac{1}{2} k x^2 + \frac{1}{2} m v^2.$$
(17.10)

Due to energy conservation, it must remain constant for the entire motion. Therefore, we can calculate E_{tot} at any given time. If the bob is at the turning point, the kinetic energy disappears and it is:

$$E_{\text{tot}} = \frac{1}{2} k x_0^2.$$
(17.11)

17.1 · Harmonic Oscillations

Alternatively, we can determine the total energy at the zero crossing, i.e., when passing through the equilibrium position, where the spring energy is zero:

$$E_{\text{tot}} = \frac{1}{2} m v_{\max}^2 = \frac{1}{2} m x_0^2 \omega_0^2. \tag{17.12}$$

To check the conservation of energy, we calculate E_{tot} at an arbitrary time of the oscillation:

$$\begin{aligned} E_{\text{tot}} &= \frac{1}{2} k x^2(t) + \frac{1}{2} m v^2(t) \\ &= \frac{1}{2} k x_0^2 \sin^2(\omega_0 t + \varphi_0) + \frac{1}{2} m x_0^2 \omega_0^2 \cos^2(\omega_0 t + \varphi_0) \\ &= \frac{1}{2} k x_0^2 \sin^2(\omega_0 t + \varphi_0) + \frac{1}{2} m x_0^2 \frac{k}{m} \cos^2(\omega_0 t + \varphi_0) \\ &= \frac{1}{2} k x_0^2 (\sin^2(\omega_0 t + \varphi_0) + \cos^2(\omega_0 t + \varphi_0)) \\ &= \frac{1}{2} k x_0^2. \end{aligned} \tag{17.13}$$

We obtain a result consistent with the above calculations.

17.1.3 The Compound Pendulum

In Experiment 17.2 to Experiment 17.5 further examples of oscillators are presented. We want to examine one of them mathematically, the so-called compound pendulum or physical pendulum. A "compound pendulum" refers to a rigid body that is pivot-mounted at a point outside its center of gravity, for example, the rod shown in ◘ Fig. 17.2. If it is deflected from its equilibrium position, a restoring torque is created:

$$M = -F_G a = -m g h \sin \varphi. \tag{17.14}$$

For small deflections, it can be approximated by

$$M \approx -m g h \varphi, \tag{17.15}$$

with φ measured in radians. We insert this into the equation of motion:

$$\begin{aligned} I \frac{d^2 \varphi(t)}{dt^2} &= M \\ I \frac{d^2 \varphi(t)}{dt^2} &\approx -m g h \varphi(t) \end{aligned} \tag{17.16}$$

$$\frac{d^2 \varphi(t)}{dt^2} + \frac{m g h}{I} \varphi(t) = 0$$

$$\frac{d^2 \varphi(t)}{dt^2} + \omega_0^2 \varphi(t) = 0 \quad \text{with} \quad \omega_0^2 = \frac{m g h}{I}.$$

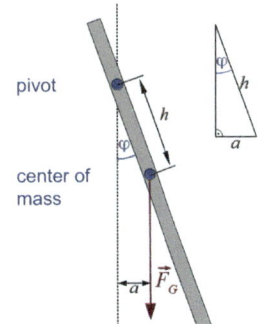

◘ Fig. 17.2 Compound pendulum

This is the same differential equation as we obtained for the spring pendulum, but now in the angle φ and not in the displacement x. Again, a sinusoidal pattern emerges as a solution (or a cosine pattern):

$$\varphi(t) = \varphi_0 \sin(\omega_0 t + \varphi'). \tag{17.17}$$

Please note that here φ_0 denotes the amplitude of the oscillation, while the phase is given by φ'. Such oscillations that follow a sinusoidal pattern are called "harmonic oscillations". They always occur when the restoring force or torque increases proportional to the displacement.

© RWTH Aachen, Physics Experiment Collection

Experiment 17.2: Compound Pendulum

This compound pendulum is a trapezoidal board with several suspension points. This allows us to study the dependence on the moment of inertia I, which changes with the distance of the suspension point to the center of gravity.

Experiment 17.3: Mach's Pendulum Apparatus

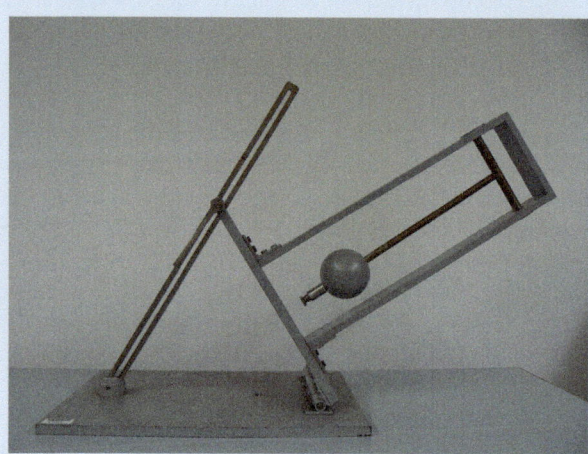

© RWTH Aachen, Physics Experiment Collection

This rod pendulum is another variant of a physical pendulum. The suspension is designed in such a way that the plane of oscillation is rigidly connected to the suspension. If we tilt the suspension, the plane of oscillation also tilts along with it. With a tilted plane of oscillation, the full gravitational force no longer contributes

17.1 · Harmonic Oscillations

to the restoring force. We have g reduced to $g \cos \alpha$. The oscillation slows down. This allows us to study the dependence of the period on the acceleration due to gravity.

Experiment 17.4: Torsion Pendulum

With a long spring we can build not only a spring pendulum, but also a torsion pendulum. You rotate the bob around its own axis, taking care not to displace it upwards or downwards from its equilibrium position. If you let it go, a torsional oscillation around the spring axis occurs. If you additionally deflect it upwards or downwards, a more complicated movement starts, which we want to discuss later in ▶ Sect. 17.4.

Resting position

Experiment 17.5: Simple Gravity Pendulum

The simple gravity pendulum is an idealized pendulum consisting of a massless suspension and a point-like bob. Although this is not fully achievable, we can come close enough to the idealization to carry out a good measurement of the standard gravity. With moderate effort, an accuracy of 0.1% can be achieved.

The illustration shows the forces acting on the bob. The horizontal deflection x and the deflection angle φ are related via $\sin \varphi = \frac{x}{l}$. The restoring force is $F_r = -\sin \varphi F_G$. For small deflections, F_r points approximately in the direction of x, resulting in a restoring force of $F_r = -\frac{mg}{l} x$. The differential equation is:

$$m \frac{d^2 x(t)}{dt^2} = -\frac{mg}{l} x(t)$$

$$\frac{d^2 x(t)}{dt^2} + \omega_0^2 x(t) = 0 \quad \text{with} \quad \omega_0^2 = \frac{g}{l},$$

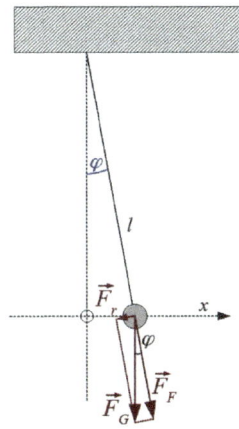

with the usual sinusoidal solution.

As a suspension, we uses a thread or a thin rod, which can be suspended almost friction-free with a cutting edge. For a pendulum length of approx. 1 m we should use a compact weight of around 1 kg, e.g., a steel ball. If it is too light, friction has too much influence. We carefully displace the bob, to ensure that it swings cleanly in a stable plane (possibly hang bifilar). The length of the suspension is measured from the suspension point to the center of gravity of the bob. This measurement limits the accuracy of determining the standard gravity. The oscillation period can be determined more accurately by measuring a larger number of oscillations.

Experiment 17.6: Simple Gravity Pendulum 2

We calculated (Experiment 17.5) that the oscillation period of a simple gravity pendulum does not depend on the mass of the bob. This initially surprising result can be beautifully demonstrated with this simple experiment. On a rod (figure), three pendulums of the same length are mounted one after the other. The three bobs are metal balls of the same diameter, but made of different materials and are therefore of different weights. At the front we have a brass ball, in the middle a ball of aluminum, and finally a lead ball at the back. The lead ball is more than four times as heavy as the aluminum ball. If we displace the three balls together and let them go at the same moment, we can observe how they oscillate synchronously over many oscillation periods. They show identical oscillation periods, determined only by the length of the thread and standard gravity.

Example 17.1: Constraints and Constraining Forces

When we introduced rigid bodies, we explained that a point mass has three degrees of freedom and a rigid body generally has six degrees of freedom. Here we describe simple pendulums with a single degree of freedom represented by its displacement and when we will switch to coupled pendulums, we will explain that a coupled pendulum has two degrees of freedom. What makes the difference?

The difference is that the bob cannot move freely, it has only one possibility of movement. For example, think about a thread pendulum. The bob can only move along the oscillation direction of the pendulum, but not per-

17.1 · Harmonic Oscillations

pendicular to it. Its weight and the suspension through the thread prevent any other motion. Therefore, we count only one degree of freedom. We say that the motion of the bob is constraint. These constraints reduce the possibilities of movement of the bob and thus reduce the number of degrees of freedom.

We want to illustrate this thinking with the example of a simple gravity pendulum. The bob, which we approximate by a point mass, initially has three degrees of freedom, which in a Cartesian coordinate system (see figure) correspond to movements in the x-, y- and z-directions. The bob is bifilarly suspended, which prevents it from moving in the z-direction. This leads to the constraint:

$$z = 0.$$

The gravitational force acting on the ball keeps the thread taut, so that the thread length and thus the distance from the origin of coordinates is always l. We can express this through a second constraint, namely:

$$x^2 + y^2 = l^2.$$

Each constraint reduces the number of degrees of freedom by one. With the two constraints, only one of the original three degrees of freedom remains. This brings us to the question of how we can calculate the motion of the system given the constraints. In addition to the restoring force $\vec{F}_r$, which drives the oscillation, other forces act on the bob, namely the other component of the weight and the force excerted by the thread. These forces ensure that the motion of the bob follows the constraints. Accordingly, we call these forces the constraining forces $\vec{F}_{\text{con}}$. They always act perpendicular to the allowed direction of motion and therefore perform no work on the system. If we want to calculate the motion from the basic law of mechanics, we must also take into account the constraining forces:

$$m\frac{d^2\vec{r}}{dt^2} = \vec{F}_r + \vec{F}_{\text{con}}.$$

This leads to the correct result, but is often cumbersome, as we need to know the constraining forces. Briefly, we want to discuss another approach. We expand the concept of the coordinate to the generalized coordinates, which are quantities that determines the position of the body, here the bob. The adjective "generalized" indicates that it is not necessarily a length, as is the case with Cartesian coordinates. The aim is now to find such a quantity that uniquely defines the position of the bob and automatically fulfills the constraints. In the case of the gravity pendulum, we may use the angle (see figure) to fulfill these conditions. The transformation from the generalized coordinate to the Cartesian coordinates is:

$x = l \sin \varphi$

$y = -l \cos \varphi$

$z = 0$.

The condition $z = 0$ is unmistakably fulfilled and $x^2 + y^2 = (l \sin \varphi)^2 + (-l \cos \varphi)^2 = l^2(\sin^2 \varphi + \cos^2 \varphi) = l^2$ shows that the second constrain is automatically fulfilled, too. The equation of motion is:

$$m \frac{d^2}{dt^2}(l \sin \varphi(t)) = 0$$

$$m \frac{d^2}{dt^2}(-l \cos \varphi(t)) = -mg,$$

where we have omitted the third component. The derivatives yield:

$$m l \left(-\sin \varphi \left(\frac{d\varphi}{dt} \right)^2 + \cos \varphi \frac{d^2\varphi}{dt^2} \right) = 0$$

$$m l \left(\cos \varphi \left(\frac{d\varphi}{dt} \right)^2 + \sin \varphi \frac{d^2\varphi}{dt^2} \right) = -mg.$$

We multiply the first equation by $\cos \varphi$ and the second by $\sin \varphi$ and add the two. Then, after a short rearrangement, we get:

$$\frac{d^2\varphi(t)}{dt^2} = -\frac{g}{l} \sin \varphi(t).$$

This is the sought-after equation of motion. The solution of the differential equation will automatically sat-

17.1 · Harmonic Oscillations

isfy the constraints, as they are valid for any values of φ. In the small angle approximation $\sin \varphi \approx \varphi \approx x/l$ the well-known differential equation of the simple gravity pendulum is recovered. For this simple example, this might not be the path to the solution, but perhaps you can imagine how helpful the reduction of degrees of freedom might be for other examples.

Finally, it should be mentioned that the choice of generalized coordinates is not unique. For example, we could have chosen the path length s (arc length) of the bob from the equilibrium position as our generalized coordinate. But keep in mind, the number of generalized coordinates, is uniquely determined by the system.

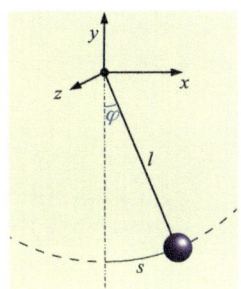

17.1.4 Representation in Phase Space

We have studied the equations of motion of harmonic oscillations using several examples. We want to revisit those oscillations from a different perspective that is important in areas of modern physics: the representation of motion in phase space.

We can unambiguously describe the state of a point mass by specifying its location and velocity at a single fixed point in time. Assuming we know the forces that act on the point mass, we can calculate from this state the location and velocity at any point in time in the future as well as in the past. In a diagram in which we plot the location and the velocity against each other, the movement of the mass point is represented by a line. This representation of the motion is called the "phase space". For a general motion in three spatial dimensions, the phase space has 6 dimensions (3 location and velocity coordinates each). In our example of a harmonic oscillation, the motion only takes place in one coordinate and the phase space therefore only has 2 dimensions.

The displacement of the oscillation is given by the quantity x. The restoring force in the harmonic case is then $F_r = -k\,x$. We determine the velocity of the bob at a time t from the total energy:

$$E_{\text{tot}} = \frac{1}{2}m\,v^2(t) + \frac{1}{2}k\,x^2(t) = \frac{1}{2}m\,v_{\text{max}}^2 = \frac{1}{2}k\,x_{\text{max}}^2 \,. \tag{17.18}$$

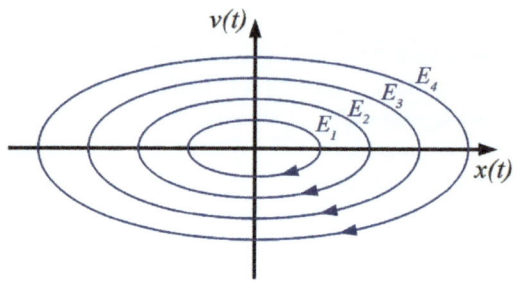

Fig. 17.3 Phase space ellipses of harmonic oscillations

From this it follows:

$$\frac{1}{2} m v^2(t) = \frac{1}{2} k x_{max}^2 - \frac{1}{2} k x^2(t)$$
$$v^2(t) = \frac{k}{m} x_{max}^2 - \frac{k}{m} x^2(t) \qquad (17.19)$$
$$v^2(t) = \frac{k}{m} x_{max}^2 \left(1 - \frac{x^2(t)}{x_{max}^2}\right).$$

From Eq. 17.18 we see that $v_{max}^2 = \frac{k}{m} x_{max}^2$ and thus:

$$v^2(t) = v_{max}^2 \left(1 - \frac{x^2(t)}{x_{max}^2}\right)$$
$$v(t) = \pm v_{max} \sqrt{1 - \frac{x^2(t)}{x_{max}^2}}. \qquad (17.20)$$

This is the parameter representation of an ellipse in the coordinates $x(t)$ and $v(t)$ with time as the parameter. Some ellipses can be seen in ◘ Fig. 17.3. The paths are each traversed in a clockwise direction. A full revolution around the ellipse corresponds to one oscillation period. The size of the ellipses is given by the total energy of the system. Ellipses with larger semi-axes describe oscillations with more energy (larger amplitude).

Laplace's Demon

According to the laws of classical mechanics, specifying the location and velocity of a body at a single point in time is sufficient to accurately predict its motion from the past and into the future. To perform this calculation, one must know the forces acting on the body. Since these are exerted by other bodies, which may themselves be moving, the movements of all involved bodies must be calculated together in a system of coupled differential equations $m_i \frac{d^2 \vec{r}_i(t)}{dt^2} = \vec{F}_i(\vec{r}_1, \ldots, \vec{r}_N)$. In practice, this is only possible for simple systems, but philosophically speaking, this technical limitation is irrelevant. From a snapshot of the world, from which the positions and velocities of all bodies at that time can be deduced, it is in principle possible for a person to predict the future exactly according to the laws of classical physics. This hypothetical person is called "Laplace's Demon" after the mathematician Pierre-Simon Laplace.

However, it should be noted that a Laplace's Demon can only exist within the framework of classical physics. Quantum physics refutes this idea, and results from chaos theory limit it even for classical physics.

17.1.5 Anharmonic Oscillations

We have discussed a number of examples of harmonic oscillations, but not all oscillations are necessarily harmonic. As already mentioned, the restoring force is key. In harmonic oscillations, it is proportional to the displacement. We have already seen other situations, for example in Experiment 11.1, the gravitational balance exhibited a torsional oscillation of the spheres, whose restoring torque is not proportional to the displacement. We presented a measurement of the oscillation over more than two periods, showing a small but clearly visible deviation from a harmonic oscillation. The differential equation is:

$$I \frac{d^2 \varphi(t)}{dt^2} = M. \tag{17.21}$$

The torque is composed of the torsion of the thread M_t and the gravitational force due to the two large spheres M_G. It is

$$\begin{aligned} M_t &= -G_T \frac{\pi r_t^4}{2 L_t} \varphi \\ M_G &= G \frac{m M}{r^2} l \quad \text{with} \quad r = r_0 - \frac{l}{2} \sin \varphi, \end{aligned} \tag{17.22}$$

where $\varphi = 0$ represents the equilibrium position, in which the small and large spheres have the distance r_0 (G is the gravitational constant, G_T the shear modulus of the wire). If you insert this into Eq. 17.21, you get a nonlinear differential equation that does not lead to a harmonic solution.

The solution to these nonlinear differential equations is often difficult. We want to end the example of the gravitational balance here and instead resume the discussion of the simple gravity pendulum. We treated it as a linear oscillation in the horizontal deflection x (Experiment 17.5). Now, we reformulate the pendulum as a rotational oscillation in the coordinate φ. It results in ($I = m l^2$):

$$I \frac{d^2\varphi(t)}{dt^2} = -mgl \sin \varphi(t)$$
$$\frac{d^2\varphi(t)}{dt^2} + \omega_0^2 \sin \varphi(t) = 0 \quad \text{with} \quad \omega_0^2 = \frac{g}{l}. \tag{17.23}$$

At this point we introduced the approximation for small displacements $\sin \varphi \approx \varphi$ and thus arrived at a harmonic differential equation. We kept only the first term from the power series expansion of the sine. Now, we consider the next larger term as a correction $\sin \varphi \approx \varphi - \frac{1}{6}\varphi^3$. The differential equation now reads:

$$\frac{d^2\varphi(t)}{dt^2} + \omega_0^2 \varphi(t) - \frac{1}{6} \omega_0^2 \varphi^3(t) = 0 \quad \text{with} \quad \omega_0^2 = \frac{g}{l}. \tag{17.24}$$

We want to try a harmonic approach again $\varphi(t) = \varphi_0 \sin(\omega' t + \tilde{\varphi})$, where we allow for a frequency that deviates from ω_0. If we insert this ansatz into the differential equation, we get

$$-\varphi_0 \omega'^2 \sin(\omega' t + \tilde{\varphi}) + \varphi_0 \omega_0^2 \sin(\omega' t + \tilde{\varphi})$$
$$- \frac{1}{6} \varphi_0^3 \omega_0^2 \sin^3(\omega' t + \tilde{\varphi}) = 0. \tag{17.25}$$

The cubic term can be converted using $\sin^3 x = \frac{1}{4}(3 \sin x - \sin 3x)$ into:

$$-\varphi_0 \omega'^2 \sin(\omega' t + \tilde{\varphi}) + \varphi_0 \omega_0^2 \sin(\omega' t + \tilde{\varphi})$$
$$- \frac{1}{6} \varphi_0^3 \omega_0^2 \frac{1}{4}(3 \sin(\omega' t + \tilde{\varphi}) - \sin(3(\omega' t + \tilde{\varphi}))) = 0. \tag{17.26}$$

This equation cannot be satisfied for arbitrary times, which shows that the differential equation cannot be exactly solved with a harmonic approach. However, we can still derive an approximation solution from it. It is the last term with the triple frequency that changes the sinusoidal motion. It averages out over a period of $\omega' t$. We can therefore omit it, resulting in the relation:

$$-\varphi_0 \omega'^2 \sin(\omega' t + \tilde{\varphi}) + \varphi_0 \omega_0^2 \sin(\omega' t + \tilde{\varphi})$$
$$-\frac{1}{6}\varphi_0^3 \omega_0^2 \frac{3}{4} \sin(\omega' t + \tilde{\varphi}) = 0$$
$$-\omega'^2 + \omega_0^2 - \frac{1}{8}\varphi_0^2 \omega_0^2 = 0 \quad (17.27)$$
$$\omega'^2 = \omega_0^2 \left(1 - \frac{\varphi_0^2}{8}\right).$$

This is the sought-after correction. With increasing amplitude, the frequency of the pendulum decreases (the period increases) compared to an infinitesimal amplitude. For example, for an amplitude of 10°, the correction is 0.38 %.

Note that in this representation of the simple gravity pendulum as a rotational oscillating, a correction due to the non-vanishing extension of the bob can easily be incorporated. Initially, we had assumed a point-like bob. If instead we assume a sphere with radius r, whose center of gravity marks the pendulum's length l, we only need to correctly calculate the moment of inertia I according to Steiner's theorem. The result is $I = m\left(l^2 + \frac{2}{5}r^2\right)$ and thus

$$\omega'^2 = \frac{g}{l\left(1 + \frac{2}{5}\frac{r^2}{l^2}\right)}. \quad (17.28)$$

If necessary, the moment of inertia of the thread could also be included here (it would then also have to be taken into account in the restoring force). But this effect is usually negligible.

We will describe a further correction due to friction in the following section.

17.2 Damped Oscillations

In the first subchapter about oscillations, we ignored friction, even though it could be observed in the experiments. Now, we turn our attention to the influence of friction on the oscillations. In Experiment 17.7, we study the effects systematically.

Experiment 17.7: Pohl's Wheel

The Pohl's wheel (figure) is a torsion pendulum build from a metal wheel (copper) mounted rotatably on an axis and connected to a spiral spring (visible behind the wheel in the picture), which causes a restoring torque towards the equilibrium position of the wheel. If we displace the wheel, i. e., if we turn it away from its equilibrium position, it performs harmonic oscillations. An electronic sensor records the rotation and displays it on a computer. The wheel is equipped with an eddy current brake (at the lowest point of the wheel), which allows to apply an adjustable friction. The two images show the wheel and a measurement of the displacement of the wheel with relatively low friction (damping).

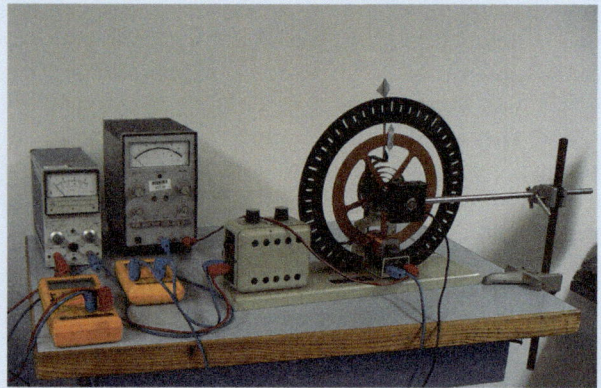

© RWTH Aachen, Physics Experiment Collection

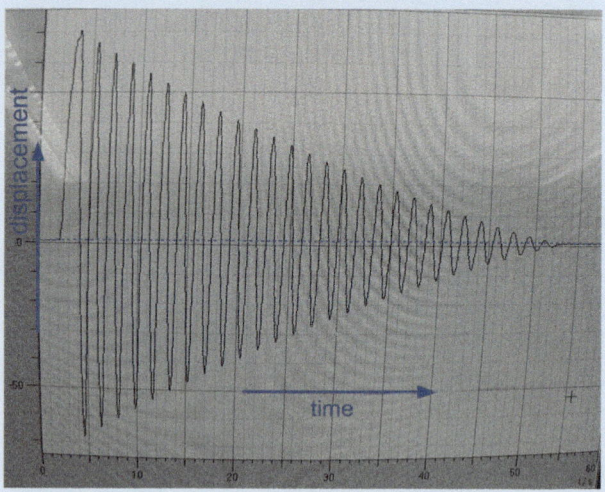

© RWTH Aachen, Physics Experiment Collection

17.2 · Damped Oscillations

As you can see, friction leads to a gradual decrease in amplitude. If the damping is increased, the amplitude decreases more rapidly. With very high damping, no periodic motion occurs at all as the wheel gradually returns to the equilibrium position, it is overdamped.

◘ Fig. 17.4 shows again the spring pendulum, where we now consider the restoring force $F_r = -k\,x$ as well as the friction $F_{\text{fric}} = -c\,v$. We assume that it is a kind of Stokes' friction, which is proportional to the velocity. The differential equation reads:

$$m\,a = F_r + F_{\text{fric}}$$

$$m\frac{d^2x(t)}{dt^2} = -k\,x(t) - c\,v(t) \quad (17.29)$$

$$\frac{d^2x(t)}{dt^2} + \frac{c}{m}\frac{dx(t)}{dt} + \frac{k}{m}x(t) = 0.$$

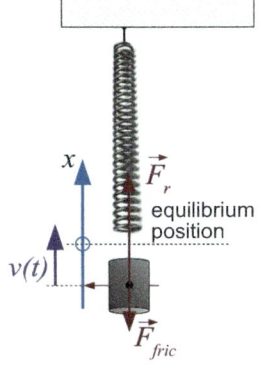

◘ Fig. 17.4 Spring pendulum with friction

We introduce two constants:

$$\frac{d^2x(t)}{dt^2} + 2\gamma\frac{dx(t)}{dt} + \omega_0^2 x(t) = 0 \quad \text{with } 2\gamma = \frac{c}{m},\ \omega_0^2 = \frac{k}{m}. \quad (17.30)$$

The constant ω_0^2 is already known to us from the frictionless case, where it indicated the circular frequency of the oscillation. It will turn out that it is closely linked to the frequency here, too. The second constant γ describes the damping of the system due to friction. The definition is $2\gamma = c/m$. However, note that other text book authors define the damping without the factor 2 ($\gamma = c/m$), which can lead to confusion.

Now, we want to find a solution for the differential equation. We expect an oscillation whose amplitude decreases over time. We choose an exponential approach, which includes the harmonic oscillation if we allow the exponent to take complex values. It is more general than the approach we used for the differential equation of the undamped oscillation, as it also includes a decrease in amplitude for a negative, real exponent. The ansatz is:

$$x(t) = a\,e^{\lambda t}$$
$$\frac{dx(t)}{dt} = a\,\lambda\,e^{\lambda t} \qquad (17.31)$$
$$\frac{d^2x(t)}{dt^2} = a\,\lambda^2\,e^{\lambda t}.$$

We insert this into the differential equation and obtain:

$$a\,\lambda^2\,e^{\lambda t} + 2\,\gamma\,a\,\lambda\,e^{\lambda t} + \omega_0^2\,a\,e^{\lambda t} = 0$$
$$a\,\lambda^2 + 2\,\gamma\,a\,\lambda + \omega_0^2\,a = 0 \qquad (17.32)$$
$$\lambda^2 + 2\,\gamma\,\lambda + \omega_0^2 = 0.$$

We obtain as a condition a quadratic equation for λ, which must be satisfied to solve the differential equation. Finally, we get two solutions for λ:

$$\lambda_{1,2} = -\gamma \pm \sqrt{\gamma^2 - \omega_0^2}. \qquad (17.33)$$

(At this point, you may understand why we inserted the factor 2 into the definition of γ.) The solution to the differential equation will be the sum of these two solutions for λ. The amplitude a does not enter into the condition, so we can choose arbitrary amplitudes a_1 and a_2 corresponding to the two values λ_1 and λ_2. The general solution is:

$$\begin{aligned}x(t) &= a_1 e^{\left(-\gamma+\sqrt{\gamma^2-\omega_0^2}\right)t} + a_2 e^{\left(-\gamma-\sqrt{\gamma^2-\omega_0^2}\right)t} \\ &= e^{-\gamma t}\left(a_1 e^{\sqrt{\gamma^2-\omega_0^2}\,t} + a_2 e^{-\sqrt{\gamma^2-\omega_0^2}\,t}\right).\end{aligned} \qquad (17.34)$$

What kind of motion this solution actually describes depends on the external conditions, in this case on the values of the two constants γ and ω_0. They determine whether the exponent is real or imaginary. We distinguish three cases $\gamma^2 < \omega_0^2$, $\gamma^2 = \omega_0^2$ and $\gamma^2 > \omega_0^2$.

17.2.1 Underdamped Oscillation ($\gamma < \omega_0$)

In this case, the frictional force (on average) is smaller than the restoring force. The radicand (the expression under the root) is negative. We separate the minus sign of the radicand by introducing a new constant $\omega^2 = -\left(\gamma^2 - \omega_0^2\right) > 0$. Then it is:

17.2 · Damped Oscillations

$$e^{\sqrt{\gamma^2-\omega_0^2}\,t} = e^{\sqrt{-\omega^2}\,t} = e^{\sqrt{-1}\sqrt{\omega^2}\,t} = e^{i\omega t} \qquad (17.35)$$
$$e^{-\sqrt{\gamma^2-\omega_0^2}\,t} = e^{-\sqrt{-\omega^2}\,t} = e^{-\sqrt{-1}\sqrt{\omega^2}\,t} = e^{-i\omega t}.$$

and the solution simplifies to

$$x(t) = e^{-\gamma t}\left(a_1 e^{i\omega t} + a_2 e^{-i\omega t}\right). \qquad (17.36)$$

However, we must respect the following consideration: The displacement $x(t)$ is a physical quantity. It can only take real values. The fact that we have used complex numbers here is merely a mathematical trick to shorten the calculations. The exponential functions will return complex numbers for some values of the time. To render the displacement real, we can only allow certain complex values for a_1 and a_2. You can easily check (determine the imaginary part of $x(t)$), that this is the case if $a_1 = a_2^* = a$. The complex phase of the number a is denoted by φ. Then,

$$\begin{aligned}
x(t) &= e^{-\gamma t}\left(a\, e^{i\omega t} + a^* e^{-i\omega t}\right) \\
&= e^{-\gamma t}\left(a\cos\omega t + ia\sin\omega t + a^*\cos(-\omega t) + ia^*\sin(-\omega t)\right) \\
&= e^{-\gamma t}\left(a\cos\omega t + ia\sin\omega t + a^*\cos\omega t - ia^*\sin\omega t\right) \\
&= e^{-\gamma t}\left((a+a^*)\cos\omega t + i(a-a^*)\sin\omega t\right) \\
&= e^{-\gamma t}\left(2\Re(a)\cos\omega t + i\,2i\,\Im(a)\sin\omega t\right) \\
&= e^{-\gamma t}\left(2\Re(a)\cos\omega t - 2\Im(a)\sin\omega t\right) \\
&= 2e^{-\gamma t}\left(|a|\cos\varphi\cos\omega t - |a|\sin\varphi\sin\omega t\right) \\
&= 2|a|e^{-\gamma t}\left(\cos\varphi\cos\omega t - \sin\varphi\sin\omega t\right) \\
&= 2|a|e^{-\gamma t}\cos(\omega t + \varphi).
\end{aligned} \qquad (17.37)$$

We obtain a harmonic oscillation with the angular frequency ω, whose amplitude is given by the absolute value of the constant a. At the time $t=0$ the oscillation starts with the amplitude $|a|$, after which the amplitude decreases exponentially due to friction. The time constant of the decrease is given by γ, i. e., in each time interval of duration $\ln 2/\gamma$ the amplitude decreases by a factor of 2.

The angular frequency of this weakly damped oscillation is reduced compared to the free oscillation. Our calculation yielded

$$\omega = \sqrt{\omega_0^2 - \gamma^2} < \omega_0 \qquad (17.38)$$

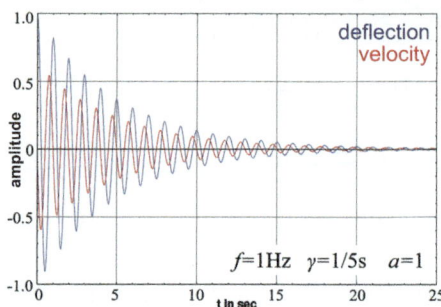

Fig. 17.5 Temporal progression of a weakly damped oscillation

The initial phase of the oscillation is given by the phase of the constants a. For example, if one chooses a to be real, then $\varphi = 0$, so that the oscillation starts at maximum displacement at $t = 0$. Choosing a purely imaginary a ($\varphi = \pi/2$) would correspond to the situation where the oscillation starts from the equilibrium position at $t = 0$. The temporal progression of such an oscillation can be seen in Fig. 17.5.

17.2.2 Overdamping ($\gamma > \omega_0$)

In this case, the frictional force is stronger than the restoring force. The radicand of the exponential is positive, so no complex numbers occur as long as we choose a_1 and a_2 to be real. We define $\alpha^2 = \gamma^2 - \omega_0^2 > 0$ and obtain:

$$e^{\sqrt{\gamma^2 - \omega_0^2}\,t} = e^{\sqrt{\alpha^2}\,t} = e^{\alpha t}$$
$$e^{-\sqrt{\gamma^2 - \omega_0^2}\,t} = e^{-\sqrt{\alpha^2}\,t} = e^{-\alpha t}.$$
(17.39)

The solution simplifies to:

$$x(t) = e^{-\gamma t}\left(a_1 e^{\alpha t} + a_2 e^{-\alpha t}\right). \tag{17.40}$$

We reparametrize the amplitudes:

$$\left.\begin{array}{l}\bar{a} = \frac{a_1 + a_2}{2} \\ \Delta a = \frac{a_1 - a_2}{2}\end{array}\right\} \Rightarrow \begin{cases} a_1 = \bar{a} + \Delta a \\ a_2 = \bar{a} - \Delta a. \end{cases} \tag{17.41}$$

17.2 · Damped Oscillations

Then it is:

$$\begin{aligned} x(t) &= e^{-\gamma t}\left((\bar{a}+\Delta a)e^{\alpha t}+(\bar{a}-\Delta a)e^{-\alpha t}\right) \\ &= e^{-\gamma t}\left((\bar{a}e^{\alpha t}+\bar{a}e^{-\alpha t})+(\Delta a\, e^{\alpha t}-\Delta a\, e^{-\alpha t})\right) \\ &= e^{-\gamma t}(2\bar{a}\cosh\alpha t+2\Delta a\sinh\alpha t) \\ &= e^{-\gamma t}((a_1+a_2)\cosh\alpha t+(a_1-a_2)\sinh\alpha t). \end{aligned} \quad (17.42)$$

We want to consider two special cases. If $a_1 = -a_2$, we get a zero amplitude at the time $t = 0$. The motion starts from the rest position with a non-vanishing velocity. The bob is slightly displaced and then returns asymptotically to the equilibrium position. An oscillation, i.e., a periodic motion, does not occur.

Another special case is obtained for $a_1 = \frac{\alpha+\lambda}{\alpha-\lambda}a_2$. As can be verified by differentiation and substitution, in this case the velocity at time $t = 0$ is zero, but not the displacement. The motion starts at maximum displacement (almost at the turning point) and returns asymptotically to the equilibrium position. This special case is illustrated in ◘ Fig. 17.6. Here too—as in all other cases—no overshoot of the equilibrium position and no periodic motion occurs.

17.2.3 Critical Damping ($\gamma = \omega_0$)

Finally, we want to consider the limiting case between under- and overdamping, called "critical damping". Note that this case only occurs approximately in nature, as the condition will never be exactly fulfilled. However, it is of mathematical interest.

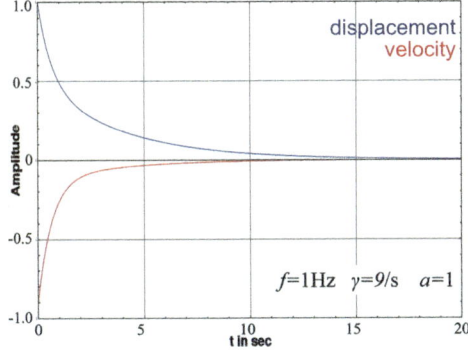

◘ **Fig. 17.6** Temporal progression in the overdamped case

The solution for this case is:

$$x(t) = a(1 + \gamma t)e^{-\gamma t}. \tag{17.43}$$

It is depicted in ◘ Fig. 17.7.

Finally, we present the phase space diagrams exemplarily for the cases of underdamping in ◘ Fig. 17.8 and critical damping in ◘ Fig. 17.9.

We can now revisit the measurement of the standard gravity with the simple gravity pendulum from the previous section. We had already indicated that we need to correct the measurement for damping. Obviously, this is a case of underdamping. The damping constant γ can be determined experimentally from the decay of the amplitude over many oscillations. We need to correct the measured frequency upwards due to the damping. We measure:

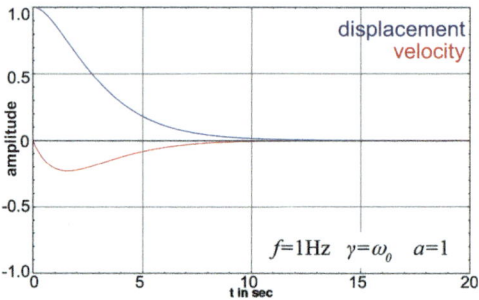

◘ **Fig. 17.7** Temporal progression in the case of critical damping

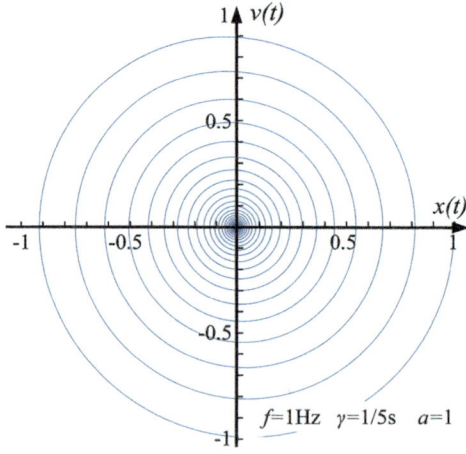

◘ **Fig. 17.8** Phase space diagram of an underdamped oscillation

17.3 · Forced Oscillations

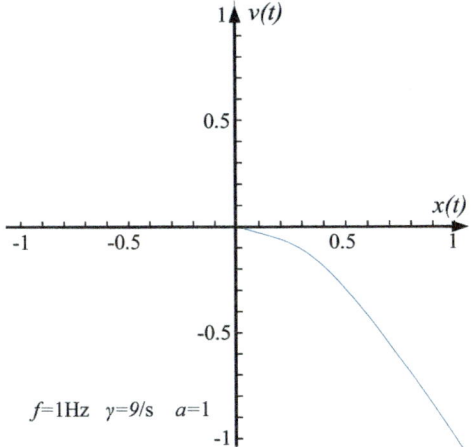

Fig. 17.9 Phase space diagram of a critically damped oscillation

$$\omega = \omega_0 \sqrt{1 - \frac{\gamma^2}{\omega_0^2}} \Rightarrow \omega \approx \omega_0 \left(1 - \frac{1}{2}\frac{\gamma^2}{\omega_0^2}\right). \qquad (17.44)$$

To summarize the results for the pendulum once again: We have now found three noticeable corrections. Combined, they result in

$$\begin{aligned}\omega^2 &= \frac{g}{l}\left(1 - \frac{\varphi_0^2}{8}\right)\frac{1}{1 + \frac{2}{5}\frac{r^2}{l^2}}\left(1 - \frac{1}{2}\frac{\gamma^2}{\omega_0^2}\right)^2 \\ &\approx \frac{g}{l}\left(1 - \frac{\varphi_0^2}{8} - \frac{2}{5}\frac{r^2}{l^2} - \frac{\gamma^2}{\omega_0^2}\right).\end{aligned} \qquad (17.45)$$

17.3 Forced Oscillations

In the first two sections, we considered oscillations that occurred free from external influences. After the initial displacement, no external force acted on the bob (if one considers the restoring force as an internal force). These are called "free oscillations". Now, we consider systems that are periodically driven from the outside, i.e., systems in which a periodic force from the outside acts on the system during the entire movement. This is called a "forced oscillation".

Experiment 17.8: Forced Oscillations on the Pohl's Wheel

We use again the Pohl's wheel, which you already know from Experiment 17.7. The wheel can be driven from the outside via the spiral spring. It is attached to the wheel on one side and to a rotatable disc on the other side, which can be moved by a motor via an eccentric. This generates a sinusoidal torque on the wheel. The amplitude of the driving force can be adjusted via the eccentric and the frequency of the driving force via the speed of the motor.

We start the motor at a very low frequency and slowly turn it up. Initially, the wheel follows the driving force exactly. It moves in synchrony with the same frequency and with the amplitude set by the eccentric. If the frequency is increased, it is observed that the wheel follows the driving frequency and that the amplitude of the oscillation gradually increases. As the frequency increases further, the amplitude continues to increase. It becomes many times larger than the amplitude of the eccentric. It can be shown that the amplitude at this point is largely determined by the damping set at the eddy current brake. If the frequency is further increased, the amplitude eventually decreases again. The resonant frequency has been exceeded. If the frequency is increased further and further, it eventually approaches zero. In addition we observe a change in the phase of the oscillation. While the wheel oscillated synchronously with the excitation at low frequency, it now oscillates in antiphase. This is visible from the pointer at the top of the wheel compared to the pointer on the eccentric, which can be seen behind the black disc.

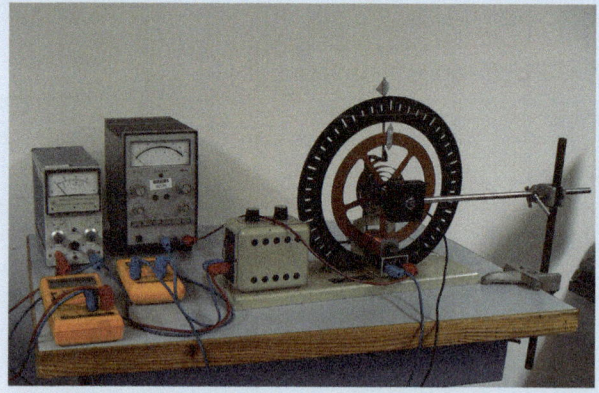

© RWTH Aachen, Physics Experiment Collection

17.3 · Forced Oscillations

Forced oscillations can be demonstrated with Pohl's wheel. The external force is periodic with the angular frequency ω:

$$F_{\text{ext}} = F_0 \cos \omega t. \tag{17.46}$$

In principle, we could insert a phase here, $F_{\text{ext}} = F_0 \cos(\omega t + \varphi_{\text{ext}})$, but without loss of generality, we can choose the zero point of time such that the exciting force has its maximum at $t = 0$ ($\varphi_{\text{ext}} = 0$).

We set up the differential equation:

$$m a = F_r + F_{\text{fric}} + F_{\text{ext}} \tag{17.47}$$

with the restoring force F_r, the frictional force F_{fric} and the external force F_{ext}:

$$m \frac{d^2 x(t)}{dt^2} = -k x(t) - c \frac{d x(t)}{dt} + F_0 \cos \omega t$$

$$\frac{d^2 x(t)}{dt^2} + \frac{c}{m} \frac{d x(t)}{dt} + \frac{k}{m} x(t) = \frac{F_0}{m} \cos \omega t \tag{17.48}$$

$$\frac{d^2 x(t)}{dt^2} + 2 \gamma \frac{d x(t)}{dt} + \omega_0^2 x(t) = K \cos \omega t,$$

with $2\gamma = c/m$, $\omega_0^2 = k/m$ and $K = F_0/m$. We have obtained an inhomogeneous differential equation. In mathematics, you learn that the general solution of such an inhomogeneous differential equation is composed of the general solution of the corresponding homogeneous differential equation ($F_{\text{ext}} = 0$) and a particular solution of the inhomogeneous differential equation.

We already discussed the solution of the homogeneous differential equation in the previous subchapter. It describes a damped oscillation, corresponding to the transient response of the system in reaction to the start of the external force at time $t = 0$. It exites a free oscillation that decays due to friction, so that after a certain time only the forced motion driven by the external force remains. This is the solution of the inhomogeneous differential equation with the term of the external force. It will turn out that this part of the solution has a constant amplitude over time and is therefore called the "stationary solution".

We try to find a stationary solution. With a sinusoidal or cosine-shaped excitation, the solution itself should also be harmonic. We therefore assume:

$$x_{\text{stat}}(t) = x_0(\omega') \cos(\omega' t - \varphi(\omega')), \tag{17.49}$$

We explicitly allow the amplitude of the motion to change with the exciting frequency ω. As the frequency of the oscillation, we initially assumed an arbitrary frequency ω'. We will see that it must match the exciting frequency ω.

Now we calculate the derivatives:

$$\frac{d x_{\text{stat}}(t)}{dt} = -x_0\, \omega'\, \sin\left(\omega' t - \varphi\right)$$

$$\frac{d^2 x_{\text{stat}}(t)}{dt^2} = -x_0\, \omega'^2 \cos\left(\omega' t - \varphi\right) \tag{17.50}$$

and insert into the differential equation:

$$-x_0\, \omega'^2 \cos\left(\omega' t - \varphi\right) - 2\gamma\, x_0\, \omega' \sin\left(\omega' t - \varphi\right)$$
$$+ \omega_0^2\, x_0 \cos\left(\omega' t - \varphi\right) = K \cos \omega t\,. \tag{17.51}$$

At this point, it is already apparent that this equation can only be satisfied for arbitrary times t, if the two frequencies ω' and ω match. So we set $\omega' = \omega$ and obtain:

$$-x_0\, \omega^2 \cos(\omega t - \varphi) - 2\gamma\, x_0\, \omega \sin(\omega t - \varphi)$$
$$+ \omega_0^2\, x_0 \cos(\omega t - \varphi) = K \cos \omega t\,. \tag{17.52}$$

Now we use the angle sum identities of the trigonometric functions to separate the phase terms $\sin(\omega t - \varphi) = \sin \omega t \cos \varphi - \cos \omega t \sin \varphi$ and $\cos(\omega t - \varphi) = \cos \omega t \cos \varphi + \sin \omega t \sin \varphi$:

$$-x_0\, \omega^2 (\cos \omega t \cos \varphi + \sin \omega t \sin \varphi)$$
$$- 2\gamma\, x_0\, \omega (\sin \omega t \cos \varphi - \cos \omega t \sin \varphi) \tag{17.53}$$
$$+ \omega_0^2\, x_0 (\cos \omega t \cos \varphi + \sin \omega t \sin \varphi) = K \cos \omega t$$

or

$$x_0 \left[\cos \omega t \left(\left(\omega_0^2 - \omega^2\right) \cos \varphi + 2\gamma\, \omega \sin \varphi \right) \right.$$
$$\left. + \sin \omega t \left(\left(\omega_0^2 - \omega^2\right) \sin \varphi - 2\gamma\, \omega \cos \varphi \right) \right] = K \cos \omega t\,. \tag{17.54}$$

However, this equation can only be satisfied if the coefficients of the $\cos \omega t$ terms and those of the $\sin \omega t$ terms match individually. Therefore, it must hold:

17.3 · Forced Oscillations

$$x_0 \left[\left(\omega_0^2 - \omega^2 \right) \cos \varphi + 2 \gamma \omega \sin \varphi \right] = K$$
$$x_0 \left[\left(\omega_0^2 - \omega^2 \right) \sin \varphi - 2 \gamma \omega \cos \varphi \right] = 0 \,. \quad (17.55)$$

Now, we can solve this system of two equations for x_0 and φ and obtain:

$$x_0 = \frac{K}{\sqrt{\left(\omega_0^2 - \omega^2\right)^2 + (2\gamma\omega)^2}}$$
$$\tan \varphi = \frac{2\gamma\omega}{\omega_0^2 - \omega^2} \,. \quad (17.56)$$

We have seen that the system oscillates with the frequency of the external excitation. The natural frequency ω_0 still has an influence on the amplitude of the oscillation after the transient response has faded (even a large one), but no longer on its frequency.

The amplitude of the oscillation strongly depends on the difference of the excitation frequency ω to the natural frequency ω_0 of the free oscillation. It has a maximum in the range where the external excitation frequency and natural frequency coincide. If the excitation frequency deviates from the natural frequency either upwards or downwards, the oscillation amplitude becomes increasingly smaller. Furthermore, the amplitude depends linearly on the strength K of the external excitation.

We observe a phase shift between the exciting force and the oscillation, i.e., the oscillation does not reach its maximum at the same time as the exciting force is at its maximum. For excitation frequencies far below the natural frequency, this phase shift is small, i.e., the oscillation follows the excitation almost without delay. If the excitation and natural frequencies coincide, the phase shift is exactly $\pi/2$, so that the oscillation lags behind the excitation by a quarter period. When the excitation reaches its maximum, the oscillation is just passing through the equilibrium position. Finally, if the excitation frequency is much greater than the natural frequency, the phase shift reaches the value of π, the system oscillates in opposition to the excitation. All these relationships can be beautifully demonstrated with Pohl's wheel (Experiment 17.8).

The progression of amplitude and phase are graphically represented in ◘ Figs. 17.10 and 17.11. Amplitude and phase are plotted against ω/ω_0. This results in graphs that are independent of the natural frequency of

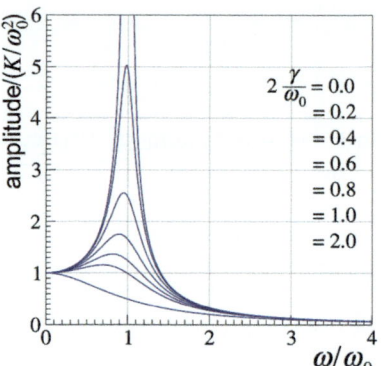

Fig. 17.10 Amplitude progression in a forced oscillation

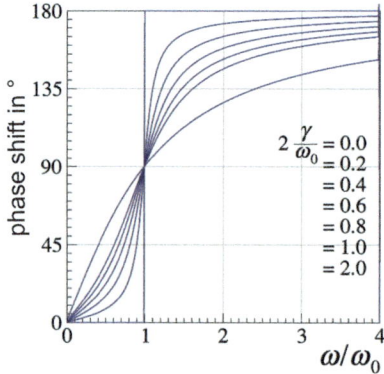

Fig. 17.11 Phase progression in a forced oscillation

the respective system. If we expresses Eq. 17.56 through the ratio ω/ω_0, we obtain:

$$x_0 = \frac{K/\omega_0^2}{\sqrt{\left(1-\left(\frac{\omega}{\omega_0}\right)^2\right)^2 + \left(\frac{2\gamma}{\omega_0}\frac{\omega}{\omega_0}\right)^2}}$$

$$\tan\varphi = \frac{\frac{2\gamma}{\omega_0}\frac{\omega}{\omega_0}}{1-\left(\frac{\omega}{\omega_0}\right)^2}\,.$$

(17.57)

As you can see, if we specify the damping through the ratio $2\gamma/\omega_0$ and further scale the amplitude with K/ω_0^2, then the same curves apply for any system. This rescaling has already been applied in the figures.

17.3 · Forced Oscillations

As the excitation frequency approaches the system's natural frequency, the amplitude of the oscillation increases sharply, referred to as a "resonance". We see from Eq. 17.56 that in the absence of damping ($\gamma = 0$) the amplitude of the oscillation tends to infinity at $\omega = \omega_0$, an artificial case that does not occur in nature. In ◘ Fig. 17.10 it can also be seen that the excitation frequency at which the maximum amplitude is reached, slightly shifts to lower frequencies with increasing damping. We want to calculate exactly where the maximum of the amplitude is. We calculate the derivative of Eq. 17.56 with respect to the excitation frequency:

$$\frac{d\,x_0(\omega)}{d\omega} = \frac{K}{\left((\omega_0^2 - \omega^2)^2 + (2\gamma\omega)^2\right)^{\frac{3}{2}}} \cdot \left((-2\omega)2(\omega_0^2 - \omega^2) + (2\gamma)2(2\gamma\omega)\right). \quad (17.58)$$

The derivative must yield zero at the maximum. Since the denominator may not become zero and we want to exclude the trivial case $K = 0$, it must hold:

$$(-2\,\omega_{\text{res}})2(\omega_0^2 - \omega_{\text{res}}^2) + (2\gamma)2(2\gamma\,\omega_{\text{res}}) = 0. \quad (17.59)$$

In the case of overdamping ($\gamma > \omega_0$), the maximum is at $\omega_{\text{res}} = 0$. This can be extrapolated in ◘ Fig. 17.10 from the case of critical damping, the strongest damping shown in the figure. We want to focus here on underdamping. In this case, the resonance frequency $\omega_{\text{res}} > 0$ and Eq. 17.59 can be further simplified by dividing by ω_{res}:

$$-4(\omega_0^2 - \omega_{\text{res}}^2) + 8\gamma^2 = 0$$
$$\Rightarrow \omega_{\text{res}} = \omega_0\sqrt{1 - \frac{2\gamma^2}{\omega_0^2}}. \quad (17.60)$$

We can see from the result, how the resonance frequency shifts towards lower frequencies with increasing damping. By substituting ω_{res} into Eq. 17.56 we obtain the resonant amplitude:

$$x_{\text{res}} = x(\omega_{\text{res}}) = \frac{K}{2\gamma\,\omega_0}\frac{1}{\sqrt{1-\gamma^2/\omega_0^2}}. \quad (17.61)$$

It increases with decreasing damping approximately like $1/\gamma$ and can reach arbitrarily large values. In this case, a lot of energy is transferred from the excitation to the system. The amplitude can increase so much that the system is destroyed, which we refer to as a "resonance catastrophe". The so-called Q-factor of the system, also known as quality factor, determines the resonant amplitude. The Q-factor is defined as:

$$Q = \frac{\omega_0}{2\gamma}. \tag{17.62}$$

As already indicated above (Figs. 17.10 and 17.11), the relationships can be represented independently of the system if we scale the system-specific quantities. We scale the frequencies to the natural frequency $\hat{\omega} = \omega/\omega_0$, use the quality factor Q to express the damping, and set $\hat{K} = K/\omega_0^2$. Then we get:

$$x_0 = \frac{\hat{K}}{\sqrt{(1-\hat{\omega}^2)^2 + \frac{\hat{\omega}^2}{Q^2}}}$$

$$\tan \varphi = \frac{\frac{\hat{\omega}}{Q}}{1-\hat{\omega}^2}. \tag{17.63}$$

The resonance is at:

$$\hat{\omega}_{res} = \frac{\omega_{res}}{\omega_0} = \sqrt{1 - \frac{1}{2Q^2}}$$

$$x_{res} = x_0(\hat{\omega}_{res}) = \frac{\hat{K}Q}{\sqrt{1 - \frac{1}{4Q^2}}}. \tag{17.64}$$

The amplitude x_0 at the natural frequency ω_0 and the width of the resonance $\Delta\hat{\omega}$, measured from the frequency to the left of the resonance, where the amplitude reaches the value $x_0(\omega_0)/\sqrt{2}$, to the corresponding frequency to the right of the resonance, are:

$$x_0(\omega_0) = \hat{K}Q$$

$$\Delta\hat{\omega} \approx \frac{1}{Q}. \tag{17.65}$$

The Q-factor determines the height and width of the resonance. The higher the Q-factor, the higher and sharper the resonance.

17.3 · Forced Oscillations

By the way, we could reparameterize the result of the free, damped oscillation through the Q-factor, too. In the case of underdamping, the amplitude of the oscillation decays like $e^{-\omega_0/(2Q)t}$, which means that it decays slower the larger the Q-factor is.

Experiment 17.9: Resonance with Spring Pendulum

With this simple setup, we demonstrate forced oscillations and the resonance phenomenon. A spring pendulum is attached to an eccentric, which is driven by a controllable motor. Damping mainly occurs through internal losses in the spring. We slowly increase the speed of the motor. When the resonance frequency is reached, the amplitude of the spring pendulum significantly exceeds the amplitude of the driving eccentric. Outside of resonance, we observe the phase shift between the eccentric and the bob.

Experiment 17.10: Tongue Frequency Meter

The tongue frequency meter is a simple apparatus for determining small frequency shifts. An old electric motor with an imbalance generates vibration in the wooden plate on which it is mounted. A yellow arrow on the axle indicates the adjustable frequency of the motor. The two weights prevent the plate from running away.
On the underside of the plate, four flat metal tongues of different lengths are mounted. A red paper strip at the end indicates any oscillations. The longest tongue has the lowest natural frequency. If we now increase the frequency of the motor starting from the lowest frequencies, we will see that the longest tongue resonates first. It begins to oscillate with a clearly visible amplitude (∼ 1 cm). When increasing the frequency further, the amplitude diminishes and the second tongue gradually resonates. Thus, by changing the frequency, we can bring one tongue after another into resonance.

© RWTH Aachen, Physics Experiment Collection

Example 17.2: Swinging

When a swing is displaced from its resting position, it swings back and forth several times until friction bring its motion to a halt. To maintain or even intensify the swinging, the child must perform work., by moving its center of gravity up and down. Only if the child moves at the swing's natural frequency can it increase the amplitude. The resonance condition must be met.

Example 17.3: Tacoma Narrows Bridge

The Tacoma Narrows Bridge was built south of Seattle over a branch of Puget Sound between 1938 and 1940. It was put into operation on July 1, 1940, and after just over 4 months of spectacular vibrations, it collapsed on November 7, 1940. The bridge had a flat roadway, which was held laterally by full-wall steel beams that unfortunately did not allow wind to pass through. On November 6/7, a strong crosswind had arisen (wind speed 67 km/h, wind force 8), which set the roadway into torsional vibrations. The torsion of the roadway is clearly visible in the first photo. The torsion transferred more and more energy from the wind flow to the roadway, which amplified the vibration to a considerable amplitude until it finally broke off (second photo). Since the amplification took hours, the bridge could be closed to traffic in time, so no one was directly harmed. Although the collapse of the bridge was strictly speaking not a resonance phenomenon (the wind blew continuously and not at the frequency of the vibrating bridge), it shows the destructive effect a continuous energy transfer can have on a vibratory system.

© University of Washington Libraries, Special Collections, UW21412

© University of Washington Libraries, Special Collections, UW21413

Experiment 17.11: Resonance Catastrophe with a Wine Glass

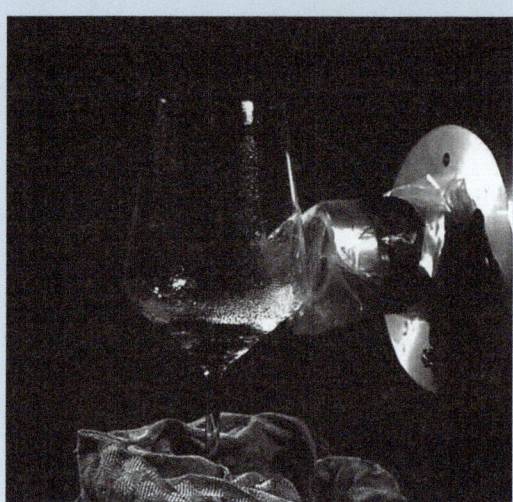

© RWTH Aachen, Physics Experiment Collection

Do you know Günter Grass's novel "The Tin Drum"? The boy Oskar Matzerath, who had decided not to grow anymore, can shatter glasses with his shrill, high voice. Here is our version: A wine glass stands in front of a speaker. The sound of the speaker is focused on the glass through a tube. The speaker is driven by a tunable sine generator and a powerful amplifier. There is a small sip of water in the glass, which helps to find the resonance frequency. If we approach the resonance frequency the glass begins to vibrate and transfers the vibration to the water. The water surface ripples and in the immediate vicinity of the resonance it is so strongly excited that it rises on the wall of the glass, as can be seen in the first picture. With appropriate power (frequency 600 to 800 Hz, the sound is now uncomfortably loud), the glass shatters upon reaching resonance (second picture).

17.4 · Coupled Oscillations

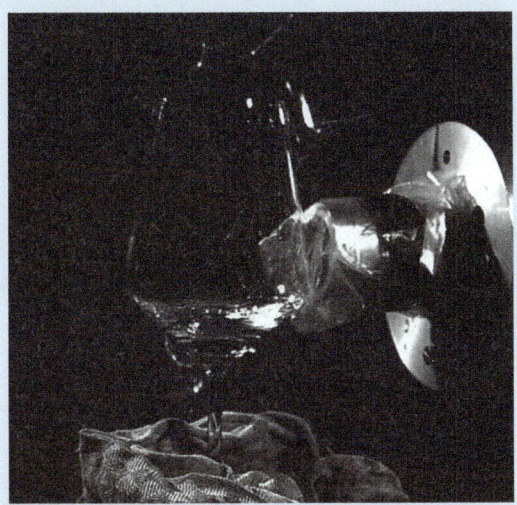

© RWTH Aachen, Physics E'xperiment Collection

17.4 Coupled Oscillations

We have concluded the discussion of oscillations of individual systems and now turn to more complex systems with several oscillating bodies. Initially, we will limit ourselves to systems with two coupled oscillations. Through the coupling, energy can be transferred from one oscillation to the other and back, which can influence the oscillations mutually. The movements that then occur are called "coupled oscillations".

> **Experiment 17.12: Coupled Pendulums**
>
> The images show coupled pendulums, which can be used to demonstrate the effects discussed in this chapter. The first is the system already outlined in ▶ Fig. 17.12, consisting of two bobs on rods, which are coupled by a (weak) spring.
> The second image shows two spring pendulums, which are mounted at the ends of a long rod. Although the rod is fixed, it moves up and down slightly due to the oscillations of the bobs, thereby coupling the two spring pendulums.

© RWTH Aachen, Physics Experiment Collection

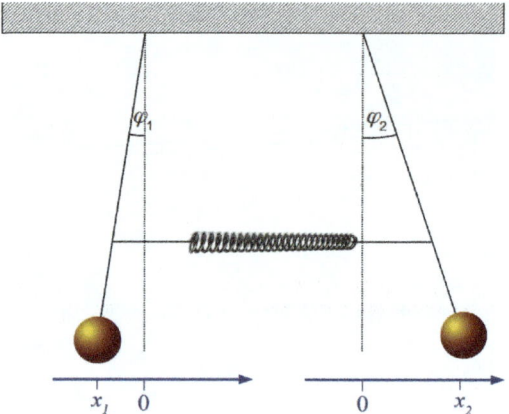

Fig. 17.12 Two coupled pendulums

© RWTH Aachen, Physics Experiment Collection

The Wilberforce pendulum is another interesting coupled pendulum. A bob is hanging on a long spring, a situation we already encountered in Experiment 17.4. The special feature of this system is that it couples two modes of oscillation in the same body. The bob on the spring can perform vertical oscillations and torsional oscillations. The spring is designed in such a way that it generates a slight torque around its longitudinal axis when stretched. This couples the two modes of oscillation.

The movements that occur as free oscillation after the bobs are released strongly depend on the initial conditions. As a rule, both bobs or the two modes of oscil-

17.4 · Coupled Oscillations

lation of the same bob swing with changing amplitude. However, under very specific conditions, a movement with constant amplitude and fixed phase relationship between the two bobs is obtained, called the "natural oscillations" or "normal modes" of the system.

Experiment 17.13: Coupled Metronomes

This experiment demonstrates a surprising effect (◨ Fig. 17.13). On a plate, 5 metronomes are mounted, which we wind up and start at maximum beat frequency (approx. 200 Hz) asynchronously. One hears a somewhat chaotic sounding clicking of the metronomes. Then, as shown in the picture, the plate is placed on rollers. If you now start the metronomes (again asynchronously!) you will see clear deflections of the plate to the right and left and after a short time the metronomes run synchronously. Through the coupling via the plate they have synchronized. The synchronization was first observed by Christiaan Huygens in 1665 on pendulum clocks. This effect also occurs in biology in some places and can be described by a model called the Kuramoto model.

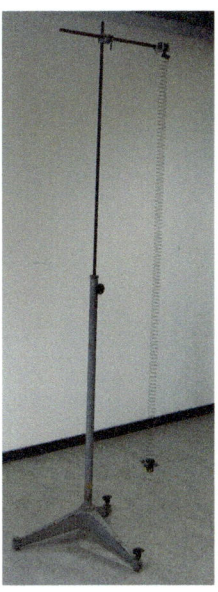

© RWTH Aachen, Physics Experiment Collection

An example of a coupled system can be seen in ◨ Fig. 17.12 and further in Experiment 17.12. Two bobs swing on a thin rod coupled by a spring between the two rods. In addition to the restoring force that acts on each of the pendulums, the spring generates a force that depends on the displacement of both pendulums. Let's call l the length of the pendulums and l' the length from the suspension to the point where the spring is attached, then

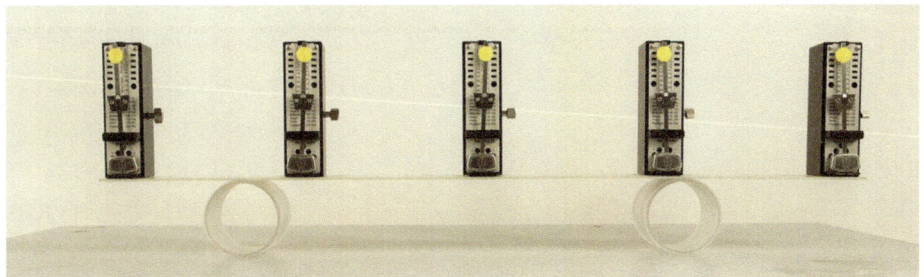

◨ Fig. 17.13 © RWTH Aachen, Physics Experiment Collection

the extension of the spring compared to the equilibrium position is:

$$\Delta s = x_2 \frac{l'}{l} - x_1 \frac{l'}{l} \qquad (17.66)$$

and the corresponding spring force is:

$$F_{12} = k_s \frac{l'}{l} (x_2 - x_1) = k'(x_2 - x_1). \qquad (17.67)$$

The second pendulum excerts this force onto the first. The corresponding reaction $F_{21} = -F_{12}$ acts from the first onto the second.

We insert the forces into Newton's second axiom and obtain two coupled differential equations:

$$\begin{aligned} m \frac{d^2 x_1(t)}{dt^2} &= -k\, x_1(t) + k'(x_2(t) - x_1(t)) \\ m \frac{d^2 x_2(t)}{dt^2} &= -k\, x_2(t) - k'(x_2(t) - x_1(t)). \end{aligned} \qquad (17.68)$$

The differential equations are called "coupled" because the solution of one equation depends on the solution of the other. The constants are $k = mg/l$ and $k' = k_s\, l'/l$.

We now have a system with two degrees of freedom, the two displacements $x_1(t)$ and $x_2(t)$. Accordingly, we need two functions to specify the position of the bobs at any given time. Again, we suspect a harmonic motion and therefore choose the following ansatz:

$$\begin{aligned} x_1(t) &= a_1 \cos(\omega_1 c t + \varphi_1) \\ x_2(t) &= a_2 \cos(\omega_2 c t + \varphi_2). \end{aligned} \qquad (17.69)$$

We substitute this into the differential equation and obtain

$$\begin{aligned} -m\, a_1 \omega_1^2 \cos(\omega_1 c t + \varphi_1) = &- k\, a_1 \cos(\omega_1 c t + \varphi_1) \\ &+ k'\, a_2 \cos(\omega_2 c t + \varphi_2) \\ &- k'\, a_1 \cos(\omega_1 c t + \varphi_1) \\ -m\, a_2 \omega_2^2 \cos(\omega_2 c t + \varphi_2) = &- k\, a_2 \cos(\omega_2 c t + \varphi_2) \\ &- k'\, a_2 \cos(\omega_2 c t + \varphi_2) \\ &+ k'\, a_1 \cos(\omega_1 c t + \varphi_1). \end{aligned}$$

$$(17.70)$$

17.4 · Coupled Oscillations

As you can see, a solution for this system of equations can only be found if $\omega_{1C} = \omega_{2C} = \omega$ and $\varphi_1 = \varphi_2 = \varphi$.[1] Then it must hold:

$$-m a_1 \omega^2 = -k a_1 + k' a_2 - k' a_1$$
$$-m a_2 \omega^2 = -k a_2 - k' a_2 + k' a_1 .$$
(17.71)

We sort the terms by a_1 and a_2

$$m \omega^2 a_1 - k a_1 - k' a_1 + k' a_2 = 0$$
$$m \omega^2 a_2 - k a_2 - k' a_2 + k' a_1 = 0$$
(17.72)

and express these conditions as a matrix equation:

$$\begin{pmatrix} m\omega^2 - k - k' & k' \\ k' & m\omega^2 - k - k' \end{pmatrix} \cdot \begin{pmatrix} a_1 \\ a_2 \end{pmatrix} = \begin{pmatrix} 0 \\ 0 \end{pmatrix}.$$
(17.73)

We obtain a two-dimensional system of linear equations. Such a system of equations has a solution only if the determinant of the matrix vanishes, i.e. if

$$\left(m\omega^2 - k - k' \right)^2 - k'^2 = 0$$
$$\left(m\omega^2 - k - k' \right)^2 = k'^2$$
$$m\omega^2 - k - k' = \pm k' .$$
(17.74)

We have obtained two solutions, which should not be surprising. When considering a system with two degrees of freedom, as is the case here, one will generally find two solutions. For a system with three degrees of freedom, we will correspondingly find a three-dimensional system of equations with generally three solutions, and so on.

We examine the two solutions more closely. Let's start with the first case with the minus sign in Eq. 17.74.

17.4.1   1. Case: $M\omega^2 - k - k' = k'$

We denote the frequency of this oscillation mode with ω_1. It is:

---

[1] One can also find solutions with $\varphi_2 = \varphi_1 + \pi$. However, these are also included in our approach, if we choose the signs of a_1 and a_2 accordingly.

$$m\omega_1^2 - k - k' = -k'$$
$$m\omega_1^2 = k$$
$$\omega_1 = \sqrt{\frac{k}{m}} = \sqrt{\frac{g}{l}} = \omega_0. \tag{17.75}$$

We substitute this back into the system of Eq. 17.72 and obtain:

$$m\frac{k}{m}a_1 - ka_1 - k'a_1 + k'a_2 = 0$$
$$-k'a_1 + k'a_2 = 0 \tag{17.76}$$
$$a_1 = a_2 = a.$$

The following motion results:

$$x_1(t) = a\cos(\omega_0 t + \varphi)$$
$$x_2(t) = a\cos(\omega_0 t + \varphi). \tag{17.77}$$

This oscillation is already known to us from the uncoupled pendulum. The two pendulums oscillate with identical amplitude and in phase with the natural frequency ω_0 of a single pendulum. As a result, the spring is always relaxed. There is no coupling between the pendulums. The motion is sketched in ◘ Fig. 17.14.

17.4.2 2. Case: $M\omega^2 - k - k' = k'$

Now we have:

$$m\omega_2^2 - k - k' = k'$$
$$m\omega_2^2 = k + 2k' \tag{17.78}$$
$$\omega_2 = \sqrt{\frac{k+2k'}{m}}.$$

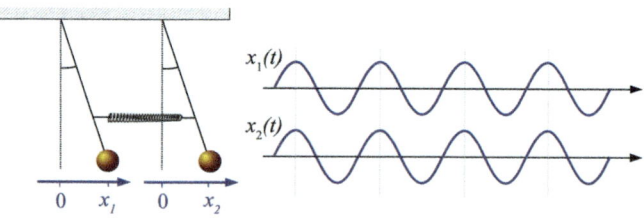

◘ **Fig. 17.14** First eigenmode of a coupled pendulum

17.4 · Coupled Oscillations

The frequency ω_2 is higher than the frequency we found in the first case. It increases with the coupling between the two pendulums. We again substitute into Eq. 17.72 to determine the amplitudes:

$$m\frac{k+2k'}{m}a_1 - k\,a_1 - k'a_1 + k'a_2 = 0$$
$$k'a_1 + k'a_2 = 0 \qquad (17.79)$$
$$a_1 = -a_2.$$

Again, we obtain oscillations of the same amplitude, but now the two pendulums oscillate in opposite directions. The spring is periodically stretched and compressed. There is coupling between the pendulums. The motion is shown in ◘ Fig. 17.15.

The movements described in case 1 and case 2 are called the "natural oscillations", "eigenmodes" or "normal modes" of a coupled pendulum. They are stationary motions insofar as the amplitude of the individual pendulums does not change over time. They can be demonstrated with the pendulums described in Experiment 17.12 by specifically displacing the bobs so that only one of the eigenmodes is excited.

In general, the motion of such a system of two coupled pendulums will be a superposition of the eigenmodes with different amplitudes.

$$\begin{aligned}x_1(t) &= A_1 \cos(\omega_1 t + \varphi_1) + A_2 \cos(\omega_2 t + \varphi_2) \\ x_2(t) &= A_1 \cos(\omega_1 t + \varphi_1) - A_2 \cos(\omega_2 t + \varphi_2).\end{aligned} \qquad (17.80)$$

We want to take a closer look at a special case. What happens if we displace one of the bobs at time $t = 0$ while the second one remains in its equilibrium position? We want to displace the first one to $x_1(0) = A$ and hold the second at $x_2(0) = 0$. We represent this by setting in Eq. 17.80 $\varphi_1 = \varphi_2 = 0$ and $A_1 = A_2 = A/2$. Then it is

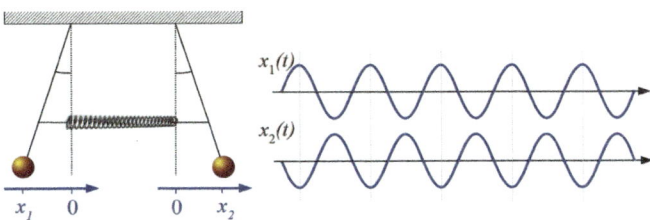

◘ **Fig. 17.15** Second eigenmode of a coupled pendulum

$$x_1(t) = \frac{A}{2}\cos(\omega_1 t) + \frac{A}{2}\cos(\omega_2 t)$$
$$x_2(t) = \frac{A}{2}\cos(\omega_1 t) - \frac{A}{2}\cos(\omega_2 t). \tag{17.81}$$

We consider the further motion of the system. With the angle sum identities $\cos x + \cos y = 2\cos\frac{x+y}{2}\cos\frac{x-y}{2}$ and $\cos x - \cos y = -2\sin\frac{x+y}{2}\sin\frac{x-y}{2}$ we get:

$$x_1(t) = A\cos\left(\frac{\omega_1+\omega_2}{2}t\right)\cos\left(\frac{\omega_1-\omega_2}{2}t\right)$$
$$x_2(t) = -A\sin\left(\frac{\omega_1+\omega_2}{2}t\right)\sin\left(\frac{\omega_1-\omega_2}{2}t\right) \tag{17.82}$$

and thus

$$x_1(t) = A\cos\left(\frac{\omega_1+\omega_2}{2}t\right)\cos\left(\frac{\omega_2-\omega_1}{2}t\right)$$
$$x_2(t) = A\sin\left(\frac{\omega_1+\omega_2}{2}t\right)\sin\left(\frac{\omega_2-\omega_1}{2}t\right). \tag{17.83}$$

The solutions each contain an oscillating term with a frequency that corresponds to the average of the two normal modes, representing the actual oscillation. The two bobs swing back and forth at this frequency. The second term oscillates much more slowly. In the case of weak coupling k', ω_1 and ω_2 are almost equal, and $\frac{\omega_2-\omega_1}{2}$ represents a much smaller frequency. This term generates a slow periodic increase and decrease of the amplitude, it modulates the amplitudes.

Please note that although the initial amplitude of the second pendulum is zero, it increases with the frequency $\frac{\omega_2-\omega_1}{2}$ and reaches the same maximum value as the amplitude of the first pendulum. The energy of the system, which was initially stored in the potential energy of the first pendulum, is gradually and completely transferred to the second pendulum and then back again. This slow increase and decrease of the amplitudes is called "beating". Our example is shown in ◘ Fig. 17.16.

17.4 · Coupled Oscillations

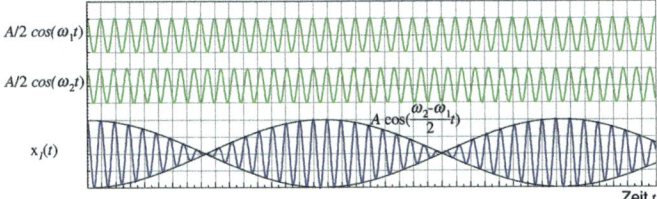

Fig. 17.16 Amplitude progression of a beating oscillation. The two green curves show the two frequencies ω_1 and $\omega_2 = 1.06\,\omega_1$, the blue curve shows the resulting beat with the envelope $\cos\left(\frac{\omega_2-\omega_1}{2}t\right)$.

Experiment 17.14: Beating

Beating can be demonstrated with two tuning forks. If you strike them at the same time, you should hear a clear tone with a constant amplitude, as they oscillate at the same frequency. But, we detune one of the two forks by attaching a small weight (in the picture on the far left prong). The frequencies are now slightly different. You hear a periodic increase and decrease of the amplitude (volume).

The effect is quite noticeable acoustically. It can be visualized with a microphone and an oscilloscope. On the screen of the oscilloscope, the rise and fall of the amplitude can be seen.

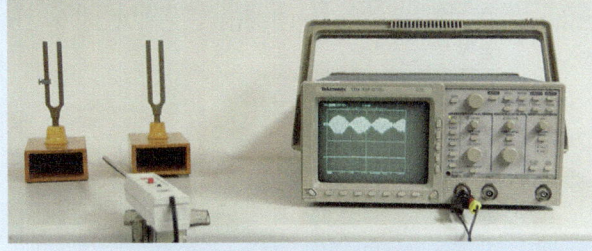

© RWTH Aachen, Physics Experiment Collection

Example 17.4: Double Pendulum

The double pendulum consists of two pendulums linked together, as can be seen in the figure. A church bell, for example, can be considered a double pendulum. The bell body represents the upper pendulum, with the clap-

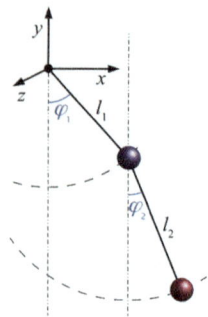

per suspended as the second pendulum. The double pendulum exhibits interesting phenomena. Before we get to that, we want to revisit the discussion of generalized coordinates from Example 17.1.

We have two bobs in this system, which we approximate as point masses. This gives us six degrees of freedom $(x_1, y_1, z_1, x_2, y_2, z_2)$. We assume again that the bobs are suspended in such a way that they can only move in the plane shown. This provides the first two constraints: $z_1 = 0$ and $z_2 = 0$. Two more constraints are necessary to maintain the lengths l_1 and l_2 to keep the suspensions constant. Imagine they were massless rods. Then the following must apply:

$$x_1^2 + y_1^2 = l_1^2$$
$$(x_2 - x_1)^2 + (y_2 - y_1)^2 = l_2^2.$$

From the original six degrees of freedom, we must therefore subtract four, leaving two degrees of freedom. We could solve the motion of the system by considering the constraining forces that enforce the conditions, but we want to switch directly to generalized coordinates. We do not want to limit ourselves to the small angle approximation, but even allow the bobs to make full rotations around their suspension (this is not possible with the church bell). It makes sense to use the two angles as generalized coordinates. The transformation to the Cartesian coordinates is:

$$x_1 = -l_1 \cos \varphi_1 \quad x_2 = -l_1 \cos \varphi_1 - l_2 \cos \varphi_2$$
$$y_1 = l_1 \sin \varphi_1 \quad y_2 = l_1 \sin \varphi_1 + l_2 \sin \varphi_2$$
$$z_1 = 0 \quad z_2 = 0.$$

Insert these quantities into the constraints and you will see that they are implicitly fulfilled. We would now have to set up the equations of motion in the generalized coordinates, which can be most easily realized with the help of the Lagrange formalism, which we do not want to introduce here. If you want to calculate the movements (numerically!), we provide you with the equations of motion:

$$\frac{d^2\varphi_1(t)}{dt^2} = -\frac{m_2}{m_1+m_2}\frac{l_2}{l_1}\left(\frac{d^2\varphi_2(t)}{dt^2}\cos(\varphi_1-\varphi_2)\right.$$
$$\left.+\left(\frac{d\varphi_2(t)}{dt}\right)^2\sin(\varphi_1-\varphi_2)\right)-\frac{g}{l_1}\sin\varphi_1$$

$$\frac{d^2\varphi_2(t)}{dt^2} = -\frac{l_1}{l_2}\left(\frac{d^2\varphi_1(t)}{dt^2}\cos(\varphi_1-\varphi_2)\right.$$
$$\left.+\left(\frac{d\varphi_1(t)}{dt}\right)^2\sin(\varphi_1-\varphi_2)\right)-\frac{g}{l_2}\sin\varphi_2.$$

This is a system of nonlinear coupled differential equations. The double pendulum is phenomenologically interesting in that it represents a simple system that exhibits chaotic behavior. As with any system, the motion that occurs is determined by the initial conditions $\varphi_1(0)$, $\varphi_2(0)$, $\frac{d\varphi_1}{dt}|_{t=0}$ and $\frac{d\varphi_2}{dt}|_{t=0}$. If you repeated the experiment with only minimally changed initial conditions, you would expect almost the same course of motion. This is the case with most systems, but not here. Even an infinitesimal change in the initial condition leads to a completely different course of motion after a short time. Such a system, in which an infinitesimal change in the initial conditions leads to a macroscopic change in motion, is called a chaotic system.

17.5 Standing Waves

In the previous section, we discussed coupled systems with two degrees of freedom. Now we want to extend these to more degrees of freedom. We start with the extension to three degrees of freedom. ◘ Fig. 17.17 shows such a system.

Instead of setting up and solving the differential equations as in Eq. 17.54, we try to find the eigenmodes of the system experimentally in Experiment 17.15. The calculation with more degrees of freedom shows nothing new, it just gets longer. We find three eigenmodes for a system with three degrees of freedom and four eigen-

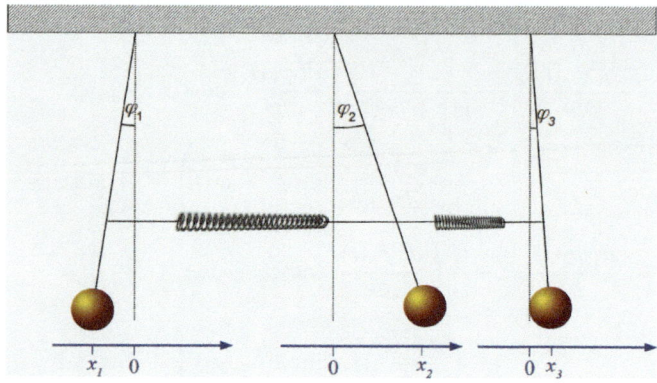

Fig. 17.17 System of three coupled pendulums

modes for the system with four degrees of freedom. This is no coincidence. In general, a system with N degrees of freedom also has N eigenmodes.

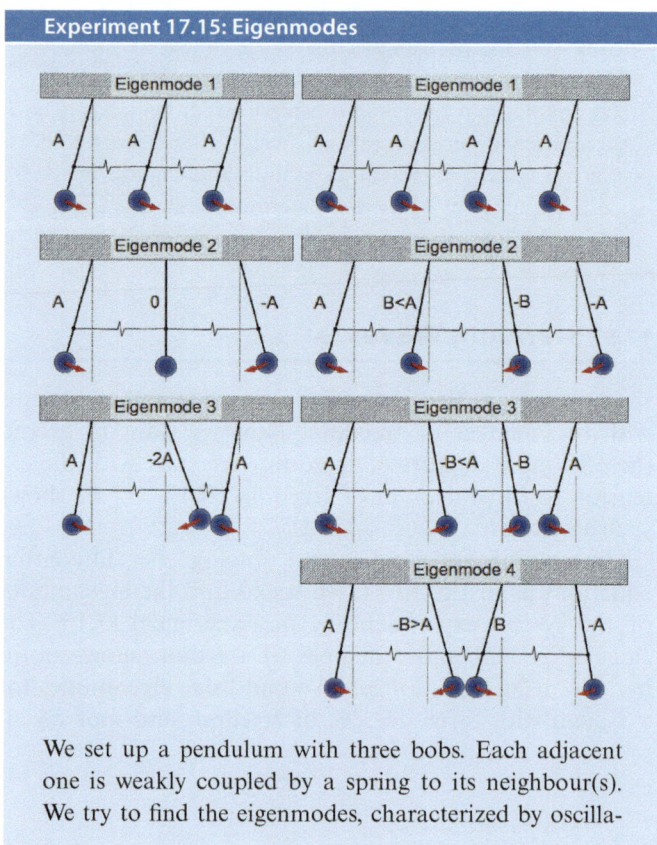

We set up a pendulum with three bobs. Each adjacent one is weakly coupled by a spring to its neighbour(s). We try to find the eigenmodes, characterized by oscilla-

17.5 · Standing Waves

tions of constant amplitudes of each individual pendulum. With a little finesse, we find the three modes that are shown in the sketches. The initial movement (red arrows) and the initial displacement of the bobs are indicated in the sketch. In the first eigenmode, all three pendulums oscillate in phase with the natural frequency of a single pendulum. The coupling springs remain relaxed. In the second eigenmode, the middle pendulum remains at rest. The two outer ones oscillate in opposite phases to each other with a frequency that is slightly higher than in the first eigenmode. In the third eigenmode, the two outer ones oscillate in phase against the middle one. The amplitude of the middle one is twice as high as that of the outer ones and the frequency is even slightly higher than in the second eigenmode.

The eigenmodes for four pendulums are also sketched. Try to find the ones for five pendulums!

Example 17.5: Calculation of Eigenmodes

In the picture, a 4-fold pendulum can be seen. The bobs move along the direction of propagation (longitudinal). The five springs all have the same strength k. Then the first bob from the left experiences a force $-k\,x_1$, from the spring to its left, if we assume that the force vanishes in the equilibrium position of the bob. The spring to its right excerts a force $k(x_2 - x_1)$ on the same bob.

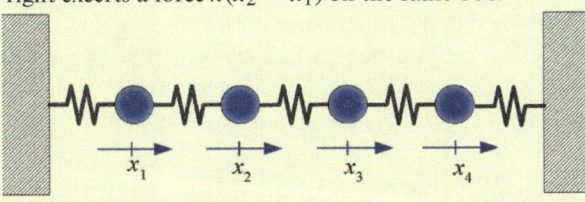

The equations of motion are:

$$m\frac{d^2 x_1(t)}{dt^2} = -k\,x_1(t) + k\,(x_2(t) - x_1(t))$$
$$m\frac{d^2 x_2(t)}{dt^2} = -k\,(x_2(t) - x_1(t)) + k\,(x_3(t) - x_2(t))$$
$$m\frac{d^2 x_3(t)}{dt^2} = -k\,(x_3(t) - x_2(t)) + k\,(x_4(t) - x_3(t))$$
$$m\frac{d^2 x_4(t)}{dt^2} = -k\,(x_4(t) - x_3(t)) - x_4(t)\,. \qquad (17.84)$$

Again, we choose a harmonic ansatz $x_i(t) = a_i \cos \omega t$, insert it and express the result in matrix notation ($k' = k/m$):

$$\begin{pmatrix} -2k' + \omega^2 & k' & 0 & 0 \\ k' & -2k' + \omega^2 & k' & 0 \\ 0 & k' & -2k' + \omega^2 & k' \\ 0 & 0 & k' & -2k' + \omega^2 \end{pmatrix} \cdot \begin{pmatrix} a_1 \\ a_2 \\ a_3 \\ a_4 \end{pmatrix} = \begin{pmatrix} 0 \\ 0 \\ 0 \\ 0 \end{pmatrix}. \quad (17.85)$$

To allow for a solution, the determinant of the matrix must vanish, which leads to the following condition:

$$\omega^8 - 8k'\omega^6 + 21k'^2\omega^4 - 20k'^3\omega^2 + 5k'^4 = 0. \quad (17.86)$$

This quartic equation has four solutions:

$$\omega_1^2 = \frac{(3-\sqrt{5})}{2} k' \approx 0.382 k'$$

$$\omega_2^2 = \frac{(5-\sqrt{5})}{2} k' \approx 1.382 k'$$

$$\omega_3^2 = \frac{(3+\sqrt{5})}{2} k' \approx 2.618 k'$$

$$\omega_4^2 = \frac{(5+\sqrt{5})}{2} k' \approx 3.618 k'. \quad (17.87)$$

Now that we have found the frequencies that solve the system of differential equations, we can determine the complete solutions by substituting the frequencies into the differential equations. This gives us the amplitudes of the four bobs, which we represent as a vector (a_1, a_2, a_3, a_4). The amplitude vector A_1 corresponds to the frequency ω_1, A_2 orresponds to ω_2, etc.:

$$A_1 = \left(+1, +\frac{\sqrt{5}+1}{2}, +\frac{\sqrt{5}+1}{2}, +1 \right) a$$

$$A_2 = \left(+1, +\frac{\sqrt{5}-1}{2}, -\frac{\sqrt{5}-1}{2}, -1 \right) a$$

$$A_3 = \left(+1, -\frac{\sqrt{5}-1}{2}, -\frac{\sqrt{5}-1}{2}, +1 \right) a \quad (17.88)$$

$$A_4 = \left(+1, -\frac{\sqrt{5}+1}{2}, +\frac{\sqrt{5}+1}{2}, -1 \right) a.$$

17.5 · Standing Waves

If the number of degrees of freedom increases, the eigenmodes become more and more complex. They no longer have the central importance in systems with many degrees of freedom as they do in simple systems, as other interesting phenomena occur. We want to investigate some of them. For example, if you deflect a bob at the end of the system, the excitation travels from one degree of freedom to the next through the system. The excitation spreads in a form referred to as a "wave". We demonstrate this behaviour with the apparatus in Experiment 17.16 and Experiment 17.17.

Experiment 17.16: Pendulum Chain

On the depicted frame, 13 bobs are mounted with thin rods. They can move in the direction of their neighbor, but also across it. Each ball is connected to its nearest neighbors by a spring. If we displace the first bob and let it go, you can observe how the deflection is transmitted from one bob to the next. A simple wave runs through the system. It doesn't matter whether we displace the first bob in the direction of its neighbor (longitudinal) or across it (transversal), a wave is created in both cases.

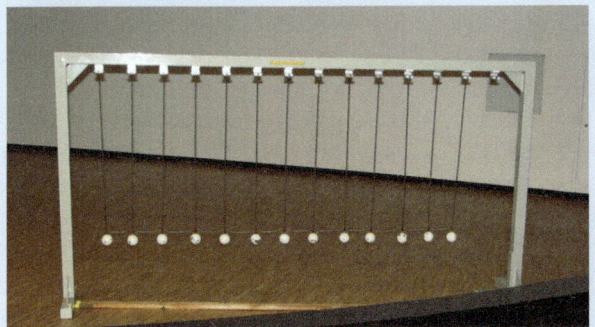

© RWTH Aachen, Physics Experiment Collection

> **Experiment 17.17: Wave Machine**
>
>
>
> © RWTH Aachen, Physics Experiment Collection
>
> This system consists of rods that are pivotally mounted in the middle on a horizontal axis. At both ends of the rods, a weight is attached, which is marked yellow for better visibility. Through an elastic attachment, the rods take a horizontal equilibrium position, around which they can perform torsional vibrations. They are also coupled to their immediate neighbors by a spring mechanism. Different excitations can now be studied. The photo shows a continuous harmonic excitation leading to a wave.

From an energy perspective, these systems can be treated uniformly. In all of them, each individual element of the system has the ability to absorb, store, and release both potential energy and kinetic energy. ◻ Figure 17.18 shows a pendulum chain, similar to Experiment 17.16, which is displaced transversally. Each element of the system consists of a mass and the springs to its neighbor(s). The springs store potential energy, the masses kinetic energy. The technical implementation of the elements varies from system to system, but the two forms of energy occur in all (mechanical) systems.

In our example, the elements that store potential energy and those that store kinetic energy were separated. This does not have to be the case. Consider, for example, a rubber band (Experiment 17.18). Potential energy

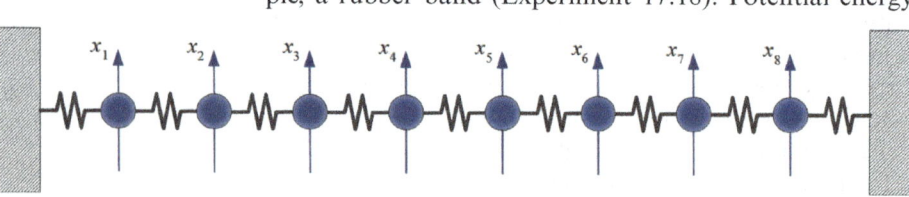

◻ **Fig. 17.18** Spring chain

is stored in the elastic stretches of the band, while kinetic energy is stored in the movement of the band. Here, both functions coincide. The example of the rubber band differs in another aspect from the pendulum chain. The pendulum chain consists of individual discrete elements (the bobs), whereas the rubber band is a continuous system. For mathematical treatment, it is divided into small discrete elements and eventually the size of the elements is reduced to zero in a limiting process. This results in the description of a continuous system. Assuming a fixed length of the rubber band, the number of elements must approach infinity in the limit, and thus also the number of degrees of freedom and indeed, a continuous system has an infinite number of degrees of freedom.

> **Experiment 17.18: Rope Waves in the Rubber Band**
>
> A rubber band is attached on one side to a rod (on the right in the picture), while the other end is held by the experimenter. He stretches the band and then displaces it with a kick. The photo, taken with a high-speed camera at 500 frames per second, shows the result of an upward displacement. You can see a step in the band that moves to the right, where it will hit the suspension and be reflected.
>
>
>
> © RWTH Aachen, Physics Experiment Collection

We distinguish two directions of displacement of oscillating systems. We speak of "longitudinal deflections" or "longitudinal waves" when the deflection has the same direction as the propagation of the excitation. If the deflection is transverse to the direction of propagation, we speak of "transverse waves". Some systems can only be excited longitudinally, others only transversely, and some both longitudinally and transversely. In Experiment 17.12, the suspensions of the coupled pendulums were designed so that only longitudinal displacements are possible. With the pendulum chain in Experiment 17.16 we were able to demonstrate both types of excitation, the rubber band in Experiment 17.18 can only oscillate transversely, etc.

In finite systems, the excitation will eventually reach the end of the system. For example, the displacement we saw in Experiment 17.18 hits the end of the rubber band at its attachment. There, the energy contained in the excitation is not lost, but it is reflected and runs back. How the reflection proceeds depends on the technical realization of the termination, namely whether the last elements of the system can be deflected themselves or whether they are restricted by the end of the system. With the rubber band, such a restriction exists. Due to the attachment, the last piece of the band can no longer move freely. Basically, two limit cases are distinguished, the reflection at the fixed end, where no displacement at the end of the system is possible, and the reflection at the loose end, where full mobility is given up to the end. Both cases can be shown on the wave machine in Experiment 17.17. They are schematically represented in ◘ Figs. 17.19 and 17.20.

Let's take a closer look at ◘ Fig. 17.19. The excitation runs from the left towards the wall, where the medium (the rubber band) is firmly attached (green curve). In our minds, we let the excitation run into the wall. Exactly the part that seems to disappear into the wall runs as a reflection in the opposite direction away from the wall (orange curve). However, the displacement is inverted, referred to

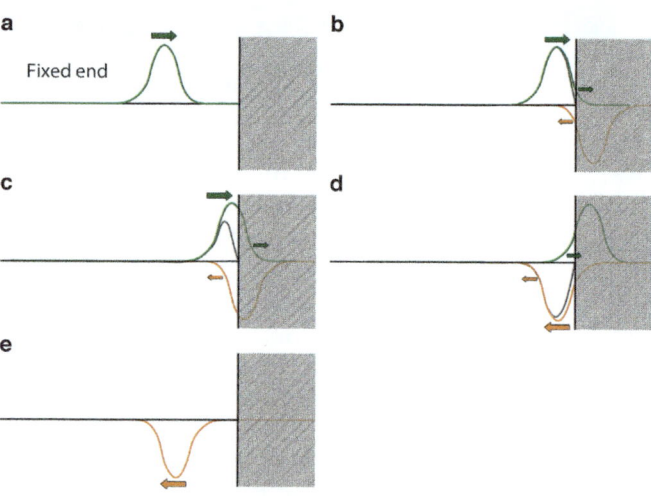

◘ **Fig. 17.19** Reflection at the fixed end

17.5 · Standing Waves

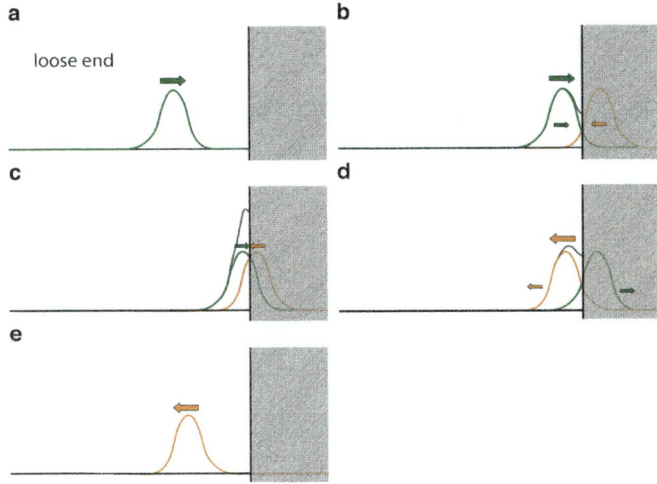

Fig. 17.20 Reflection at the loose end

as a "phase jump of 180°". The deflection that is actually observed is the superposition of the incoming and outgoing excitation (gray curve). Due to the phase jump, the incoming and outgoing excitation always add up to zero at the wall, so that this element does not move.

In the case of reflection at the loose end, on the other hand, there is no phase jump. As a result, the incoming and outgoing excitation add up constructively and there is a maximum deflection at the end of the medium.

Entirely new phenomena can be observed when the medium of the vibrations is not only limited at one end, but at both ends. Think of the string of a musical instrument. You create an excitation by striking or plucking the string. The excitation initially runs against one end, is reflected there and then runs in the opposite direction. It reaches the other end, is reflected there again and so on. You get many incoming and outgoing waves that overlap (superpose). A noticeable vibration only occurs when wave crests meet wave crests and wave troughs meet wave troughs. Such a constructive superposition only occurs at certain frequencies. Then standing waves form, whose associated wavelengths are in a geometric ratio to the length of the string. We will now take a closer look at these. The string of a musical instrument can serve as an illustration. However, standing waves occur in many other situations. But first, we want to show standing waves in an experiment (Experiment 17.19), with the wave machine already known from Experiment 17.17.

> **Experiment 17.19: Standing Waves with the Wave Machine**
>
> The wave machine from Experiment 17.17 can be periodically excited on the left side with a tunable frequency. The excitation runs to the right towards the end and is reflected there. The elements of the wave machine, including the last one, are freely movable, creating a loose end, which can be reconfigured into a fixed end by clamping the final element. The reflected excitation overlaps with the now continuously incoming excitation. This superposition generates a clearly visible deflection only at certain frequencies. If we slowly tune through the frequenicies, we will see only irregular trembling of the elements at most frequencies. But at some specific frequencies, patterns of a stable oscillation on the wave machine emerge, representing standing waves. The amplitude of the oscillations of the individual elements varies periodically along the axis. In ◘ Fig. 17.21, some examples of these patterns can be seen, which are obtained for either a loose or fixed end. The wavelength λ of these standing waves is linked to the exciting frequency f via $c = \lambda f$, where c is the speed of propagation of the excitation on the wave machine. The images in ◘ Fig. 17.21 show the deflections of the elements at eight equidistant points in time within half an oscillation period. Points at which the elements of the system do not move are clearly recognizable. They are called the nodes. The distance between two nodes corresponds to half a wavelength.

> **Experiment 17.20: Standing Waves on Rubber Rope**
>
> On the rubber rope, which you already know from Experiment 17.18, a standing wave can be generated by periodic excitation. In ◘ Fig. 17.22 you can see the first overtone with a node in the middle of the image and a fixed end on the right edge of the image.

17.5 · Standing Waves

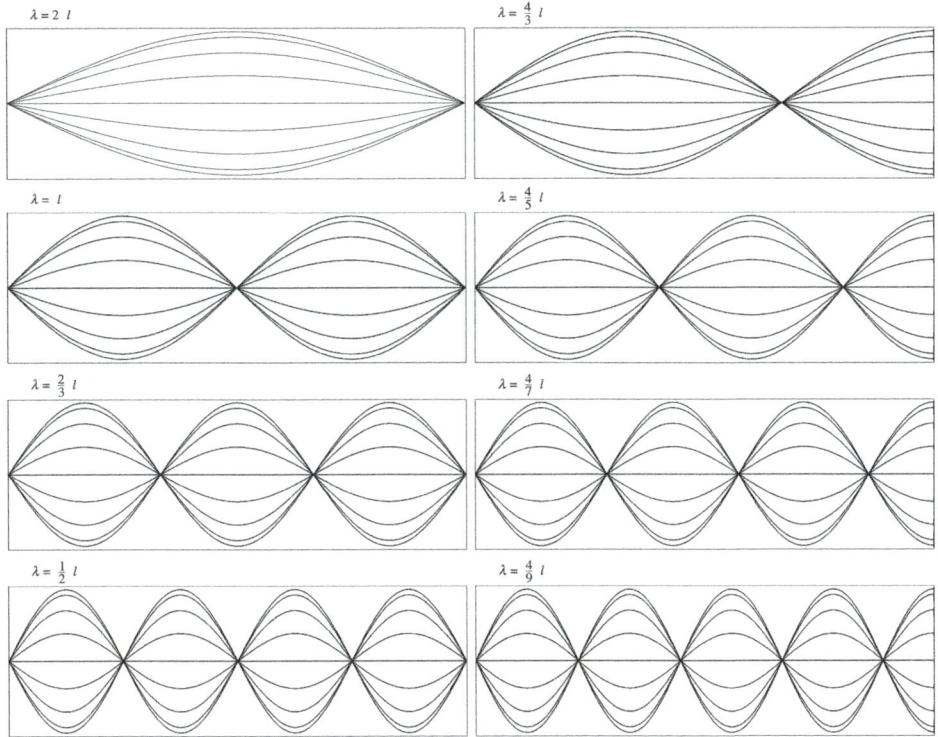

Fig. 17.21 Displacement of the elements at eight equidistant points in time within half an oscillation period

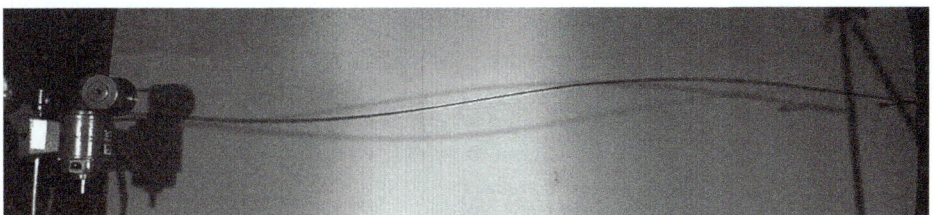

Fig. 17.22 Standing wave on a rubber rope. © RWTH Aachen, Physics Experiment Collection

Experiment 17.21: Kundt's Tube

We partially filled a glass tube, closed at the right end, with fine sand. A dog whistle - a tunable Galton's whistle - is attached to the left. At a suitable frequency, standing waves are created in the tube. At the nodes, the air is at rest and the sand remains in place, while at the vibrating antinodes, the sand is stirred up by the high amplitudes. If it falls accidentally into the area of a node, it stays there, if it falls into the area of an antinode, it is stirred up again, until it eventually reaches the quiet area of a node. Thus, the area of the antinodes is cleared of sand and the sand accumulates at the nodes.

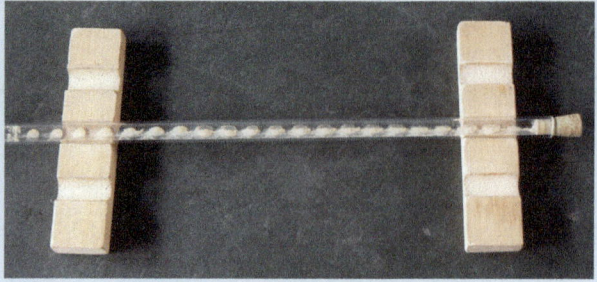

© RWTH Aachen, Physics Experiment Collection

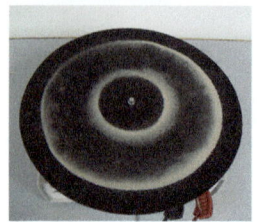

© RWTH Aachen, Physics Experiment Collection

Experiment 17.22: Chladni Patterns

Chladni patterns are vibration patterns of a plate. They are named after Ernst Florens Friedrich Chladni, who described them in 1787 in his treatise *Discoveries on the Theory of Sound*. They represent two-dimensional standing waves in the plate.

We use a round metal plate. It can be excited with a violin bow, but it works better with a vibration generator, whose frequency can be tuned. The plate is connected in the middle with a screw to the exciter of the vibration generator. The plate is sprinkled with fine sand to make the waves visible. If we tune the frequency, complicated but regular patterns appear in the sand, which reflect the vibrations of the plate. The sand follows the distribution of the vibration nodes due to the mechanism that we already know from Kundt's tube (Experiment 17.21).

© RWTH Aachen, Physics Experiment Collection

© RWTH Aachen, Physics Experiment Collection

17.5 · Standing Waves

Experiment 17.23: Vibrations of a Guitar String

© RWTH Aachen, Physics Experiment Collection

With a stroboscope, we can make the vibrations of a guitar visible. It illuminates the string at fixed intervals. If the frequency of the stroboscope is synchronized with that of the string, we can see the string always in the same phase of the vibration and can recognize the nodes and antinodes. In the picture, you can see the guitar and the lamps of the stroboscope at the bottom right. The lowest string (E-string, here at the top of the photo) is struck. The stroboscope illuminates the maximum deflection downwards. For comparison, the rest position of the string is indicated with a red dashed line.

❓ Problems

1. An unknown mass m_0 is hanging on a spring. When a piece of mass $m_1 = 50$ g is added to this mass, the oscillation period increases by 20%. Calculate the original mass m_0!
2. Unfortunately, on a hydrometer the scale for determining the liquid density is not readable anymore. Therefore, it is submerged for a short time deeply into the liquid and the duration of the resulting oscillations is determined. The hydrometer has a diameter of 1.1 cm at the liquid surface and a mass of 0.12 kg. Calculate the density of the liquid in which the hydrometer performs oscillations with a duration of 2.3 s!

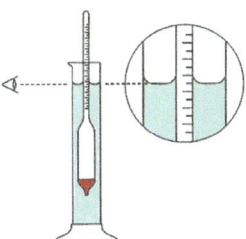

3. Imagine a tube from one side of the Earth to the other, passing through the Earth's center. In this tube, gravity increases linearly from the Earth's center to the Earth's surface. How long would a body need in free fall from one side of the Earth to the other (radius of the Earth $R_E = 6.4 \cdot 10^6$ m)? Compare the result with the time the body would need on a fictitious orbit directly above the Earth's surface.

4. In a U-shaped tube, a column of liquid of length l oscillates back and forth. Calculate the oscillation frequency!

5. A rod of length $2l$, which is assumed to be massless, is suspended to rotate around its center. At its ends the point masses m_1 and m_2 are attached. Determine the oscillation frequency of this physical pendulum at small deflections around the equilibrium position!

6. A solid cylinder with radius R and mass m is rotatably mounted about a horizontal axis at a distance s parallel to the symmetry axis of the cylinder. What is the reduced length l_r of this physical pendulum? What is the oscillation period T for small deflections? At what value of s is T minimal?

7. A torsion pendulum consists of a round plate suspended from a wire of length 1 m and thickness 1 mm. The oscillation period of the pendulum is 1 s. If a cylindrical disc of mass 0.5 kg and diameter 15 cm is centered on the plate, the oscillation period increases to 5.4 s. What is the torsional constant D^* of the suspension? Use the result to calculate the shear modulus of the wire material!

8. A pendulum of mass m and length l should be used for measuring the acceleration due to gravity. In case the oscillation amplitude φ_0 is not negligible, an anharmonic component arises in the oscillation, which leads to the following correction of g with respect to the pure mathematical pendulum

$$g = \frac{4\pi^2 l}{T^2}\left(1 + \frac{\varphi_0^2}{8}\right)$$

a) Determine the correction of g, when the pendulum's point mass m is changed to a sphere with radius R!

b) Determine the correction of g, when you take into account the mass of the pendulum string m_{string}!

17.5 · Standing Waves

c) Determine the correction of g, which results from considering the buoyancy of the sphere with density ρ_K in the surrounding air of density ρ_{air}!

d) Determine the correction of g, when you take into account the damping constant γ of the oscillation due to friction effects (air resistance of the sphere and bearing friction)!

e) Calculate the contribution of each of the above corrections for a pendulum, in which a sphere with a radius of 2 cm and a mass of 0.38 kg on a string of length 1 m and mass 5 g performs oscillations after an initial deflection of 10 cm. The oscillation amplitude decreases to half after 100 oscillations. Compare the corrections with each other and draw conclusions with regard to a most precise measurement of g!

9. A pendulum body of mass $m = 0.2$ kg is attached to a spring with the spring constant $k = 20\,\text{N/m}$. The pendulum body is immersed in an oil bath and the spring is set into motion by a sinusoidal periodic force with the maximum value $F = 3\,\text{N}$. When varying the excitation frequency, the maximum oscillation amplitude observed is $x_r = 0.2$ m. What is the damping constant γ? What is the resonance frequency ω_r? What is the phase shift φ_r between excitation and oscillation at the resonance?

10. Two identical pendulum bodies are fixed at the end of two identical rods. The two pendulums are coupled together by a spring. If the two pendulums are set in anti-phase oscillations, they perform a total of 612 oscillations within 10 min. In the case of in-phase oscillation, one counts 625 oscillations. Now the system is brought to rest and then one of the two pendulums is set in motion. How long does it take until the second pendulum reaches the maximum oscillation amplitude and the first pendulum is at rest again?

11. On the occasion of the completion of the Cologne Cathedral, the largest church in Germany, Emperor Wilhelm I donated in 1872 a total of 22 captured French cannons to manufacture the largest free-hanging bell to that date. When the Emperor's bell was to be rung for the first time, the bell, despite the efforts of two dozen "Deutz cuirassiers" who were assigned to start the swinging, made no sound at all.

a) A bell forms a double pendulum, consisting of two physical pendulums, the bell body and the

clapper (see figure). Create a substitute sketch in which you replace the bell body and clapper with mathematical pendulums with the respective reduced pendulum length. Let l be the reduced pendulum length of the bell body, l_1 that of the clapper and a the distance between the pivot points of the bell body and clapper. Justify without calculation using a physical argument why the bell body and clapper oscillate synchronously, if $a = l - l_1$ applies.

b) At the initiative of the secondary school teacher W. Veltmann from the town Düren near Cologne, investigations were carried out on the bell[2] and the oscillation period of the pure bell body was determined to be 3.64 s, and that of the clapper to be 3.25 s. For the distance a between the pivot points, a value of 66.7 cm was measured. Now justify why the bell did not ring, and consider how a remedy could have been provided.

After changes had been made to the suspension of the bell and clapper, the Emperor's Bell could be reasonably rung. However, it was still referred to as the "Mute of Cologne" by the population. In June 1918, it was melted down again, this time for German cannons.

---

[2] W. Veltmann, On the movement of a bell; Polytechnic Journal 1876, Volume 220 (pp. 481–495).

Waves

Contents

18.1 Harmonic Waves – 534

18.2 Wave Equation – 542

18.3 Wave Packets – 544

18.4 Energy Density and Energy Transport – 551

18.5 Reflection and Interference – 553

© The Author(s), under exclusive license to Springer-Verlag GmbH, DE, part of Springer Nature 2025
S. Roth and A. Stahl, *Mechanics*,
https://doi.org/10.1007/978-3-662-68079-7_18

18.1 Harmonic Waves

In the previous chapter, you were introduced to initial wave phenomena. You saw waves propagating in one direction on a medium, such as on a rubber band or on the string of a guitar. This is referred to as linear or one-dimensional waves. Waves can also exist in two or three spatial dimensions. In Experiment 18.1, we show water waves that propagate in two spatial directions. An example of a three-dimensional wave would be the propagation of sound in air. In this chapter, we will systematically describe the wave phenomena.

Experiment 18.1: Water Waves

Waves are generated in a shallow water tray by an exciter that moves up and down periodically. The exciter can be point-shaped, as shown in the figure, or line-shaped, generating either circular waves or plane waves. The waves are optically projected onto a screen. The water waves are illuminated from above. The water tray has a transparent bottom, under which there is a tilted mirror. This deflects the light onto a matte screen, where the wave pattern can then be observed from the outside.

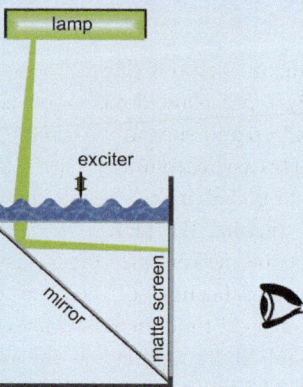

When light hits a wave crest, it acts like a focusing lens. It concentrates the light onto the matte screen, while the wave troughs defocus it. As a result, the wave crests appear as bright and the wave troughs as dark structures on the matte screen. The exciter can be adjusted in frequency. The lamp is operated at the same frequency as a stroboscope to facilitate observation. The first photograph shows a circular wave at 20 Hz.

18.1 · Harmonic Waves

The frequency and wavelength of the waves are linked via their speed of propagation[1] c : $c = \lambda f$. The speed, in turn, depends on the depth of the water in the basin. This can be seen nicely from the plane wave in the second photograph. In the middle, a long plexiglass plate was placed at the bottom of the basin, thus reducing the water depth and hence the speed. This reduces the wavelength, and the wave crests move closer together in this area.

© RWTH Aachen, Physics Experiment Collection

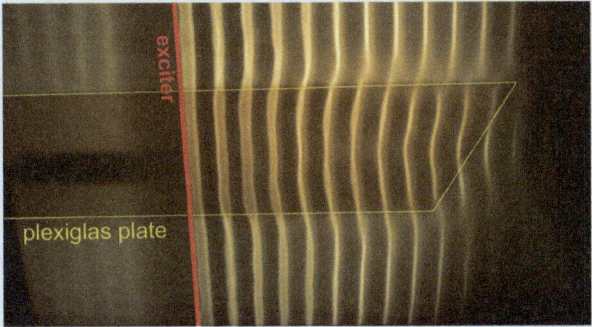

© RWTH Aachen, Physics Experiment Collection

When a wave propagates, oscillations occur in two different types (see ◻ Fig. 18.1). If we create a snapshot of the wave, we will see that the displacement changes periodically along the direction of propagation. In this chapter, we want to assume again that all oscillations are harmonic. Then we can describe the displacement at a fixed point in time t_i by

---

1   As we will see later, speed refers here to the phase velocity.

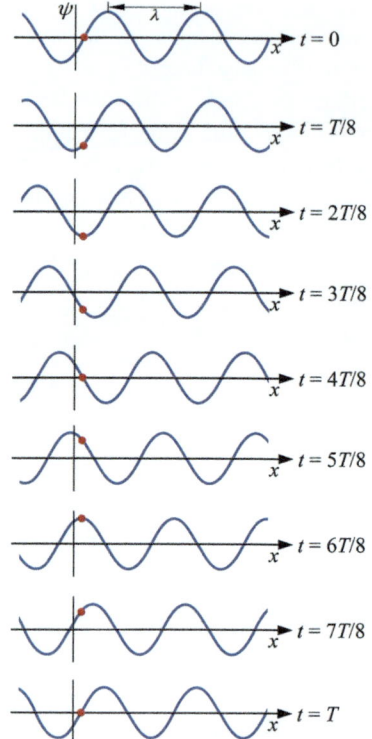

Fig. 18.1 Snapshots of a 1-dimensional wave at different times. The times are given as fractions of the oscillation period

$$\psi_{t_i}(x) = \psi_0 \sin\left(2\pi \frac{x}{\lambda}\right) \tag{18.1}$$

with the wavelength λ, which indicates the spatial distance between two wave crests, and the amplitude ψ_0 of the wave.

However, if we observe a fixed point in the wave field (location x_i) over some period of time, we observe a temporal rise and fall of the point, as indicated by the red point in ◘ Fig. 18.1. If this movement is also harmonic, then we can write

$$\psi_{x_i}(t) = \psi_0 \sin\left(2\pi \frac{t}{T}\right), \tag{18.2}$$

with the temporal oscillation period T. We merge the two oscillations into:

18.1 · Harmonic Waves

$$\psi(x,t) = \psi_0 \sin\left(2\pi \frac{x}{\lambda} \mp 2\pi \frac{t}{T}\right). \tag{18.3}$$

The upper sign describes a wave that moves in a positive x direction, the lower one a wave in negative x direction.

Let's look again at ◻ Fig. 18.1. The red marked point goes through exactly one oscillation period in the time T. In the same time, the position of maximum displacement marked by a vertical line has shifted to the right by one wavelength λ. The oscillation period T indicates the temporal period with which a fixed point in the wave field oscillates. It is associated with the frequency of excitation via

$$\omega = 2\pi f = 2\pi \frac{1}{T}. \tag{18.4}$$

The wavelength λ indicates the spatial period of the wave. It is associated with the wave number k, which indicates the number of wave trains (multiplied by 2π) per unit length in the direction of propagation:

$$k = 2\pi \frac{1}{\lambda}. \tag{18.5}$$

With these quantities, we can rewrite a one-dimensional wave as:

$$\psi(x,t) = \psi_0 \sin(kx \mp \omega t). \tag{18.6}$$

Experiment 18.2: Water Waves 2

With this experiment, we want to focus on an important aspect of wave propagation. Waves usually propagate in a medium.[2] The medium moves locally, but it is by no means transported in the direction of propagation. The sketch shows this for a water wave. The black dots on the dashed line represent water molecules in the equilibrium state (without wave). When a wave passes through, they move on an approximate circular path around their equilibrium position, as indicated by the blue dots in the sketch and the red arrows. One revolution on the circle corresponds to the passage of a wave crest over a wave trough to the next wave crest (one pe-

---

[2] You will also get to know waves without a medium, e.g. light waves or gravitational waves.

riod). Once the wave has passed, the radius of the circular path decreases, and in the end, the molecules are (approximately) back in the same place as at the beginning. They are *not* carried along in the direction of propagation by the wave.

This can be demonstrated with the wave tray from Experiment 18.1 by placing small styrofoam balls in the water. They float on the surface. If a wave is excited, the styrofoam balls tremble a little back and forth, but essentially stay in their place.

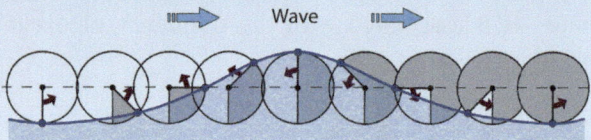

Example 18.1: Sound Waves

Sound waves are periodic compressions and decompressions of air. The sketch shows the excitation of a sound wave by a speaker. The diaphragm of the speaker, which moves with the frequency of the excitation, compresses the air in front of the speaker and then decompresses it again. This excitation is then passed on from layer of air to layer of air. The vertical lines in the sketch symbolize layers of air that would be equidistant in their resting state.

Example 18.2: Gravitational Waves

Not every wave requires a medium for propagation. Gravitational waves, which were first described by Albert Einstein in his General Theory of Relativity[3], have no material basis at all. They are periodic shortenings and lengthenings of space. The image shows the

---

3 Gravitational waves were first detected in 2015 by the LIGO experiment.

18.1 · Harmonic Waves

simulation of a gravitational wave as it should be radiated from a binary system of two stars.

Example 18.3: Sand Waves

Waves exist in various areas of nature. In the picture, wind has created waves in the sand.

© NASA

Waves can propagate in one, two, or three dimensions. One-dimensional waves include, for example, the rope waves we learned about in Experiment 17.18. We have seen two-dimensional waves that spread out in a plane as water waves in the wave tray (Experiment 18.1). These waves are further distinguished by the shape of the wavefront. The two most important are the circular wave and the plane wave. They are shown in ◘ Fig. 18.2. The blue lines correspond to the wave crests.

The displacement of a plane wave that propagates in x- or $-x$-direction, can be written as

$$\psi(\vec{r},t) = \psi_0 \sin(kx \mp \omega t), \qquad (18.7)$$

where $\vec{r} = (x, y)$ is a two-dimensional position vector. For a plane wave propagating in any direction, we introduce the wave vector $\vec{k}$. It has the magnitude of the wave number $\left|\vec{k}\right| = 2\pi \frac{1}{\lambda}$ and points in the direction of propagation. Then the wave is described by

$$\psi(\vec{r},t) = \psi_0 \sin\left(\vec{k} \cdot \vec{r} \mp \omega t\right). \qquad (18.8)$$

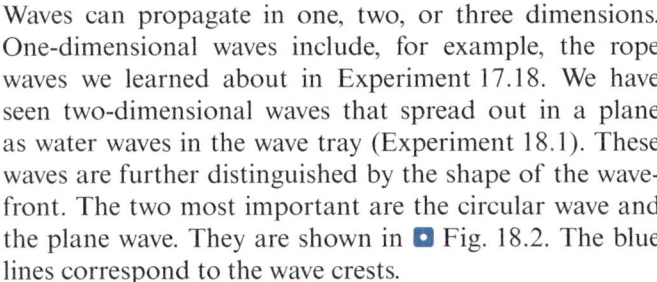

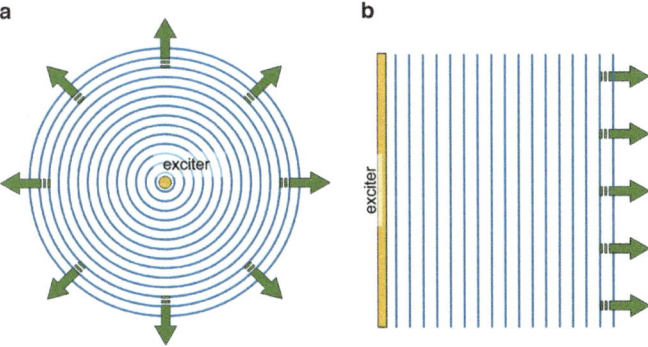

◘ **Fig. 18.2** The wave field of a circular wave (**a**) and a plane wave (**b**) in two dimensions

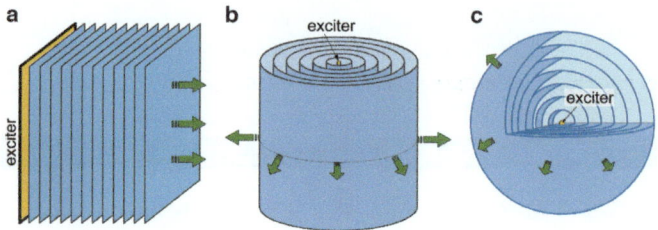

Fig. 18.3 The wave field of a plane wave (**a**), a cylindrical wave (**b**) and a spherical wave (**c**) in three dimensions

A circular wave, which spreads starting from the origin of the coordinate system, is described by the following relation:

$$\psi(\vec{r}, t) = \frac{\psi_0}{r} \sin(kr - \omega t). \tag{18.9}$$

When considering waves that spread in three dimensions in space, there are a multitude of waveforms. The most important ones are shown in ◘ Fig. 18.3. These are the plane wave, the cylindrical wave and the spherical wave. Equations 18.8 and 18.9 also apply to three-dimensional waves. The position and wave vectors are now three-dimensional.

Already in ▶ Sect. 17.5 we pointed out the different directions in which oscillating systems can be displaced. In the case of waves, we refer to these directions as the "polarization" of the waves. If the displacement occurs in the direction of propagation, we call it a "longitudinal polarization". This is the case, for example, with sound waves (Example 18.1). The air molecules move back and forth in the direction of the wave. In the case of water waves, however, the movement of the elements is perpendicular to the direction of the wave. This is referred to as "transverse polarization". The sand waves from Example 18.3 are also transverse. Please note that there are two possible directions of transverse polarization in three-diemnsional space. If a wave propagates in the z direction, for example, a transverse wave can displace either in the x-direction or in the y-direction. This is referred to as the two "polarization states of the transverse wave". Mathematically, the polarization directions can be captured by specifying the amplitude of the wave as a vector pointing in the direction of the deflection.

18.1 · Harmonic Waves

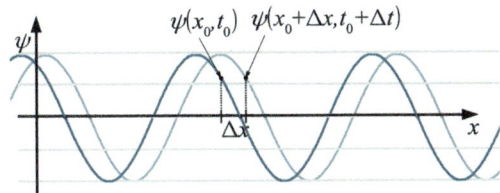

◘ **Fig. 18.4** For the determination of the phase velocity

Finally, in this introduction to waves, we want to determine their velocity. To do this, we consider a point on the wave with a fixed phase and track it over time. In ◘ Fig. 18.4, we have indicated such a point with a phase ϕ slightly beyond the maximum. Let's assume that at time t_0 this phase has reached the location x_0. The wave moves to the right. At a later time $t_0 + \Delta t$, the position with phase ϕ has reached the location $x_0 + \Delta x$. The wave function is

$$\psi(x,t) = \psi_0 \sin\left(2\pi \frac{x}{\lambda} - 2\pi \frac{t}{T}\right). \qquad (18.10)$$

An identical phase at (x_0, t_0) and $(x_0 + \Delta x, t_0 + \Delta t)$ implies

$$2\pi \frac{x_0}{\lambda} - 2\pi \frac{t_0}{T} = 2\pi \frac{x_0 + \Delta x}{\lambda} - 2\pi \frac{t_0 + \Delta t}{T},$$

$$0 = 2\pi \frac{\Delta x}{\lambda} - 2\pi \frac{\Delta t}{T},$$

$$\frac{\Delta t}{T} = \frac{\Delta x}{\lambda},$$

$$v_{ph} = \frac{\Delta x}{\Delta t} = \frac{\lambda}{T} = \lambda f = \frac{\omega}{k}. \qquad (18.11)$$

There are several ways to assign a velocity to a wave. This is the quantity that indicates the velocity at which a point with a fixed phase propagates. Therefore, this quantity is called the "phase velocity of the wave". We will get to know other definitions of wave velocities that have different meanings.

18.2 Wave Equation

Now, we have learned about the relationship between frequency and wavelength of a wave: $v_{Ph} = \lambda f$. The phase velocity in a specific case depends on the medium that transports the wave. We want to calculate the phase velocity in a simple example, the example of a wave on a string of a musical instrument (Experiment 17.23).

To calculate the oscillations, we need to determine the restoring force. It originates from the tensile force $\vec{F}$ with which the string is strained (see ◘ Fig. 18.5). We make an approximation to determine this force: We assume that the deflection of the string is small compared to the wavelength. Then the mass elements of the string move exclusively transverse to the direction of the string. We choose our coordinate system so that the string is strained in the z-direction. The displacement is ψ. We consider a piece of the string of length Δz. It has the mass $m = \mu \Delta z$, where μ is the mass per unit length of the string. At the beginning and end of the mass element, the tensile forces $\vec{F}(z)$ and $\vec{F}(z + \Delta z)$ act. The components in the x-direction correspond to the restoring force. These are

$$F_x(z) = \sin(\phi(z))(-|\vec{F}(z)|)$$
$$F_x(z + \Delta z) = \sin(\phi(z + \Delta z))|\vec{F}(z + \Delta z)|, \qquad (18.12)$$

where the minus sign in the first equation takes into account that the tensile force at z acts to the left. So we have

$$F_R = [\sin(\phi(z + \Delta z)) - \sin(\phi(z))]F. \qquad (18.13)$$

For small deflections, $\sin \phi \approx \tan \phi$ applies, which can be represented by the derivative of the deflection:

$$F_R \approx \left[\frac{\partial \psi(z + \Delta z)}{\partial z} - \frac{\partial \psi(z)}{\partial z}\right] F. \qquad (18.14)$$

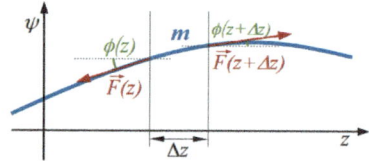

◘ **Fig. 18.5** On the calculation of wave propagation on a string

18.2 · Wave Equation

We now insert this into Newton's second axiom:

$$\left[\frac{\partial \psi(z+\Delta z, t)}{\partial z} - \frac{\partial \psi(z,t)}{\partial z}\right] F = \mu \Delta z \frac{\partial^2 \psi(z,t)}{\partial t^2}. \quad (18.15)$$

Finally, we let the mass element become infinitesimally small. Then we have:

$$\lim_{\Delta z \to 0} \frac{\left[\frac{\partial \psi(z+\Delta z,t)}{\partial z} - \frac{\partial \psi(z,t)}{\partial z}\right]}{\Delta z} F = \frac{\partial^2 \psi(z,t)}{\partial z^2} F = \mu \frac{\partial^2 \psi(z,t)}{\partial t^2} \quad (18.16)$$

or

$$\frac{\partial^2 \psi(z,t)}{\partial t^2} = \frac{F}{\mu}\frac{\partial^2 \psi(z,t)}{\partial z^2} = v_{Ph}^2 \frac{\partial^2 \psi(z,t)}{\partial z^2}. \quad (18.17)$$

This is called the "wave equation". It describes the motion of the elements of the medium. It is a second-order differential equation. It relates the second derivative of the displacement with respect to time to the second derivative with respect to position. This form of the wave equation is not a special case for the string wave, it applies to all harmonic waves.

In ▶ Sect. 18.1 we had already provided an equation that described the propagation of a harmonic wave, without deriving it from the basic laws of mechanics (there, however, with x as the direction of propagation). We had stated (Eq. 18.6):

$$\psi(z,t) = \psi_0 \sin(kz - \omega t + \phi_0). \quad (18.18)$$

Now we can show that this approach actually solves the wave equation and thus follows from the basic laws. We determine the derivatives and substitute:

$$\frac{\partial^2}{\partial z^2}\psi(z,t) = -k^2 \psi_0 \sin(kz - \omega t + \phi_0)$$
$$\frac{\partial^2}{\partial t^2}\psi(z,t) = -\omega^2 \psi_0 \sin(kz - \omega t + \phi_0) \quad (18.19)$$

and thus

$$-\omega^2 \psi_0 \sin(kz - \omega t + \phi_0) = \frac{F}{\mu}\left(-k^2 \psi_0 \sin(kz - \omega t + \phi_0)\right)$$

$$v_{Ph}^2 = \frac{\omega^2}{k^2} = \frac{F}{\mu}.$$

$$(18.20)$$

We see that we indeed provided a correct solution to the wave equation in ▶ Sect. 18.1. Incidentally, we have managed to trace the phase velocity of the wave back to the properties of the medium. The phase velocity is

$$v_{Ph} = \sqrt{\frac{F}{\mu}}. \tag{18.21}$$

It increases with the tension of the string. Since the wavelength in a musical instrument is predetermined by the geometry of the instrument, the tone pitch (frequency) increases with the tension of a string due to $v_{Ph} = \lambda f$. This is exploited when tuning the instrument.

18.3 Wave Packets

In ▶ Sect. 17.4 you learned about the phenomenon of beating. When two oscillations with almost the same frequency overlap, the amplitude of the oscillation is modulated:

$$x(t) = 2x_0 \sin\left(\frac{\omega_1 - \omega_2}{2}t\right) \sin\left(\frac{\omega_1 + \omega_2}{2}t\right). \tag{18.22}$$

The second term describes the rapid oscillation with the average of the two original frequencies. The first term represents a slow modulation of the amplitude of the oscillation.

Beating also occurs with waves, when two waves of the same propagation direction and similar frequency overlap. If we superimpose the two waves:

$$\begin{aligned}\psi_1(x,t) &= \psi_0 \sin(k_1 x - \omega_1 t) \\ \psi_2(x,t) &= \psi_0 \sin(k_2 x - \omega_2 t),\end{aligned} \tag{18.23}$$

we obtain

$$\begin{aligned}\psi(x,t) = {}&2\psi_0 \sin\left(\frac{k_1 - k_2}{2}x - \frac{\omega_1 - \omega_2}{2}t\right) \\ &\cdot \sin\left(\frac{k_1 + k_2}{2}x - \frac{\omega_1 + \omega_2}{2}t\right).\end{aligned} \tag{18.24}$$

The second term corresponds to that of a wave, as we already know it. The frequency of the wave is $\omega = (\omega_1 + \omega_2)/2$ and the wave number is $k = (k_1 + k_2)/2$.

18.3 · Wave Packets

If the frequency and thus also the wave number of the two original waves are similar, then in the first term $\omega_1 - \omega_2$ and $k_1 - k_2$ are significantly smaller than ω or k. The first term then changes much more slowly than the rapidly oscillating second term. It represents an amplitude modulation, as we have already learned in the case of beating in oscillations.

The output waves ψ_1 or ψ_2 have a constant amplitude. They are infinitely extended. By superimposing, the waves were divided into packets, between which the amplitude drops to zero. If infinitely many waves are superimposed, we can achieve a situation where the amplitude of the resulting wave is different from zero only in a single spatially limited area. This is called a "wave packet".

◘ Figure 18.6 shows the propagation of a Gaussian-modulated wave packet. Snapshots are shown at successive points in time, each one oscillation period later. The red dot marks a displacement with a fixed phase (maximum displacement). In the first shot, this point is at the maximum of the wave packet. During the propagation, it gradually falls back. In the bottom picture, it

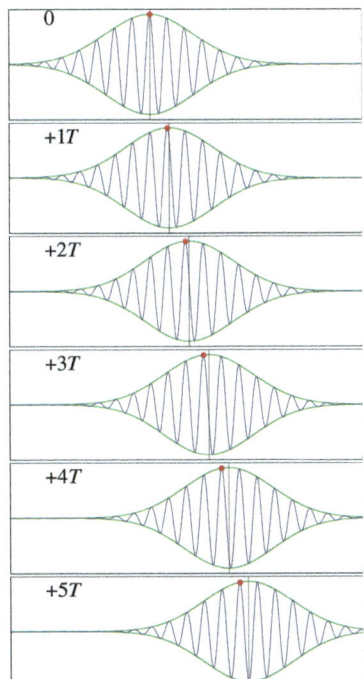

◘ **Fig. 18.6** Propagation of a wave packet

is already significantly to the left of the maximum of the wave packet. Here, the phase propagates slower than the wave packet.

We return to the overlay of two waves:

$$\psi(x,t) = 2\psi_0 \sin(k_{mod}x - \omega_{mod}t) \sin(kx - \omega t)$$

$$\text{with} \quad k_{mod} = \frac{k_1 - k_2}{2}, \quad \omega_{mod} = \frac{\omega_1 - \omega_2}{2},$$

$$k = \frac{k_1 + k_2}{2}, \quad \omega = \frac{\omega_1 + \omega_2}{2}. \tag{18.25}$$

We determine the propagation speed of the wave packet by following a point of fixed amplitude of the modulation. Analogous to Eq. 18.11, it must hold

$$k_{mod}x - \omega_{mod}t = k_{mod}(x + \Delta x) - \omega_{mod}(t + \Delta t)$$
$$0 = k_{mod}\Delta x - \omega_{mod}\Delta t \tag{18.26}$$
$$\frac{\Delta x}{\Delta t} = \frac{\omega_{mod}}{k_{mod}} = \frac{\omega_1 - \omega_2}{k_1 - k_2} = \frac{\Delta \omega}{\Delta k}.$$

For a real wave packet, which arises from infinitely closely spaced initial waves, one then obtains

$$v_g = \frac{d\omega}{dk}. \tag{18.27}$$

This is called the "group velocity of the wave packet". It indicates how fast the wave packet moves. As can be seen from the example in ◘ Fig. 18.6, it can differ from the phase velocity v_{ph} of the wave.

We can insert the relation $v_{ph} = \omega/k$ for the phase velocity, which we found in Eq. 18.11, here:

$$\omega = k v_{ph}$$
$$v_g = \frac{d\omega}{dk} = \frac{d}{dk}(k v_{ph}) = v_{ph} + k \frac{dv_{ph}}{dk}. \tag{18.28}$$

If the phase velocity is independent of the wavelength of the wave, then the phase and group velocities are the same. If this is not the case, the individual waves that make up a wave packet move at different velocities. This changes the propagation speed of the wave packet. This phenomena is referred to as "dispersion".

18.3 · Wave Packets

Example 18.4: Dispersion of Electromagnetic Waves

An example of dispersion-free waves is the propagation of light in a vacuum. The phase velocity is the speed of light in vacuum ($c_{\text{vac}} = 299\,792\,458$ m/s), which is the same for all electromagnetic waves. Therefore,

$$v_g = v_{ph} = c_{\text{vac}}.$$

When electromagnetic waves propagate through a medium, this is no longer the case. If we consider the propagation of radio waves in the atmosphere as an example, the following applies approximately:

$$\omega = \sqrt{\omega_p^2 + c_{\text{vac}}^2 k^2} \quad \text{with} \quad \omega_p = 3\,\text{MHz}.$$

This results in

$$v_{ph} = \frac{\omega}{k} = \frac{\sqrt{\omega_p^2 + c_{\text{vac}}^2 k^2}}{k} = \sqrt{c_{\text{vac}}^2 + \frac{\omega_p^2}{k^2}} > c_{\text{vac}}$$

$$v_g = \frac{d\omega}{dk} = \frac{1}{2}\frac{1}{\sqrt{\omega_p^2 + c_{\text{vac}}^2 k^2}}\left(2 c_{\text{vac}}^2 k\right) = \frac{c_{\text{vac}}^2}{v_{ph}} < c_{\text{vac}}.$$

Please note: The statement of the theory of relativity that nothing moves faster than light in vacuum refers to the group velocity. The phase velocity can be greater than c_{vac}.

A wave packet is created by the superposition of waves of different frequencies. It can be written as:

$$\psi(x,t) = \psi_0 \int_{-\infty}^{\infty} A(k,\omega) \sin(kx - \omega t)\, dk. \quad (18.29)$$

The function $A(k,\omega)$ indicates the strength and phase with which a wave of frequency ω contributes to the wave packet. This allows the creation of wave packets with any amplitude profile. However, $A(k,\omega)$ then contains arbitrary frequencies between zero and infinity.

The function $A(k,\omega)$ can be determined from $\psi(x,t)$ through a Fourier transformation. We do not want to go into the mathematical formalism here, but rather demonstrate the frequency analysis in an experiment (Experiment 18.3).

Experiment 18.3: Fourier Analysis with Microphone

Software that performs automatic Fourier analyses is widely used today. We show here the Fourier spectra of a tuning fork and a human voice. The sound is converted into an electrical signal by a microphone and digitized by an analog-to-digital converter (flash-ADC). The software then performs the Fourier analysis over an adjustable time interval, i.e., it numerically determines the inverse integral to Eq. 18.29.

The Fourier spectrum of the tuning fork can be seen in (a). It emits the concert pitch a' with a frequency of 440 Hz. It can be seen in the Fourier spectrum as a narrow high peak. It is almost a purely harmonic oscillation. Overtones are hardly recognizable.

Unlike the human voice (Figure b). The experimenter sings a note *a* into the microphone, an octave lower than the tuning fork's pitch. The fundamental frequency of the note is at 220 Hz. Overtones are clearly recognizable, the strongest at 440 Hz (a'), 660 Hz (e'') and 880 Hz (a'').

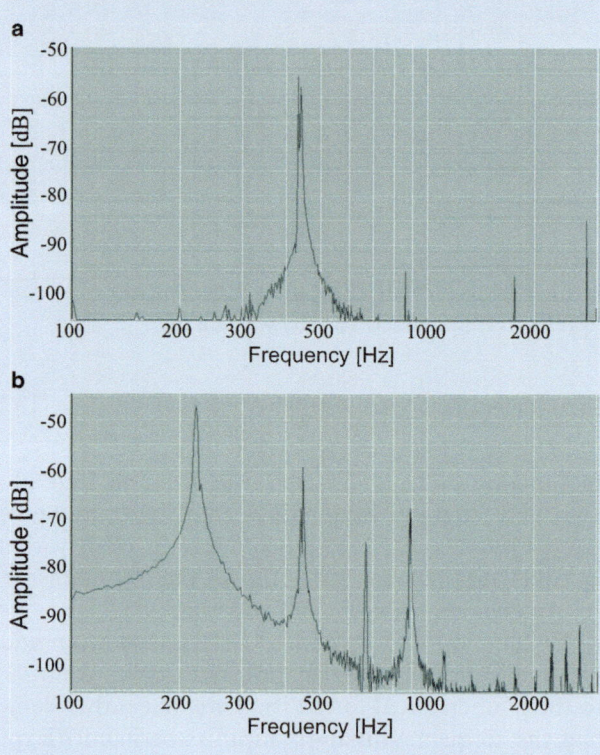

© RWTH Aachen, Physics Experiment Collection

18.3 · Wave Packets

Example 18.5: Fourier Representation

In the main text, we discussed the Fourier representation of wave packets. We want to give you a simple example to illustrate the Fourier representation. The wave packet moves in the z-direction and has the following form:

$$\psi(z,t) = \psi_0 \sin(2\pi f t - kz) \sin^2(2\pi f_{\text{mod}} t - kz).$$

We have chosen $\psi_0 = 1$ and $f_{\text{mod}} = \frac{1}{40} f$. The figure shows the temporal course of the displacement at a fixed point along the wave propagation. Imagine it was an acoustic wave, then this could be the temporal course with which the wave packet hits a microphone.

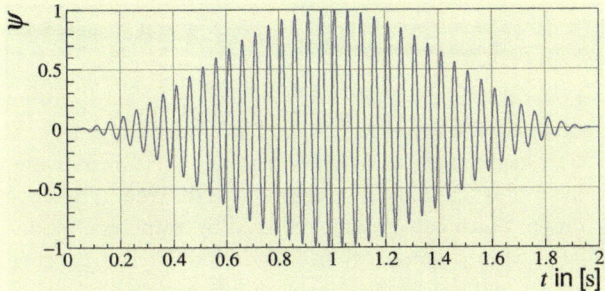

We can represent this wave packet by a Fourier series:

$$\psi(t) = \frac{a_0}{2} + \sum_{l=1}^{\infty} (a_l \cos(2\pi l f_0 t) + b_l \sin(2\pi l f_0 t)).$$

The coefficients a_l and b_l can be calculated using the following formulas:

$$a_l = \frac{1}{\pi} \int_{-\pi}^{\pi} \psi(t) \cos(2\pi l f_0 t)\, dt$$

$$b_l = \frac{1}{\pi} \int_{-\pi}^{\pi} \psi(t) \sin(2\pi l f_0 t)\, dt.$$

We have evaluated the integrals numerically. In our example, only a few coefficients are different from zero, namely $b_{39} = -1/4$, $b_{40} = 1/2$ and $b_{41} = -1/4$. The frequency f_0 corresponds to the frequency f_{mod} and b_{40} belongs to the carrier frequency f of our wave packet. In general, however, we only get an approximate description of the wave packet for a Fourier series of finite length.

Furthermore, you should note that our example for $\psi(z,t)$ does not represent a single wave packet. If we had continued our illustration to the right or left, you would recognize that the wave packet repeats regularly. The frequency f_0 in the Fourier series is the frequency of this repetition. Only periodic functions can be represented by Fourier series. If we wanted to represent a single wave packet, we would have to switch from the Fourier series to a Fourier integral. Then, not only the discrete frequencies lf_0 appear in the spectrum. Instead, we get a continuous spectrum of frequencies.

Example 18.6: Amplitude Modulation

Low-frequency signals such as speech or music often cannot be transmitted directly over desired transmission media such as a radio channel. For transmission, the useful signal must be shifted to a different frequency range, which can be accomplished by amplitude modulation. Furthermore, by shifting into different frequency ranges, multiple useful signals can be transmitted simultaneously and without mutual interference.

In amplitude modulation, the amplitude of a high-frequency carrier wave is varied depending on the low-frequency (modulating) signal to be transmitted. Amplitude modulation is used, for example, in radio broadcasting on the longwave (LF), medium wave (MF), shortwave frequency (HF) bands, in television, depending on the television standard used, or in CB radio.

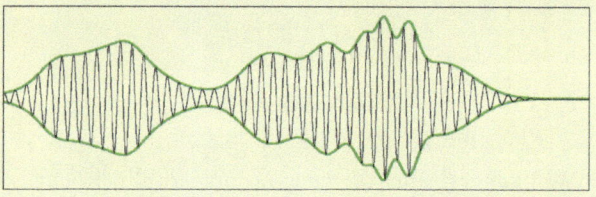

In the case of dispersion-free transmission, the shape of the wave packet will not change over time, but in the case of non-vanishing dispersion the shape of the wave packet changes. Usually, it spreads out. This is a technical problem especially for the transmission of digital signals, which consist of rectangular pulses. Due to the dispersion, the pulses become wider and wider increasing

18.4 · Energy Density and Energy Transport

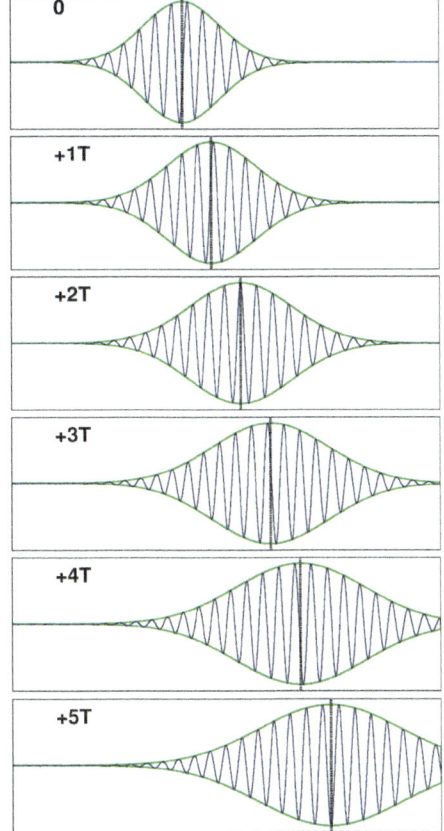

◘ **Fig. 18.7** Spreading of a wave packet through dispersion

with the length of the transmission path. A minimum distance between successive pulses must be maintained so that they do not merge into each other. This limits the transmission capacity.

In ◘ Fig. 18.7, a Gaussian wave packet is depicted, which spreads under dispersion. You can see how the wave packet widens with propagation.

18.4 Energy Density and Energy Transport

In a mechanical wave, masses move against a restoring force. Kinetic energy is continuously converted into potential energy and vice versa, from which we conclude that energy is stored in a wave. To determine the energy content, we go back to a system of coupled oscillations.

In ▶ Sect. 17.1 we saw that the energy of an oscillator can be expressed as

$$E = \overline{E}_{kin} + \overline{E}_{pot} = E_{kin}^{max} = E_{pot}^{max}. \tag{18.30}$$

On average, the energy is equally distributed between kinetic and potential components. At the zero crossing, the entire energy is present as kinetic energy. It has reached its maximum. At the turning point, on the other hand, the potential energy is maximum and equal to the total energy. If the total energy is expressed via the maximum kinetic energy, the result is:

$$E = \frac{1}{2} m v_0^2 = \frac{1}{2} m x_0^2 \omega^2. \tag{18.31}$$

For a wave in a continuous medium, one cannot specify individual oscillators. Instead, one considers the energy density. It is obtained from the formula for the oscillator by replacing the mass with the mass density

$$\varepsilon = \frac{1}{2} \rho \psi_0^2 \omega^2. \tag{18.32}$$

We called the amplitude in the oscillator x_0, for a wave we now use ψ_0.

From the energy density, we determine the intensity of a wave. It is defined as the energy transported by the wave per unit of time through a unit area:

$$\text{Intensity} = \frac{\text{Energy}}{\text{Area} \cdot \text{Time}}. \tag{18.33}$$

Intensity is also referred to as "energy flux density". The wave passes through an area of size A. The intensity is determined from (see ◉ Fig. 18.8):

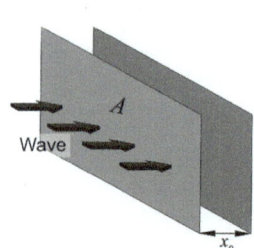

◉ **Fig. 18.8** To determine the energy flux density

$$I = \frac{E}{A\,t} = \frac{\varepsilon\, \Delta V}{A\,t}. \tag{18.34}$$

The volume of the wave field that penetrates the area t over time A is given by:

$$\Delta V = A\, \Delta x = A\, v_g\, t \tag{18.35}$$

and thus:

$$I = \frac{\varepsilon\, \Delta V}{A\,t} = \varepsilon\, v_g = \frac{1}{2} \rho\, v_g\, \psi_0^2\, \omega^2. \tag{18.36}$$

The intensity of the wave is proportional to the square of its amplitude and to the square of its frequency. This relationship, which we have derived here for mechanical waves, applies generally to all waves.

We conclude the section with a final note on the energy transport of a wave: For plane waves, the amplitude is the same everywhere in space and it also does not change over time. As a result, despite wave propagation, the energy density does not change at any point in the wave field. Therefore, one cannot speak of energy transport by a plane wave. This is fundamentally different with wave packets. Here, the energy is localized in the area of the wave packet. If the wave packet moves through the medium at the speed v_g, the energy stored in it also moves. The wave transports energy.

18.5 Reflection and Interference

In ▶ Sect. 17.5 we learned how waves are reflected at the end of a string. This phenomenon is by no means limited to one-dimensional waves. Waves are reflected at the boundaries of the medium. We can demonstrate this with water waves in the wave tray (Experiment 18.4).

Experiment 18.4: Reflection of Water Waves

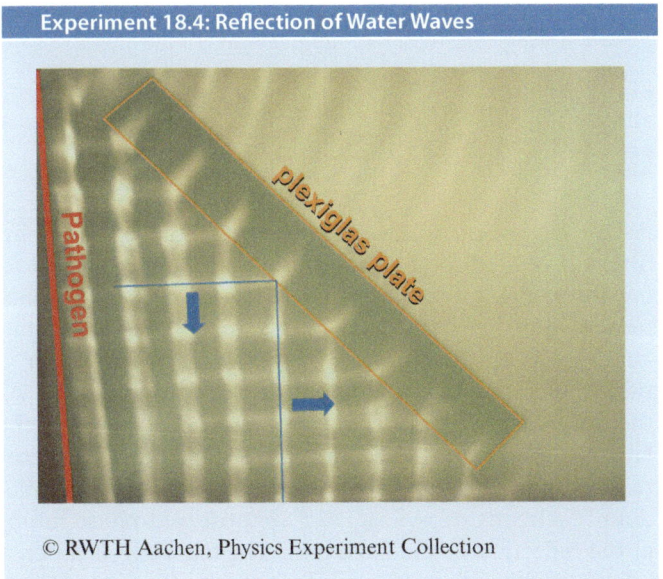

© RWTH Aachen, Physics Experiment Collection

Again, we use the wave tray, which you already know from Experiment 18.1. Before the exciter, we place a rectangular plexiglass plate in the water. It is just barely covered with water. The wave coming from the left is reflected at the boundary to the shallow water over the plexiglass plate. The blue line shows the snapshot of an already partially reflected wavefront. The arrows indicate the direction of movement.

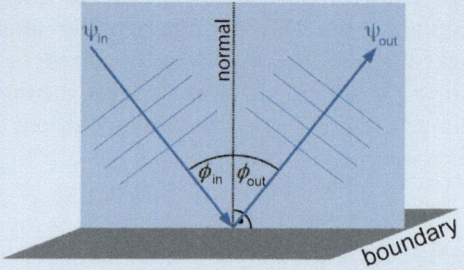

Upon closer inspection, we can recognize the law of reflection. The wave is reflected so that the angle of incidence ϕ_{in} and the angle on exit (reflection angle) ϕ_{out} are equal: $\phi_{in} = \phi_{out}$. This is called the "law of reflection". In the sketch, the angles are defined. They are always measured against the normal to the boundary. For three-dimensional waves, the angle of reflection against the normal does not yet uniquely determine the direction of the outgoing wave. It must also be noted that the wave is always reflected in a plane that includes the normal to the boundary surface.

As we have seen, waves can be represented by many individual oscillators. For each oscillator, there is a restoring force in the form $F_r = -kx$ that enables the oscillation. (Note that here k does not indicate the wave number, but the elastic constant of the system). The larger k, the smaller the amplitudes of the oscillators will be. At the boundary of a medium, k changes. The side with the larger value of k is called the "denser medium" and correspondingly the side with the smaller k the "thinner medium". Already in ▶ Sect. 17.5 we observed phase jumps at the reflection at the fixed end. The fixed end is a dense medium with infinitely large k. In general, a phase jump of π occurs whenever a wave is reflected on a denser medium. No phase jump occurs during a reflection on the thinner medium.

18.5 · Reflection and Interference

Example 18.7: Echo

An echo is created by the reflection of sound waves on the surrounding mountain walls.

At the end of the chapter on waves, we briefly touch on a phenomenon that is characteristic of waves: interference. We will delve deeper into the topic in wave optics. Whenever you observe interference effects, you can be sure that you are dealing with waves.

"Interference" refers to the superposition of two or more waves. The amplitudes of the waves add up in:

$$\Psi_{\text{tot}}(\vec{r},t) = \Psi_1(\vec{r},t) + \Psi_2(\vec{r},t) + \ldots . \quad (18.37)$$

This is another example of a superposition principle. The statement of Eq. 18.37, that the amplitudes add up, may seem self-evident, but it is not trivial at all. One could have equally expected that the intensities of the waves would add up. But this is not the case. The amplitudes add up. The intensity of the resulting wave is obtained for waves of the same frequency from:

$$I_{\text{tot}}(\vec{r},t) = \frac{1}{2} \rho \, v_g \, \omega^2 (\Psi_1(\vec{r},t) + \Psi_2(\vec{r},t) + \ldots)^2 . \quad (18.38)$$

Since amplitudes can also take negative values, it can certainly happen that the intensity of the wave resulting from the superposition is less than the intensities of the individual waves. Everything depends on the phase between the individual waves. In Experiment 18.5 we show some interference effects.

> **Experiment 18.5: Interference with Water Waves**
>
> We use the wave tray from Experiment 18.1 one last time. A simple example of interference is created by two circular waves that originate from neighboring exciters. This can be seen in the first example. In the right area, you can clearly see directions in which wave crests and wave troughs alternate. Here, the wave is spreading. In between are directions where you see a uniform, structureless brightness, where the water surface is at rest. The two waves cancel each other in these areas due to interference.
>
> The second example shows a classic interference experiment: the double slit. In the water tray, an obstacle (marked in orange) is placed, which only has two narrow openings. The double slit is excited from the left with plane water waves. To the right of the slits, in their immediate vicinity, you can see circular wavelets emanating. Further away from the two slits, these overlap and lead to an interference pattern similar to the one in the first example. Again, there are directions from the slit where no excitations can be seen. Here, the wavelets from the two slits cancel each other. In between are interference maxima.
>
>
>
> © RWTH Aachen, Physics Experiment Collection

18.5 · Reflection and Interference

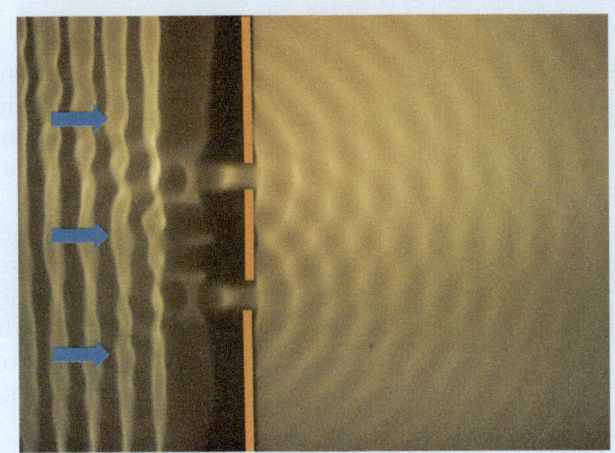

© RWTH Aachen, Physics Experiment Collection

❓ Problems

1. An airplane is flying over an observer at the zenith and the observer estimates that the direction from which he hears the sound of the airplane deviates by about 30° from the direction in which he sees the airplane. What is the speed of the airplane, assuming a speed of sound of 330 m/s?

2. For the propagation speed v of a transverse wave along a tensioned string, the following applies $v = \sqrt{\sigma/\rho}$, where σ is the string tension and ρ is the density of the string material. A violin has four steel strings ($\rho_{steel} = 7.86 \cdot 10^3$ kg/m³), which can freely oscillate between the saddle and the bridge over a length of 33 cm. Their diameters and the frequencies to which they are tuned are given as follows:

E-string	0.25 mm	664.50 Hz
A-string	0.45 mm	443.00 Hz
D-string	0.70 mm	295.33 Hz
G-string	0.75 mm	196.89 Hz

What is the total force with which the four strings of the violin are tensioned?

3. The speed of sound in solids with Young's modulus E and density ρ is $v = \sqrt{E/\rho}$. The speed of sound in a gas with bulk modulus κ and density ρ is

$v = \sqrt{\kappa/\rho}$. In a Kundt's tube filled with oxygen under normal pressure, standing waves are generated using a 0.75 m long steel rod. What is the distance between adjacent nodes?
- Elastic modulus and density of steel:
  $E = 2.0 \cdot 10^{11}$ N/m², $\rho = 7.8 \cdot 10^3$ kg/m³;
- Bulk modulus and density of oxygen:
  $\kappa = 1.4 \cdot 10^5$ N/m², $\rho = 1.4$ kg/m³.

4. Two loudspeakers are placed 2 m apart and emit the same sound. An observer is equidistant from both loudspeakers and 3 m away from the connecting line between the loudspeakers. The observer now starts walking parallel to this connecting line and notices a minimum in volume after a distance of 1.45 m. What frequency does the emitted sound have?

Acoustics

Contents

19.1 Sound Waves – 560

19.2 Perception of Sound – 570

19.3 Moving Sound Sources – 576

19.4 Musical Instruments – 583

© The Author(s), under exclusive license to Springer-Verlag GmbH, DE, part of Springer Nature 2025
S. Roth and A. Stahl, *Mechanics*,
https://doi.org/10.1007/978-3-662-68079-7_19

19.1 Sound Waves

In this chapter we want to specifically deal with sound waves. They are an example of the waves that we discussed in the previous chapter. However, due to the great importance of sound in everyday life—just think of speech or music—we want to dedicate a separate chapter to it.

Sound propagates in air and other gases or liquids as periodic compression and decompression of the medium. Sound can also propagate in solids, although the mechanism of sound propagation in solids can differ from that in liquids and gases. We want to focus primarily on the propagation of sound in fluids.

A sound wave is created when an exciter compresses (or decompresses) the air at one point. For example, if the membrane of a speaker moves forward, it compresses the layer of air directly in front of the membrane (see ◘ Fig. 19.1). The pressure in this layer increases. It expands and compresses the next layer of air in front of it. In this way, the excitation spreads from layer to layer. Air molecules move back and forth between the compressions and decompressions. A longitudinal wave is created. Transverse sound waves can only occur in solids. Every sound wave needs a medium. It is based on the movement of the molecules of the medium. A sound wave cannot propagate in vacuum (see Experiment 19.4).

We demonstrate the propagation of sound first by looking at some experiments (Experiment 19.1 to Experiment 19.5).

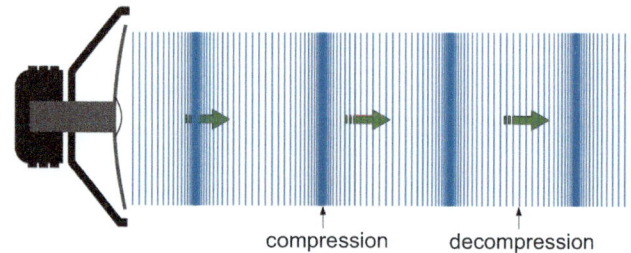

◘ **Fig. 19.1** Schematic representation of a sound wave

19.1 · Sound Waves

Experiment 19.1: Flickering Candle Flame

With a candle we visualize the movement of the air in a sound wave. The burning candle is placed directly in front of the speaker's membrane. A tunable sine generator is used as the signal source. At low frequencies (10 Hz), the movement of the air can be clearly seen by the flickering of the candle. With higher frequencies, the flickering becomes faster (don't be misleaded, you do not directly see the frequency of the sound). The flickering can still be seen even when the frequency of the sound is already above the human hearing range.

© RWTH Aachen, Physics Experiment Collection

Experiment 19.2: Rubens's Flame Tube

This experiment demonstrates that sound waves indeed build up from pressure differences. In the middle of the tube, there is a connection through which the tube is slowly flooded with propane gas. On the upper side, there are small holes in a regular pattern in the tube through which the gas escapes again. Now we seal the right end of the tube with a plug and attach a loudspeaker as airtight as possible on the left side. It transmits a sinusoidal tone adjustable in frequency. Then we ignite the gas at the top and adjust the frequency of the tone. At certain frequencies, standing waves occur, which can be recognized by a regular distribution of flames, as seen in the picture.

© RWTH Aachen, Physics Experiment Collection

At the nodes of the standing wave, the gas is almost at rest. The gas flow is reduced to such an extent that no visible flame is produced at these points. The situation is different at the antinodes. Here, the pressure periodically changes between positive and negative pressure compared to the surroundings. At negative pressure, air is sucked in from the outside and mixed with the gas. At positive pressure, this gas is then expelled. A clear flame is produced, the height of which is proportional to the amplitude of the wave.

For those who like to play: With music from the speaker, you can get impressive visualizations of the music.

Experiment 19.3: Sound Propagation in Air

We had explained that the propagation of a sound wave is associated with a longitudinal movement of air molecules. This simple experiment demonstrates this. We set up a transmission path for sound with two tambourines. We strike the left tambourine forcefully with a drum stick. The right one will pick up the sound. In between is an empty tube. It merely serves to focus the sound on the second tambourine, so that the intensity along the transmission path decreases only insignificantly. The movement of the molecules in the sound wave deflects the drum skin of the right tambourine. We make this deflection visible with a ping-pong ball that hangs in front of the tambourine, in contact with the drum skin. It is unfortunately obscured by the tambourine in the picture. If we strike the left tambourine, it is slightly repelled by the right tambourine. Since the movement is only slight, we illustrate it with a minature figure that we place behind the ping-pong ball. With the left tambourine, we generate a sound wave. The air transfers it to the right tambourine. This pushes the ping-pong ball which then knocks over the figure.

© RWTH Aachen, Physics Experiment Collection

Experiment 19.4: Sound in a Vacuum

Sound requires a medium - usually air - to propagate. If we remove the air, the sound can no longer propagate. This experiment demonstrates this impressively. We hang a loud siren on a stand under a glass bell. The simple stand suppresses the propagation of sound to the outside via the floor. Initially, there is air under the bell and the sound of the siren is clearly audible. Then we pump out the air with a simple vacuum pre-pump. The sound of the siren becomes quieter and quieter. At the same time, we observe the pressure under the bell. From about 30 mbar on, the siren can no longer be heard.

In the end we reverse the experiment. We turn off the pump and slowly let air back into the bell through a valve. With the air, the sound returns.

© RWTH Aachen, Physics Experiment Collection

Experiment 19.5: Sound Propagation in Wood

This experiment is intended to show that sound also propagates in a solid through the movement of molecules. We use a long wooden rod as the medium. On the right side, we strike it with a rubber mallet. At the left end, a small wooden ball, which is suspended as a pendulum, indicates a deflection. This could have been caused by the sound wave. However, it could also be due to the fact that we slightly move the rod when striking it, thus transferring momentum. To suppress this second effect, we attach the wooden rod to a very heavy stone block using strong clamps. This should prevent movement of the rod. But in the end, it is not entirely clear whether the deflection is really due to the sound wave.

Experiment 19.6: Pin Hole Siren

The pin hole siren is a simple device for generating a loud noise. In a pinhole siren, holes are placed on a disc at a fixed radius. A strong air stream blows against the rotating disc and is alternately let through the holes or hits the disc. This creates periodic pressure fluctuations and therefore sound. The amplitude progression is rather rectangular, which leads to a shrill unpleasant tone. Our sirene consists of a metal disc with several rings of pin holes (see illustration), which is rotated at a high frequency. We blow compressed air (a few bar) from a bottle at the disk. The higher the pressure, the louder the tone.

The main frequency of the tone depends on the rotation frequency of the disc and on the number of holes in a ring. On our disc, the number of pin holes per ring is chosen for the various rings in such a way that a major scale is created. You can play a little tune on the disc.

© RWTH Aachen, Physics Experiment Collection

The fact that sound is actually a wave can be demonstrated by interference. Only waves show interference effects. In Experiment 19.7, you can see an example of interference with sound waves.

Experiment 19.7: Interference with Sound

We set up two speakers a few meters apart. Both speakers are driven by the same signal. We choose a frequency in the lower range, e.g., 220 Hz (wavelength $\lambda \approx 1.5$ m). In the picture, the crests of the sound waves from both

19.1 · Sound Waves

speakers are depicted. At some points, the lines cross. There, the wave crests from both speakers amplify each other. However, there are also points where the wave crests of one speaker meet the wave troughs of the other and the two waves cancel each other. If you walk a few meters in front of the two speakers, you can clearly hear that the volume is louder at some points than at others.
The following setup is also impressive. The two speakers are placed directly next to each other and one of the speakers is reversed in polarity. Now its wave is out of phase with the other. If you stand some distance in front of the speakers, you only hear a faint sound, as the waves of the two speakers cancel each other. If you now unplug one of the two speakers, the sound surprisingly becomes much louder.

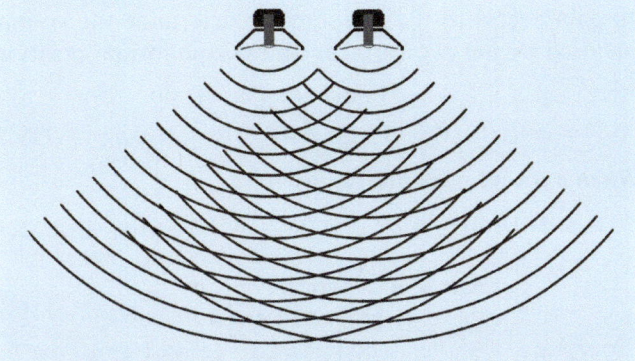

To capture the propagation of sound mathematically, we start from a sound channel with imaginary walls and a rectangular cross-section, as indicated in ◘ Fig. 19.2. The coordinate x denotes the position along the channel. The sound wave propagates in this direction. Two planes of air molecules are indicated in the channel, which, in the

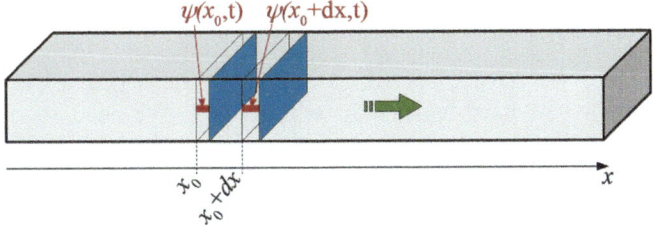

◘ **Fig. 19.2** On the calculation of the propagation of a sound wave

equilibrium state, i.e., without a sound wave, are at the positions x_0 and $x_0 + dx$. The sound wave displaces these molecules from their equilibrium position in the x-direction. The displacement from the equilibrium position is $\psi(x_0, t)$ or $\psi(x_0 + dx, t)$. The displacement of the molecules locally changes the pressure, leading to the necessary restoring forces for the propagation of the wave.

We express the displacement at $x_0 + dx$ through a Taylor expansion from the displacement at x_0. With the notation dx we have already indicated that we will make this distance very small, so that a first-order expansion will suffice:

$$\psi(x_0 + dx, t) = \psi(x_0, t) + \left[\frac{\partial}{\partial x}\psi(x, t)\right]_{x=x_0} dx. \quad (19.1)$$

To gain access to the pressure, we determine the volume enclosed by the two surfaces. In the equilibrium position, it is

$$V = A(x_0 + dx) - A x_0 = A\, dx. \quad (19.2)$$

When a sound wave arrives, we have:

$$V + dV = A(x_0 + dx + \psi(x_0 + dx, t)) - A(x_0 + \psi(x_0, t))$$

$$= A\, dx + A\left[\frac{\partial}{\partial x}\psi(x, t)\right]_{x=x_0} dx \quad (19.3)$$

$$\Rightarrow dV = A\left[\frac{\partial}{\partial x}\psi(x, t)\right]_{x=x_0} dx = V\left[\frac{\partial}{\partial x}\psi(x, t)\right]_{x=x_0}.$$

With the arrival of a sound wave, the volume changes, which was originally occupied by the air molecules that were between x_0 and $x_0 + dx$. This changes the pressure in this area. We denote the pressure that prevailed before the arrival of the sound wave with p_0. With the sound wave, we have $p = p_0 + dp$, with

$$dp = -\bar{\kappa}\frac{dV}{V} = -\bar{\kappa}\left[\frac{\partial}{\partial x}\psi(x, t)\right]_{x=x_0}, \quad (19.4)$$

where we have resorted to a relation here (Eq. 14.15), which you have learned in ▶ Sect. 14.3. We need to relate the compression module $\bar{\kappa}$ to the pressure p. We use a relation that we can only derive within the framework of kinetic gas theory. It will result in $dp = -\gamma p_0 \frac{dV}{V}$, i.e., $\bar{\kappa} = \gamma p_0$, with a constant $\gamma = C_p/C_V$ called the heat capacity ratio. With this, we have

19.1 · Sound Waves

$$dp = -\gamma p_0 \left[\frac{\partial}{\partial x}\psi(x,t)\right]_{x=x_0}. \qquad (19.5)$$

The force that acts on the displaced air layer results from the pressure (again we use a first-order Taylor expansion in x):

$$\begin{aligned}
dF &= pA - \left(p + \frac{\partial p}{\partial x}dx\right)A \\
&= -\frac{\partial p}{\partial x}dxA \\
&= -\frac{\partial}{\partial x}(p_0 + dp)dxA \\
&= -\frac{\partial}{\partial x}(dp)dxA \\
&= -\frac{\partial}{\partial x}\left(-\gamma p_0 \left[\frac{\partial}{\partial x}\psi(x,t)\right]_{x=x_0}\right)dxA \\
&= \gamma p_0 A \frac{\partial^2 \psi(x,t)}{\partial x^2}dx.
\end{aligned} \qquad (19.6)$$

To determine the movement of the air layer, we use Newton's second law of mechanics. The displacement of the air layer is $\psi(x,t)$, we calculate its acceleration as the second derivative with respect to time. With the air layer, a mass is $dm = \rho\, dV = \rho A\, dx$ connected. Then:

$$dF = dm\frac{\partial^2 \psi(x,t)}{\partial t^2} = \rho A \frac{\partial^2 \psi(x,t)}{\partial t^2}dx. \qquad (19.7)$$

We substitute Eq. 19.6 and obtain:

$$\begin{aligned}
\gamma p_0 A \frac{\partial^2 \psi(x,t)}{\partial x^2}dx &= \rho A \frac{\partial^2 \psi(x,t)}{\partial t^2}dx \\
\gamma p_0 \frac{\partial^2 \psi(x,t)}{\partial x^2} &= \rho \frac{\partial^2 \psi(x,t)}{\partial t^2} \\
\frac{\partial^2 \psi(x,t)}{\partial t^2} &= \frac{\gamma p_0}{\rho}\frac{\partial^2 \psi(x,t)}{\partial x^2}.
\end{aligned} \qquad (19.8)$$

The differential equation represents the equation of motion for a mass element of the medium. It is the wave equation, as we have already seen from Eq. 18.17. It has the solution:

$$\psi(x,t) = \psi_0 \sin(kx - \omega t + \phi_0) \qquad (19.9)$$

with the speed of sound:

$$v_{\text{ph}} = \frac{\omega}{k} = \sqrt{\frac{\gamma p_0}{\rho}}. \tag{19.10}$$

The group velocity follows from:

$$\omega = k\sqrt{\frac{\gamma p_0}{\rho}}$$
$$\Rightarrow v_g = \frac{d\omega}{dk} = \sqrt{\frac{\gamma p_0}{\rho}} = v_{\text{ph}}. \tag{19.11}$$

As we can see, group and phase velocity are the same. Therefore, the propagation of a sound wave in air is dispersion-free. This is important for human communication. If the sound wave packets were to disperse during propagation, acoustic communication over longer distances would not be possible.

The speed of sound in air under normal conditions is:

$$\left.\begin{array}{l} p_0 = 1.01 \cdot 10^5 \text{ Pa} \\ \rho = 1.29 \dfrac{\text{kg}}{\text{m}^3} \\ \gamma = \dfrac{c_p}{c_V} = 1.40 \end{array}\right\} \Rightarrow v_{\text{Ph}} = 331 \, \frac{\text{m}}{\text{s}}. \tag{19.12}$$

Sound takes about 3 s to travel 1 km.

Experiment 19.8: Measurement of the Speed of Sound

The measurement of the speed of sound in air can be easily achieved using an electronic stopwatch. The start and stop signals for the clock are generated by two microphones, which are set up at a distance l apart from each other. We generate a bang with a toy gun, the propagation of which we then measure. The result is approximately 330 m/s.

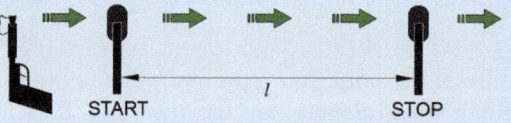

Experiment 19.9: Speed of Sound in Helium

As we have seen, the speed of sound depends on the properties of the medium. In air, it was 331 m/s. In

19.1 · Sound Waves

helium, the speed of sound is significantly higher. Under normal conditions, it is 981 m/s. If you fill organ pipes with helium instead of air, it results in higher tones. In the organ pipe, a standing wave is excited by the flowing gas. The wavelength of the excitation is determined by the geometry of the pipe. The frequency of the vibration, which we perceive as pitch, is then proportional to the speed of sound in the medium according to $v = \lambda f$ and is correspondingly higher in helium than in air.

We excite the organ pipes by blowing them with air from a compressed air bottle. If we switch to a helium bottle, you will hear an increase in pitch as the air in the pipe is replaced by helium.

© RWTH Aachen, Physics Experiment Collection

Experiment 19.10: Voice Pitch in Helium

This is a simple experiment that always leads to amusement. You inhale helium from a bottle and then speak. As with the organ pipes in Experiment 19.9, the pitch shifts in the larynx. You get a squeaky, unnaturally high voice. After a few breaths, the effect disappears. However, you should not repeat the experiment too often as it is not beneficial for the vocal cords.

Experiment 19.11: Speed of Sound in a Solid

With this experiment we determine the speed of sound in metal rods. The rods are 1.5 m long and 1.2 cm in diameter. We use rods made of copper, brass, aluminum, and steel. As a time sensor, we use a piezocrystal. The sensor is placed on the ground. It is protected by a thin felt mat. The rods are carefully placed on the sensor. They are held vertically by a clamp. The clamp should only be loosely attached, as it otherwise dampens the sound too much. Now the rod is lightly struck at the top with a small hammer. From the signal of the sensor, it can be seen that the sound is reflected up and down in the rod multiple times. From the time interval between two reflections, the speed of sound is determined as $v = 2L/\Delta t$. The results are values from 4000 to 5000 m/s.

Sensor

© U. S. National Oceanic and Atmospheric Administration

Example 19.1: Thunderstorm

Lightning and thunder occur simultaneously. The thunder takes about 3 s to cover a kilometer. In contrast, the time it takes for the light from the lightning to spread the same distance is negligible (a few µs). The distance of the lightning can be estimated from the time difference between the arrival of the flash of light and the sound of thunder at the observer as 1 km per 3 s.

19.2 Perception of Sound

Sound is a subjective sensation. We describe it with terms such as "loud" and "quiet", "low" and "high" or "harmonic" and "dissonant". Because the perception of sound is so important for our everyday life, we want to at least try to explain how it is related to the physical properties of sound waves. Let's start with a brief look into the ear (◘ Fig. 19.3).

We perceive sound through our ears. The sound reaches the eardrum via the external auditory canal and sets it vibrating. The hammer transmits the vibrations to the anvil, where they are led by the stirrup to the cochlea (snail). In the fluid-filled cochlea, they are converted into electrical signals. They reach the brain via the auditory nerve and are processed there. The frequency of the incoming sound waves determines the pitch, their intensity the volume.

◘ Figure 19.4 shows the sensitivity of the ear depending on the frequency. There is no clear lower limit for hearing. Below about 20 Hz, sound waves are no longer perceived as tones, but as vibration. Above 10 to 20 kHz (depending on age), the sensitivity drops sharply and sounds can no longer be sensed.

In music, the pitch is expressed through the notes (see ◘ Fig. 19.5). The tones of the different octaves are designated from bottom to top with uppercase letters sometimes with a number attached that indicates the corresponding octave. ◘ Table 19.1 gives the frequencies of the tones from C4 to C5 (well tempered tuning). This octave contains the concert pitch A4, which usually serves as a reference.

Sounds, as they are produced by a musical instrument or by singing, do not consist of a single sinusoidal

19.2 · Perception of Sound

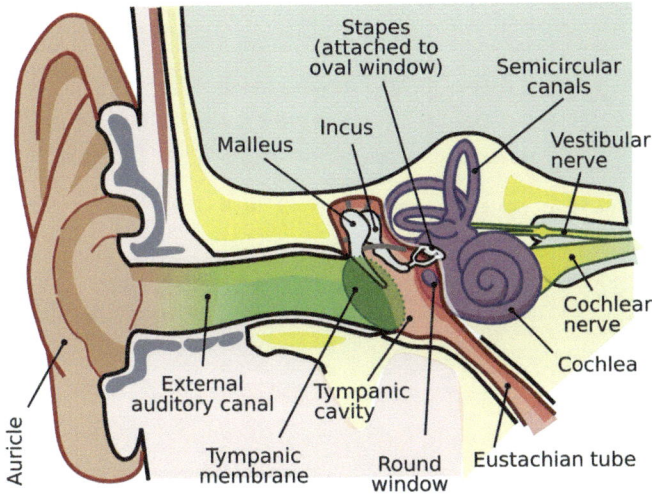

 Fig. 19.3 Anatomy of the ear. © Wikimedia: Lars Chittka; Axel Brockmann[1]

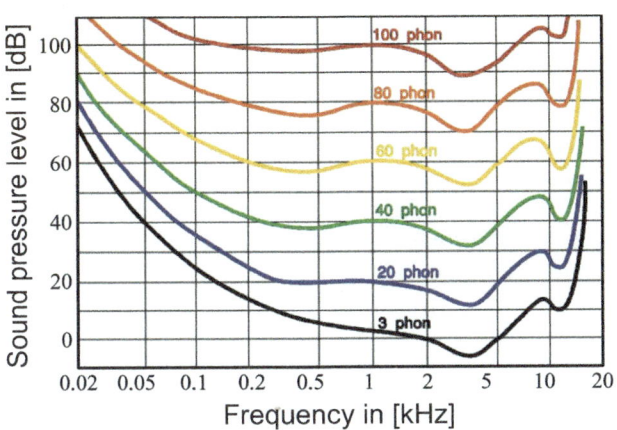

 Fig. 19.4 Sensitivity of the ear

wave with a fixed frequency. They are composed of a fundamental tone and a series of more or less intense overtones. The fundamental tone is perceived as pitch, the overtones determine the timbre. For example, if we con-

---

1 Perception Space—The Final Frontier, A PLoS Biology Vol. 3, No. 4, e137 doi:10.1371/journal.pbio.0030137 (Fig. 1A/Large version), vectorised by Inductiveload, CC BY 2.5, ▶ https://commons.wikimedia.org/w/index.php?curid=5957984

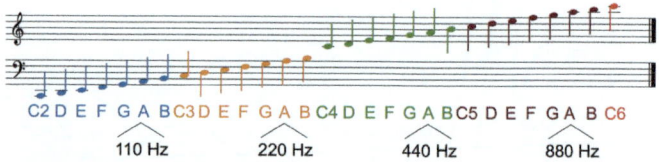

• **Fig. 19.5** Scale of tones

Table 19.1	Frequencies of the tones of the C4 octave
C4	262 Hz
D4	294 Hz
E4	330 Hz
F4	349 Hz
G4	392 Hz
A4	440 Hz
B4	494 Hz
C5	524 Hz

sider a string instrument with a string length of L, standing waves with the wavelengths $\frac{n}{2}\lambda = L$ can propagate on it. As a rule, all these waves are excited together. They correspond to tones with the frequencies

$$f = \frac{n}{2}\frac{v}{L} \tag{19.13}$$

Assuming a string with the concert pitch A4 (440 Hz) as the fundamental tone, the overtones are A5 (880 Hz), E5 (1320 Hz), A5 (1760 Hz), $C^{\#}6$ (2200 Hz), E6 (2640 Hz), $F^{\#}6$ (3080 Hz), A6 (3520 Hz), etc. These overtones are in integer ratios to the fundamental tone. They define the intervals in our scales. • Table 19.2 lists the different intervals. In the overtone series of the concert pitch, an octave to A5 appears first, then a fifth to E5, and then again an octave to the fundamental tone. The $C^{\#}$ corresponds to a major third, the $F^{\#}$ to a major sixth (each in relation to an A). We perceive these intervals as particularly harmonious. The dissonant intervals, such as the seconds or the sevenths, only appear in the overtone series in very high orders. For other fundamental tones, the frequencies are to be shifted accordingly, while the frequency ratios remain the same.

19.2 · Perception of Sound

Table 19.2 Tone intervals

Prime	C/C	1 : 1
Minor second	C#/C	10 : 9
Major second	D/C	9 : 8
Minor third	D#/C	6 : 5
Major third	E/C	5 : 4
Fourth	F/C	4 : 3
Excessive fourth	F#/C	7 : 5
Fifth	G/C	3 : 2
Minor sixth	G#/C	8 : 5
Major sixth	A/C	5 : 3
Minor seventh	A#/C	9 : 5
Major seventh	B/C	15 : 8
Octave	C/C	2 : 1

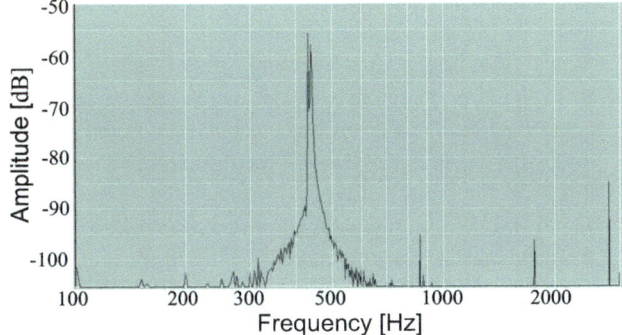

Fig. 19.6 Frequency spectrum of a tuning fork (concert pitch A4). © RWTH Aachen, Physics Experiment Collection

Figure 19.6 shows the Fourier spectrum of a tuning fork with the concert pitch A4, recorded with the apparatus from Experiment 18.3. The fundamental tone and the first overtones at 880 Hz, 1320 Hz, 1760 Hz and 2200 Hz are clearly visible.

Example 19.2: Equal Temperament

G4	391.1 Hz	G4	391.1 Hz
A4	440 Hz	A4	440 Hz
B4	495 Hz	B4	488.9 Hz
C#5	550 Hz	C5	521.5 Hz
D5	586.7 Hz	D5	586.7 Hz
E5	660 Hz	E5	651.9 Hz
F#5	733.3 Hz	F#5	733.3 Hz
G#5	825 Hz	G5	782.2 Hz
A5	880 Hz		

In the text, we have explained the tone intervals. They are given in ◨ Table 19.2. Now imagine you have to tune a keyboard instrument like the one shown. You start with the concert pitch A4 and set the tones according to the integer ratios. By changing the tension of the strings, you can change the propagation speed of the wave on the string and thus its frequency. You have now tuned the instrument in a so-called pure tuning in A. The frequencies are given in the table in the two left columns for the tones of the A major scale.

Now you want to play a piece in G major. You go down from the concert pitch A4 by a major second and get G4 (391, $\overline{1}$ Hz). Now you want to build the scale on this tone with the integer ratios of the intervals. You get the tones of the right columns. Compare the frequencies! The fifth D5 in G major matches the fourth (also D5) in A major exactly. But this is not true for all tones. The B4 has a frequency of 495 Hz in A major according to your pure tuning. But in the G major scale, it should have a frequency of 488.9 Hz. It will sound impure if you want to play in G major. Actually, you would have to retune the instrument to G scale beforehand. But this is too much effort. Therefore, other tunings have been considered, such as the well tempered tuning most commonly used today. The octave, which comprises 12 semitones, is divided into 12 equal frequency steps. The frequency ratios are $\sqrt[12]{2^n}$: 1. This levels out the differences in tuning. The instrument sounds a little impure in all keys, but is equally useful for all. Again starting from the concert pitch A4, the problematic B4 mentioned above would now be 493.9 Hz, while it should be 495 Hz in A

19.2 · Perception of Sound

> major and 490 Hz in G major (the G4 is 392 Hz in this tuning). There are now deviations in all keys, but they are overall smaller.

Like the pitch, the volume is also a subjective perception. It is linked to the intensity of the sound wave. In ▶ Sect. 18.4 we had defined the intensity of a wave as the power transmitted by a wave through a unit area. As we had seen, the intensity is proportional to the square of the wave amplitude.

The human ear has an enormous dynamic range. It can perceive sound waves in a range from a minimum of 10^{-12} W/m² to a maximum of 1 W/m² and even beyond. However, sound above 1 W/m² is painful and can cause damage to the ear. The pain-free range covers 12 orders of magnitude, a dynamic range that hardly any technical measuring device achieves. The sensitivity of the human hearing depends on the frequency of the tones. It is highest in the range of 1 kHz to 4 kHz. This frequency range includes the fundamental tones of speech. Music covers a significantly larger frequency range. If you want to fully capture the sound of music, the tones must be louder than those of normal conversation. The spectral sensitivity of the ear is given in ◘ Fig. 19.4.

It is probably due to the enormous dynamic range of our hearing that we do not perceive sound intensity as linear. What we perceive as volume depends more logarithmically on the intensity. To produce a sound that we perceive as twice as loud, a sound wave with approximately ten times the intensity is necessary. For example, a human perceives a sound wave with the intensity 10^{-2} W/m² as twice as loud as one whose intensity is 10^{-3} W/m². ◘ Table 19.3 lists the intensities of some common sources of sounds.

Due to the nearly logarithmic relationship between the subjective perception of volume and the objective sound intensity, it is usually expressed through a logarithmic scale. The unit of this scale is named after Graham Bell. It is called 1 bel. The derived unit decibel (dB) is more well-known. There are 10 dB = 1 bel. We define the sound level L of a sound wave as

$$L[\text{dB}] = 10 \log \frac{I}{I_0} \tag{19.14}$$

Table 19.3 Sound intensity and sound level of some sound sources

	Intensity [W/m²]	Sound level [dB]
Airplane turbine, 10 m distance	1000	150
Pain threshold jackhammer Rock concert	1	120
Screaming	10^{-3}	90
Car interior at 120 km/h	10^{-5}	70
Conversation (50 cm)	$3 \cdot 10^{-6}$	65
Minimal whisper	10^{-10}	20
Threshold of hearing	10^{-12}	0

with the intensity I_0 as a reference, for which the lower hearing limit of 10^{-12} W/m² is typically used.

Example 19.3: Threshold of hearing

To illustrate the high sensitivity of the human ear, one can convert the minimal perceivable intensity $I_0 = 10^{-12}$ W/m² into an amplitude. From Eq. 18.36 it follows

$$\psi_0 = \frac{1}{\omega}\sqrt{\frac{2 I_0}{\rho v_g}}.$$

At a frequency of 2 kHz ($\omega = 12\,500$ Hz), with a density of $\rho = 1.29\,\frac{\text{kg}}{\text{m}^3}$ and $v_g = 331$ m/s we get an amplitude of just $5 \cdot 10^{-12}$ m. That's less than the diameter of an atom!

19.3 Moving Sound Sources

When sound sources or the observer move, the frequencies and wavelengths of the waves change. We want to investigate these effects here. We first deal with the most important effect, the so-called Doppler effect. It has nothing to do with the German word for doubling (doppeln). The effect is named after the Austrian mathemati-

19.3 · Moving Sound Sources

cian and physicist Christian Doppler (◘ Fig. 19.7), who theoretically predicted the effect.

The Doppler effect always occurs when the sound source or observer moves relative to the medium of the wave. It occurs not only with sound, but with all waves. Even with electromagnetic waves (light), there is a comparable effect. However, the situation with electromagnetic waves is somewhat different, as electromagnetic waves do not need a medium of propagation. Here, we limit the discussion to sound waves.

We compare the emission of a sound wave from a stationary source with that of a moving source (◘ Fig. 19.8). With a stationary source, the distance between two wave peaks is just one wavelength λ. This changes when the sound source moves. Let's assume the observer is on the right in the figure, i.e., the sound source is moving towards him. Then the distance between two wave crests shortens for the observer, as the source has moved towards him by the distance s between the emission of two crests:

$$\lambda' = \lambda - s = \lambda - v_s T. \tag{19.15}$$

◘ **Fig. 19.7** Portrait of Christian Doppler (1803–1853). © Austrian Archives/IMAGNO/picture-alliance

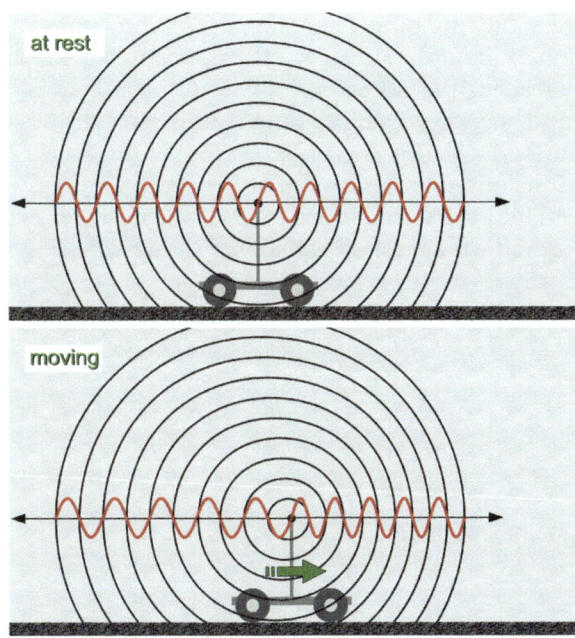

◘ **Fig. 19.8** Sound waves from a stationary and a moving source

Here, v_s is the velocity of the source and T the oscillation period of the exciter. The velocity of the sound wave is $v_{\text{ph}} = \lambda/T$. This results in:

$$\lambda' = \lambda - v_s \frac{\lambda}{v_{\text{ph}}} = \lambda\left(1 - \frac{v_s}{v_{\text{ph}}}\right)$$
$$f' = \frac{v_{\text{ph}}}{\lambda'} = \frac{f}{1 - \frac{v_s}{v_{\text{ph}}}}.$$
(19.16)

Therefore, the observer sees a wave whose wavelength is shortened and whose frequency is increased compared to the stationary case.

Nov, let's examine the case where the sound source is stationary, but the observer is moving. Your first reaction might be that this should lead to the same result, as it only depends on the relative velocity, but this is not correct. The two cases are objectively distinguishable by the movement in the medium. In the case discussed above, the observer is stationary in the medium, while now the sound source is stationary in the medium.

We proceed from the following consideration: When the observer is stationary, a time T passes between two consecutive wave crests reaching him. If he moves towards the source, this time appears shortened to T'. We address the situation from the reference system of the stationary sound source. At the moment the first wave crest reaches the observer, he is a distance λ away from the next wave crest. In the time T', this second wave crest has moved towards the observer by the distance $T' \cdot v_{\text{Ph}}$, and the observer has moved at the speed v_{obs} by the distance $T' v_{\text{obs}}$ the second wave crest. If the observer reaches the second wave crest after the time T', the following must hold true:

$$T' v_{\text{ph}} + T' v_{\text{obs}} = \lambda$$
$$T' v_{\text{ph}} \left(1 + \frac{v_{\text{obs}}}{v_{\text{ph}}}\right) = \lambda$$
$$\lambda' = \frac{\lambda}{1 + \frac{v_{\text{obs}}}{v_{\text{ph}}}},$$
(19.17)

from which follows:

$$f' = f\left(1 + \frac{v_{\text{obs}}}{v_{\text{ph}}}\right).$$
(19.18)

19.3 · Moving Sound Sources

We summarize all the results ($\Rightarrow\Leftarrow$ indicates that the sound source or observer are moving towards each other):

$$\begin{array}{ccc} & \text{moving source} & \text{moving observer} \\ \Rightarrow\Leftarrow & \lambda' = \lambda\left(1 - \frac{v_S}{v_{\text{ph}}}\right) & \lambda' = \frac{\lambda}{1 + \frac{v_{\text{obs}}}{v_{\text{ph}}}} \\ \Leftarrow\Rightarrow & \lambda' = \lambda\left(1 + \frac{v_S}{v_{\text{ph}}}\right) & \lambda' = \frac{\lambda}{1 - \frac{v_{\text{obs}}}{v_{\text{ph}}}} \end{array}$$

$$\begin{array}{ccc} & \text{moving source} & \text{moving observer} \\ \Rightarrow\Leftarrow & f' = \frac{f}{1 - \frac{v_S}{v_{\text{ph}}}} & f' = f\left(1 + \frac{v_{\text{obs}}}{v_{\text{ph}}}\right) \\ \Leftarrow\Rightarrow & f' = \frac{f}{1 + \frac{v_S}{v_{\text{ph}}}} & f' = f\left(1 - \frac{v_{\text{obs}}}{v_{\text{ph}}}\right). \end{array}$$

(19.19)

When the observer and the source move towards each other, the observer registers an increased frequency and a shortened wavelength. If they move away from each other, the frequency is lowered and the wavelength is increased. However, the extent of the changes differs slightly (in the second order in v/v_{Ph}), depending on whether the source or the observer is moving.

Experiment 19.12: Doppler Effect with Whistle

You can perform this experiment yourself. All you need is a string, a small electric sound source, e.g., a buzzer, a small siren, or a whistle, and a partner. You tie the sound source to the string and rotate it as fast as possible over your head. You will not hear any change in pitch, but the external partner will hear a clear rise and fall in pitch, depending on whether the whistle is coming towards him or moving away from him.

Example 19.4: Redshift

The Doppler effect also occurs with electromagnetic waves. If a transmitter is moving towards us, the wavelength is shortened, the frequency is shifted into the blue, if it moves away from us, the wave is redshifted. Due to the much higher propagation speed of electromagnetic waves compared to sound, however, a noticeable effect only occurs at relative velocities approaching the speed of light.

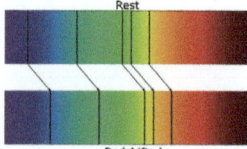

An example is the shift of spectral lines from stars moving towards or away from us. The illustration shows an example. The spectral lines can be seen as black lines against a background that is illustrated in the color of the respective wavelength. On the top you see the lines of a star at rest, on the bottom of a distant star recessing from us. The redshift is clearly visible. However, it should be noted that for such distant objects, the cosmological redshift, which is caused by the expansion of the universe, must be taken into account. It predominates at large red shifts. The contribution of the Doppler effect is no longer discernible on this scale.

Example 19.5: Doppler Effect on a Siren

If a police car with flashing lights and siren passes us, we can hear how the pitch of the siren or the sound of the engine changes as it passes. As the car approaches us, the siren sounds higher. The reason is simple: The sound waves of the siren are "compressed", leading to a higher tone. If it moves away, the sound waves are stretched, which in turn causes a lower tone. The change in pitch is therefore only an observational effect. For the driver, his siren sounds the same all the time.

Now, we have discussed moving sound sources. We implicitly assumed that the velocities of the sources are less than the speed of sound. When the velocity approaches the speed of sound, something new arises from the Doppler effect: the sonic boom.

When deriving the Doppler effect, we considered the distance between two wave crests with a moving source. We obtained (Eq. 19.16)

$$\lambda' = \lambda - s = \lambda\left(1 - \frac{v_s}{v_{\text{ph}}}\right). \tag{19.20}$$

19.3 · Moving Sound Sources

If the speed of the source reaches the speed of sound ($v_s = v_{\text{ph}}$), then λ' becomes zero. The wave crests overlap. A wavefront with particularly high amplitude is created, which is perceived as a bang. It is called the "sound barrier" or the "sonic boom".

In ◘ Fig. 19.9, three situations are shown. In the first, the sound source is moving at 60% of the speed of sound, in the second exactly at the speed of sound, and in the third at 1.4 times the speed of sound. Each shows the emission of a wave crest at a distance of one oscillation period. The positions of the sound source are numbered in time with the numbers 1 (now) to 6 (5*T* previ-

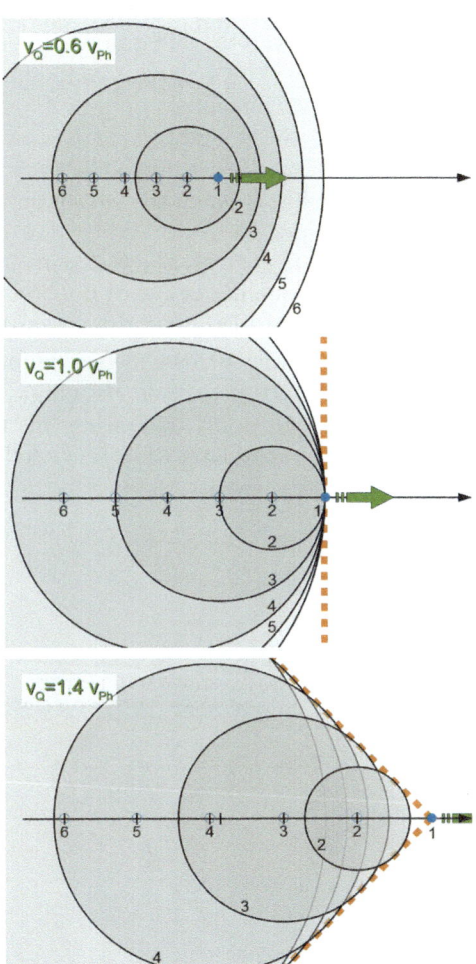

◘ **Fig. 19.9** Sound propagation at high speeds

ously). The wave crests spread out spherically from the point where they were generated.

In the first image at 60% of the speed of sound, you can recognize the Doppler effect. In front of the sound source (to the right), the wave crests are closer together. The wavelength is shortened. Behind the sound source, the distance between the wave crests is significantly larger. The wave fronts of the individual wave crests are clearly separated from each other. There is no overlap.

This is different in the second situation, where the sound source itself is moving at the speed of sound. At the level of the source, the wave crests meet and overlap to form a sonic boom. The wave front that is created is straight (a plane in space). It is indicated by the dashed, orange line. In the image, this seems to be true only near the sound source. However, this is because only the most recent 5 wave crests are shown. If you include all earlier wave crests, a perfect plane results.

In the last image, the sound source is moving at a speed greater than the speed of sound. The pressure wave forms a cone (orange line), which the source carries with it. The sound source appears to sit on the tip of the cone. It is called the "Mach front" or "Mach wave", named after the German physicist Ernst Mach.

We want to determine the opening angle of the Mach front. The sketch in ◘ Fig. 19.10 shows a sound source moving at supersonic speed. In time t it covers the distance $v_s t$ (from point 4 to point 1). During this time, the sound wave that was emitted at point 4 has reached a radius $v_{ph} t$. The wavefronts have overlapped to form a cone,

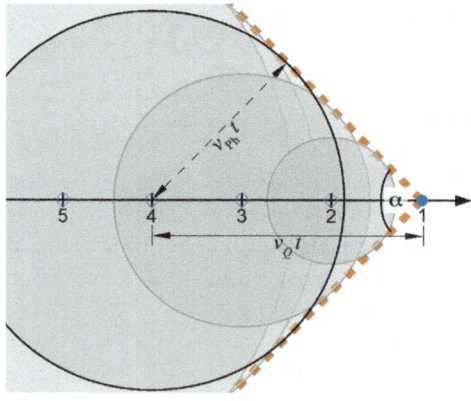

◘ **Fig. 19.10** Opening angle of the Mach front

19.4 · Musical Instruments

the opening angle of which is marked as α in the sketch. The distance $\overline{14}$, the dashed line $v_{ph}t$ and the wavefront from a right-angled triangle. We read off:

$$\sin\frac{\alpha}{2} = \frac{v_{ph}t}{v_s t} = \frac{1}{\frac{v_s}{v_{ph}}} = \frac{1}{M}. \tag{19.21}$$

The quantity M is called the "Mach number". It indicates the speed of the source as a multiple of the speed of sound. For the speed of sound (Mach 1), the half opening angle of the cone is 90°. The lateral surface of the cone has degenerated into a plane, the sound barrier. As the speed increases, the cone becomes increasingly pointed.

Example 19.6: Sonic Boom in Aircraft

The speed of sound in air is about 330 m/s. It depends on humidity, temperature, and pressure and reduces at 10 km altitude to around 290 m/s or 1050 km/h. When an aircraft reaches this speed, it breaks the sound barrier. The Mach front is formed, which becomes visible under specific weather conditions due to condensation. The pressure wave disrupts the flow conditions on the aircraft. The aircraft must penetrate it, which requires additional thrust. The aircraft must be designed for this, like the fighter jet shown in the photo.

© Photo U. S. Navy

Example 19.7: Mach Cone in Water Waves

Sonic boom and Mach cone do not only occur in sound waves, as can be seen here with a duck moving at multiple water wave speeds.

© Prof. Dr. Friedrich Balck

19.4 Musical Instruments

At the end of this chapter on acoustics, we want to take a closer look at musical instruments. They can be roughly divided into three groups:
- **String instruments**, where the sound is produced by a vibrating string and amplified by a resonating body. This group includes the string instruments violin, viola, cello, and double bass (strings are played with

a bow), as well as the guitar, harp, and harpsichord, where the string is plucked. Also, the piano or grand piano, where the string is struck.
- **Wind instruments**, where the sound is produced by a vibrating column of air. A resonating body is usually not necessary. There are woodwind instruments, which stimulate the vibration via a wooden reed (saxophone, clarinet, oboe, bassoon), and brass instruments, where the vibration is stimulated with the lips (trumpet, trombone, tuba, horn), and the organ with different pipes.
- **Percussion instruments** with different techniques, e.g., drums, where a membrane vibrates, or xylophones, where bars made of various materials are struck.

We examine the monochord as an example of a simple string instrument (Experiment 19.13). Some strings are stretched over a resonating body. The pitch can be changed by movable bridges. The string is firmly clamped at both ends. When the string is struck, an excitation is generated. It runs to the end of the string, is reflected there (fixed end), and superimposes itself with the incoming wave. We call ψ_{in} the wave before the reflection and ψ_{out} the wave after. We want to assume harmonic waves. Then it is:

$$\psi_{in}(x,t) = \psi_0 \cos(\omega t - kx)$$
$$\psi_{out}(x,t) = \psi_0 \cos(\omega t + kx + \pi) = -\psi_0 \cos(\omega t + kx).$$
(19.22)

The outgoing wave differs from the incoming one by the direction of propagation ($+kx$ instead of $-kx$) and the phase jump by π due to the reflection at the fixed end. We use the angle sum identities of the cosine function to rewrite and add the waves:

$$\psi_{in}(x,t) = \psi_0 [\cos(\omega t) \cos(kx) + \sin(\omega t) \sin(kx)]$$
$$\psi_{out}(x,t) = -\psi_0 [\cos(\omega t) \cos(kx) - \sin(\omega t) \sin(kx)]$$
$$\Rightarrow \psi_\Sigma(x,t) = 2\psi_0 \sin(\omega t) \sin(kx).$$
(19.23)

The superposition of incoming and outgoing waves ψ_Σ results in a standing wave. The first sine term describes the oscillation of the wave over time, the second sine term indicates the variation of the oscillation along the strings. At the point of reflection, the standing wave has a node. On the monochord, this reflection does not only occur at

19.4 · Musical Instruments

one end of the string, but at both ends. Only waves that have a node at both ends can form. We choose the x-axis so that the left end of the string is at $x = 0$. Then there is automatically a node there ($\psi_\Sigma(0, t) = 0$). For the right end of the string at $x = L$ it must apply ($\psi_\Sigma(L, t) = 0$):

$$kL = n\pi \Rightarrow \frac{2\pi}{\lambda}L = n\pi \Rightarrow L = \frac{n}{2}\lambda. \qquad (19.24)$$

The vibration modes for $n = 1, 2, 3$ are shown in Fig. 19.11 at the times $t = 0$, $t = \frac{\pi}{8}T$, $t = \frac{2\pi}{8}T$ etc. The vibrations with $n = 2, 3, \ldots$ have additional nodes. The wavelengths and frequencies of the modes are

$$\lambda = \frac{2L}{n} \quad \text{and} \quad f = \frac{v_{\text{ph}}}{\lambda} = v_{\text{ph}}\frac{n}{2L}. \qquad (19.25)$$

If a string is, for example, adjusted so that the fundamental mode ($n = 1$) corresponds to a tone C4, then the overtones ($n = 2, 3, \ldots$) are the tones C5, G5, C6, E6, etc.

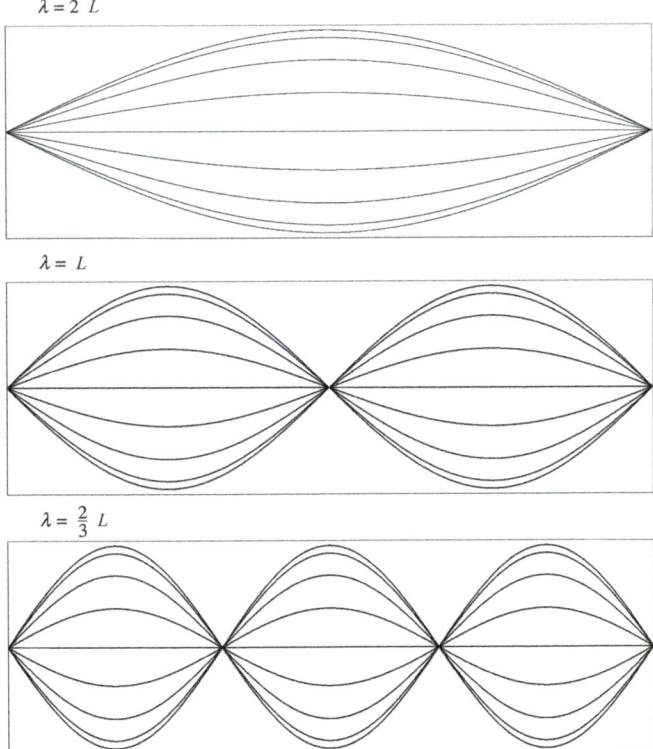

Fig. 19.11 The first vibration modes of the monochord

In addition to the fundamental tone, the pure intervals sound.

The vibrations of the strings are transmitted via the bridge to the resonating body of the instrument and stimulate it to vibrate on its own. Only through this amplification does a clearly audible tone emerge. However, the resonating body also has a significant influence on the sound of the instrument. It does not amplify all frequencies equally. Some frequencies stimulate stronger resonances in the resonating body and are thus emphasized, while others are not.

> **Experiment 19.13: Monochord**
>
> A monochord consists of several strings stretched over a resonating body. Our example has only two strings. The strings can be bowed, plucked, or struck. They are both tuned to the same note. If we play one string, the other string is stimulated to resonate and amplify the tone. With a movable bridge, we can shorten the length of the individual strings. If we set them to integral partial lengths, we can specifically amplify overtones of the played string. This creates a full, overtone-rich sound.
>
>
>
> © RWTH Aachen, Physics Experiment Collection

> **Example 19.8: Violin**
>
> The amplification of tones by the resonating body significantly determines the sound of a violin. Material, shape, and thickness of the bottom and top have a noticable influence. The top contains the two F-holes and is reinforced on the inside by a bar, the bass bar. Unlike the box-shaped resonating body of the monochord, which forms sharp resonances, a violin has broad, closely spaced resonances due to its rounded shapes. The illustration shows Chladni sound figures (see Experiment 17.22) of the top (above) and bottom (below) of

a violin. The resonance frequencies of the top (147 Hz, 222 Hz, 304 Hz–349 Hz) and the bottom (116 Hz, 167 Hz, 222 Hz, 230 Hz, 349 Hz, 403 Hz) are slightly offset (in the ratio 5 : 4). The fourth mode is missing in the top. The resonances must be well adjusted to produce a full, pleasant sound. This is the art of violin making.

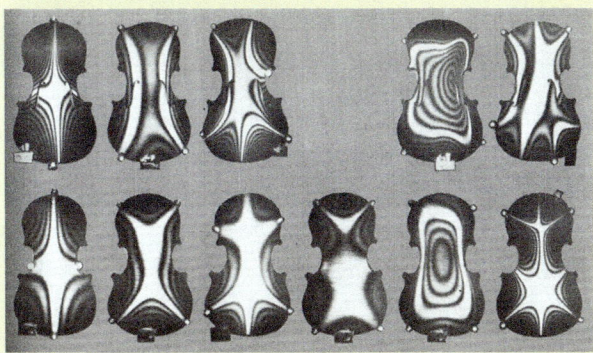

© Spectrum Academic Publishing, Physics of Musical Instruments, Article Carleen Maley Hutschins, Sound and Acoustics of the Violin, Fig. 5

In wind instruments, a column of air vibrates in the tube. Let's consider a trumpet as an example. The tone is stimulated at the mouthpiece by pressure variations with the lips. The pressure in the air column has a loose end at the mouthpiece, i.e., a vibration antinode. In contrast, the pressure of the air column in the bell has a fixed end. In the simplest model, one could describe the trumpet with a straight tube with a fixed and a loose end. This would lead to vibrations as shown in ◖ Fig. 19.12. The frequencies of the modes are given by

$$f_n = \frac{(2n-1)}{4} \frac{v_{\text{ph}}}{l}, \qquad (19.26)$$

where l is the length of the tube. The trumpeter can stimulate different modes. With the valves of the trumpet, he can change the tube length and thus also reach intermediate tones between the fundamental modes.

This model of a one-sided open, cylindrical tube greatly simplifies the describption of the instrument. Clarinets have a straight tube and only a small bell. For them, the formula is approximately correct. For trum-

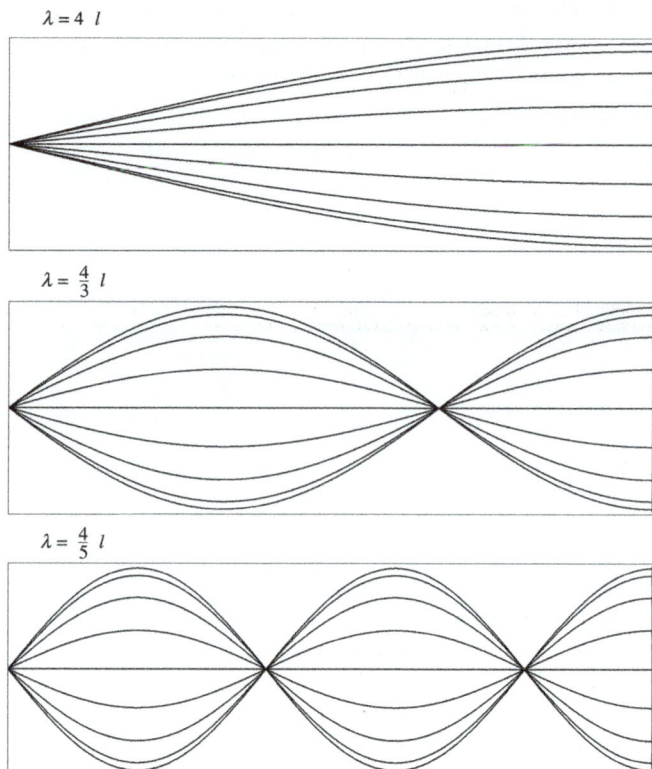

◻ **Fig. 19.12** Vibration modes in a simple tube

pets, one can try to achieve a somewhat more realistic modeling of a real instrument by at least roughly considering the gradual increase in tube diameter along the instrument. In a bit more realistic case, we describe the sound tube as composed of a cylindrical and a conical part. This case can still be treated analytically. At the entrance of the cylindrical tube is the mouthpiece, through which the player stimulates the vibration with his lips. At the end of the conical tube follows the bell. The various brass instruments differ in the length ratio of the cylindrical and conical parts. A tuba consists almost entirely of a conical part, while a trombone predominates in the cylindrical part.

One can construct a simplified model of a trumpet with a mouthpiece, a rubber hose, and a plastic funnel. With a setup similar to the one from Experi-

19.4 · Musical Instruments

ment 18.3, the resonance frequencies can be determined. Figure 19.13[2] shows the results of a calculation of the fundamental tone and the first five overtones of the hose trumpet as a function of the length ratio $L_1/(L_1 + L_2)$ (blue lines) compared to some measurements (red crosses). As you can see, the length ratio has a significant influence on the location of the resonances.

Starting from this model, one can try to approach a real instrument step by step by using the actual course of the tube diameter, including the bell, considering the excitation by the mouthpiece, and so on.

We have now examined the monochord as an example of a string instrument and the trumpet as a wind instrument in some more detail. At the end of this chapter, we want to take a closer look at a percussion instrument, the timpani, a type of kettledrum. With the timpani, the sound is produced by vibrations on a skin that is stretched over the drum. Underneath the skin is a resonating body of semispherical shape that amplifies the tones. The vibrations on the skin can be made visible as Chladni sound figures. Figure 19.14 shows some vibration modes as an example. Due to the high symmetry of the timpani, the vibrations modes are particularly regular. When you strike the timpani, you hear a spectrum of vibration modes that make up the sound. Although you cannot isolate individual modes in striking the instrument, you can change the relative strength of the individual modes by striking it in different positions and thus change the timbre.

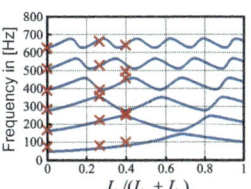

Fig. 19.13 Resonance frequencies of a hose trumpet (total length $L_1 + L_2 = 1.5$ m, hose diameter 19 mm). © From Bachelor's thesis Jonathan Parlasca, RWTH Aachen University

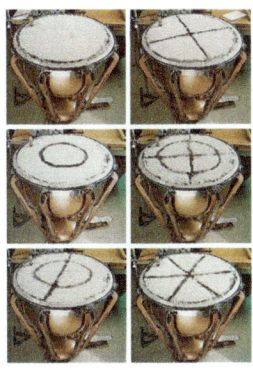

Fig. 19.14 Chladni sound figures on a timpani

Example 19.9: Eigenmodes of a circular membrane

At the end of the volume, we would like to present you with a mathematically somewhat more challenging example. We want to calculate the eigenmodes of a circular membrane, as they occur, for example, in drums. We will use polar coordinates, which we have indicated in the figure. First, we need the wave equation, which we will extend from Eq. 18.17, which deals with the one-dimensional case, to our two-dimensional example. It then reads:

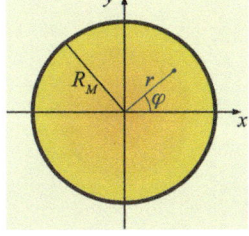

---

2 From the bachelor's thesis of Thomas Jonathan Parlasca, RWTH Aachen, September 25, 2012.

$$v_{ph}^2\left[\frac{\partial^2\psi(x,y,t)}{\partial x^2}+\frac{\partial^2\psi(x,y,t)}{\partial y^2}\right]=\frac{\partial^2\psi(x,y,t)}{\partial t^2},$$

which we will now represent in polar coordinates:

$$v_{ph}^2\left[\frac{\partial^2\psi(r,\varphi,t)}{\partial r^2}+\frac{1}{r}\frac{\partial\psi(r,\varphi,t)}{\partial r}+\frac{1}{r^2}\frac{\partial^2\psi(r,\varphi,t)}{\partial\varphi^2}\right]$$
$$=\frac{\partial^2\psi(r,\varphi,t)}{\partial t^2}.$$

As a first step towards the solution, we separate the time-dependent part. It can be captured by the following approach:

$$\psi(r,\varphi,t)=v(r,\varphi)\cdot\sin(\omega t+\phi_0),$$

The sine term describes the temporal oscillation, while the function $v(r,\varphi)$, which no longer depends on time, describes the spatial pattern of the standing waves, including the location of the nodes and antinodes. If we insert this approach into the differential equation, we get:

$$\sin(\omega t+\phi_0)\left[\frac{\partial^2 v(r,\varphi)}{\partial r^2}+\frac{1}{r}\frac{\partial v(r,\varphi)}{\partial r}+\frac{1}{r^2}\frac{\partial^2 v(r,\varphi)}{\partial\varphi^2}\right]$$
$$=-\frac{\omega^2}{v_{ph}^2}v(r,\varphi)\sin(\omega t+\phi_0).$$

This differential equation is fulfilled when the spatial function $v(r,\varphi)$ satisfies the following differential equation:

$$\frac{\partial^2 v(r,\varphi)}{\partial r^2}+\frac{1}{r}\frac{\partial v(r,\varphi)}{\partial r}+\frac{1}{r^2}\frac{\partial^2 v(r,\varphi)}{\partial\varphi^2}=-\lambda v(r,\varphi),$$

where we have introduced a new constant $\lambda=\omega^2/v_{ph}^2$. We can now proceed to sorting the r- and φ-dependencies:

$$r^2\frac{\partial^2 v(r,\varphi)}{\partial r^2}+r\frac{\partial v(r,\varphi)}{\partial r}+\lambda r^2 v(r,\varphi)$$
$$=-\frac{\partial^2 v(r,\varphi)}{\partial\varphi^2}.$$

This differential equation can be solved with a factorized approach:

$$v(r, \varphi) = \psi_0 R(r) \Phi(\varphi).$$

We have explicitly written out the amplitude ψ_0 so that we no longer have to worry about the scaling of $R(r)$ and $\Phi(\varphi)$. We insert the ansatz into the differential equation and obtain after a short transformation:

$$\frac{1}{R(r)}\left[r^2 \frac{\partial^2 R(r)}{\partial r^2} + r\frac{\partial R(r)}{\partial r} + \lambda r^2\right] = \sigma = -\frac{1}{\Phi(\varphi)}\frac{\partial^2 \Phi(\varphi)}{\partial \varphi^2}.$$

Since the left side only depends on r and the right side only depends on φ, this equation can only be satisfied if each side, taken separately, yields something that depends neither on r nor on φ. We have called this constant σ and have already entered it into the equation. This gives us two uncoupled differential equations, which after a further short transformation we can write as follows:

1. $r^2 \dfrac{\partial^2 R(r)}{\partial r^2} + r\dfrac{\partial R(r)}{\partial r} + (\lambda r^2 - \sigma)R(r) = 0$

2. $\dfrac{\partial^2 \Phi(\varphi)}{\partial \varphi^2} + \sigma \Phi(\varphi) = 0.$

The second differential equation should look familiar to you. It is the differential equation of an undamped free oscillation with the solution:

$$\Phi(\varphi) = \cos\left(\sqrt{\sigma}\varphi + \Phi_\varphi\right).$$

However, note that the variable here is not time t, but the polar coordinate φ. Therefore, $\Phi(\varphi) = \Phi(\varphi + 2\pi)$ must hold, as these coordinates correspond to the same point on the membrane. The condition is fulfilled when $\sqrt{\sigma} = n$ is true, with an integer, non-negative number n [3]. Since φ represents the angle to the x-axis, we can make Φ_φ zero by suitable rotation of the coordinate system around the center of the membrane. We will therefore omit the parameter in the following.

---

[3] Negative numbers n also fulfill the condition. However, these solutions are already included in the solutions with a positive n with a suitable choice of ϕ_φ.

We now turn to the remaining differential equation. We eliminate the constant λ by a substitution. $\rho = \sqrt{\lambda}\,r$, which leads us to the final form:

$$\rho^2 \frac{\partial^2 R(\rho)}{\partial \rho^2} + \rho \frac{\partial R(\rho)}{\partial \rho} + \left(\rho^2 - n^2\right) R(\rho) = 0.$$

This is the Bessel's differential equation, named after the astronomer, physicist, and mathematician Friedrich Wilhelm Bessel, who derived many properties of the solution functions. The solution cannot be represented by trigonometric or related functions, but that should not bother you. The solutions are called Bessel functions of the first kind $J_n(\rho)$. For each number n there is a unique solution function. It can be specified, for example, by a power series:

$$J_n(\rho) = \left(\frac{\rho}{2}\right)^v \sum_{k=0}^{\infty} \frac{(-1)^k}{k!(n+k)!} \left(\frac{\rho}{2}\right)^{2k}.$$

The first four are shown in the figure below. The zeros of the Bessel functions can be found in the following table. We will need them in the following.

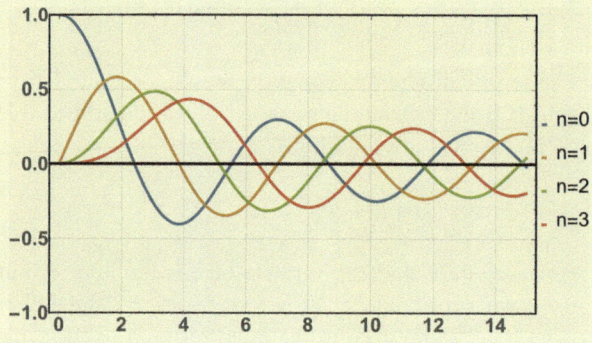

Zero	$J_0(\rho)$	$J_1(\rho)$	$J_2(\rho)$	$J_3(\rho)$	$J_4(\rho)$
0.	2.4048	3.8317	5.1356	6.3802	7.5883
1.	5.5201	7.0156	8.4172	9.7610	11.0647
2.	8.6537	10.1735	11.6198	13.0152	14.3725
3.	11.7915	13.3237	14.7960	16.2235	17.6160
4.	14.9309	16.4706	17.9598	19.4094	20.8269

19.4 · Musical Instruments

So we have found $R(r) = J_n(\sqrt{\lambda} r)$. Now it's time to address the boundary condition of the problem. We are calculating the vibration of a circular membrane, which is clamped at the outer edge to the body of the instrument. The deflection must vanish at the outer edge, which leads us to the condition $R(R_M) = 0$ or $J_n(\sqrt{\lambda} R_M) = 0$. If we insert the constant λ, we get the condition:

$$J_n\left(\frac{\omega_{nl} R_M}{v_{ph}}\right) = 0.$$

The frequencies ω_{nl} are the eigenfrequencies of the membrane. The index n denotes the order of the Bessel function and the index l labels the different zeros of the function. The fundamental mode is obtained, for example, for $(n = 0, l = 0)$:

$$\frac{\omega_{00} R_M}{v_{ph}} = 2.4048 \Rightarrow \lambda_{00} = \frac{2\pi R_M}{2.4048}.$$

To convert the wavelength λ_{00} into a frequency, we still need the phase velocity v_{ph}, which depends on the tension of the membrane (tuning). For the D-bass timpani with a diameter of $2R_M = 73$ cm this results in fundamental tones between D2 (73.3 Hz) and A2 (110 Hz) depending on the tune. Some of the modes are depicted in the figures. Compare with the Chladni sound figures in ▶ Fig. 19.14!

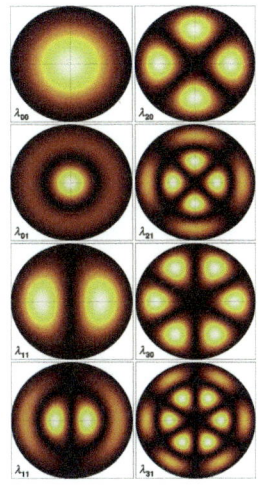

❓ Problems

1. If you close the upper end of an organ pipe while playing, its pitch changes. What musical interval separates the fundamental tone before and during closing. What is the result for the first overtone? The frequency of the first overtone decreases by 131 Hz due to the closing of the pipe's end. In which tone is the organ pipe tuned and how long is it?
2. In the mechanical workshop of the Physics Institute, a pillar drill with a sound level of 72 dB, a lathe with 78 dB, and a milling machine with 74 dB are operated simultaneously. What is the total sound level of the three machines?

3. A train passes a railway crossing, with the engine driver sounding a whistle signal as a warning. A musician waiting at the railway barrier observes that the whistle tones differ by a major third (frequency ratio 5/4) when the locomotive is approaching and moving away. How fast is the train?

4. A ship is stationary on the surface of the sea and sends out an sonar signal with a frequency of 50 kHz. After a time of 2.6 s, it receives the signal reflected from a submarine at an angle of 65° to the vertical. How far away is the submarine ($v_{sound} = 1475\,\text{m/s}$)? At what speed is the submarine moving in horizontal direction towards the ship if the frequency of the received signal is 380 Hz above that of the transmitted signal?

5. A sound source is moving at a speed of 2 m/s towards a wall and emits a tone of frequency 262 Hz. Determine the beat frequency that a stationary observer behind the transmitter perceives.

Supplementary Information

A1 List of Symbols – 596

A2 Solutions to the Problems – 599

A3 Mathematical Introduction – 621

© The Editor(s) (if applicable) and The Author(s), under exclusive license to Springer-Verlag GmbH, DE, part of Springer Nature 2025
S. Roth and A. Stahl, *Mechanics*, https://doi.org/10.1007/978-3-662-68079-7

A1 List of Symbols

work	W	▶ Sect. 7.1
Avogadro constant	N_A	
orbital or instantaneous velocity	$\vec{v}_O$	▶ Sect. 5.8
acceleration	$\vec{a}$	▶ Sect. 5.5
decibel	dB	▶ Sect. 19.2
density	ϱ	
angular momentum	$\vec{L}$	▶ Sect. 13.1
torque	$\vec{M}$	▶ Sect. 12.2
angle of rotation	$\vec{\varphi}$	▶ Sect. 12.2
pressure	p	▶ Sect. 15.1
diameter	d	
Young's modulus	E	▶ Sect. 14.1
energy	E	▶ Sect. 7.2
energy density	ε	▶ Sect. 18.4
standard acceleration	$\vec{g}$	▶ Sect. 5.6
stiffness	k	▶ Sect. 14.1
area	A	
frequency	f	▶ Sect. 17.1
velocity	$\vec{v}$	▶ Sect. 5.3
gravitational force	$\vec{F}_G$	▶ Sect. 11.3.2
gravitational field strength	$\vec{G}$	▶ Sect. 11.5
gravitational constant	G	▶ Sect. 11.3.1
gravitational potential	Φ_G	▶ Sect. 11.5
group velocity	v_g	▶ Sect. 18.3
Q-factor	Q	▶ Sect. 17.4
momentum	$\vec{p}$	▶ Chap. 8
Intensity	I	▶ Sect. 18.4
Joule	J	▶ Sect. 7.1
Kelvin	K	▶ Sect. 2.2
kilogram	kg	▶ Sect. 2.5
kinetic energy	E_{kin}	▶ Sect. 7.2
compressibility	κ	▶ Sect. 15.2
bulk modulus	$\overline{\kappa}$	▶ Sect. 15.2

A1 List of Symbols

force	F	▶ Sect. 6.2
impulse	T	▶ Sect. 8.5
power	P	▶ Sect. 7.1
velocity of light	c	
Mach number	M	▶ Sect. 19.3
mass	m, M	▶ Sect. 6.2
area density	μ	▶ Sect. 18.2
mass flow rate	J	▶ Sect. 16.2
mass flux	j	▶ Sect. 16.2
meter	m	▶ Sect. 3.2
Newton	N	▶ Sect. 6.2
surface tension		▶ Sect. 15.5.1
Pascal	Pa	▶ Sect. 14.1
period	T	▶ Sect. 17.1
phase velocity	v_{ph}	▶ Sect. 18.3
potential	φ	▶ Sect. 17.4
frequency of precession	$\vec{\omega}_P$	▶ Sect. 13.5
Poisson's ratio	μ	▶ Sect. 14.1
radian	rad	▶ Sect. 2.6
radius	r	
solid angle	Ω	▶ Sect. 2.6
reduced mass	m'	▶ Sect. 11.2
coefficient of friction	μ_s, μ_k	▶ Chap. 9
friction	$\vec{F}_{fric}$	▶ Sect. 9.1
relative elongation		▶ Sect. 14.1
Reynolds number	Re	▶ Sect. 16.7
restoring force	$\vec{F}_r$	▶ Sect. 17.1
sound pressure	L	▶ Sect. 19.2
shear stress	τ	▶ Sect. 14.4
center of gravity	$\vec{r}_{cm}$	▶ Sect. 12.3
period	T	▶ Sect. 17.1
second	s	▶ Sect. 2.4
steradian	sr	▶ Sect. 2.6
distance	s	
shear modulus	G	▶ Sect. 14.4

moment of inertia	I	▶ Sect. 13.2
inertia matrix	$\tilde{I}$	▶ Sect. 13.6
viscosity	η	▶ Sect. 16.4
volume	V	
volumetric flow rate	I	▶ Sect. 16.2
wave function	$\psi(\vec{r},t)$	▶ Sect. 18.1
wave length	λ	▶ Sect. 18.1
wave vector	$\vec{k}$	▶ Sect. 18.1
wave number	k	▶ Sect. 18.1
drag coefficient	C_D	▶ Sect. 16.7
angle	$\alpha, \beta, \gamma, \theta, \phi$	
angular acceleration	$\vec{\alpha}$	▶ Sect. 13.2
angular velocity	$\vec{\omega}$	▶ Sect. 10.3
time	t	▶ Sect. 2.4
centrifugal force	$\vec{F}_{CF}$	▶ Sect. 10.3
centripetal force	$\vec{F}_{CP}$	▶ Sect. 10.3
tension	σ	▶ Sect. 14.1

A2 Solutions to the Problems

Chapter 1
No problems

Chapter 2
1. Scalars: time, temperature, mass, energy, work
   Vectors: momentum, force, acceleration, velocity
2. $\eta = \frac{\pi}{8} \frac{r^4}{l} \Delta p \frac{1}{l}$
   $\Rightarrow [\eta] = \frac{m^4 Pa}{m \frac{m^3}{s}} = Pa\,s = \frac{N}{m^2}s = \frac{kg\,m\,s}{m^2 s^2} = \frac{kg}{m\,s}$
3. a) $[c_1] = m; [c_2] = \frac{m}{s}; [c_3] = \frac{m}{s^2}$
   b) $[c_1] = \frac{m}{s}; [c_2] = \frac{1}{s^2}; [c_3] = m$
   c) $[c_1] = \frac{m}{s^2}$
   d) $[c_1] = m; [c_2] = \frac{1}{s}; [c_3] = \frac{m}{s}$
4. Volume: $V = \frac{\pi}{4} d^2 h \Rightarrow h = \frac{4V}{\pi d^2}$. Surface $A = 4\frac{V}{d} + \frac{\pi}{2} d^2$ should be minimal. $\frac{\partial A}{\partial d} = 0$ results in the relationship: $-4\frac{V}{d^2} + \pi d = 0 \Rightarrow 4V = \pi d^3 \Rightarrow h = d$

Chapter 3
1. The difference is 124.7 g. Within the given accuracy, this agrees with 125 g.
2. It should be reset at the latest after $\Delta t = \frac{30\,s}{10^{-5}} = 3 \cdot 10^6$ s, so once a month.
3. Absolute accuracy of the distance:
   $\Delta s = 3 \cdot 10^{-10} \cdot 3.3 \cdot 10^8$ m $= 0.1$ m.
   Absolute accuracy of the runtime:
   $\Delta t = \frac{\Delta s}{c} = \frac{0.1\,m}{3 \cdot 10^8 \frac{m}{s}} = 3 \cdot 10^{-10}$ s $= 0.3$ ns.
4. With the given values for l and T results in $g = 9.89438\ldots \frac{m}{s^2}$. From the error propagation, one obtains a relative error of $\frac{\sigma_g}{g} = 0.021$, so in total $g = 9.89 \pm 0.21 \frac{m}{s^2}$. The uncertainty of T is the dominant contribution. The uncertainty of l can be neglected.

Chapter 4
No problems

Chapter 5
1. (a) $9.5 \cdot 10^{-9} \frac{m}{s} = 3.4 \cdot 10^{-8} \frac{km}{h}$
   (b) $2.98 \cdot 10^4$ m/s $= 1.07 \cdot 10^5$ km/h
   (c) $1.02 \cdot 10^{-4}$ m/s $= 3.67 \cdot 10^{-4}$ km/h
   (d) $333 \frac{m}{s} = 1200 \frac{km}{h}$

2. $\bar{v} = 2\frac{v_1 v_2}{v_1+v_2} = 23\,\frac{\text{km}}{\text{h}} < \frac{1}{2}(v_1 + v_2)$

3. $v = \sqrt{2gh} = 44\,\frac{\text{m}}{\text{s}}$

$t = \sqrt{\frac{2s}{g}} = 4.5\,\text{s}$

4. $t_2 = \frac{a_1}{|a_2|}t_1 = \frac{0.8}{1.2}t_1$

$2s = a_1 t_1^2 + |a_2|\left(\frac{a_1}{a_2}\right)^2 t_1^2 \Rightarrow t_1 = 38.7\,\text{s}$

$t = t_1 + t_2 = 64.5\,\text{s}$

$v_{\max} = a_1 t_1 = 31.0\,\frac{\text{m}}{\text{s}} = 111\,\frac{\text{km}}{\text{h}}$

$t' = t_1 + t_v + t_2 = 34.7\,\text{s} + 7.1\,\text{s} + 23.1\,\text{s} = 64.9\,\text{s}$

$t' - t = 0.4\,\text{s}$

5. $v - v_0 = at$

$s - s_0 = \frac{1}{2}at^2 + v_0 t$

$v^2 - v_0^2 = 2a(s - s_0)$

6. $s = \frac{v^2}{2a}$

$\frac{s_2}{s_1} = \frac{v_2^2}{v_1^2} = 2.0$

$v_2^2 - v_z^2 = 2as = v_1^2$

$v_z = \sqrt{v_2^2 - v_1^2} = 49\,\frac{\text{km}}{\text{h}}$

7. $\frac{1}{2}at_1^2 = s_1 = 25\,\text{m}$

$vt_2 = at_1 t_2 = s_2 = 75\,\text{m}$

$t = t_1 + t_2 = 9.58\,\text{s}$

$\frac{2s_1}{t_1^2} = a = \frac{s_2}{t_1(t - t_1)}$

$t_1 = \frac{2s_1}{2s_1 + s_2} t = 3.83\,\text{s}$

$a = \frac{2s_1}{t_1^2} = 3.4\,\frac{\text{m}}{\text{s}^2}$

$v = at_1 = 13\,\frac{\text{m}}{\text{s}}$

A2 Solutions to the Problems

8. $\dfrac{19\,\text{m}}{360\,\frac{\text{m}}{\text{s}}} \cdot 19\,\dfrac{\text{m}}{\text{s}} = 1.0\,\text{m}$

9. (a) No wind: $t_a = \dfrac{2s}{v}$

   (b) Side wind: $t_b = \dfrac{2s}{\sqrt{v^2-w^2}} = \dfrac{2s}{v}\dfrac{1}{\sqrt{1-\frac{v^2}{w^2}}}$

   (c) Headwind and tailwind: $t_c = \dfrac{s}{v-w} + \dfrac{s}{v+w}$

   $= \dfrac{2s}{v}\dfrac{1}{1-\frac{w^2}{v^2}}$

   $t_a < t_b < t_c$

10. $s = vt = v\sqrt{\dfrac{2h}{g}} = 12\,\dfrac{\text{m}}{\text{s}}\cdot\sqrt{\dfrac{2\cdot 0.5\,\text{m}}{9.81\,\frac{\text{m}}{\text{s}^2}}} = 3.8\,\text{m}$

   $\tan\phi = \dfrac{v_{\text{vert}}}{v_{\text{hori}}} = \dfrac{gt}{v} = \dfrac{g}{v}\sqrt{\dfrac{2h}{g}} = \dfrac{9.81\,\frac{\text{m}}{\text{s}^2}}{12\,\frac{\text{m}}{\text{s}}}\sqrt{\dfrac{2\cdot 0.5\,\text{m}}{9.81\,\frac{\text{m}}{\text{s}^2}}} = 0.262$

   $\Rightarrow \phi = 15°$

11. $t = \sqrt{\dfrac{2}{g}h} = \sqrt{\dfrac{2\,l}{g\sqrt{2}}}$

    $v = \dfrac{s}{t} = \dfrac{l/\sqrt{2}}{t} = \sqrt{\dfrac{g\,l}{2\sqrt{2}}} = \sqrt{\dfrac{9.81\,\frac{\text{m}}{\text{s}^2}\cdot 10\,\text{m}}{2\sqrt{2}}} = 5.9\,\dfrac{\text{m}}{\text{s}}$

12. $t = 2\dfrac{v/\sqrt{2}}{g} = \sqrt{2}\dfrac{v}{g} = \sqrt{2}\dfrac{167\,\frac{\text{m}}{\text{s}}}{9.81\,\frac{\text{m}}{\text{s}^2}} = 24\,\text{s}$

13. a) $x = (v\cos\beta)t$

    $y = (v\sin\beta)t - \dfrac{1}{2}gt^2$

    $y(t_{\text{end}}) = 0 \Rightarrow t_{\text{end}} = \dfrac{2v\sin\beta}{g}$

    $R = x(t_{\text{end}}) = \dfrac{2v^2}{g}\sin\beta\cos\beta = 2h\sin 2\beta$

    b) R is maximum for $\sin 2\beta = 1 \Rightarrow \beta_{\max} = 45°$

    c) $x = (v\cos\beta)t$

    $y = H + (v\sin\beta)t - \dfrac{1}{2}gt^2$

    $y(t_{\text{end}}) = 0 \Rightarrow \dfrac{1}{2}gt_{\text{end}}^2 - v\sin\beta\, t_{\text{end}} - H = 0$

    $\Rightarrow t_{\text{end}} = \dfrac{1}{g}\left(v\sin\beta + \sqrt{v^2\sin^2\beta + 2gH}\right)$

    $R = x(t_{\text{end}}) = 2h\cos\beta\left(\sin\beta + \sqrt{\sin^2\beta + \dfrac{H}{h}}\right)$

d) $\dfrac{d}{d\beta}\left(\dfrac{R}{h}\right) = -2\sin^2\beta + 2\cos^2\beta$

$\qquad - 2\sin\beta\sqrt{\sin^2\beta + H/h}$

$\qquad + 2\cos\beta \cdot \dfrac{2\sin\beta\cos\beta}{2\sqrt{\sin^2\beta + H/h}}$

With $s = \sin\beta$ and $a = H/h$ follows from $\dfrac{d}{d\beta}\left(\dfrac{R}{h}\right) = 0$:

$(1 - 2s^2)\sqrt{s^2 + a} - s(s^2 + a) + s(1 - s^2) = 0$

$\sqrt{s^2 + a} = \dfrac{2s^3 + as - s}{1 - 2s^2}$

$(s^2 + a)(1 - 2s^2)^2 = (2s^3 + as - s)^2$

$s^2 = \dfrac{1}{2+a} \Rightarrow \sin\beta_{max} = \dfrac{1}{\sqrt{2 + H/h}}$

14. $\sin\beta_{max} = \dfrac{1}{\sqrt{2 + H/h}}$

$h = \dfrac{v^2}{2g} = 8.61\,\text{m}$

$\sin\beta_{max} = \dfrac{1}{\sqrt{2 + \frac{2.1\,\text{m}}{8.61\,\text{m}}}} = 0.668 \Rightarrow \beta_{max} = 42°$

Normally, strikes are made at smaller angles (30°–40°) because, due to human anatomy, one can impart a greater speed to the ball at these angles.

15. $\sin 2\beta_1 = \sin 2\beta_2$

$\beta_1 = 60° \Rightarrow \beta_2 = 30°$

$R = \dfrac{v^2}{g}\sin 2\beta \Rightarrow v = \sqrt{\dfrac{gR}{\sin 2\beta}} = 499\,\dfrac{\text{m}}{\text{s}}$

$\Delta t = t_1 - t_2 = \dfrac{R}{v\cos\beta_1} - \dfrac{R}{v\cos\beta_2}$

$= \dfrac{R}{v}\left(\dfrac{1}{\cos\beta_1} - \dfrac{1}{\cos\beta_2}\right)$

$= 44\,\text{s} \cdot \left(\dfrac{1}{\cos 60°} - \dfrac{1}{\cos 30°}\right)$

$= 37\,\text{s}$

16.

$\omega^2 r = a = \sqrt{(4g)^2 - g^2} = \sqrt{15}g$

$f = \dfrac{\omega}{2\pi} = \dfrac{1}{2\pi}\sqrt{\dfrac{\sqrt{15}g}{r}} = 0.40\dfrac{1}{s}$

$\cos\alpha = \dfrac{g}{4g} = \dfrac{1}{4} \Rightarrow \alpha = 76°$

17. $a = \omega^2 r = \left(\dfrac{2\pi}{24 \cdot 3600\,s}\right)^2 \cdot 6.4 \cdot 10^6\,m = 0.034\,\dfrac{m}{s^2}$

$\Rightarrow \dfrac{a}{g} = 0.0035$

Chapter 6

1. $a = \dfrac{F}{m} = \dfrac{300\,kN}{87\,t + 100\,t} = 1.6\,\dfrac{m}{s^2}$

    $a = \dfrac{F}{m} = \dfrac{300\,kN}{87\,t + 4000\,t} = 0.073\,\dfrac{m}{s^2}$

    $F = m \cdot a = 100\,t \cdot 1.6\,\dfrac{m}{s^2} = 160\,kN$

    $F = m \cdot a = 4000\,t \cdot 0.073\,\dfrac{m}{s^2} = 294\,kN$

2. $F = m \cdot a = m \cdot \dfrac{v^2}{2s} = 507\,kN$
3. $F_{up} = m(g + a) = 1100\,kg\left(9.81\,\dfrac{m}{s^2} + 0.9\,\dfrac{m}{s^2}\right) = 11.8\,kN$

    $F_{down} = m(g - a) = 9.8\,kN$

4. $a = g\sin\beta = g\sin(\arctan 0.12) = 1.2\,\dfrac{m}{s^2}$
5. $F_1 = m_1 g \cos\theta, \quad F_2 = m_2 g \sin\theta$

    $F_1 = F_2 \Leftrightarrow \dfrac{m_1}{m_2} = \tan\theta$

    $a = \dfrac{F_1 - F_2}{m_1 + m_2} = g\dfrac{m_1\cos\theta - m_2\sin\theta}{m_1 + m_2}$

    $m_1 = m_2 \Rightarrow a = \dfrac{g}{2}(\cos\theta - \sin\theta)$

6. $F = m\dfrac{s}{L}g \Rightarrow a = \dfrac{F}{m} = \dfrac{g}{L}s \Rightarrow \ddot{s} = \dfrac{g}{L}s$

    Ansatz: $s(t) = Ae^{\alpha t} + Be^{-\alpha t} \Rightarrow \dot{s}(t) = A\alpha e^{\alpha t} - B\alpha e^{-\alpha t}$
    Substitute into ODE: $\ddot{s} = \alpha^2 s \Rightarrow \alpha = \sqrt{g/L}$
    1. Initial condition: $\dot{s}(0) = 0 = A\alpha - B\alpha \Rightarrow A = B$
    This results in: $s(t) = A(e^{\alpha t} + e^{-\alpha t}) = 2A\cosh\alpha t$

2. Initial condition: $s(0) = l = 2A \Rightarrow A = l/2$
Solution: $s(t) = l\cosh\sqrt{\frac{g}{L}}t$
End at: $s(\tau) = L \Rightarrow \tau = \sqrt{\frac{L}{g}}\operatorname{arccosh}\frac{L}{l}$

7. $ma = -kv \Rightarrow \dot{v} = -\frac{k}{m}v \Rightarrow v(t) = v_0 e^{-\frac{k}{m}t}$

$$k = \frac{F}{v} = \frac{5\,\text{kN}}{10\,\frac{\text{m}}{\text{s}}} = 500\,\frac{\text{Ns}}{\text{m}}$$

$$t_1 = \frac{m}{k}\ln\frac{v_0}{v_1} = \frac{400\,\text{kg}}{500\,\frac{\text{Ns}}{\text{m}}}\ln\frac{50}{5} = 1.8\,\text{s}$$

8. $\tan 8° = \frac{a}{g} \Rightarrow a = g\tan 8° = 1.4\,\frac{\text{m}}{\text{s}^2}$

9. $F = \sqrt{(mg + ma\sin(20°))^2 + (ma\cos(20°))^2} = 874\,\text{N}$

10. $\dfrac{\frac{1}{2}mg}{F_{\text{rope}}} = \dfrac{d}{\sqrt{(l/2)^2 + d^2}} \approx \dfrac{d}{l/2} \Rightarrow d = \dfrac{lmg}{4F_{\text{rope}}}$

$$= \frac{10\,\text{m} \cdot 25\,\text{kg} \cdot 9.81\,\frac{\text{m}}{\text{s}^2}}{4 \cdot 10\,\text{kN}} = 0.061\,\text{m}$$

▪▪ Chapter 7

1. $W = mgh = 441\,\text{J}$

   $E_{\text{kin}} = \frac{1}{2}mv^2 = 130\,\text{J}$

   $E_{\text{pot}} = mgh = 4.2 \cdot 10^{10}\,\text{J}$

   $E_{\text{kin}} = \frac{1}{2}mv^2 = 1.4 \cdot 10^{10}\,\text{J}$

   $P_S = E_0 4\pi r^2 = 3.9 \cdot 10^{26}\,\text{W}$

   $E_{\text{pot}} = mgh = 1.7 \cdot 10^{13}\,\text{J}$

   $P_P = \dfrac{dm}{dt}gh = 8.6 \cdot 10^8\,\text{W}$

2. $2\pi rF = GF' \Rightarrow F' = \frac{2\pi r}{G}F = \frac{2\pi \cdot 1\,\text{m}}{0.008\,\text{m}}200\,\text{N} = 157\,\text{kN}$

3. $U_{\text{rear}}F_{\text{rear}} = U_{\text{front}}F_{\text{front}}$
   $U'_{\text{rear}}F'_{\text{rear}} = U'_{\text{front}}F'_{\text{front}}$

   $F_{\text{rear}} = F'_{\text{rear}} \Rightarrow \dfrac{F'_{\text{front}}}{F_{\text{front}}} = \dfrac{U_{\text{front}}}{U_{\text{rear}}} \cdot \dfrac{U'_{\text{rear}}}{U'_{\text{front}}} = \dfrac{45}{13} \cdot \dfrac{34}{29} = 4$

4. $\frac{1}{2}mv_3^2 = \frac{1}{2}mv_2^2 - \frac{1}{2}mv_1^2 \Rightarrow v_3 = \sqrt{v_2^2 - v_1^2} = 83\,\frac{\text{km}}{\text{h}}$

5. $\frac{1}{2}mv_x^2 = \frac{1}{2}mgh \Rightarrow v_x = \sqrt{gh}$

$\frac{v_y}{g} = t = \sqrt{\frac{2h}{g}} \Rightarrow v_y = \sqrt{2gh}$

$\tan \beta = \frac{v_y}{v_x} = \sqrt{2} \Rightarrow \beta = 55°$

6. $R = 2h \cos \beta \left(\sin \beta + \sqrt{\sin^2 \beta + \frac{H}{h}} \right)$

with $H = L(1 - \cos \beta)$ and $h = \frac{v^2}{2g}$

$mgL - mgH = \frac{1}{2}mv^2 \Rightarrow mgL \cos \beta = \frac{1}{2}mv^2$

$\Rightarrow L \cos \beta = \frac{v^2}{2g} = h$

$S = L \sin \beta + R$

$= L \left(\sin \beta + 2 \cos^2 \beta \left(\sin \beta + \sqrt{\sin^2 \beta + \frac{1 - \cos \beta}{\cos \beta}} \right) \right)$

Angle for maximum S is $\beta = 41°$

7. $F = \frac{P}{v} = 125\,\text{N}$

8. $v = 14\,\frac{\text{km}}{\text{h}} = 3.9\,\frac{\text{m}}{\text{s}}$; $m = 12\,400\,\text{kg} + 80\,\text{places} \approx 20\,\text{t}$;
   Slope: $\sigma = 20\% = 0.2$

$P = \frac{\Delta E}{\Delta t} = \frac{mg\Delta h}{\Delta t} = mg\frac{\Delta h}{\Delta s}\frac{\Delta s}{\Delta t} \approx mg\sigma v = 160\,\text{kW} < 175\,\text{kW}$

9. Force due to vehicle weight: $F_G \approx mg \cdot 4\%$
   Force due to friction: $F_R = kv^2$

$P = F \cdot v = F_R(v_1) \cdot v_1 + F_G \cdot v_1 = F_R(v_2) \cdot v_2 - F_G \cdot v_2$

$\Rightarrow kv_1^3 + F_G v_1 = kv_2^3 - F_G v_2 \Rightarrow k = F_G \frac{v_1 + v_2}{v_2^3 - v_1^3}$

$P = kv_1^3 + F_G v_1 = F_G v_1 \left(\frac{v_1 + v_2}{v_2^3 - v_1^3} v_1^2 + 1 \right)$

$= F_G v_1 \left(\frac{\frac{v_2}{v_1} + 1}{\left(\frac{v_2}{v_1}\right)^3 - 1} + 1 \right) = 142\,\text{kW}$

Chapter 8

1. $\vec{r}_S = (0.0, z_s)$

$$z_S = \frac{1}{M} \int_0^H z\, dm = \frac{\rho \pi R^2}{M} \int_0^H z\left(1 - \frac{z}{H}\right)^2 dz$$

$$= \frac{\rho \pi R^2}{\frac{1}{3}\rho \pi R^2 H} \frac{1}{12} H^2 = \frac{1}{4} H$$

2. $x_S = \frac{\rho \pi r_1^2 (r_2 - r_1)}{M} = \frac{r_1^2(r_2 - r_1)}{r_2^2 - r_1^2} = \frac{r_1^2}{r_1 + r_2}$

3. $\frac{h}{a} \approx \frac{1}{2}\frac{a}{l} \Rightarrow h \approx \frac{a^2}{2l} = 1.4 \cdot 10^{-4}$ m

$\frac{1}{2} M u^2 = Mgh \Rightarrow u = \sqrt{2gh}$

$mv = mv_0 - Mu \Rightarrow v = \frac{1}{m}\left(mv_0 - M\sqrt{\frac{ga^2}{l}}\right) = 173\,\frac{\text{m}}{\text{s}}$

4. $v_2 = \sqrt{\frac{2E}{m_2\left(1 + \frac{m_2}{m_1}\right)}} \approx \sqrt{\frac{2E}{m_2}} = 710\,\frac{\text{m}}{\text{s}}; \quad v_1 = \frac{m_2}{m_1}v_2 = 1.9\,\frac{\text{m}}{\text{s}}$

5. $v' = \frac{\sqrt{p_1^2 + p_2^2}}{m_1 + m_2} = \frac{\sqrt{m_1^2 v_1^2 + m_2^2 v_2^2}}{m_1 + m_2} = 17\,\frac{\text{km}}{\text{h}}$

$\tan \varphi = \frac{p_2}{p_1} = \frac{m_2 v_2}{m_1 v_1} \Rightarrow \varphi = 41°$

Before the impact: $E_{\text{kin}} = \frac{1}{2} m_1 v_1^2 + \frac{1}{2} m_2 v_2^2$
After the impact: $E'_{\text{kin}} = \frac{1}{2}(m_1 + m_2) v'^2$

$\frac{\Delta E_{\text{kin}}}{E_{\text{kin}}} = \frac{E_{\text{kin}} - E'_{\text{kin}}}{E_{\text{kin}}} = 1 - \frac{E'_{\text{kin}}}{E_{\text{kin}}} = 0.52$

6. a) φ_b undefined; $\varphi_w = \frac{\pi}{2} - \varphi_b$; $v_b = v \cos \varphi_b$; $v_w = v \sin \varphi_b$
   b) $\sin \varphi_b = \frac{b}{d}$; $\varphi_w = \frac{\pi}{2} - \varphi_b$
   c) From geometry, it follows: $\varphi_b = 1.455 < \frac{\pi}{2}$
   For extended spheres: $\varphi'_b = \varphi_b + \arcsin \frac{b}{r_{\text{wb}}}$;
   $b = d \sin (\varphi'_b)$
   Iterative substitution gives: $b = 56.8$ mm;
   $\varphi'_b = 1.595 > \frac{\pi}{2}$ Kinematically not possible!
   Taking into account the hole width means the
   angle can be $\frac{D/2}{r_{\text{bL}}} = 0.056$ smaller.
   Iterative substitution gives: $b = 57.17$ mm;
   $\varphi'_b = 1.53$
   The blue sphere would have to be hit exactly on
   0.03 mm.

A2 Solutions to the Problems

d) From geometry it follows: $\varphi_b = 0.185$
   For extended spheres: $b = 12.2$ mm; $\varphi'_b = 0.215$
   With hole width: $b = 15.8$ mm; $\varphi'_b = 0.279$
   Required accuracy: 3.6 mm

7. $F = \frac{dm}{dt} v = 150$ N

8. $F = \frac{dm}{dt} v \Rightarrow P_1 = Fv = \frac{dm}{dt} v^2$

   $m = \frac{dm}{dt} \frac{L}{v} \Rightarrow P_2 = m \cdot g \cdot \frac{dh}{dt} = \frac{dm}{dt} \frac{L}{v} \cdot g \cdot v \frac{h}{L} = \frac{dm}{dt} gh$

   $P = \frac{dm}{dt} \left(gh + v^2 \right)$

9. a) Ansatz: $\frac{d}{dt}(mv) = -\frac{dm}{dt}(v_{rel} - v) - mg$

   b) Condition for takeoff: $\left| \frac{dm}{dt} \right| v_{rel} > mg$

   Time of takeoff:
   $\left| \frac{dm}{dt} \right| v_{rel} = \left(m_0 - \left| \frac{dm}{dt} \right| t \right) g \Rightarrow t = \frac{m_0}{\left| \frac{dm}{dt} \right|} - \frac{v_{rel}}{g} = 6.5$ s

   Immediate takeoff:
   $\left| \frac{dm}{dt} \right| v_{rel} > m_0 g \Rightarrow m_0 < \left| \frac{dm}{dt} \right| \frac{v_{rel}}{g} = 4.08$ kg
   Maximum payload: 4.08 kg $- 4.0$ kg $= 0.08$ kg

   c) $v(t) = -v_{rel} \ln \left(\frac{m(t)}{m_0} \right) - gt$

   d) Maximum speed at $t_{end} = 7$ s with $m(t_{end}) = 0.5$ kg
   Substituted in solution of c: $v_{max} = v(t_{end}) = 98 \frac{m}{s}$

   e) $s(t) = -v_{rel} \frac{m_0}{\frac{dm}{dt}} \left(\frac{m}{m_0} \left(\ln \frac{m}{m_0} - 1 \right) + 1 \right) - \frac{1}{2} g t^2$

   f) Flight altitude at end of burn: $s(t_{end}) = 153$ m
   Afterwards, the rocket continues to rise in free fall until it reaches maximum altitude:

   $s_{max} = s(t_{end}) + \frac{v_{max}^2}{2g} = 153$ m $+ 490$ m $= 643$ m

10. $p = \int F dt = 2.5$ N $\cdot 0.2$ s $= 0.5 \frac{kg\,m}{s} \Rightarrow v = 5 \frac{m}{s} \Rightarrow s = \frac{v^2}{2g} = 1.3$ m

∎∎ Chapter 9

1. $a = \frac{F_R}{m} = \mu g$; $\quad s = \frac{v^2}{2a} = \frac{v^2}{2\mu g} = 1.6$ km

2. $F_{brems} \leq F_R = \mu_H mg \Rightarrow t = \frac{v}{\mu_H g} = 6.1$ s

   $P_{brems} = F_{brems} v = \mu_H mgv = 319$ kW

3. $a = \frac{2s}{t^2}$; $\quad ma = mg\mu_G \Rightarrow \mu_G = \frac{2s}{gt^2} = 0.04$

4. Acceleration: $ma = mg \sin\alpha - \mu mg \cos\alpha$
   Deceleration: $-ma' = -(mg \sin\alpha + \mu mg \cos\alpha)$

   $2aL = v^2 = 2a'x$

   $x = L\dfrac{\tan\alpha - \mu}{\tan\alpha + \mu} = 6.3\,\text{m}$

5. $F \cos\alpha = F_R = \mu F_N = \mu(mg - F\sin\alpha)$

   $\Rightarrow F = \dfrac{\mu mg}{\cos\alpha + \mu \sin\alpha}$

   $\dfrac{dF}{d\alpha} = 0 \Rightarrow \sin\alpha = \mu \cos\alpha \Rightarrow \tan\alpha = \mu$

   $\alpha = 17°; \quad F = 56\,\text{N}$

∎∎ Chapter 10

1. $mg = m\omega^2 r \sin 45° \Rightarrow f = \dfrac{1}{2\pi}\sqrt{\dfrac{g}{r \sin 45°}} = 0.5\,\dfrac{1}{\text{s}}$

2. $\dfrac{mv^2}{r} = mg \Rightarrow v = \sqrt{gr} = 8 \cdot 10^3\,\dfrac{\text{m}}{\text{s}}$

   $m\omega^2 r = mg \Rightarrow T = 2\pi\sqrt{\dfrac{r}{g}} = 5 \cdot 10^3\,\text{s} = 1.4\,\text{h}$

3. $\tan\alpha = \dfrac{F_Z}{F_G} = \dfrac{m\omega^2 l \sin\alpha}{mg}$

   $\alpha = \arccos\left(\dfrac{g}{l}\dfrac{1}{4\pi^2 f^2}\right)$

4. $\tan\phi = \dfrac{F_Z}{F_G} = \dfrac{v^2}{gr}$

   $F_Z = F_R \Rightarrow \dfrac{mv^2}{r} = \mu mg \Rightarrow v_{\max} = \sqrt{\mu gr}$

   $\tan\phi_{\max} = \dfrac{v_{\max}^2}{gr} = \mu \Rightarrow \phi_{\max} = 45°$

5. $\vec{a}_C = -2\vec{\omega} \times \vec{v} \Rightarrow a_C = 2\omega v \cos\varphi$

   $\vec{a}_Z = -\vec{\omega} \times (\vec{\omega} \times \vec{r}) \Rightarrow a_Z = \omega^2 R_E \cos\varphi$

   $\ddot{x} = 2\omega \cos\varphi\, gt \Rightarrow \begin{array}{l} x = \omega \cos\varphi \dfrac{1}{3} gt^3 \\ = \dfrac{1}{3}\omega \cos\varphi \sqrt{\dfrac{8h^3}{g}} = 14\,\text{mm} \end{array}$

   $\ddot{y} = \omega^2 R_E \cos\varphi \sin\varphi \Rightarrow \begin{array}{l} y = \omega^2 R_E \cos\varphi \sin\varphi \dfrac{1}{2} t^2 \\ = \omega^2 R_E \cos\varphi \sin\varphi \dfrac{h}{g} \\ = 170\,\text{mm} \end{array}$

A2 Solutions to the Problems

■■ Chapter 11

1. $G\dfrac{Mm}{r^2} = m\omega^2 r \Rightarrow r = \left(\dfrac{GM}{\omega^2}\right)^{\frac{1}{3}} = \left(\dfrac{gR_E^2}{\omega^2}\right)^{\frac{1}{3}}$

   $= 4.2 \cdot 10^7$ m

   $\Rightarrow d = r - R_E = 36 \cdot 10^3$ km

   $\Delta\omega = \dfrac{\Delta\phi}{T} \Rightarrow \dfrac{\Delta\omega}{\omega} = 2.3 \cdot 10^{-6}$

   $\dfrac{\Delta r}{r} = -\dfrac{2}{3}\dfrac{\Delta\omega}{\omega} \Rightarrow |\Delta r| = \dfrac{2}{3}r\dfrac{\Delta\omega}{\omega} = 64$ m

2. $\dfrac{a_H^3}{a_E^3} = \dfrac{T_H^2}{T_E^2} \Rightarrow a_H = \left(\dfrac{T_H}{T_E}\right)^{\frac{2}{3}} a_E = \left(\dfrac{76}{1}\right)^{\frac{2}{3}} \cdot 1$ AE

   $= 17.9$ AE

   $r_{max} = 2a_H - r_{min} = 35.2$ AE

3. $G\dfrac{Mm}{r^2} = \dfrac{mv^2}{R_A} \Rightarrow 2R_A = \dfrac{2v}{\sqrt{G\frac{4}{3}\pi\rho}} = 2.4$ km

   $f = \dfrac{1}{2\pi}\omega = \dfrac{1}{2\pi}\dfrac{v}{R_A} = 2 \cdot 10^{-4}$ s^{-1}

   $\Rightarrow T = \dfrac{1}{f} = 5 \cdot 10^3$ s $= 1.4$ h

4. $\dfrac{Mm}{M+m}\omega^2 r = G\dfrac{Mm}{r^2}$

   $\Rightarrow r^3 = \dfrac{G(m+M)}{\omega^2} = \dfrac{gR_E^2}{\omega^2}\left(1+\dfrac{m}{M}\right) \Rightarrow r = 3.9 \cdot 10^8$ m

   $r_E = r\dfrac{m}{M+m} = 4.7 \cdot 10^6$ m

5. $\varepsilon^2 = \dfrac{2L^2 E_{\text{tot}}}{G^2 m^3 M^2} + 1 = \dfrac{a^2 - b^2}{a^2}$;

$$p = \dfrac{L^2}{Gm^2 M} = \dfrac{b^2}{a} \Rightarrow E_{\text{tot}} = -G\dfrac{Mm}{2a}$$

$$E_{\text{tot}} = \dfrac{1}{2}mv^2 - G\dfrac{Mm}{r} = -G\dfrac{Mm}{2a}$$

$$\Rightarrow v = \sqrt{2GM\left(\dfrac{1}{r} - \dfrac{1}{2a}\right)} = v_F\sqrt{r_E\left(\dfrac{1}{r} - \dfrac{1}{2a}\right)}$$

$$v_{\text{Start}} = v - v_E = v_F\sqrt{r_E\left(\dfrac{1}{r_E} - \dfrac{1}{r_E + r_J}\right)} - \dfrac{v_F}{\sqrt{2}}$$

$$= \left(\sqrt{1 - \dfrac{1}{1 + 5.3}} - \dfrac{1}{\sqrt{2}}\right)v_F = 0.21 v_F$$

$$= 8.8\,\dfrac{\text{km}}{\text{s}}$$

$$v_{\text{direkt}} = \sqrt{2GM\left(\dfrac{1}{r_E} - \dfrac{1}{r_J}\right)} = v_F\sqrt{1 - \dfrac{r_E}{r_J}} = \sqrt{1 - \dfrac{1}{5.3}}\,v_F$$

$$= 0.9 v_F = 38\,\dfrac{\text{km}}{\text{s}}$$

■■ **Chapter 12**

1. $m_{\text{ladder}}\, g\dfrac{d}{2} + m_{\text{person}}\, gd - F_w h = 0$

$$\left(\dfrac{1}{2}m_{\text{ladder}} + m_{\text{person}}\right) g\dfrac{d}{h} < \mu_H\left(m_{\text{ladder}} + m_{\text{person}}\right)g$$

$$\dfrac{d}{h} < \dfrac{m_{\text{ladder}} + m_{\text{person}}}{\frac{1}{2}m_{\text{ladder}} + m_{\text{person}}}\mu_H \approx \mu_H \quad \text{für} \quad m_{\text{person}} \gg m_{\text{ladder}}$$

2. $\lambda dx = \mu ds \Rightarrow \lambda = \mu\dfrac{ds}{dx} = \mu\sqrt{1 + y'^2}$

$$y'' = \dfrac{1}{F_H}\mu\sqrt{1 + y'^2} \Rightarrow y(x) = \dfrac{F_H}{\mu}\cosh\left(\dfrac{\mu}{F_H}(x - x_0)\right) + y_0$$

3. Coordinate system with x axis and y axis through the catheti. Weight must be maintained and torques around x axis and y axis must disappear:

$$mg = F_1 + F_2; \quad mg\dfrac{a}{3} = F_1 x_1; \quad mg\dfrac{b}{3} = F_2 y_2$$

$$\dfrac{a}{x_1} + \dfrac{b}{x_2} = 3$$

$$x_1 = a \Rightarrow y_2 = \dfrac{b}{2}; \quad y_2 = b \Rightarrow x_1 = \dfrac{a}{2};$$

$$x_1 = \dfrac{2}{3}a, \quad y_2 = \dfrac{2}{3}b$$

A2 Solutions to the Problems

4. Coordinate system with fixed points on the x axis. Weight is distributed over three points and torques around the x axis and y axis must disappear:

$$mg = F_1 + F_2 + F_3; \quad mg\frac{a}{2} = F_3 x + F_2 a; \quad mg\frac{a}{2} = F_3 a$$

$$F_3 = \frac{1}{2}mg; \quad F_2 = \frac{1}{2}mg\left(1 - \frac{x_3}{a}\right); \quad F_1 = \frac{1}{2}mg\frac{x_3}{a}$$

5. $M = \rho V = \rho \frac{1}{3}\pi R^2 H$

$$z_S = \frac{1}{M}\int_0^H z\,dm = \frac{1}{M}\rho \pi R^2 \int_0^H z\left(1 - \frac{z}{H}\right)^2 dz$$

$$= \frac{1}{M}\rho \pi R^2 \frac{1}{12}H^2 = \frac{1}{4}H$$

$$M' = \rho_0 \pi R^2 \int_0^H \left(1 - \frac{z}{2H}\right)\left(1 - \frac{z}{H}\right)^2 dz = \rho_0 \frac{7}{24}\pi R^2 H$$

$$z'_S = \frac{1}{M'}\rho_0 \pi R^2 \int_0^H z\left(1 - \frac{z}{2H}\right)\left(1 - \frac{z}{H}\right)^2 dz$$

$$= \frac{1}{M'}\rho_0 \pi R^2 \frac{1}{15}H^2 = \frac{8}{35}H$$

6. a) $\cos\alpha = \frac{d_i}{d_a} \Rightarrow \alpha = 34°$

b) $F\cos\alpha \leq \mu(mg - F\sin\alpha) \Rightarrow F \leq \frac{\mu}{\cos\alpha - \mu\sin\alpha}mg$

$$\Rightarrow \frac{F}{mg} \leq 2.7$$

■■ Chapter 13

1. $a = \frac{v}{t} = 1.7\,\frac{m}{s}$ Rotation around hub:

$$\alpha = \frac{a}{r} = 5.4\,\frac{1}{s^2}$$

$$F = ma + \frac{M}{r} = ma + \frac{I\alpha}{r} = ma + \frac{mr^2\alpha}{r} = 2ma = 3.3\,N$$

Rotation around the tire's support point:

$$F = \frac{M}{r} = \frac{1}{r}(mr^2 + I)\alpha = 2mr\alpha = 2ma = 3.3\,N$$

2. $I_{\text{Sakai}} = m_{\text{arc}} R^2 + 2 \cdot \dfrac{1}{3} m_{\text{spoke}} R^2$

$= 2\rho R^3 \left(\pi - \arctan \dfrac{1}{2} + \dfrac{1}{3} \right) = 6.02 \rho R^3$

$I_{\text{ring}} = m_{\text{ring}} R^2 = 2\pi \rho R^3 = 6.28 \rho R^3$

3. $dF_R = \mu g \rho \, 2\pi \, r \, dr$ with the mass occupancy per unit area $\rho = \dfrac{m}{\pi R^2}$

$$M = \int dM = \int r \, dF_R = 2\pi \mu g \rho \int_0^R r^2 \, dr = \dfrac{2}{3} \mu m g R$$

$-M = I\alpha = \dfrac{1}{2} m R^2 \alpha \Rightarrow \alpha = -\dfrac{4}{3} \dfrac{\mu g}{R}$

$\mu m g s = \dfrac{1}{2} m v_0^2 \Rightarrow \mu = \dfrac{v_0^2}{2gs} \Rightarrow \alpha = -\dfrac{2}{3} \dfrac{v_0^2}{sR}$

$t = \dfrac{\omega_0}{-\alpha} = \dfrac{3\pi f_0 s R}{v_0^2} = 8.8 \, \text{s}$

4. $\Delta E_{\text{pot}} = m g \dfrac{l}{2}; \quad \Delta E_{\text{rot}} = \dfrac{1}{2} I \omega^2 = \dfrac{1}{6} m l^2 \omega^2$

$\Delta E_{\text{pot}} = \Delta E_{\text{rot}} \Rightarrow \omega = \sqrt{\dfrac{3g}{l}} \Rightarrow v = \omega \cdot l = \sqrt{3gl}$

5. $mg - ma = Ma + \dfrac{Ia}{r^2} \Rightarrow a = \dfrac{mg}{(M+m) + \dfrac{M}{2} \dfrac{R^2}{r^2}}$

$t = \sqrt{\dfrac{2l}{a}} = \sqrt{2l \dfrac{m + M + \dfrac{M}{2} \dfrac{R^2}{r^2}}{mg}}$

6. $a = \dfrac{mg \sin \beta}{m + \dfrac{I}{R^2}} g$

solid: $a = \dfrac{2}{3} g \sin \beta = 0.65 \, \dfrac{m}{s^2}$

hollow: $a = \dfrac{1}{2} g \sin \beta = 0.49 \, \dfrac{m}{s^2}$

sliding: $a = g \sin \beta = 0.98 \, \dfrac{m}{s^2}$

$\mu m g \cos \beta > m g \sin \beta - ma = m g \sin \beta \left(1 - \dfrac{m}{m + I/R^2} \right)$

$\mu > \tan \beta \dfrac{1}{1 + \dfrac{mR^2}{I}} = \begin{cases} \dfrac{1}{3} \tan \beta = 0.03 & \text{(solid)} \\ \dfrac{1}{2} \tan \beta = 0.05 & \text{(hollow)} \end{cases}$

7. During the rebound, momentum is converted into angular momentum by the brief force impulse. Since

A2 Solutions to the Problems

no kinetic energy is lost and the amount of momentum remains the same, the amount of angular momentum also remains the same, only the direction of rotation changes.

Bouncy ball: $F\delta t = \Delta p_x = 2mv_0 \cos\theta$;

$$F\delta t = \frac{\Delta L}{R} = \frac{I \Delta \omega}{R} = \frac{4}{5} mR\omega_0$$

$$\Rightarrow \omega_0 = \frac{5}{2} \frac{v_0 \cos\theta}{R}$$

Soccer ball: $\omega_0 = \frac{3}{2} \frac{v_0 \cos\theta}{R}$; $v_0 = 33.3 \frac{m}{s}$;

$$\tan\theta \approx \frac{0.244 \text{ m}}{0.05 \text{ m} + 0.11 \text{ m}};$$

$$R = 0.11 \text{ m} \Rightarrow f_0 = \frac{\omega_0}{2\pi} = 5\frac{1}{s}$$

8. Rolling condition: $\omega r = \omega_P R \Rightarrow \omega = \omega_P \frac{R}{r}$

Precession: $\vec{M} = \vec{R} \times \vec{F}_P = \vec{L} \times \vec{\omega}_P$; $\vec{L} = I\vec{\omega} = \frac{1}{2} mr^2 \vec{\omega}$

$$F_P = \frac{L\omega_P}{R} = \frac{I\omega\omega_P}{R} = \frac{I\omega_P^2}{r} = \frac{1}{2} mr(2\pi f)^2 = 4.9 \text{ kN}$$

$$F_{ges} = F_P + F_G = F_P + mg = F_P + 4.9 \text{ kN} = 9.8 \text{ kN}$$

9. $\Delta L_E = -\frac{2}{5} m_E R_E^2 \omega_E < 0$

$$\Delta L_{EM} = \frac{m_E m_M}{m_E + m_M} \left(r'^2_{EM} \omega'_M - r^2_{EM} \omega_M \right) > 0$$

$$\frac{m_E m_M}{m_E + m_M} \omega_M^2 r_{EM} = G \frac{m_E m_M}{r_{EM}^2} \Rightarrow r_{EM}^3 = \left(1 + \frac{m_M}{m_E} \right) \frac{gR_E^2}{\omega_M^2}$$

$$\Delta L_{EM} = m_E \left(1 + \frac{m_E}{m_M} \right)^{-1} \left(1 + \frac{m_M}{m_E} \right)^{\frac{2}{3}} g^{\frac{2}{3}} R_E^{\frac{4}{3}} \left(\omega_M'^{-\frac{1}{3}} - \omega_M^{-\frac{1}{3}} \right)$$

$$= -\Delta L_E = \frac{2}{5} m_E R_E^2 \omega_E$$

$$T'_M = \frac{2\pi}{\omega'_M}$$

$$= 2\pi \left[\frac{2}{5} R_E^{\frac{2}{3}} \omega_E \left(1 + \frac{m_E}{m_M} \right) \left(1 + \frac{m_M}{m_E} \right)^{-\frac{2}{3}} g^{-\frac{2}{3}} + \omega_M^{-\frac{1}{3}} \right]^3$$

$$= 4.6 \cdot 10^6 \text{ s} = 53 \text{ d}$$

$$r'_{EM} = \left[\left(1 + \frac{m_M}{m_E} \right) \frac{gR_E^2}{\omega'^2_M} \right]^{\frac{1}{3}} = 6.0 \cdot 10^8 \text{ m}$$

Chapter 14

1. $\dfrac{F}{A} = E\dfrac{\Delta l}{l} \Rightarrow A = \dfrac{F}{E\frac{\Delta l}{l}} = 1.2 \cdot 10^{-4}\,\text{m}^2 = 1.2\,\text{cm}^2$

   $\mu = \dfrac{E}{2G} - 1 = 0.375;\quad \dfrac{\Delta A}{A} \approx 2\dfrac{\Delta d}{d} = -2\mu\dfrac{\Delta l}{l} = -5.5 \cdot 10^{-4}$

2. $\dfrac{\Delta \rho}{\rho} = \dfrac{\Delta\frac{1}{V}}{\frac{1}{V}} = \dfrac{\frac{1}{(d+\Delta d)^2(l+\Delta l)} - \frac{1}{d^2 l}}{\frac{1}{d^2 l}} \approx -2\dfrac{\Delta d}{d} - \dfrac{\Delta l}{l}$

   $= 2\mu\dfrac{\sigma}{E} - \dfrac{\sigma}{E} = (2\mu - 1)\dfrac{\sigma}{E} = -4.3 \cdot 10^{-4}$

3. $\sigma_{max} = \dfrac{F}{A} = \dfrac{g\rho A l_{max}}{A} \Rightarrow l_{max} = \dfrac{\sigma_{max}}{g\rho} = 6.5\,\text{km}$

4. $d(\Delta l) = \dfrac{\sigma}{E}dl = \dfrac{F}{A(l)E}dl;$

   $A(l) = \dfrac{\pi}{4}\left(d_1 - (d_1 - d_2)\dfrac{l}{L}\right)^2$

   $\Delta l = \dfrac{F}{E}\int_0^L \dfrac{1}{A(l)}dl = \dfrac{4}{\pi}\dfrac{FL}{E}\dfrac{1}{d_1 d_2}$

5. a) $dF = \omega^2 r\, dm = \omega^2 r \rho A\, dr$

   $\Rightarrow \sigma(r) = \rho\omega^2 \int_r^L r'\,dr' = \dfrac{1}{2}\rho\omega^2\left(L^2 - r^2\right)$

   $d(\Delta L) = \dfrac{\sigma}{E}dr \Rightarrow \Delta L = \dfrac{1}{E}\int_0^L \sigma(r)\,dr = \dfrac{\rho\omega^2 L^3}{3E}$

   b) $F = \omega^2 R\rho AR\int_0^{\pi/2}\cos\varphi\,d\varphi = \rho A\omega^2 R^2 \Rightarrow \sigma = \rho\omega^2 R^2$

   $\dfrac{\Delta R}{R} = \dfrac{\Delta(2\pi R)}{2\pi R} = \dfrac{\sigma}{E} \Rightarrow \Delta R = \dfrac{\sigma}{E}R = \dfrac{\rho\omega^2 R^3}{E}$

   c) The ring expands three times more than the spoke

6. $F = AE\dfrac{\Delta l}{l} \Rightarrow \dfrac{F_a}{F_m} = \dfrac{d_a^2 E_a}{d_m^2 E_m}$

   $mg = F_m + 2F_a \Rightarrow F_m = \dfrac{mg}{1 + 2\frac{d_a^2 E_a}{d_m^2 E_m}};\quad F_a = \dfrac{mg}{2 + \frac{d_m^2 E_m}{d_a^2 E_a}}$

   $\dfrac{d_m^2 E_m}{d_a^2 E_a} = 1 \Leftrightarrow F_m = F_a = \dfrac{mg}{3} \Rightarrow \dfrac{d_a}{d_m} = \sqrt{\dfrac{E_m}{E_a}}$

A2 Solutions to the Problems

7. x: Position of the mass along the rod of length L

$$F_1 = \frac{L-x}{L}; \quad F_2 = \frac{x}{L}$$

$$\frac{F_1}{A_1 E_1} = \frac{\Delta l}{l} = \frac{F_2}{A_2 E_2} \Rightarrow \frac{L-x}{A_1} = \frac{x}{A_2} \Rightarrow x = \frac{L}{1 + A_1/A_2}$$

8. Shear angle: α

$$\alpha \approx \frac{\varphi R}{L} \Rightarrow F = GA\alpha = G 2\pi R d \frac{\varphi R}{L}$$

$$\Rightarrow M = RF = \frac{2\pi \varphi G}{L} R^3 d$$

$$\Rightarrow D = \frac{M}{\varphi} = \frac{2\pi G}{L} R^3 d$$

Solid wire

$$M = \varphi \frac{2\pi G}{L} \int_0^R r^3 dr = \varphi \frac{\pi G}{2L} R^4 \Rightarrow D = \frac{\pi G}{2L} R^4$$

9. $x = r\cos\varphi$;

$$dA = r\, d\varphi\, dr \Rightarrow \int x^2 dA = \int_{R_i}^{R_a} \int_0^{2\pi} r^3 \cos^2\varphi\, d\varphi\, dr$$

$$= \frac{\pi}{4}\left(R_a^4 - R_i^4\right)$$

$$\rho = \frac{E}{F_0 L_0} \frac{\pi}{4}\left(R_a^4 - R_i^4\right)$$

▪▪ Chapter 15

1. Before: $E_{pot} = -mgd$
   After: $E_{pot} = mgh - \frac{4}{3}\pi r^3 \rho g d$

$$h = \left(\frac{4\pi r^3 \rho}{3m} - 1\right)d = 1.1\,\text{m}$$

2. $V_{ice}\rho_{ice} = V_{oil}\rho_{oil} + V_{water}\rho_{water}$
   $V_{ice} = V_{oil} + V_{water}$

$$\Rightarrow \frac{V_{water}}{V_{ice}} = \frac{\rho_{ice} - \rho_{oil}}{\rho_{water} - \rho_{oil}} = 0.68$$

Water level rises by: $\frac{V_{ice}}{A_{glas}}\left(\frac{\rho_{ice}}{\rho_{water}} - \frac{V_{water}}{V_{ice}}\right) = 1.5\,\text{mm}$

Oil level drops by: $\frac{V_{ice}}{A_{glas}}\left(\frac{\rho_{ice}}{\rho_{water}} - 1\right) = -0.5\,\text{mm}$

3. a) $m + V\rho_{He} = V\rho_{air}$

$$V(h) = \frac{m}{\rho_{air}(h) - \rho_{He}(h)} = \frac{m}{\rho_{air}(0) - \rho_{He}(0)}\frac{p(0)}{p(h)}$$

$$V(0) = V(h)\frac{p(h)}{p(0)} = \frac{m}{\rho_{air}(0) - \rho_{He}(0)}$$

$$= 1.1 \cdot 10^3\,\text{m}^3$$

$$V(h) = V(0)\exp\left(\frac{\rho_0 g}{p_0}h\right) = V(0)\exp\left(\frac{h}{8\,\text{km}}\right)$$

$$= 7.2 \cdot 10^4\,\text{m}^3$$

$$\Rightarrow d = 2\left(\frac{V(h)}{4\pi/3}\right)^{\frac{1}{3}} = 52\,\text{m}$$

b) Balloon mass:
$m_B = Ad\rho = 4\pi\left(\frac{d}{2}\right)^2 d\rho = 187\,\text{kg} \Rightarrow \frac{\Delta V}{V} = \frac{m_B}{m} = +13\%$
Temperature: $\frac{\Delta V}{V} = -\frac{\Delta \rho}{\rho} \approx \frac{\Delta T}{T} = -15\%$

c) The balloon should be filled with less helium so that it does not burst when it slightly ascends above the target altitude.

4. $p(h) = \rho g(H - h) \Rightarrow F_{dam} = L\int_0^H p(H)dh = \frac{1}{2}\rho g L H^2$

$F_{mountain} = \frac{1}{2\pi}F_{ring} = \frac{1}{2\pi}\frac{2\pi R}{L}F_{dam} = \frac{R}{L}F_{dam} = \frac{1}{2}\rho g R H^2$

5. $p = \frac{4\sigma}{r}$; $p_1 = 6.0\,\text{Pa}$; $p_2 = 4.0\,\text{Pa}$;

$V_{tot} = V_1 + V_2 \Rightarrow r_{tot} = \sqrt[3]{r_1^3 + r_2^3} \Rightarrow$

$p_{tot} = \frac{4\sigma}{\sqrt[3]{r_1^3 + r_2^3}} = 3.7\,\text{Pa}$

6. $h = \frac{2\sigma}{\rho g r} = 2.9\,\text{cm}$

Tree: $r = 0.005\,\text{mm} \Rightarrow h = 2.9\,\text{m}$

▪▪ Chapter 16

1. $v = \sqrt{2gh} = 68\,\frac{\text{m}}{\text{s}}$; $P = p\dot{V} = \frac{1}{2}\rho v^2 Av = \frac{1}{2}\rho A v^3$

$\Rightarrow A = \frac{P_{el}}{\eta}\frac{2}{\rho v^3} = 5\,\text{m}^2$

A2 Solutions to the Problems

2. $v = \sqrt{2gh}; \quad \dot{h} = -\dfrac{\dot{V}}{A_{tank}} = -\dfrac{A_{hole} v}{A_{tank}} = -\dfrac{d^2}{D^2}\sqrt{2gh}$

   $\Rightarrow T = \dfrac{D^2}{d^2}\sqrt{\dfrac{2H}{g}}$

3. $\dot{V} = \dfrac{\pi R^4}{8\eta L}\rho g h = 1.0 \cdot 10^{-6} \dfrac{m^3}{s}$

   $Re = \dfrac{\rho v R}{\eta} = \dfrac{\rho \dot{V}}{\eta \pi R} = 0.57$

4. $\rho V g = 6\pi \eta v r \Rightarrow d = 2r = \sqrt{\dfrac{18\eta v}{\rho g}} = 0.6\,\text{mm}$

5. $mg = \tfrac{1}{2}\rho A c_A v^2 \Rightarrow s = \dfrac{v^2}{2a} = \dfrac{m^2 g}{\rho A c_A F}$

   Frankfurt: $\rho = 1.3\,\text{kg/m}^3 \Rightarrow s = 1.9\,\text{km}$

   Mexico City: $\rho = 0.96\,\text{kg/m}^3 \Rightarrow s = 2.5\,\text{km}$

   Maximum flight height at $v_{max} = 1100\,\text{km/h}$:

   $\rho = \dfrac{2mg}{A c_A v_{max}^2} = \rho_0 \exp\left(-\dfrac{h_{max}}{8\,\text{km}}\right) \Rightarrow h_{max} = 19\,\text{km}$

6. $F = \tfrac{1}{2} c_W \rho A v^2 \Rightarrow P = \tfrac{1}{2} c_W \rho A v^3$

   With $\rho = 1.3\,\text{kg/m}^3 \approx 1\,\text{kg/m}^3$ and $c_W \approx 0.4$ the given formula is derived.

   Substituting the values results in a rotor diameter of approximately 140 m.

▪▪ Chapter 17

1. $1.2 \cdot 2\pi\sqrt{\dfrac{m_0}{D}} = 2\pi\sqrt{\dfrac{m_0+m_1}{D}} \Rightarrow m_0 = \dfrac{m_1}{(1.2)^2-1} = 114\,\text{g}$

2. $F_A = g\rho A h = D \cdot h \Rightarrow D = g\rho A$

   $T = 2\pi\sqrt{\dfrac{m}{D}} = \dfrac{4}{d}\sqrt{\dfrac{\pi m}{g\rho}} \Rightarrow \rho = \dfrac{16\,\pi m}{d^2\,gT^2} = 0.96 \cdot 10^3\,\dfrac{\text{kg}}{\text{m}^3}$

3. $DR_E = mg; \quad T = 2\pi\sqrt{\dfrac{m}{D}} = 2\pi\sqrt{\dfrac{R_E}{g}} \Rightarrow \dfrac{T}{2} = 42\,\text{min}$

   $m\omega^2 R_E = mg; \quad T = \dfrac{2\pi}{\omega} = 2\pi\sqrt{\dfrac{R_E}{g}} \Rightarrow \dfrac{T}{2} = 42\,\text{min}$

4. $m = lA\rho; \; F = 2xA\rho g \Rightarrow D = 2A\rho g \Rightarrow f = \dfrac{1}{2\pi}\sqrt{\dfrac{2g}{l}}$

5. $f = \frac{1}{2\pi}\sqrt{\frac{mgs}{I}} = \frac{1}{2\pi}\sqrt{\frac{mg\left(\frac{m_1-m_2}{m_1+m_2}\right)l}{(m_1+m_2)l^2}} = \frac{1}{2\pi}\sqrt{\frac{m_1-m_2}{m_1+m_2}\frac{g}{l}}$

6. $l_r = s + \frac{R^2}{2s}$

$$T = 2\pi\sqrt{\frac{I+ms^2}{mgs}} = 2\pi\sqrt{\frac{R^2+2s^2}{2gs}}$$

$$\frac{dT}{ds} = 0 \Leftrightarrow s = \frac{R}{\sqrt{2}}$$

7. $\omega = \sqrt{\frac{D^*}{I}} \Rightarrow \frac{1}{\omega_1^2} - \frac{1}{\omega_2^2} = \frac{I_1-I_2}{D^*}$

$\Rightarrow D^* = 4\pi^2 \frac{\frac{1}{2}mR^2}{T_2^2 - T_1^2} = 2.0\cdot 10^{-3}$ Nm

$D^* = \frac{\pi}{32}\frac{G}{L}d^4 \Rightarrow G = \frac{64\pi mR^2 L}{(T_2^2-T_1^2)d^4} = 2.0\cdot 10^{10}\,\frac{\text{N}}{\text{m}^2}$

8. a) $g = \frac{4\pi^2 l}{T^2}\left(1 + \frac{2}{5}\frac{R^2}{l^2}\right)$

b) $g = \frac{4\pi^2 l}{T^2}\left(1 - \frac{1}{6}\frac{m_F}{m}\right)$

c) $g = \frac{4\pi^2 l}{T^2}\left(1 + \frac{\rho_L}{\rho_K}\right)$

d) $g = \frac{4\pi^2 l}{T^2}\left(1 + \frac{\gamma^2 T^2}{4\pi^2}\right)$

e) $\frac{\varphi_0^2}{8} = 1.3\cdot 10^{-3}$

$\frac{2}{5}\frac{R^2}{l^2} = 1.6\cdot 10^{-4}$

$\frac{1}{6}\frac{m_F}{m} = 2.2\cdot 10^{-3}$

$\frac{\rho_L}{\rho_K} = 1.1\cdot 10^{-4}$

$\frac{\gamma^2 T^2}{4\pi^2} = 1.2\cdot 10^{-6}$

A2 Solutions to the Problems

9. $\omega_r^2 = \omega_0^2 - 2\gamma^2 = \dfrac{D}{m} - 2\gamma^2$

$x_{res} = x(\omega_r) = \dfrac{F}{2\gamma m} \dfrac{1}{\sqrt{\dfrac{D}{m} - \gamma^2}}$

$\gamma^4 - \dfrac{D}{m}\gamma^2 + \left(\dfrac{F}{2mx_{res}}\right)^2 = 0$

$\Rightarrow \gamma = 4.1\,\dfrac{1}{s}; \quad \omega_r = 8.1\,\dfrac{1}{s}$

$\tan\varphi_r = \dfrac{2\gamma\omega_r}{\dfrac{D}{m} - \omega_r^2}$

$\Rightarrow \varphi_r = 63°$

10. $\Delta t = \dfrac{T_{beat}}{4} = \dfrac{1}{4}\dfrac{1}{f_{beat}} = \dfrac{1}{4}\dfrac{1}{\frac{f_2 - f_1}{2}} = 23\,\text{s}$

11. a) The mathematical pendulums of the bell body and the clapper have the pendulum body in the same position. If the static friction at the pivot point of the clapper is sufficiently large and the force when swinging is sufficiently small, the clapper remains in sync with the bell body.

b) $l = g\left(\dfrac{T}{2\pi}\right)^2 \Rightarrow l - l_1 = \dfrac{g}{4\pi^2}(T^2 - T_1^2) = 67\,\text{cm} \approx a$

Remedy e.g. by changing the suspension, i.e. changing of a.

▪▪ Chapter 18

1. $v_{plane} = v_{sound}\sin\alpha \approx 600\,\text{km/h}$
2. Fundamental tone: $\lambda = 2l$

$F = \sigma A = v^2\rho A = (\lambda f)^2 \rho \dfrac{d^2}{4}\pi = l^2 f^2 \rho d^2 \pi$

$F = F_E + F_A + F_D + F_G$

$= l^2 \rho \pi \left(f_E^2 d_E^2 + f_A^2 d_A^2 + f_D^2 d_D^2 + f_G^2 d_G^2\right) = 355\,\text{N}$

3. $f_0 = \dfrac{v_0}{\lambda_0} = f_{Fe} = \dfrac{v_{Fe}}{\lambda_{Fe}} \Rightarrow L_0 = \dfrac{v_0}{v_{Fe}}L_{Fe} = \sqrt{\dfrac{\kappa_0 \rho_{Fe}}{E_{Fe}\rho_0}}L_{Fe} = 47\,\text{mm}$

4. $\dfrac{\lambda}{2} = \Delta s = 0.84\,\text{m} \Rightarrow f = \dfrac{v_{sound}}{2\Delta s} \approx 200\,\text{Hz}$

Chapter 19

1. Fundamental tone: $f^0_{open}/f^0_{closed} = 2$ (Octave)
   1st overtone: $f^1_{open}/f^1_{closed} = \frac{4}{3}$ (Fourth)

   $$\Delta f^1 = f^1_{open} - f^1_{closed} = \frac{v_{sound}}{4L} \Rightarrow f^0_{open} = \frac{v_{sound}}{2L} = 2\Delta f^1$$
   $$= 262 \text{ Hz } (c')$$
   $$L = \frac{v_{sound}}{2f^0_{open}} = 0.63 \text{ m}$$

2. $L_{tot} = 10 \log \left(\sum_i 10^{\frac{L_i}{10}} \right) = 80 \text{ dB}$

3. $\frac{5}{4} = \frac{1+\frac{v}{v_s}}{1-\frac{v}{v_s}} \Rightarrow v = \frac{v_s}{9} = 132 \frac{\text{km}}{\text{h}}$

4. $l = \frac{1}{2} v_{sound} \Delta t = 1.9 \text{ km}$

   $$f' = f \frac{v_{sound} + v_{rel}}{v_{sound} - v_{rel}}$$
   $$\Rightarrow v_{rel} = v_{sound} \frac{f'-f}{f'+f} \approx v_{sound} \frac{\Delta f}{2f}$$
   $$\Rightarrow v_{submarine} = \frac{v_{rel}}{\sin 65°} = 22 \frac{\text{km}}{\text{h}}$$

5. $f_{direct} = f_{sender} \frac{1}{1+\frac{v_{sender}}{v_{sound}}}; \quad f_{reflected} = f_{sender} \frac{1}{1-\frac{v_{sender}}{v_{sound}}}$

   $$f_{beat} = (f_{reflected} - f_{direct}) = 2f_{sender} \frac{\frac{v_{sender}}{v_{sound}}}{1 - \left(\frac{v_{sender}}{v_{sound}}\right)^2}$$
   $$\approx 2f_{sender} \frac{v_{sender}}{v_{sound}} = 3 \text{ Hz}$$

A3 Mathematical Introduction

A3.1 Calculating with Powers

Quantities in physics can take on a wide range of values. For example, the distance from Earth to the nearest galaxy is 23,650,000,000,000,000,000,000 m, while the diameter of an atom is only about 0.0000000001 m. To avoid having to constantly write this large number of zeros, powers are often used in physics. The prefixes you have learned (◘ Tab. 2.3) are an example. To be able to use powers, you have to master their calculation rules.

▪▪ Definitions
(In the following n and m are integer numbers $1, 2, 3, \ldots$)

$$a^n = a \cdot a \cdot a \cdot \ldots \cdot a \quad \text{(in total } n \text{ times)}$$

$$a^{-n} = \frac{1}{a} \cdot \frac{1}{a} \cdot \frac{1}{a} \cdot \ldots \cdot \frac{1}{a} \quad \text{(int total } n \text{ times)}$$

$$a^0 = 1 \quad \text{(if } a \text{ is not equal zero)}$$

$$0^0 = 0$$

$$a^{\frac{1}{n}} = \sqrt[n]{a}$$

$$a^{\frac{n}{m}} = \sqrt[m]{a^n} = \left(\sqrt[m]{a}\right)^n$$

▪▪ Examples

$$7^4 = 7 \cdot 7 \cdot 7 \cdot 7 = 2401$$

$$10^3 = 10 \cdot 10 \cdot 10 = 1000$$

$$2^{-2} = \frac{1}{2} \cdot \frac{1}{2} = \frac{1}{4}$$

$$10^{-6} = \frac{1}{10} \cdot \frac{1}{10} \cdot \frac{1}{10} \cdot \frac{1}{10} \cdot \frac{1}{10} \cdot \frac{1}{10} = 0.000001$$

$$27^{\frac{1}{3}} = \sqrt[3]{27} = 3$$

▪▪ The most important calculation rules are
(The calculation rules for positive and negative exponents as well as fractions are the same. We can summarize them by allowing rational numbers for the exponents n and m. Rational numbers are all numbers that can be represented as a fraction. This also includes integers, as you can write 4 as $\frac{4}{1}$, for example.

a, b and p can, on the other hand, be any real number.)

$$pa^n = p(a^n)$$
$$a^n \cdot a^m = a^{(a+m)}$$
$$\frac{a^n}{a^m} = a^{(n-m)}$$
$$a^n \cdot b^n = (a \cdot b)^n$$
$$\frac{a^n}{b^n} = \left(\frac{a}{b}\right)^n$$
$$(a^n)^m = a^{(n \cdot m)}$$

■■ **Examples**

$$4\left(a^2 \cdot b^2\right) = (2ab)^2$$
$$\left(\frac{13}{17}\right)^{-4} = 83.521$$
$$10^2 \cdot 10^{-5} \cdot 10 = 10^{-2} = 0.01$$
$$a^2 \cdot b^3 \quad \text{cannot be simplified any further}$$

A3.2 Vectors

Vectors describe directed quantities, i.e., those quantities that possess a direction in addition to a value. Imagine a penalty kick in football. Certainly, it is important at what speed the shooter shoots the ball during the kick. Was it a hard shot or rather a soft one? But at least equally important is the direction of the ball. Is the shot placed in one of the corners of the goal, is it low, does it go into one of the upper corners or even over the goal?

In physics, the velocity of the ball is described by a vector. How high the speed is is indicated by its magnitude, measured in m/s. The direction indicates where the ball is moving. In sketches, velocity vectors are represented as arrows, each pointing in the direction of movement. The length of the arrow indicates the magnitude of the velocity. The longer the arrow, the faster the movement.

Mathematically, vectors are objects in multidimensional spaces. They can be specified by their components in the respective dimensions. However, they exist independently of a component representation. We distinguish

A3 Mathematical Introduction

vectors from simple numbers by a small arrow above the symbol. For example,

$$\vec{a} = (a_1, a_2, a_3, \ldots, a_n).$$

This vector has n components. So it is a n-dimensional vector, where n is a natural number $(1, 2, 3, 4, \ldots)$. Initially, in physics, you will almost exclusively encounter three-dimensional vectors, where the three dimensions represent the three directions in space (up-down, right-left, front-back).

If we want to specify the components of a vector, we must first define a coordinate system (see ▶ Sect. 5.2). Here is an illustration of our penalty kick example. The figure shows the ball flying into the top right corner of the goal. It is a hard shot with a speed of about 81 km/h. The velocity vector in the indicated Cartesian coordinate system is

$$\vec{v} = \left(14\,\frac{\text{km}}{\text{h}}, 21\,\frac{\text{km}}{\text{h}}, 77\,\frac{\text{km}}{\text{h}}\right) = (14, 21, 77)\,\frac{\text{km}}{\text{h}}$$

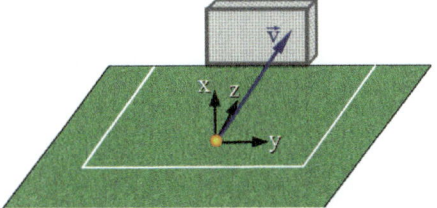

Note that physical vectors usually carry units.
The most important calculation rules for vectors are:

$$\vec{a} + \vec{b} = (a_1 + b_1, a_2 + b_2, a_3 + b_3)$$
$$\vec{a} - \vec{b} = (a_1 - b_1, a_2 - b_2, a_3 - b_3)$$
$$c \cdot \vec{a} = (c \cdot a_1, c \cdot a_2, c \cdot a_3)$$

with the two vectors

$$\vec{a} = (a_1, a_2, a_3) \quad \vec{b} = (b_1, b_2, b_3)$$

and c an arbitrary real number. The absolute value of a vector is determined according to the following rule:

$$|\vec{a}| = \sqrt{a_1^2 + a_2^2 + a_3^2}.$$

In our example, the absolute value of the velocity is

$$|\vec{v}| = \sqrt{14^2 (\text{km/h})^2 + 21^2 (\text{km/h})^2 + 77^2 (\text{km/h})^2}$$
$$|\vec{v}| = \sqrt{196 (\text{km/h})^2 + 441 (\text{km/h})^2 + 5929 (\text{km/h})^2}$$
$$|\vec{v}| = \sqrt{6566 (\text{km/h})^2}$$
$$|\vec{v}| = 81.03 \,\text{km/h}.$$

The rules are written here for three dimensions. They apply to n dimensions accordingly.

Sometimes there are situations where only the direction of a vector is relevant, not its absolute value. In these cases, we use unit vectors. They are scaled such that the absolute value is 1. Instead of a vector arrow above the symbol, a unit vector is usually denoted with a small hat.

$$\hat{a} = \frac{\vec{a}}{|\vec{a}|} = \frac{1}{\sqrt{a_1^2 + a_2^2 + a_3^2}} (a_1, a_2, a_3).$$

The direction of our soccer ball during a penalty kick would be:

$$\hat{v} = \frac{1}{81.03 \,\text{km/h}} (14, 21, 77) \,\text{km/h} = (0.1728, 0.2592, 0.9503).$$

Vectors can also be multiplied together. The most important multiplication is the dot product. You multiply the vectors component by component and add the results:

$$\vec{a} \cdot \vec{b} = a_1 \cdot b_1 + a_2 \cdot b_2 + a_3 \cdot b_3.$$

The result is a number that is associated with the direction between the two vectors. If ϕ is the angle between the two vectors, then:

$$\vec{a} \cdot \vec{b} = |\vec{a}| |\vec{b}| \cos \phi.$$

Using the dot product, you can determine the angle between two directions

$$\cos \phi = \frac{\vec{a} \cdot \vec{b}}{|\vec{a}| |\vec{b}|}.$$

For example, you could ask about the difference in direction of a penalty kick to the upper right corner and the upper left corner:

left: $\vec{v}_l = (14, 21, 77)$ km/h
right: $\vec{v}_r = (14, -21, 77)$ km/h

$$\vec{v}_l \cdot \vec{v}_r = (14 \cdot 14 - 21 \cdot 21 + 77 \cdot 77)\left(\frac{\text{km}}{\text{h}}\right)^2 = 5684 \left(\frac{\text{km}}{\text{h}}\right)^2$$

$$\cos \phi = \frac{5684 \,(\text{km/h})^2}{(81.03 \,\text{km/h})^2} = 0.8657 \rightarrow \phi = 30.0°.$$

You should remember some special cases:
- If the two vectors point in the same direction (angle zero), the scalar product is maximum. The result is the product of the absolute values of the two vectors.
- If there is a right angle between the vectors, the scalar product vanishes.
- If the vectors point in opposite directions, the scalar product is negative. The result is the negative of the product of the absolute values of the two vectors.

It is important to realize that the scalar product does not depend on the coordinate system. Although the components of both vectors change when the coordinate system is changed, the scalar product always yields the same value.

In conclusion, an important relation for the lengths of vectors, the so-called triangle inequality, should be mentioned.

$$||\vec{a}| - |\vec{b}|| \leq |\vec{a} + \vec{b}| \leq |\vec{a}| + |\vec{b}|$$

A3.3 Sequences

Sequences are mappings from natural numbers to a range of values W. Each natural number is assigned a number in the range of values. The sequence can be represented by a table, by listing the sequence members, or by a calculation rule. Here is an example:

1 2 3 4 5 6 7 8 9
0 1 4 9 16 25 36 49 64

The same sequence could also be given in this way

$a_n = 0, 1, 4, 9, 16, 25, 36, 49, 64, \ldots$

or via a calculation rule

$$a_n = (n-1)^2.$$

In this sequence, the number 0 is assigned to 1, 1 to 2, 4 to 3, and so on. The range of values could be the whole positive numbers (including 0).

Here are three more examples:

$$b_n = -1, +1, -1, +1, -1, +1, -1, \ldots \quad b_n = (-1)^n$$

$$c_n = 1, \frac{1}{2}, \frac{1}{3}, \frac{1}{4}, \frac{1}{5}, \frac{1}{6}, \frac{1}{7}, \ldots \quad c_n = \frac{1}{n}$$

$$d_n = 1, 1, 1, 1, 1, 1, 1, 1, \ldots \quad d_n = 1$$

It is interesting to see how the sequences develop for increasing n. The first sequence (a_n) for example, continues to grow. This means that one can specify an arbitrarily large number – called a bound – but the sequence still grows beyond this bound. One just has to choose n large enough. In this case we say the sequence (a_n) diverges or it grows towards infinity. This is how the concept of infinity is defined in mathematics (although the definition in mathematics is usually formulated in a more compact way).

In contrast to the first sequence, the third sequence (c_n) converges. It is getting closer and closer to zero. The difference between c_n and zero becomes smaller as n increases. Zero is then called the limit of the sequence. This can be formulated similarly to divergence: A sequence converges towards a limit if one can specify an arbitrarily small number (bound), such that the distance between sequence member a_n and the limit becomes smaller than this number, provided that n is chosen large enough. Then, for this a_n and all subsequent members, the distance to the limit must be smaller than the bound. The limit does not necessarily have to be zero. Sequences can converge towards any number. For example, the fourth sequence d_n converges towards the limit 1.

The second sequence b_n is also interesting. The value of the sequence constantly jumps between $+1$ and -1. The sequence obviously does not diverge. For example, one could choose 100 as a bound. The sequence never exceeds 100. But it also does not converge. One might be tempted to specify zero as the limit, but the distance to zero never becomes less than 1. Neither $+1$ nor -1 are limits. Although the distance to $+1$ sometimes even becomes zero, e.g., at $n = 42$, but this does not apply to

A3 Mathematical Introduction

all subsequent elements. This sequence does not converge. It has no limit.

Symbolically, the limit values are written as (the overturned eight is the symbol for infinity):

$$\lim_{n \to \infty} a_n = \infty$$

$$\lim_{n \to \infty} b_n \text{ does not exist}$$

$$\lim_{n \to \infty} c_n = 0$$

$$\lim_{n \to \infty} d_n = 1.$$

The example of a stroboscopic recording of a falling ball discussed in Experiment 5.2 can be described by sequences. There, the averagevelocity between t_3 and t_4 was given as the following term, where $s(t_i)$ is the height of the ball at time t_i:

$$v_4 = \frac{s_4}{\Delta t} = \frac{s(t_4) - s(t_3)}{\Delta t} = \frac{s(t_4) - s(t_4 - \Delta t)}{\Delta t}$$

This is less than the speed at time t_4. Now the following sequence can be given:

$$(v_4)_n = \frac{s(t_4) - s(t_4 - \Delta t/n)}{\Delta t/n}$$

With increasing n, the time difference between the beginning and end of the interval at which the positions of the ball are determined, is getting smaller. The velocity of the ball changes less for shorter intervals, so that the average velocity in the interval approaches the actual velocity at the end of the interval. In fact, this sequence converges to the velocity at time t_4.

In the following we give the elements of a sequence for an acceleration due to gravity of 9.81 m/s^2, a time point $t_4 = 10 \text{ s}$ and $\Delta t = 1 \text{ s}$. $\Delta t/n$ is the time interval over which the average is calculated.

n	$\Delta t/n$	$(v_4)_n$
1	1 s	93.195 m/s
2	0.5 s	95.648 m/s
3	0.25 s	96.874 m/s
4	0.125 s	97.587 m/s
5	0.0625 s	97.793 m/s
...	...	...
∞		98.1 m/s

A3.4 Calculation Rules for Limits

Without further derivation, we present the calculation rules for limits. They are given for the case of sequences. They can be extended accordingly to series and functions.

$$\lim_{n \to \infty} c \cdot a_n = c \cdot \lim_{n \to \infty} a_n$$

Here, c can be any number, assuming that the sequence converges. If the sequence (a_n) diverges, then $c \cdot (a_n)$ also diverges, so the calculation rule above symbolically applies (then on the right stands $c \infty$, which is strictly speaking not defined). However, one must pay attention to the signs. For example, if the sequence diverges positively and one chooses c negatively, then $c \cdot (a_n)$ diverges negatively, etc. In the still missing case, where the limit of the sequence does not exist, $c \cdot (a_n)$ also has no limit. Here too, the calculation rule must be understood symbolically.

$$\lim_{n \to \infty} (a_n + b_n) = \lim_{n \to \infty} a_n + \lim_{n \to \infty} b_n$$

$$\lim_{n \to \infty} (a_n - b_n) = \lim_{n \to \infty} a_n - \lim_{n \to \infty} b_n$$

These calculation rules are also to be understood symbolically in the case of divergences or non-existing limits.

$$\lim_{n \to \infty} (a_n \cdot b_n) = \lim_{n \to \infty} a_n \cdot \lim_{n \to \infty} b_n$$

$$\lim_{n \to \infty} \left(\frac{a_n}{b_n} \right) = \frac{\lim_{n \to \infty} a_n}{\lim_{n \to \infty} b_n}$$

For the last calculation rule, one must exclude that sequence members b_n or their limit yield zero, to prevent a division by zero.

A3.5 Derivative of a Function

We start with an example from physics. We want to describe a car journey. Here, s is the distance traveled by the car. We plot the distance versus the time

A3 Mathematical Introduction

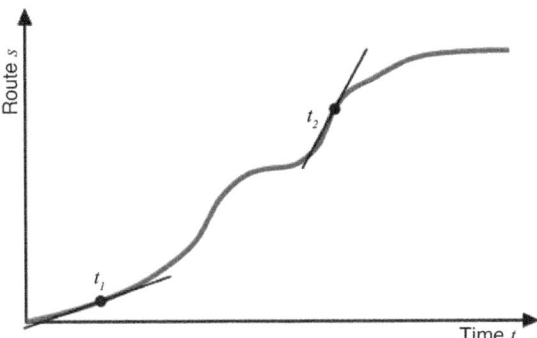

At the beginning, the curve is quite flat. In a fixed time interval, the distance travelled increases only slightly. This means that the car is driving slowly. If the curve is steeper, as in point t_2, the distance covered per time interval is larger, i.e., the car is driving faster. The slope of the curve indicates the velocity. It can be determined by drawing a tangent to the curve and determining its slope. This procedure is indicated at the time points t_1 and t_2.

The slope of a curve is positive when the value of the function increases with time. In our example, this is almost always the case. The car moves away from the starting point ($s = 0$). At the very end, the curve becomes horizontal. The distance no longer increases. The slope and thus the velocity of the car are zero (the car has reached its destination).

We call it a negative slope when the function decreases over time. In our example, this could be interpreted as the car driving backwards.

If the slope is always positive, we call it a monotonically increasing function. The slope must never be negative, but zero is allowed. In our example, this means that the car is allowed to stop, but not to reverse. If the slope is everywhere greater than zero, we call it a strictly monotonically increasing function. Now the car must always drive. Stopping is no longer allowed. For monotonically or strictly monotonically decreasing functions the same holds accordingly.

Let's now leave our example and first consider the simple case of a straight line. We describe it as a function $f(x)$. Here, x can be time, but it doesn't have to be. The figure shows an example.

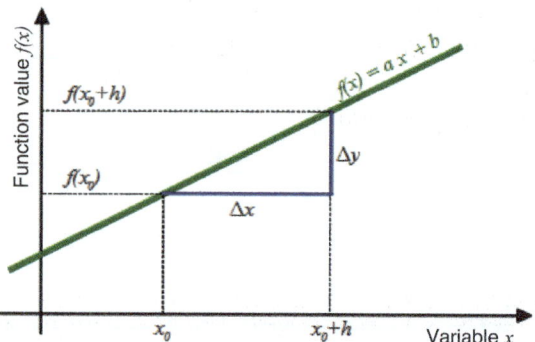

The slope of the function is denoted by $f'(x)$. It is determined by the slope triangle (see figure):

$$f'(x) = \frac{\Delta y}{\Delta x}.$$

This corresponds to the tangent of the angle that the line forms with the x axis. For a straight line $f(x) = ax + b$ we get:

$$f'(x) = \frac{\Delta y}{\Delta x} = \frac{f(x_0 + h) - f(x_0)}{(x_0 + h) - x_0}$$
$$= \frac{(a(x_0 + h) + b) - (ax_0 + b)}{h} = \frac{ah}{h} = a.$$

Now we can introduce the derivative of any function: The derivative of a function $f'(x)$ is defined as the slope of the tangent to the curve at the point x. Another notation is:

$$f'(x) = \frac{d}{dx} f(x).$$

In physics, a dot is often used instead of a dash for derivatives with respect to time:

$$\frac{d}{dt} f(t) = \dot{f}(t).$$

But how do you determine the tangent? For this, one must perform a limit process. It is outlined in the following figures.

A3 Mathematical Introduction

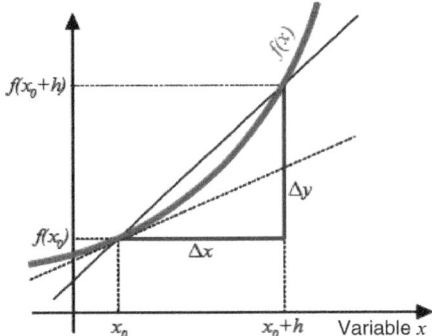

The first figure shows a function at the point x_0 where a slope triangle is depicted. The red line shows the slope of the straight line indicated by the triangle. It is not the required tangent. The tangent is indicated by a dashed red line.

The slope, which is determined by the slope triangle, depends on the choice of h. If we reduce h, it approaches the tangent. For illustration, in the following figure h is halved.

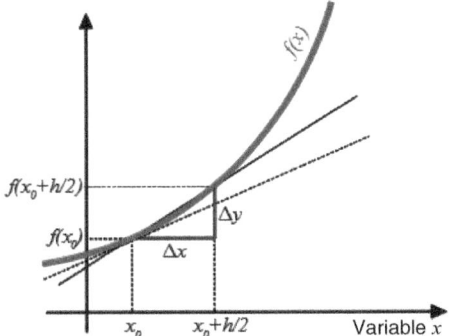

We observe that the line, which is now determined by the slope triangle, is already significantly closer to the tangent. In the next figure, a further halving of h is carried out.

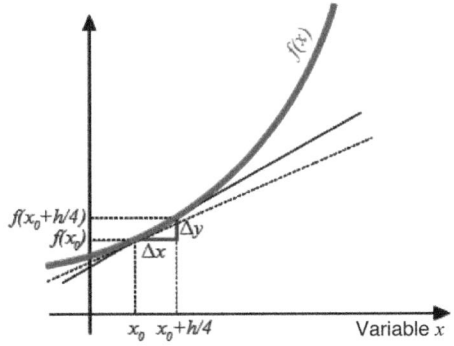

Therefore, the slope of the tangent is obtained from the slope triangle through the limit transition $h \to 0$:

$$f'(x) = \lim_{h \to 0} \frac{f(x_0 + h) - f(x_0)}{h}.$$

Here is a brief note of caution. The derivative is a limit process. As we have seen in ▶ Sect. 3 of the mathematical appendix A3 when introducing the limit with sequences, the limit does not exist in all cases. One must first ensure that the derivative exists before trying to calculate it. The function is then called differentiable. In classical physics, cases of non-differentiable functions are rather rare. Here, one is usually on the save side by the fact that nature does not make jumps (*natura non facit saltus*). However, one should not forget this difficulty.

For starting experimental physics, this may suffice with respect to abstract aspects of the mathematical concept of derivation. Mathematicians may forgive us for the casual approach.

A3.6 Calculation Rules for Derivatives

We will first calculate some derivatives as examples:

$$f(x) = ax^2.$$

Then holds:

$$f'(x) = \lim_{h \to 0} \frac{f(x+h) - f(x)}{h} = \lim_{h \to 0} \frac{(a(x+h)^2) - (ax^2)}{h}$$

$$= \lim_{h \to 0} \frac{ax^2 + 2axh + ah^2 - ax^2}{h}$$

$$= \lim_{h \to 0} \left(\frac{2axh}{h} + \frac{ah^2}{h} \right) = \lim_{h \to 0} (2ax) + \lim_{h \to 0} (ah)$$

$$= 2ax + 0 = 2ax$$

and therefore

$$\frac{d}{dx}(ax^2) = 2ax.$$

Similarly, we calculate the derivative of a third order polynomial:

A3 Mathematical Introduction

$$f(x) = ax^3$$

$$f'(x) = \lim_{h \to 0} \frac{f(x+h) - f(x)}{h} = \lim_{h \to 0} \frac{(a(x+h)^3) - (ax^3)}{h}$$

$$= \lim_{h \to 0} \frac{ax^3 + 3ax^2h + 3axh^2 + ah^3 - ax^3}{h}$$

$$= \lim_{h \to 0} \left(3ax^2\right) + \lim_{h \to 0} (3axh) + \lim_{h \to 0} \left(ah^2\right) = 3ax^2.$$

The figure shows on the left (a) the graph of the function and on the right (b) the graph of the derivative.

In general, for polynomials:

$$\frac{d}{dx}(ax^n) = nax^{n-1}.$$

A third example uses the angle sum identities of trigonometric functions. We are looking for the derivative of the function $f(x) = \sin x$:

$$f'(x) = \lim_{h \to 0} \frac{f(x+h) - f(x)}{h} = \lim_{h \to 0} \frac{\sin(x+h) - \sin x}{h}.$$

Now $\sin \alpha - \sin \beta = 2 \cos \frac{\alpha+\beta}{2} \sin \frac{\alpha-\beta}{2}$ and therefore

$$f'(x) = \lim_{h \to 0} \frac{2 \cos \frac{x+h+x}{2} \sin \frac{x+h-x}{2}}{h}$$

$$= 2 \lim_{h \to 0} \frac{\cos(x + h/2) \sin h/2}{h}$$

$$= \lim_{h \to 0} \cos(x + h/2) \cdot \lim_{h \to 0} \frac{\sin h/2}{h/2} = \cos x \cdot 1 = \cos x$$

Hence, we get $\frac{d}{dx}(\sin x) = \cos x$.

In practice, the derivatives of elementary functions can be looked up in a mathematical reference book (unless one knows them by heart). The most important ones are listed here:

$f(x)$	$f'(x)$	remark
c	0	
x	1	
x^n	nx^{n-1}	$n \in Q$
$\frac{1}{x^n}$	$\frac{-n}{x^{n+1}}$	$x \neq 0$
$\sqrt{x}$	$\frac{1}{2\sqrt{x}}$	$x \neq 0$
$\sqrt[n]{x}$	$\frac{1}{n\sqrt[n]{x^{n-1}}}$	$x \neq 0$
e^x	e^x	
a^x	$a^x \ln a$	
$\ln x$	$\frac{1}{x}$	$x \neq 0$
$\log_a x$	$\frac{1}{x} \log_a e$	$x \neq 0$
$\sin x$	$\cos x$	
$\cos x$	$-\sin x$	
$\tan x$	$\frac{1}{\cos^2 x}$	$\cos x \neq 0$
$\cot x$	$\frac{-1}{\sin^2 x}$	$\sin x \neq 0$
$\sinh x$	$\cosh x$	
$\cosh x$	$\sinh x$	
$\tanh x$	$\frac{1}{\cosh^2 x}$	$\cosh x \neq 0$
$\arcsin x$	$\frac{1}{\sqrt{1-x^2}}$	$x \neq 1$
$\arccos x$	$\frac{-1}{\sqrt{1-x^2}}$	$x \neq 1$

With the help of the following calculation rules, the derivatives of more complicated functions can then be determined from the elementary functions in the table:

$$\frac{d}{dx}(c \cdot f(x)) = c \cdot \frac{d}{dx}f(x)$$

$$\frac{d}{dx}(f(x) + g(x)) = \frac{d}{dx}f(x) + \frac{d}{dx}g(x)$$

$$\frac{d}{dx}(f(x) - g(x)) = \frac{d}{dx}f(x) - \frac{d}{dx}g(x)$$

■■ **Product Rule**

$$\frac{d}{dx}(f(x) \cdot g(x)) = f(x) \cdot \frac{d}{dx}g(x) + g(x)\frac{d}{dx}f(x)$$

A3 Mathematical Introduction

■■ **Quotient Rule**

$$\frac{d}{dx}\left(\frac{f(x)}{g(x)}\right) = \frac{g(x)\frac{d}{dx}f(x) - f(x)\frac{d}{dx}g(x)}{(g(x))^2}$$

■■ **Chain Rule**

$$\frac{d}{dx} = f(g(x)) = \frac{d}{dz}f(z) \cdot \frac{d}{dx}g(x) \quad \text{with} \quad z = g(x)$$

A3.7 Indefinite Integrals

In the description of the movements, we initially started from the position function $s(t)$ and asked ourselves what velocity and what acceleration the body has at the given time. We learned that we need to differentiate the position function:

$$s(t) \to v(t): \quad v(t) = \frac{d}{dt}s(t)$$

$$v(t) \to a(t): \quad a(t) = \frac{d}{dt}v(t).$$

Then the problem appeared that we wanted to proceed in the opposite direction, i.e. the acceleration is given and we wanted to derive the velocity and then the position of the object from it.

Mathematically, one can describe it like the following. We know a function $f(x)$ and are looking for a function $F(x)$, which fulfills the following condition:

$$\frac{d}{dx}F(x) = f(x).$$

The function $F(x)$ is called an antiderivative or primitive function to $f(x)$. In our example, we knew $a(t)$ and searched for $v(t)$, so that:

$$\frac{d}{dt}v(t) = a(t).$$

There is no general calculation method with which we could find an antiderivative. However, from the table of derivatives that you have seen in the previous section, one can generate a table of antiderivatives. For example, the table gives:

$$\frac{d}{dx}x^n = nx^{n-1}.$$

Hence, x^n is an antiderivative of nx^{n-1}:

$$f(x) = nx^{n-1} \rightarrow F(x) = x^n.$$

Since constants are preserved during differentiation, both functions can be multiplied by $\frac{1}{n}$ and we obtain:

$$f(x) = x^{n-1} \rightarrow F(x) = \frac{1}{n}x^n.$$

This can be written a bit more elegantly by replacing $n-1$ by m:

$$f(x) = x^m \rightarrow F(x) = \frac{1}{m+1}x^{m+1}.$$

In this way, one can generate tables of antiderivatives, as found in mathematical reference books. In mathematics, you will learn calculation methods on how to trace the search for the antiderivatives back to the entries in the table.

The antiderivative of a function f is often written as the so-called indefinite integral:

$$F(x) = \int f(x)dx.$$

Later, you will get to know the definite integrals, which you can recognize by the integration limits given at the top and bottom of the integration symbol.

Concluding this very short introduction, we would like to point out that the antiderivative is never unique. This means that there is more than one antiderivative for $f(x)$ (even infinitely many). If $F_0(x)$ is an antiderivative of $f(x)$, then $F_k(x) = F_0(x) + k$ is also an antiderivative with an arbitrary number k, as

$$\frac{d}{dx}F_k(x) = \frac{d}{dx}(F_0(x) + k) = \frac{d}{dx}F_0(x) + \frac{d}{dx}k = f(x),$$

since the derivative of a constant k is zero. You must not forget this constant when searching for an antiderivative. Its value can usually be determined by the boundary conditions in physical applications. In our example, we could use the initial condition that at the time $t = 0$ the speed should be v_0. Therefore, $k = v_0$ must apply.

A3.8 The Second Derivative

You had learned the derivative of a function as the slope of the curve. We now want to investigate this further. We use the example of the distance travelled by a car once again.

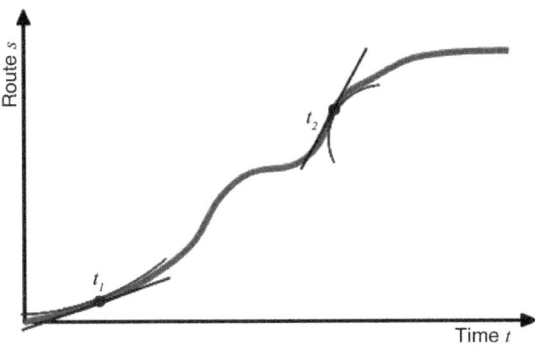

We had seen that the tangents to the curve indicate the velocity. The velocity is not constant. It changes during the travel. If the slope (from left to right) increases, the velocity increases. The car accelerates. If it decreases, the car decelerates. You can determine this from the curve by fitting a circle into the curve that has the same curvature as the curve at the respective time. At the times t_1 and t_2 the circles are drawn. At t_1 the radius of the circle is large, the curvature is low. The speed increases only slightly. At t_2 the circle is on the lower side of the curve. The curvature is negative. The speed decreases over time. The car decelerates. Also, the radius of the circle is much smaller than at t_1 and thus the curvature is greater.

The curvature of the curve indicates the acceleration. It is obtained by differentiating the function twice:

$$f''(x) = \frac{d^2}{dx^2}f(x) = \frac{d}{dx}\left(\frac{d}{dx}f(x)\right).$$

In physics, when differentiating with respect to time, we write again:

$$\frac{d^2}{dt^2}f(t) = \ddot{f}(t).$$

From the behavior of the slope and the curvature, the behavior of the function itself can be derived. In particular, extreme values (these are the points where the function reaches a maximum or a minimum) and the saddle and inflection points can be determined. These are sketched in the figure below.

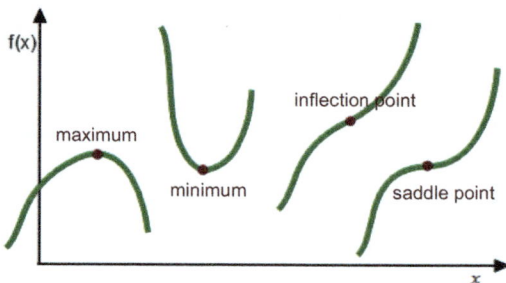

Let's first consider the maximum. It is a point x_{max} for which in a neighborhood of x_{max} the following applies: $f(x_{\text{max}}) > f(x)$. If the neighborhood is bounded, we speak of a local maximum (a function can have several of these). If the neighborhood extends over the entire domain of the function, it is a global maximum.

From the figure, it can be seen that the slope to the left of the maximum is positive (function is increasing) and to the right of the maximum is negative (function is decreasing). At the maximum itself, the slope is zero. This must be the case, because if there were a positive slope, for example, it would mean that the function value to the right of the maximum would be even greater, which cannot be the case.

A minimum is defined accordingly. Here, the slope to the left of the minimum is negative (function decreases) and to the right of the minimum is positive (function increases). Again, the slope disappears at the minimum. You can find the minima and maxima by looking for the points where the slope becomes zero:

$$\frac{d}{dx} f(x) = 0.$$

Now you can distinguish between maxima and minima by considering the curvature (second derivative). In the case of the maximum, it is negative, because the slope decreases in the area of the maximum (it changes from positive to negative). For the minimum, on the other hand, the second derivative is positive.

A3 Mathematical Introduction

So, we conclude:

maximum: $\quad \dfrac{d}{dx}f(x) = 0 \quad \dfrac{d^2}{dx^2}f(x) < 0$

minimum: $\quad \dfrac{d}{dx}f(x) = 0 \quad \dfrac{d^2}{dx^2}f(x) > 0.$

In the remaining case, where the curvature also reduces to zero, a saddle point is formed (unless the third derivative also disappears). The saddle point can be seen in the image on the right. At the saddle point, the curve has a horizontal tangent (slope zero).

saddle point: $\quad \dfrac{d}{dx}f(x) = 0 \quad \dfrac{d^2}{dx^2}f(x) = 0.$

The saddle point is a special case of the inflection point, which is also shown in the image:

inflection point: $\quad \dfrac{d}{dx}f(x) \neq 0 \quad \dfrac{d^2}{dx^2}f(x) = 0.$

At the inflection point, the curvature changes sign. If you have a positive slope (as in the image), the slope gradually decreases to the left of the inflection point. However, it does not reach zero as in the saddle point. To the right of the inflection point, it then increases again. The same applies for a negative slope.

A3.9 Inverse Functions

In Example 5.10 the function arccos appeared (written out *arcuscosinus*). This is the inverse function of the cosine function.

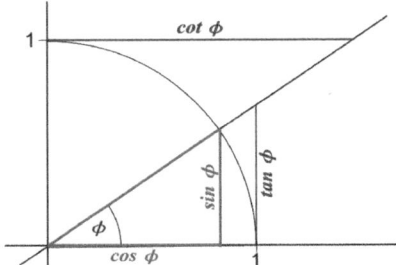

Let's first consider the trigonometric functions themselves before we turn to the inverse functions in general.

The figure shows the definition of the trigonometric functions on the unit circle. Here, the hypotenuse of the triangle has a length of 1. In the general case,

$$\sin \phi = \frac{\text{opposite}}{\text{hypotenuse}}$$

applies and correspondingly for the other trigonometric functions. Often an angle is given, and we wonder how large the projection of a line under this angle onto the cathetus is. The answer is obtained from the trigonometric functions. As an example, the graph of the sine function is shown in the next figure. If you want to read the value of the function, you look for the angle ϕ on the x-axis, find the associated point on the curve, and read off its y-coordinate (see arrows).

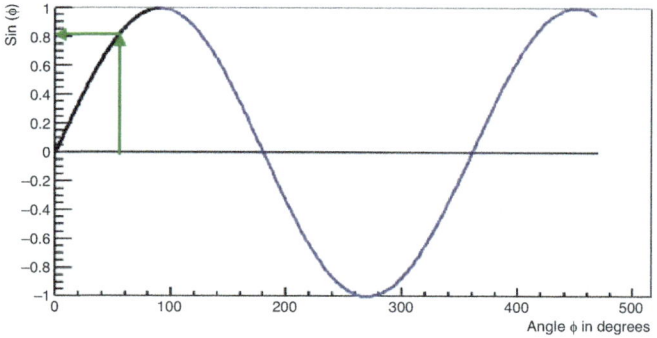

Now we want to follow the opposite direction. With the value of the sine function given, we want to know for which angle the sine function takes this value. This is shown in the next figure.

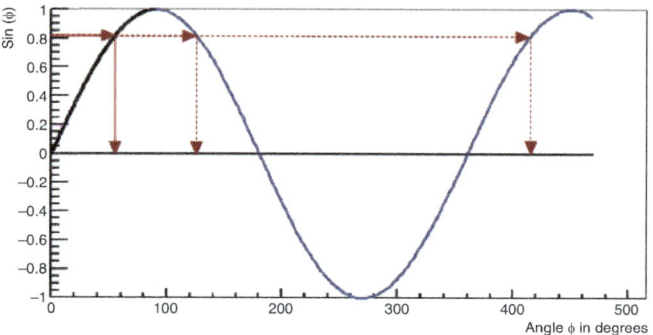

The result can be represented as a new function under certain conditions, which we will discuss further below,

A3 Mathematical Introduction

by swapping the x- and y-axes. This new function is called the "inverse function of $f(x)$" and is written as $f^{-1}(x)$, in our example $\sin^{-1}\phi$. Some common inverse functions have their own names. For the trigonometric functions, these are the already mentioned functions $\arcsin x$, $\arccos x$, $\arctan x$ and $\mathrm{arccot}\, x$.

In Example 5.10 at the end of the calculation of the angle with the maximum range, we arrived at:

$$\cos\beta_{max} = \frac{\sqrt{2gh_0 + v_0^2}}{\sqrt{2gh_0 + 2v_0^2}}$$

from where we can find the desired angle through the inverse function of the cosine:

$$\beta_{max} = \arccos(\cos\beta_{max}) = \arccos\left(\frac{\sqrt{2gh_0 + v_0^2}}{\sqrt{2gh_0 + 2v_0^2}}\right).$$

The graph of the inverse function of the sine can be seen in the next figure.

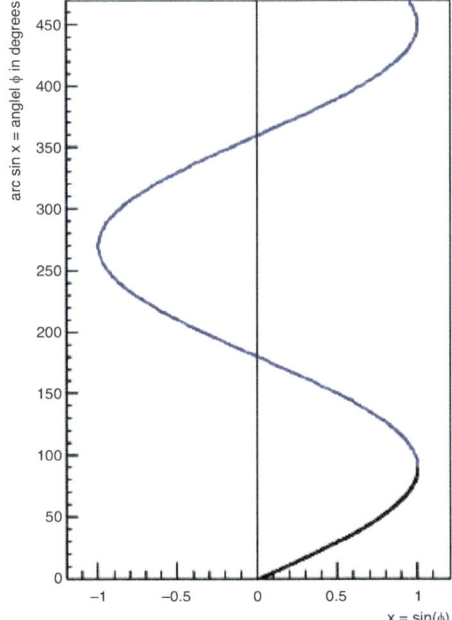

When looking at the inverse function in the last two images, it is noticeable that the inversion is not unique.

If, as in the example, a value of 0.8 is given for the sine, there is more than one angle (actually infinitely many if you extend the graph to arbitrary angles) that leads to a sine of 0.8. In the last image, you can see this by the fact that to each x more than one function value is associated. However, this does not result in a meaningful function. In order to be able to meaningfully invert a function, one must limit the domain of the function to the extent that the inversion becomes unique, i.e., to the extent that each value in the range only appears once. For the sine, this can be done by, for example, focusing on the black area of the graph for angles between 0° and 90° degrees. Usually, for the inversion of the sine, angles between $-90°$ and 90° degrees are chosen. This is the maximum range in which the sine is uniquely invertible. It is said that one restricts oneself to a branch of the function. The choice of the branch is initially arbitrary. In applications in physics, one must be careful whether this is actually the branch that provides the correct solution. In our example with the football, another extremum of the shooting range is found in the range 90° to 180° degrees at the value $180° - \beta_{max}$. This corresponds to a shot in the opposite direction (in the picture to the left), which would have the same range.

A3.10 Differential Equations

The fundamental law of mechanics is a differential equation. It specifies a condition that every motion $\vec{r}(t)$ in mechanics has to fulfill. The conditional equation not only contains $\vec{r}(t)$ itself, but also the second derivative of the motion $\vec{a}(t) = \frac{d^2}{dt^2}\vec{r}(t)$. Therefore, it is called a differential equation:

$$\frac{d^2}{dt^2}\vec{r}(t) = \frac{1}{m}\vec{F}(\vec{r}(t), t).$$

It relates the second derivative of the motion to the motion itself. The task is to find one or all possible motions that fulfill this relation.

Solving such differential equations is often not simple. In the examples, you have learned the method of separation of variables (Examples 6.3 and 6.4) and a numerical solution method (Example 6.5). With a little experience, you will succeed in many cases in guessing the solution by trial and error. It is important that you always check whether the assumed solution is actually a solution to the differential equation. You do this by substituting the assumed solution into the differential equation. This is illustrated by the following example. A constant force results in a constant acceleration:

$$m \frac{d^2}{dt^2} s(t) = F_0 .$$

For this case, we had calculated the motion in ▶ Sect. 5.5. The result was:

$$s(t) = \frac{1}{2} a_0 t^2 + v_0 t + s_0 .$$

To use this, we need to calculate the derivatives:

$$\frac{d}{dt} s(t) = a_0 t + v_0$$

$$\frac{d^2}{dt^2} s(t) = a_0 .$$

Substituted into the differential equation, we get:

$$m \, a_0 = F_0 .$$

This is fulfilled if we set the acceleration to $a_0 = F_0/m$.

Now, you will rightly say that it is difficult to guess the exact form of the motion. But this is not necessary. An approximate idea is sufficient. For example, if you suspect that the solution to our example is a polynomial in time, then this is enough. You set up a general polynomial:

$$s(t) = \sum_{n=0}^{N} c_n t^n .$$

You calculate the derivatives:

$$\frac{d}{dt} s(t) = \sum_{n=1}^{N} n c_n t^{n-1}$$

$$\frac{d^2}{dt^2} s(t) = \sum_{n=2}^{N} n(n-1) c_n t^{n-2} .$$

Note that the constant term disappears from the sum in each differentiation. Now insert this into the differential equation:

$$m\frac{d^2}{dt^2}s(t) = m\sum_{n=2}^{N} n(n-1)c_n t^{n-2} = F_0.$$

This relation must be fulfilled for any time. On the right side F_0 is independent of time, so the left side must also be time-independent. However, on the left, there is only a single term that is time-independent, namely the term for $n = 2$. It is proportional to $t^0 = 1$. All other terms must disappear. So $c_n = 0$ for $n > 2$. This reduces everything to:

$$m\left(2(2-1)c_2 t^{2-2}\right) = F_0$$
$$2\,m\,c_2 = F_0$$
$$c_2 = \frac{1}{2}\frac{F_0}{m}.$$

For the coefficients c_0 and c_1 no condition applies. They can initially be chosen arbitrarily. This results in the solution of the differential equation:

$$s(t) = \frac{1}{2}\frac{F_0}{m}t^2 + c_1 t + c_0.$$

This is already very similar to our result. The coefficients c_0 and c_1 are not determined from the differential equation. They are defined by the boundary conditions, e.g. location and velocity at the time $t = 0$. This results in a single unique solution, which is generally the case in mechanics. If the position and velocity of a body are determined at a certain point in time, its motion is uniquely determined for all times. Under these boundary conditions, there can only be a single solution to the differential equation.

A3.11 The Integral

When calculating the work, we faced the problem of a varying force along the path. In this case, we divide the path into individual sections. These do not necessarily have to be of the same length. However, for simplicity, we will assume here that they are. The diagram below shows an example of how to calculate the work. The evolution of the force along the path is shown in green. The force changes with the path. The distance is divided into 7 equidistant sections Δs. For each section of the path, the work can be determined as $W_i = F_i \Delta s$, where we assume that the force points along the path. The total work is then:

$$W = \sum_{i=1}^{7} W_i \approx \sum_{i=1}^{7} F_i \Delta s.$$

In the last step, we specified the force on each section by a representative value F_i. We still have to determine these values. We consider two different possibilities. We take once the maximum of the force in the section Δs and once the minimum. These two possibilities are indicated in the diagram. $F_i \Delta s$ corresponds to the area of a rectangle in this diagram (height × width). The red rectangles show the calculation with the maximum force, the blue rectangles the result for the minimum force. The red rectangles are partially covered by the blue ones. They also extend down to the x-axis. The sum of the red areas is called an "upper sum" of the function and that of the blue ones a "lower sum".

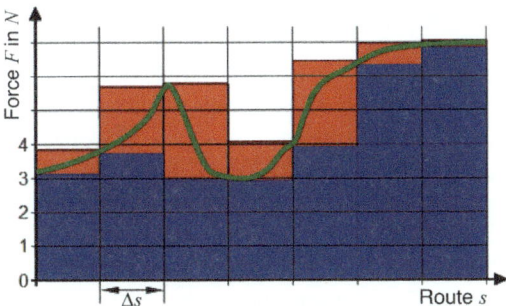

The upper sum as well as the lower sum represent an approximate calculation of the work performed. This approximation can be improved by dividing the total distance into smaller sections. The nextfigure shows a subdivision into twice as many (14) equidistant segments.

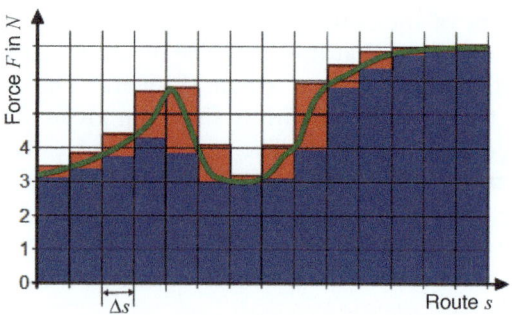

The upper sum overestimates the actual work performed, while the lower sum underestimates it. The difference is particularly large in areas where the function is steep. In the flat area on the right, however, the upper and lower sums have already approximated the function quite well. If you make the division into the path sections finer and finer, you can get arbitrarily close to the actual work performed. Therefore, we divide the distance into n equidistant path segments $\Delta s = s/n$ and let n increase:

$$W = \lim_{n \to \infty} \sum_{i=1}^{n} F_i \frac{s}{n}.$$

This is called the (definite) integral and is written as:

$$W = \lim_{n \to \infty} \sum_{i=1}^{n} F_i \frac{s}{n} = \int_0^s F(s')ds'.$$

The distance should start at 0 and end at s. These limits of the integration path are indicated on the integral sign. Under the integral, we have written s' to distinguish this variable from the length s of the distance.

Geometrically speaking, the integral corresponds to the area F_{ab} under the curve in the units specified on the axes. This is shown in the next figure. Here we have—as is customary in mathematics—chosen x instead of s as the integration variable and generalized the integral in the sense that we no longer start the integration path at 0, but at any value a:

$$F_{ab} = \int_a^b f(x)dx.$$

A3 Mathematical Introduction

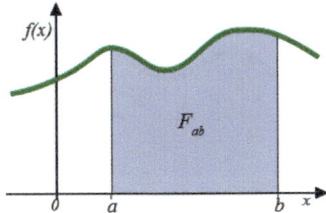

From this geometric interpretation, some simple calculation rules can be derived:

Areas can be composed additively. Therefore, it must hold:

$$\int_a^b f(x)dx = \int_0^b f(x)dx - \int_0^a f(x)dx$$

and further (see sketch below):

$$\int_a^b f(x)dx = \int_a^c f(x)dx + \int_c^b f(x)dx.$$

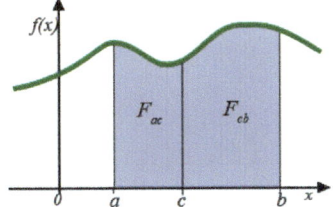

From this, we finally see that:

$$\int_a^b f(x)dx = -\int_b^a f(x)dx.$$

We now want to turn our attention to the important fundamental theorem of calculus. We will once again use the sketch with a, b and c, but make F_{bc} very small:

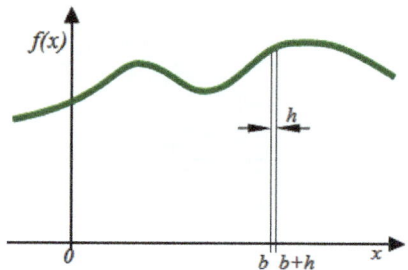

We consider the area under the curve, starting from 0 and ending at b. We want to consider b as a variable, i.e., write down how the area changes as a function of b:

$$F_{0b} = F(b) = \int_0^b f(x)dx.$$

To understand how this area changes with b, we calculate its derivative with respect to b.

$$\frac{dF(b)}{db} = \lim_{h \to 0} \frac{F(b+h) - F(b)}{h}$$

$$= \lim_{h \to 0} \frac{1}{h} \left[\int_0^{b+h} f(x)dx - \int_0^b f(x)dx \right]$$

$$= \lim_{h \to 0} \frac{1}{h} \left[\int_0^b f(x)dx + \int_b^{b+h} f(x)dx - \int_0^b f(x)dx \right]$$

$$= \lim_{h \to 0} \frac{1}{h} \left[\int_b^{b+h} f(x)dx \right]$$

$$= \lim_{h \to 0} \frac{1}{h} \left[hf(b) \right] = \lim_{h \to 0} f(b) = f(b).$$

In the second last line, we took advantage of the fact that for a very small h the area can be approximated by a rectangle. Hence, the integral is $h \cdot f(b)$. We get as a result that the derivative of the integral is the function itself :

$$\frac{d}{db} \int_0^b f(x)dx = f(b).$$

A3 Mathematical Introduction

The function over which one integrates is called the "integrand". The result can also be expressed as follows: The integral is an antiderivative of the integrand.

Now, if we use the geometric rules we have already learned, we get:

$$F_{ab} = \int_a^b f(x)dx = \int_0^b f(x)dx - \int_0^a f(x)dx = F(b) - F(a).$$

This finally results in the fundamental theorem of calculus:

It is $\int_a^b f(x)dx = F(b) - F(a)$ where $\dfrac{d}{dx}F(x) = f(x)$,

i.e., $F(x)$ is an antiderivative of $f(x)$.

We have already discussed how to determine an antiderivative in the section on indefinite integrals and we do not want to repeat it here. For definite integrals, you therefore have to insert the upper and lower limit into the antiderivative and form the difference. A possible constant in the antiderivative (which is not unique) is eliminated when forming the difference, so that the definite integral provides a unique result.

Here—without explanation—we give some important rules for integrals (they also apply to indefinite integrals):

a) Constants can be "pulled out":

$$\int_a^b cf(x)dx = c\int_a^b f(x)dx.$$

b) Integrals of sums:

$$\int_a^b (f(x) + g(x))dx = \int_a^b f(x)dx + \int_a^b g(x)dx.$$

c) Integrals of products. From

$$\int_a^b \frac{d}{dx}(f(x) \cdot g(x))dx = \int_a^b \left(\frac{d}{dx}f(x)\right) \cdot g(x)dx$$

$$+ \int_a^b f(x) \cdot \left(\frac{d}{dx}g(x)\right)dx$$

by rearranging, we get a rule called partial integration

$$\int_a^b f(x) \cdot \left(\frac{d}{dx}g(x)\right) dx = f(x) \cdot g(x)|_a^b$$
$$- \int_a^b \left(\frac{d}{dx}f(x)\right) \cdot g(x) dx.$$

The middle term is to be understood as follows:

$$f(x) \cdot g(x)|_a^b = f(b)g(b) - f(a)g(a).$$

d) Substitution, where the integration variable is changed:

$$\int_a^b f(g(x)) dx = \int_{g^{-1}(a)}^{g^{-1}(b)} f(u) \frac{d}{du} g^{-1}(u) du,$$

where $g^{-1}(u)$ is the inverse function to $g(x)$.

A3.12 Improper Integrals

In some cases (e.g., Example 7.6), integrals appear that extend to infinity. Such integrals are called "improper", as they are not actually an integral, but the limit of an integral. In the solution, when inserting the limits, the term "$1/\infty$" also appeared in Example 7.6. This is a casual notation that is quite common in physics. Here we want to try to explain the meaning.

We take this integral as an example again:

$$W = \int_R^\infty k \frac{1}{r^2} dr.$$

The idea is to first calculate the integral as a definite integral up to a fixed limit, and then let the limit grow to infinity. So:

$$W = \lim_{b\to\infty} \int_R^b k \frac{1}{r^2} dr$$

$$= \lim_{b\to\infty} \left(-k \frac{1}{r}\bigg|_R^b\right)$$

$$= \lim_{b\to\infty} \left(-k\left(\frac{1}{b} - \frac{1}{R}\right)\right)$$

$$= k \lim_{b\to\infty} \left(\frac{1}{R} - \frac{1}{b}\right)$$

$$= k\left(\lim_{b\to\infty}\left(\frac{1}{R}\right) - \lim_{b\to\infty}\left(\frac{1}{b}\right)\right) = k\left(\frac{1}{R} - 0\right) = \frac{k}{R}.$$

If both limits are infinite (integration from $-\infty$ to $+\infty$), then two limit transitions have to be performed after the integration.

Finally, we would like to point out that both differentiation and integration, as well as the improper limits described here, represent limit processes. These limit processes may only be interchanged in their order if all limits exist and are finite!

A3.13 Multiple Integrals

Initially, we had always integrated in just one variable. In the definition of the center of mass (▶ Sect. 8.2), integrals over three dimensions appear for the first time. This is evident from the fact that the integral sign no longer specifies just an interval in one dimension (on the *x*-axis), but a three-dimensional volume over which one has to integrate. We first consider the integral of the mass. We will discuss the vector integral in the numerator of the definition of the center of mass later. The integral is:

$$M = \int_V dm = \int_V \rho(\vec{r}) dV.$$

This integral can be written out as three integrals, but then a coordinate system must be chosen. In Cartesian coordinates, we get:

$$M = \int_{z_{min}}^{z_{max}} \int_{y_{min}}^{y_{max}} \int_{x_{min}}^{x_{max}} \rho(x, y, z) dx\, dy\, dz,$$

where the integration limits denote the geometric boundaries of the body. For complicated shapes, the body may need to be broken down into partial volumes and these added up at the end. The three integrals are performed one after the other. The order chosen here is first x, then y and finally z:

$$M = \int_{z_{min}}^{z_{max}} \left(\int_{y_{min}}^{y_{max}} \left(\int_{x_{min}}^{x_{max}} \rho(x,y,z) dx \right) dy \right) dz.$$

The brackets are usually omitted. Note that this is only one possible order of integration. The order can be swapped if all integrals in the original and new order exist and are finite.

If the integration limits are constants, then one integrates over a cuboid. An example can be seen in the figure. For simplicity, we set $x_{max} = 0$ and $y_{min} = 0$.

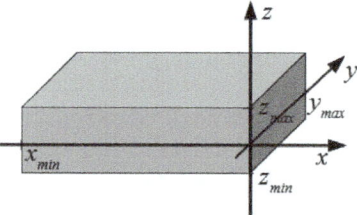

The previous example with constant integration limits yields cuboids. In the general case, the limits depend on the coordinates. This allows any arbitrary body to be integrated. As an example, we want to determine the mass of a triangular smoothing file (without handle). The file has a triangular cross-section and becomes slightly thinner towards the tip. The two sketches show the cross-section and longitudinal profile.

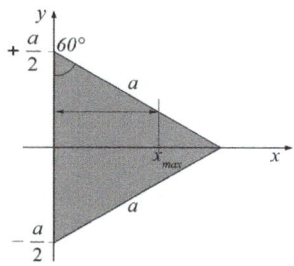

A3 Mathematical Introduction

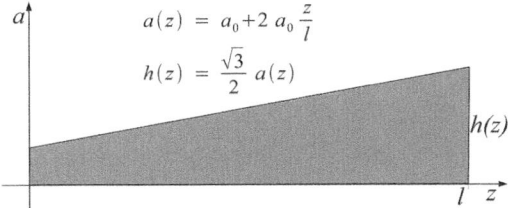

In the sketches, the coordinate system is chosen such that the file points along the z-axis. The tip is located at $z = 0$. The file has a length l. The cross-section is an equilateral triangle with the edge length a. The edge length at the tip of the file is a_0 and increases towards the shaft to $3a_0$. The height is h. The formulas for a and h are given in the sketch.

First, we integrate over x. The integration limit x_{max} depends on the postion in y. It can be determined from the right-angled (half) triangle. The angle in an equilateral triangle is $60°$. The adjacent side has the length $\frac{a}{2} - y$ and the opposite side is x_{max}. The tangent of $60°$ is $\sqrt{3}$. This yields for positive y

$$x_{max} = \sqrt{3}\left(\frac{a(z)}{2} - y\right).$$

We assume the density ρ_0 to be constant. The mass is:

$$M = 2 \int_0^l \int_0^{+a/2} \int_0^{x_{max}} \rho_0 \, dx \, dy \, dz.$$

We have exploited the symmetry in y to limit the integral to positive values of y. Now we have to solve the integrals. We can pull the constant in front:

$$M = 2\rho_0 \int_0^l \int_0^{+a/2} \int_0^{x_{max}} dx \, dy \, dz.$$

The x-integral is simple. The integrand is a 1. The antiderivative is x:

$$M = 2\rho_0 \int_0^l \int_0^{+a/2} x\big|_0^{x_{max}} \, dy\, dz$$

$$= 2\rho_0 \int_0^l \int_0^{+a/2} (x_{max} - 0)\, dy\, dz$$

$$= 2\rho_0 \int_0^l \int_0^{+a/2} \sqrt{3}\left(\frac{a}{2} - y\right) dy\, dz.$$

In the last step, we have to introduce x_{max}. This is necessary because x_{max} depends on y. Neglecting that, we would make a mistake in the following y-integral. The integrand is a first-degree polynomial in y, whose antiderivative can also be easily found:

$$M = 2\sqrt{3}\,\rho_0 \int_0^l \left(\frac{a}{2}y - \frac{1}{2}y^2\right)\bigg|_0^{+a/2} dz$$

$$= 2\sqrt{3}\,\rho_0 \int_0^l \left(\left(\frac{a^2}{4} - \frac{a^2}{8}\right) - 0\right) dz$$

$$= 2\sqrt{3}\,\rho_0 \int_0^l \frac{a(z)^2}{8} dz$$

$$= \frac{1}{4}\sqrt{3}\,\rho_0 \int_0^l \left(a_0 + 2a_0\frac{z}{l}\right)^2 dz$$

$$= \frac{1}{4}\sqrt{3}\,\rho_0 a_0^2 \int_0^l \left(1 + 4\frac{z}{l} + 4\frac{z^2}{l^2}\right) dz$$

In the final step, a second-degree polynomial has to be integrated. This results in:

$$M = \frac{1}{4}\sqrt{3}\,\rho_0 a_0^2 \left(z + 2\frac{z^2}{l} + \frac{4z^3}{3l^2}\right)\bigg|_0^l$$

$$= \frac{1}{4}\sqrt{3}\,\rho_0 a_0^2 \left(l + 2l + \frac{4}{3}l\right)$$

$$= \frac{13}{12}\sqrt{3}\,\rho_0 a_0^2\, l.$$

In this way, the masses of any body can be calculated.

A3.14 Vector Integrals

We once again return to the formula for determining the center of mass:

$$\vec{r}_{MM} = \frac{\int_V \vec{r}\, dm}{\int_V dm} = \frac{1}{M}\int_V \rho(\vec{r})\vec{r}\, dV$$

The integral in the numerator differs from the integrals considered so far. The integrand is not a simple function, but a vector (more precisely, a vector-valued function). Consequently, the result of the integration is not a number, but a vector. It is called a "vector integral".

Vector integrals are calculated component-wise. The integral over the x-component of the integrand yields the x-component of the resulting vector. The same applies for y and z, respectively. In our example, this would be

$$\vec{r}_{MM} = \begin{pmatrix} \frac{1}{M}\int_V \rho(\vec{r})x\, dV \\ \frac{1}{M}\int_V \rho(\vec{r})y\, dV \\ \frac{1}{M}\int_V \rho(\vec{r})z\, dV \end{pmatrix} = \frac{1}{M}\begin{pmatrix} \int_V \rho(\vec{r})x\, dV \\ \int_V \rho(\vec{r})y\, dV \\ \int_V \rho(\vec{r})z\, dV \end{pmatrix}$$

A3.15 Solving Systems of Equations

We revisit the example of the billiard shot with two balls on a plane (▶ Sect. 8.3.6). In Eq. 8.34 we had written down the conditions:

$$\frac{1}{2}m_1 u_1^2 = \frac{1}{2}m_1 v_1^2 + \frac{1}{2}m_2 v_2^2$$
$$m_1 u_1 = m_1 v_1 \cos\theta_1 + m_2 v_2 \cos\theta_2$$
$$0 = m_1 v_1 \sin\theta_1 + m_2 v_2 \sin\theta_2$$

We simplify the problem by assuming that the masses are equal. Then

$$\begin{cases} u_1^2 = v_1^2 + v_2^2 \\ u_1 = v_1 \cos\theta_1 + v_2 \cos\theta_2 \\ 0 = v_1 \sin\theta_1 + v_2 \sin\theta_2 \end{cases}$$

applies. Such a set of several equations is called a "system of equations". Here, the three determining equations are opposed by four unknown quantities, namely v_1, v_2,

θ_1 and θ_2. We call such a system of equations underdetermined. The set of solution consists not only of individual values, but is a one-dimensional set. We can freely choose one quantity. For a certain range of this quantity, a solution can then be found. We want to specify the angle θ_1.

To solve a system of equations, three so-called equivalence transformations are available:

a) Substituting: This means we solve one of the equations for one of the variables and then replace it in the other equations.
b) Equating: If two equations contain an identical side, we can equate the other sides ($a = b$, $a = c$ then also applies $b = c$).
c) Adding/Subtracting: We can add or subtract two equations (both sides of the equations).

We start solving our example by applying the equivalence transformation a) to the third equation:

$$0 = v_1 \sin \theta_1 + v_2 \sin \theta_2$$

$$\rightarrow \sin \theta_2 = -\frac{v_1}{v_2} \sin \theta_1$$

$$\rightarrow \cos \theta_2 = \sqrt{1 - \left(\frac{v_1}{v_2}\right)^2 \sin^2 \theta_1}.$$

Please note that the angles are constrained between 0° and 90° and therefore the cosine and the sine are always positive. We now insert this into the second equation. We no longer need to consider the third one. We have spent it:

$$\begin{cases} u_1^2 = v_1^2 + v_2^2 \\ u_1 = v_1 \cos \theta_1 + v_2 \sqrt{1 - \left(\frac{v_1}{v_2}\right)^2 \sin^2 \theta_1} \end{cases}$$

We isolate the root in the second equation and square it:

$$u_1 - v_1 \cos \theta_1 = v_2 \sqrt{1 - \left(\frac{v_1}{v_2}\right)^2 \sin^2 \theta_1}$$

$$u_1^2 + v_1^2 \cos^2 \theta_1 - 2 v_1 u_1 \cos \theta_1 = v_2^2 - v_1^2 \sin^2 \theta_1$$

$$u_1^2 + v_1^2 \cos^2 \theta_1 + v_1^2 \sin^2 \theta_1 - 2 v_1 u_1 \cos \theta_1 = v_2^2$$

$$u_1^2 + v_1^2 - 2 v_1 u_1 \cos \theta_1 = v_2^2.$$

A3 Mathematical Introduction

If we now also isolate v_2^2 in the first equation, we obtain:

$$\begin{cases} u_1^2 - v_1^2 = v_2^2 \\ u_1^2 + v_1^2 - 2v_1 u_1 \cos\theta_1 = v_2^2 \end{cases}$$

Now, in both equations we find v_2^2 on the right side, hence we can apply method b) and set the two equations equal. We obtain:

$$u_1^2 - v_1^2 = u_1^2 + v_1^2 - 2v_1 u_1 \cos\theta_1$$
$$-2v_1^2 = -2v_1 u_1 \cos\theta_1$$
$$v_1 = u_1 \cos\theta_1 .$$

By this, we have found the first part of the solution. The equation gives v_1 as a function of the two known quantities u_1 and θ_1.

Now we insert this result into the original system of equations and start from the beginning:

$$\begin{cases} u_1^2 = u_1^2 \cos^2\theta_1 + v_2^2 \\ u_1 = u_1 \cos^2\theta_1 + v_2 \cos\theta_2 \\ 0 = u_1 \cos\theta_1 \sin\theta_1 + v_2 \sin\theta_2 \end{cases}$$

The next part of the solution can be directly determined from the first equation:

$$u_1^2 = u_1^2 \cos^2\theta_1 + v_2^2$$
$$u_1^2 \left(1 - \cos^2\theta_1\right) = v_2^2$$
$$u_1^2 \sin^2\theta_1 = v_2^2$$
$$v_2 = u_1 \sin\theta_1 .$$

It remains to determine θ_2. For this, we use the previously found relation for $\cos\theta_2$:

$$\cos\theta_2 = \sqrt{1 - \left(\frac{v_1}{v_2}\right)^2 \sin^2\theta_1}$$

$$\cos\theta_2 = \sqrt{1 - \left(\frac{u_1 \cos\theta_1}{u_1 \sin\theta_1}\right)^2 \sin^2\theta_1}$$

$$\cos\theta_2 = \sqrt{1 - \cos^2\theta_1}$$
$$\cos\theta_2 = \sin\theta_1$$
$$\theta_2 = \arccos\sin\theta_1 .$$

Using this, we have found the desired result. It is:

$$v_1 = u_1 \cos\theta_1$$
$$v_2 = u_1 \sin\theta_1$$
$$\theta_2 = \arccos\sin\theta_1 \,.$$

The initial velocity u_1 can take any values (positive, as it is an absolute value). The angle θ_1 on the other hand has a restricted range. It must be in the interval 0° to 90°. It is easy to convince oneself that for angles $\theta_1 > 90°$ no solution can be found.

A3.16 The Cross Product

In rotating reference systems, a new class of vectors appears, the so-called axial vectors. These include, for example, the angular velocity $\vec{\omega}$, the torque $\vec{M}$ or the angular momentum $\vec{L}$. These vectors point along the axes of rotation, as can be seen in the sketch for $\vec{\omega}$.

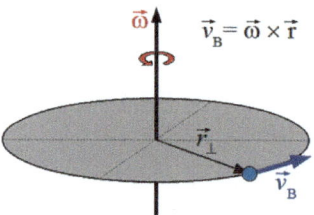

We consider the orbital velocity $\vec{v}_B$ of a rotating body. It is obviously proportional to the angular velocity $\vec{\omega}$. The faster the reference system rotates, the faster the body moves. The orbital velocity of the body also depends on the distance of the body from the axis of rotation, more precisely, on the perpendicular distance to the axis of rotation $\vec{r}_\perp$. The larger it is, the greater the orbital velocity.

The number of revolutions per second (the frequency) is given by:

$$f = \frac{\omega}{2\pi}\,.$$

With each revolution, the body covers the distance $s = 2\pi\, r_\perp$, so the orbital velocity is:

$$v_B = \frac{\omega}{2\pi}\, 2\pi\, r_\perp = \omega\, r_\perp\,.$$

A3 Mathematical Introduction

The second sketch shows the general case with a position vector, whose origin does not lie in the orbital plane of the body. We read from the figure:

$v_B = \omega r \sin \varphi$.

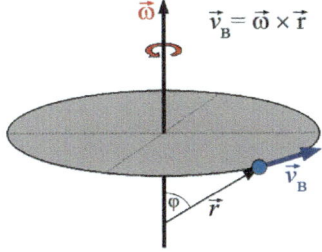

This sets the absolute value of the orbital velocity. The direction of the orbital velocity is tangential. As can be seen from the sketch, i.e., the direction is perpendicular to the position vector and perpendicular to the angular velocity.

To express this mathematically, we introduce the cross or vector product. It is written as $\vec{a} \times \vec{b} = \vec{c}$. The result is again a vector that is perpendicular to both initial vectors. The absolute value of the cross product is:

$|\vec{a} \times \vec{b}| = |\vec{a}| \cdot |\vec{b}| \sin \varphi$,

as we need it in our case. The angle φ is the angle between the two vectors $\vec{a}$ and $\vec{b}$. The absolute value is the area of a parallelogram spanned by the two vectors.

With the above rule, the sign of the vector product is not yet determined. The sign of the cross product is chosen such that the vectors $\vec{a}, \vec{b}, \vec{c}$ form in this order a right-handed tripod. For illustration, one can use the right(!) hand (see figure). Point your thumb in the direction of the first vector of the cross product and your index finger in the direction of the second. If you then spread your middle finger, as seen in the picture, perpendicular to the thumb and index finger, it indicates the direction of the resulting vector of the cross product. The rule in the second picture is also useful. It applies to rotational movements. Point your right thumb in the direction of the angular velocity. Then the four fingers indicate the direction of rotation.

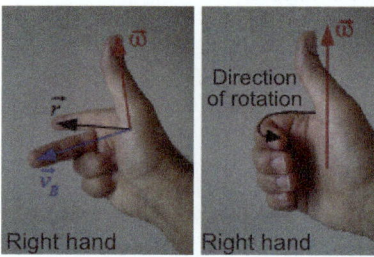

The cross product can be calculated from the components of the vectors as:

$$\vec{a} \times \vec{b} = \begin{pmatrix} a_y b_z - a_z b_y \\ a_z b_x - a_x b_z \\ a_x b_y - a_y b_x \end{pmatrix}.$$

Here are the most important calculation rules:

$$\vec{a} \times \vec{a} = 0$$
$$\vec{a} \times \vec{b} = -\vec{b} \times \vec{a}.$$

A3.17 Geometry of the Ellipse

Ellipses play a central role in celestial mechanics. The ellipse is a closed line, similar to a circle. It is defined by two focal points as the set of all points P, for which the sum of the distances to the two given focal points F_1 and F_2 equals $2a$. The center C of the ellipse is the center of the line connecting the focal points. The size a is the major semi-axis of the ellipse. The sizes aregiven in the sketch. For all points on the ellipse, $r_1 + r_2 = 2a$ holds.

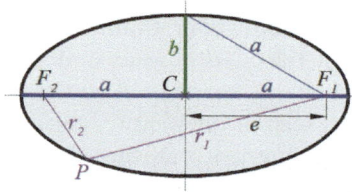

As the two focal points move closer together, the ellipsemore and more resembles a circle. When the two foci finally coincide, a circle is formed. Then $r_1 = r_2 = r$ and the major semi-axis becomes the radius of the circle.

A3 Mathematical Introduction

The minor semi-axis has the length b. It is already uniquely determined by the major semi-axis and the position of the foci.

One calls $e = \sqrt{a^2 - b^2}$ the eccentricity of the ellipse and $\epsilon = e/a$ the numerical eccentricity. The numerical eccentricity is dimensionless. Its value lies between 0 and 1. A numerical eccentricity of 0 corresponds to a circle. Then $a = b$. As the value approaches one, the ellipse becomes flatter and eventually degenerates into a line.

In astronomy, the point of a planet's closest approach to the sun is called the perihelion. If the sun is at the focal point F_1, this is the point on the right where the major axis touches the ellipse. The point of greatest distance from the sun (opposite) is called the aphelion.

In polar coordinates, the ellipse can be represented as follows:

$$r(\varphi) = \frac{p}{1 + \epsilon \cos \varphi},$$

with the semi-parameter $p = b^2/a$, the numerical eccentricity ϵ and the angle φ of the radius vector to the direction of the perihelion.

The ellipse belongs to the conic sections. These are the curves that are obtained as the intersection of a plane with a cone. The sketch shows an example that leads to an ellipse. Circles (plane perpendicular to the axis of the cone), parabolas (plane parallel to the axis of the cone), or hyperbolas can result.

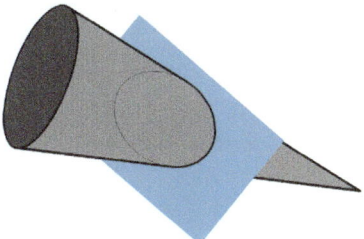

A3.18 Taylor Expansion

In calculating the gravitational force, we used an approximation method that is often used in physics (▶ Sect. 11.3.2). It is called a Taylor expansion. We want to examine it a bit more closely here.

The exact form of the gravitational force is:

$$F_G = G \frac{m m_\oplus}{r^2}.$$

We are interested in the behaviour at distances near the Earth's radius. This is outlined in the figure.

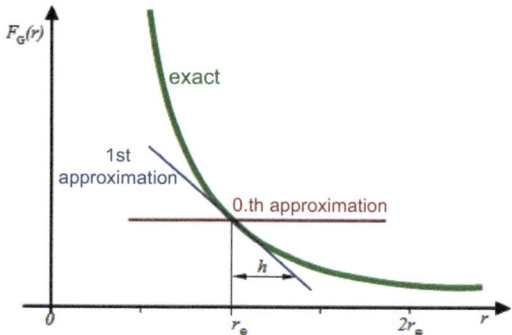

As a very first rough approximation—this is referred to as the 0th approximation—we can assume that the gravitational force near the Earth's radius is constant. In the sketch, this corresponds to the horizontal red line. In practice, you often use this approximation. Whenever you write $F_G = mg$, you assume that the weight is independent of the height h. Only at heights that go far beyond 1 km, the change in gravitational force becomes noticeable.

You will get a better description of the gravitational force if you approximate the exact curve in the sketch with a straight line. What slope should this straight line have? Well, its slope should correspond to the slope of the curve at the starting height (Earth's radius). This slope is the derivative of the exact function at this point. This is called the 1st approximation. It has the mathematical form:

$$F_G(r_\oplus + h) \approx F_G(r_\oplus) + \left.\frac{d}{dr} F_G(r)\right|_{r=r_\oplus} \cdot h.$$

It is a straight line described by a linear function of the height h with the slope given above.

This approximation can be further improved by adding a parabola to the straight line. This results in an approximation of the form:

$$F_G(r_\oplus + h) \approx F_G(r_\oplus) + c_1 \cdot h + c_2 \cdot h^2.$$

With the choice of the coefficient c_1 as above, the approximation has the same slope as the exact form. If we choose the coefficient c_2 such that the curvature between the approximation and the exact function also match, then this is the best choice. The second approximation is therefore:

$$F_G(r_\oplus + h) \approx F_G(r_\oplus) + \left.\frac{d}{dr}F_G(r)\right|_{r=r_\oplus} \cdot h + \left.\frac{d^2}{dr^2}F_G(r)\right|_{r=r_\oplus} \cdot h^2.$$

This procedure can be extended to terms of higher and higher order and results in a series:

$$F_G(r_\oplus + h) = F_G(r_\oplus) + \sum_{n=1}^{\infty} \left.\frac{d^n}{dr^n}F_G(r)\right|_{r=r_\oplus} \cdot h^n$$

$$= \sum_{n=0}^{\infty} \left.\frac{d^n}{dr^n}F_G(r)\right|_{r=r_\oplus} \cdot h^n,$$

where the 0-th derivative in the infinite sum is the value of the function itself. This is called the Taylor series or Taylor expansion of a function.

Not only functions can be approximated using the Taylor expansion. It can also be used to estimate how good the approximation is. You need to consider the value of the first term that has been neglected. This is exactly what we did in the calculation for F_G. We used the zeroth approximation $F_G = mg$ and determined the next term as an estimate of the approximation. For $h = 100\,\text{m}$, a relative error of 31 ppm (parts per million $= 0.000031$) results.

Before applying the Taylor series, however, you must be aware of an important characteristic. The Taylor series of a function generally only converges within a limited range of values, the so-called convergence range. Outside of this range, the Taylor series is meaningless. Practically, this means that the point at which one wants to use the Taylor series should be as close as possible to the value around which the expansion is performed. An often used example for a Taylor series is:

$$\frac{1}{\sqrt{1-x}} = 1 + \frac{1}{2}x + \frac{3}{4}x^2 + \ldots$$

Here, one obtains rapid convergence when x is significantly smaller than 1. If x becomes larger, the convergence worsens, i.e., one must consider more terms to achieve the same accuracy. Finally for $x \geq 1$ the series no longer converges at all.

A3.19 Volume Integrals

We have used volume integrals several times, for example, for determining the mass of a body or for determining its center of gravity. We represent these integrals over the volume of the body through triple integrals. We had already discussed this in ▶ Sect. 13 of this Appendix A3. Here again is the example of an integral over a cuboid.

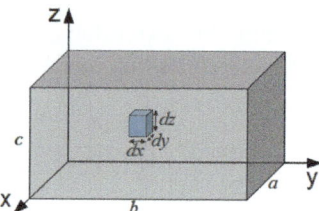

Geometrically speaking, we are breaking down the cuboid into infinitely small cuboids with edge lengths dx, dy and dz. An infinitesimal cuboid at the location (x, y, z) provides the contribution $f(x, y, z)dx\, dy\, dz$ to the integral. In the example of calculating the mass of the cuboid the function f would be the density ρ at the location (x, y, z). The integral is then:

$$I = \int_0^c \int_0^b \int_0^a f(x, y, z) dx\, dy\, dz\, .$$

For a cuboid, this is in most cases an appropriate way to determine the integral. But what should we do if we want to integrate a sphere. Now the Cartesian coordinates that we used above are no longer suitable. We want to use spherical coordinates. One might naively assume that the integral might be:

$$\int_0^R \int_0^\pi \int_0^{2\pi} f(\varphi, \theta, r) d\varphi\, d\theta\, dr\, .$$

But this is not correct. The second sketch shows why. The volume of the small elements, from which the integral is composed, depends on the position of the given $d\varphi\, d\theta\, dr$. This was not the case with Cartesian coordinates. You can see in the sketch that the volume increases with increasing r, and it also depends on θ. The lengths of the

A3 Mathematical Introduction

edges of the element can be extracted from the sketch. They are dr, $r\,d\theta$ and $\sin\theta\,r\,d\varphi$. For the integration the volume of the element must be taken into account. The integral in spherical coordinates is therefore:

$$I = \int_0^R \int_0^\pi \int_0^{2\pi} f(\varphi,\theta,r) r^2 \sin\theta\,d\varphi\,d\theta\,dr.$$

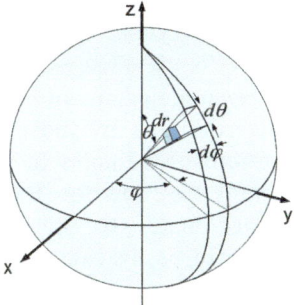

We have given a geometric justification for the form of the integral in spherical coordinates. More systematically speaking, one must consider that in a coordinate transformation of the integration variables, not only x, y and z need to be transformed, but also dx, dy and dz. For one-dimensional integrals, we have already seen this in ▶ Sect. 11 of this appendix A3. We called it "substitution" there.

In three dimensions, something similar applies. If u_1, u_2 and u_3 are the coordinates in which the integral is given (in our case these were x, y and z), and v_1, v_2 and v_3 are new coordinates (in our example these were φ, θ and r), then you must use the so-called Jacobian matrix. It reads:

$$J = \begin{pmatrix} \frac{\partial u_1}{\partial v_1} & \frac{\partial u_1}{\partial v_2} & \frac{\partial u_1}{\partial v_3} \\ \frac{\partial u_2}{\partial v_1} & \frac{\partial u_2}{\partial v_2} & \frac{\partial u_2}{\partial v_3} \\ \frac{\partial u_3}{\partial v_1} & \frac{\partial u_3}{\partial v_2} & \frac{\partial u_3}{\partial v_3} \end{pmatrix}$$

The determinant of this matrix (Jacobian determinant) must be included in the integral:

$$I = \int_0^R \int_0^\pi \int_0^{2\pi} f(\varphi,\theta,r) \det J\,d\varphi\,d\theta\,dr.$$

In our example we get:

$$u_1 = x = r\cos\varphi \sin\theta \quad \frac{\partial u_1}{\partial v_1} = \frac{\partial u_1}{\partial \varphi} = -r\sin\varphi \sin\theta$$

$$\frac{\partial u_1}{\partial v_2} = \frac{\partial u_1}{\partial \theta} = r\cos\varphi \cos\theta$$

$$\frac{\partial u_1}{\partial v_3} = \frac{\partial u_1}{\partial r} = \cos\varphi \sin\theta$$

$$u_2 = y = r\sin\varphi \sin\theta \quad \frac{\partial u_2}{\partial v_1} = \frac{\partial u_2}{\partial \varphi} = r\cos\varphi \sin\theta$$

$$\ldots$$

from which the Jacobian determinant is derived. The two most important determinants are those for the transformation from Cartesian coordinates to cylindrical coordinates $\det J = \rho$ and the one for the transformation from Cartesian coordinates to spherical coordinates $\det J = r^2 \sin\theta$. You can easily calculate other Jacobian determinants yourself.

A3.20 Matrices

In the equations in ▶ Sect. 13.6 we first encountered a matrix ($\tilde{I}$). We want to briefly address the most important calculation rules for matrices. A matrix has parallels to a vector, however, the numbers are not sorted in one, but in two dimensions, in the rows and columns of the matrix. A matrix with n lines and m columns is called a $n \times m$-Matrix. A single element of a $n \times m$-matrix $\tilde{X}$ then has two indices X_{ij}. The first index indicates the row to which the element belongs, the second the column. This matrix is written as:

$$\tilde{X} = \begin{pmatrix} X_{11} & X_{12} & \cdots & X_{1m} \\ X_{21} & X_{22} & \cdots & X_{2m} \\ \vdots & \vdots & \ddots & \vdots \\ X_{n1} & X_{n2} & \cdots & X_{nm} \end{pmatrix}.$$

In this book, we mostly use 3×3-matrices, such as the inertia tensor. You can add and subtract matrices by adding or subtracting the individual elements of the matrix. To do this, the dimensions of the matrices must match. For example:

A3 Mathematical Introduction

$$\tilde{X} - \tilde{Y} = \begin{pmatrix} X_{11} - Y_{11} & X_{12} - Y_{12} & \cdots & X_{1m} - Y_{1m} \\ X_{21} - Y_{21} & X_{22} - Y_{22} & \cdots & X_{2m} - Y_{2m} \\ \vdots & \vdots & \ddots & \vdots \\ X_{n1} - Y_{n1} & X_{n2} - Y_{n2} & \cdots & X_{nm} - Y_{nm} \end{pmatrix}.$$

You can multiply matrices with a constant:

$$c\tilde{X} = \begin{pmatrix} cX_{11} & cX_{12} & \cdots & cX_{1m} \\ cX_{21} & cX_{22} & \cdots & cX_{2m} \\ \vdots & \vdots & \ddots & \vdots \\ cX_{n1} & cX_{n2} & \cdots & cX_{nm} \end{pmatrix}.$$

Furthermore, you can multiply two matrices with each other. For example, $\tilde{Z} = \tilde{X} \cdot \tilde{Y}$, with the elements of $\tilde{Z}$ given by $Z_{ij} = \sum_k X_{ik} Y_{kj}$. As a matrix

$$\tilde{X} \cdot \tilde{Y} = \begin{pmatrix} \sum_k X_{1k} Y_{k1} & \sum_k X_{1k} Y_{k2} & \cdots & \sum_k X_{1k} Y_{km} \\ \sum_k X_{2k} Y_{k1} & \sum_k X_{2k} Y_{k2} & \cdots & \sum_k X_{2k} Y_{km} \\ \vdots & \vdots & \ddots & \vdots \\ \sum_k X_{nk} Y_{k1} & \sum_k X_{nk} Y_{k2} & \cdots & \sum_k X_{nk} Y_{km} \end{pmatrix}.$$

As can be seen from the multiplication rule, a multiplication of two matrices is only possible if the dimensions of the two matrices match. If one wants to multiply a $n \times m$-matrix with a $k \times l$-matrix this is only possible if $m = k$. The product results is a $n \times l$-matrix. The case of multiplying two 3×3-matrices results in another 3×3-matrix.

In calculations with matrices, the so-called transposed matrix might appear. The transposed matrix to $\tilde{X}$ is denoted by $\tilde{X}^T$. Written out, it is:

$$\tilde{X}^T = \begin{pmatrix} X_{11} & X_{21} & \cdots & X_{n1} \\ X_{12} & X_{22} & \cdots & X_{n2} \\ \vdots & \vdots & \ddots & \vdots \\ X_{1m} & X_{2m} & \cdots & X_{mn} \end{pmatrix}.$$

You can imagine this as a mirroring of the matrix elements along its diagonal. The transposed matrix of a $n \times m$ matrix is a $m \times n$ matrix.

Vectors can also be considered as matrices. In this sense a vector $\vec{a}$ would be a $n \times 1$ matrix:

$$\vec{a} = \tilde{a} = \begin{pmatrix} a_{11} \\ a_{21} \\ \vdots \\ a_{n1} \end{pmatrix},$$

where the second index is usually omitted. The transposed vector corresponds to a $1 \times n$ matrix:

$$\vec{a}^T = \tilde{a}^T = \begin{pmatrix} a_{11} & a_{12} & \cdots & a_{1n} \end{pmatrix}.$$

Our scalar product can be represented as a matrix multiplication:

$$\vec{a} \cdot \vec{b} = \tilde{a}^T \cdot \tilde{b} = \left(\sum_k a_{1k} b_{k1} \right).$$

We have discussed about vectors as one-dimensional objects $\vec{a} = (a_i)$ and matrices as two-dimensional objects $\tilde{a} = (a_{ij})$. This can be extended to higher dimensions. A three-dimensional object would then have three indices $\tilde{a} = (a_{ijk})$, etc. In general we then speak of tensors. The vector is a first-order tensor, the matrix a second-order tensor, etc. This should clarify the origin of the name "inertia tensor". In principle, tensors can be introduced up to any order. In classical mechanics, you will hardly encounter tensors with an order higher than two. However, in general relativity, tensors up to the fourth order often appear.

GPSR Compliance

The European Union's (EU) General Product Safety Regulation (GPSR) is a set of rules that requires consumer products to be safe and our obligations to ensure this.

If you have any concerns about our products, you can contact us on ProductSafety@springernature.com

In case Publisher is established outside the EU, the EU authorized representative is:

Springer Nature Customer Service Center GmbH
Europaplatz 3
69115 Heidelberg, Germany

Batch number: 08400034

Printed by Printforce, the Netherlands